Microbial Solutions for Sustainable Agriculture

This book explains microbial roles for agricultural sustainability, especially under climate changes, including the role of microorganisms in sustainable agriculture. Major coverage entails microbial enzymatic roles in agriculture, microbial phytohormones as promising sustainable plant bioagents, and phosphate-, potassium-, and calcium-solubilizing microbes as alternative biofertilizers. It introduces microbial solutions to mitigate the biotic and abiotic stress, with high attention to rising temperature and water limitation due to the climate changes.

This book

- Presents a summarized insight into important roles of microbes in facing the climate changes;
- Describes plant–microbe interaction in sustainability;
- Highlights roles of microorganisms in limited agriculture water sources;
- Provides an insight into microbial phytohormones as promising plant-sustainable bioagents;
- provides comprehensive overviews of microbial bioremediation roles for soil reclamation.

This book is aimed at researchers and graduate students in microbiology, agriculture, and sustainability.

Climate Change Resilience through Biorefinery Technologies

Series Editors: Bikash Kumar and Kiran Bala

Algal Biorefineries in Circular Economy for Climate Change Resilience
Edited by Fayaz A. Malla and Suhaib A. Bandh

Microbial Solutions for Sustainable Agriculture
Edited by Ghada Abd-Elmonsef Mahmoud and Bikash Kumar

For more information about this series, please visit: www.routledge.com/Climate-Change-Resilience-through-Biorefinery-Technologies/book-series/CCRBT

Microbial Solutions for Sustainable Agriculture

Edited by

Ghada Abd-Elmonsef Mahmoud
Bikash Kumar

CRC Press
Taylor & Francis Group
Boca Raton London New York

CRC Press is an imprint of the
Taylor & Francis Group, an **informa** business

Designed cover image: Shutterstock

First edition published 2026
by CRC Press
2385 NW Executive Center Drive, Suite 320, Boca Raton FL 33431

and by CRC Press
4 Park Square, Milton Park, Abingdon, Oxon, OX14 4RN

CRC Press is an imprint of Taylor & Francis Group, LLC

ISBN: 978-1-032-86354-2 (hbk)
ISBN: 978-1-041-19986-1 (pbk)
ISBN: 978-1-003-71448-4 (ebk)

DOI: 10.1201/9781003714484

Typeset in Times
by SPi Technologies India Pvt Ltd (Straive)

Contents

About the Editors

Dr. Ghada Abd-Elmonsef Mahmoud is an associate professor of applied microbiology in the Botany and Microbiology Department, Faculty of Science, Assiut University. She also, the head of Applied Biotechnology Department at the Molecular Biology Researches and Studies Institute, Assiut University, Assiut 71516, Egypt. She got her M.Sc. in microbiology, titled "Biodegradation of Hydrocarbons by Fungi," in 2013. She got her Ph.D. in microbiology, titled "Isolation and Characterization of Microorganisms Producing Vitamins," in 2016. Her main research interests include plant growth-promoting microorganisms, isolation and identification of different microorganisms, taxonomic ranking of bioactive material-producing microorganisms, enhancement of metabolic production efficiency of microorganisms, biodegradation of the environmental pollutants, functional impact of genetically modified microorganisms, molecular biology and metabolic profiling of pathogenic microorganisms, biosynthesis of nanoparticles from microorganisms, and their use as novel active antifungal agents. She has authored/co-authored more than 88 scientific publications, 13 book chapters, six review articles, and one book in the microbiology field. She has also been part of more than 80 workshops and training programs. She has reviewed more than 161 research papers as an academic reviewer in high-impact factor journals. She completed 13 M.Sc. and Ph.D. theses during the last five years and supervised more than 10 M.Sc. and Ph.D. students. She has been a scientific judge of the Egypt Science and Engineering Fair (ISEF) since 2022. She is a member of the American Society of Microbiology, the Egyptian Soil Science Society, the Society of Basic and Applied Mycology, and Applied Microbiology International. Her work has attracted significant impact, which is evident from 1200 citations with the h-index of 19 and the i-10 index of 35. She is an editor for the Frontiers in Energy Research Section (in Bioenergy and Biofuels), and she is a reviewer of 40+ peer-reviewed research journals for Elsevier, Wiley, MDPI, Springer, Taylor, and Francis.

Dr. Bikash Kumar is currently working as an Assistant Professor at Amity Institute of Biotechnology, Amity University Chhattisgarh, Raipur, Chhattisgarh, India. He also served a National Post-Doctoral Fellow at the Department of Biosciences and Biomedical Engineering, Indian Institute of Technology, Indore, Madhya Pradesh, and as a Post-Doctoral Research Associate at the Department of Biosciences and Bioengineering at the Indian Institute of Guwahati, Guwahati, Assam. He received his B.Sc. (Hons) in 2012 and M.Sc. (Distinction) in 2014 in biotechnology and bioinformatics from Guru Ghasidas University, Bilaspur (a central university), and Pondicherry Central University, respectively. Later, in 2014, he worked as a junior research fellow in the Department of Microbiology, Central University of Rajasthan, India. In 2015, he enrolled for Ph.D. and started his research at the Bioprocess and Bioenergy Laboratory under the research supervision of Prof. Pradeep Verma from the Department of Microbiology, Central University of Rajasthan, India. His work focuses on the development of biomass pretreatment technologies, biomass valorization, and microbial enzyme production (cellulases, laccase, and xylanases) for a greener energy and environment. His other research areas are bioprocess technology, food technology, and microbial biotechnology. The work has resulted in the publication of more than 50+ research articles, review papers, and book chapters. He has edited four books for Elsevier, Springer, and CRC Press. He has provided editorial assistance in 11 other books for CRC Press, Elsevier, and Springer and has been acknowledged. His work has attracted significant impact, which is evident from 3051+ citations with the h-index of 26 and the i-10 index of 30. He is a life member of the Biotech Research Society of India (BRSI), the Association of Microbiologists of India (AMI), the Mycological Society of India (MSI), and the Association of Carbohydrate Chemists and Technologists, India (ACCTI). He has received several awards from the Council of Scientific & Industrial Research-Senior Research Fellowship (CSIR-SRF), the Jawaharlal Nehru Memorial Fund (JNMF) scholarship, the Department of Science and Technology and Engineering Research Board (DST-SERB) National Post Doctoral Fellowship(N-PDF), and the Association of Microbiologists of India (AMI)-Young Scientist Award. Dr. Bikash Kumar is an associate editor in the Frontiers in Energy Research, a guest editor for special issue in the Frontiers in Bioengineering and Biotechnology, BMC Biotechnology for Sustainable Materials, and Green Technology, Resilience, and Sustainability, and a review editor for the Frontiers in Chemistry and Frontiers in Energy Research. He is a reviewer for 24+ international journals and has reviewed more than 82+ papers for Elsevier, Wiley, MDPI, Springer, Taylor, and Francis. Dr. Kumar is the recipient of DST-SERB NPDF project funding of ~3.5 million INR. He is included in the top 2% of the Scientist List by Stanford University in the list for the year 2023, and 2025.

List of Contributors

Alaa Z. El-Bably
Department of Water Requirement and Field Irrigation Research
Soils, Water and Environment Research Institute
Agricultural Research Center
Giza, Egypt

Arezoo Dadrasnia
Tech Center
TECNIO Network
University of Vic – Central University of Catalonia
Spain

Ashutosh Kumar Choudhary
School of Science and Technology
Swami Rama Himalayan University
Dehradun, India

Aya A.M. Abdellatif
Central Lab of Organic Agriculture
Agricultural Research Center
Giza, Egypt

Bindu Naik
Department of Food Science and Technology, Graphic Era (Deemed to be University)
Dehradun, India

Bruno Rafael de Almeida Moreira
Center of Crop Science
Queensland Alliance of Agriculture and Food Innovation (QAAFI)
The University of Queensland
Brisbane, Australia

Charu Sharma
School of Biosciences
Swami Rama Himalayan University
Dehradun, India

Eshraq A. AL-Amery
Microbiology Department
Faculty of Applied Science
Taiz University
Taiz, Yemen
and
Al Rowad Center for Scientific Research, Consultation, and Community Development
Yemen

Ghada Abd-Elmonsef Mahmoud
Botany and Microbiology Department
Faculty of Science
Assiut University
Assiut, Egypt

Hanaa Metwally Barakat
Microbiology and Chemistry Department
Faculty of Science
Al-Azhar University
Cairo, Egypt

Hanan S Khalefa
Veterinary Hygiene and Management Department
Faculty of Veterinary Medicine
Cairo University
Giza, Egypt

Hassan H.A. Mostafa
Central Lab of Organic Agriculture
Agricultural Research Center
Giza, Egypt

Ibrahim M.K. Elmaghraby
Central Lab of Organic Agriculture
Agricultural Research Center
Giza, Egypt

Imtiaz Ahmed
State Key Laboratory of Efficient Utilization of Arid and Semiarid Arable Land in Northern China
Institute of Agricultural Resources and Regional Planning
Chinese Academy of Agricultural Sciences
Beijing, China

Jobira Defera Gonfa
State Key Laboratory of Efficient Utilization of Arid and Semiarid Arable Land in Northern China
Institute of Agricultural Resources and Regional Planning
Chinese Academy of Agricultural Sciences
Beijing, China
and
Department of Plant Sciences
College of Agriculture and Natural Resources
Dilla University
Dilla, Ethiopia

Kiya Adare Tadesse
State Key Laboratory of Efficient Utilization of Arid and Semiarid Arable Land in Northern China
Institute of Agricultural Resources and Regional Planning
Chinese Academy of Agricultural Sciences
Beijing, China
and
School of Plant Sciences
College of Agriculture and Environmental Science
Haramaya University
Dire Dawa, Ethiopia

Lamiaa Ibrahim Ahmed
Department of Food Hygiene and Control
Faculty of Veterinary Medicine
Cairo University
Giza, Egypt

Mahesh Pattabhiramaiah
Centre for Applied Genetics
Department of Zoology
Bangalore University
Bangalore, India

Mahmoud Abdelaziz
State Key Laboratory of Efficient Utilization of Arid and Semiarid Arable Land in Northern China
Institute of Agricultural Resources and Regional Planning
Chinese Academy of Agricultural Sciences
Beijing, China
and
Department of Soils and Water
Faculty of Agriculture
Assiut University
Assiut, Egypt

Maissara M.K. Elmaghraby
Central Lab of Organic Agriculture
Agricultural Research Center
Giza, Egypt

Manjunatha Reddy G. R
Department of Botany
Bangalore University
Bangalore, India

Matheus Mantuanelli Roberto
University Center of Hermínio Ometto Foundation (FHO)
São Paulo, Brazil

Md Ashraful Alam
State Key Laboratory of Efficient Utilization of Arid and Semiarid Arable Land in Northern China
Institute of Agricultural Resources and Regional Planning
Chinese Academy of Agricultural Sciences
Beijing, China
and
Bangladesh Sugarcrop Research Institute
Pabna, Bangladesh

Mohamed A. Abdel-Sater
Botany and Microbiology Department
Faculty of Science
Assiut University
Assiut, Egypt

Mohamed Hefzy
Department of Water Requirement and Field Irrigation Research
Soils, Water and Environment Research Institute
Agricultural Research Center
Giza, Egypt

Mona Shaban E. M. Badawy
Department of Microbiology and Immunology
Faculty of Pharmacy (Girls),
Al-Azhar University
Cairo, Egypt

Mostafa A.S. Abdelgalil
Central Lab of Organic Agriculture
Agricultural Research Center
Giza, Egypt

Nano Alemu Daba
State Key Laboratory of Efficient Utilization of Arid and Semiarid Arable Land in Northern China
Institute of Agricultural Resources and Regional Planning
Chinese Academy of Agricultural Sciences
Beijing, China
and
School of Plant Sciences
College of Agriculture and Environmental Science
Haramaya University
Dire Dawa, Ethiopia

Neama A. Shalaby
Food Industry Technology Program
Faculty of Technology of Industry and Energy
Samannoud Technological University
Gharbia, Egypt

Nemmat A. Hussein
Botany and Microbiology Department
Faculty of Science
Assiut University
Assiut, Egypt

Nisha Dhillon
School of Biosciences
Swami Rama Himalayan University
Dehradun, India

Omnia Karem M. Riad
Department of Microbiology and Immunology
Faculty of Pharmacy (Girls)
Al-Azhar University
Cairo, Egypt

Paulo Henrique Frata Ferreira
Department of Plant Production
College of Agricultural and Technological Sciences
São Paulo State University (Unesp)
São Paulo, Brazil

Paulo Renato Matos Lopes
Department of Plant Production
College of Agricultural and Technological Sciences
São Paulo State University (Unesp)
São Paulo, Brazil

Pujawati Suryatmana
Soil Biology Laboratory
Department of Soil Science and Land Resources
Faculty of Agriculture
Padjadjaran University
Indonesia

Qomarudin Helmy
Water and Wastewater Engineering Research Group
Faculty of Civil and Environmental Engineering
Institute of Technology Bandung
Indonesia

Reda K. Darwesh
Department of Water Requirement and Field Irrigation Research, Soils
Water and Environment Research Institute
Agricultural Research Center
Giza, Egypt

Rosazlin Abdullah
Institute of Biological Sciences
Faculty of Science
Universiti Malaya
Kuala Lumpur, Malaysia

Sayed A. Abd El Hafez
Department of Water Requirement and Field Irrigation Research
Soils, Water and Environment Research Institute
Agricultural Research Center
Giza, Egypt

Shanthala Mallikarjunaiah
Centre for Applied Genetics
Department of Zoology
Bangalore University
Bangalore, India

Soma A. El Mogy
Materials Testing and Surface Chemical Analysis Lab
National Institute of Standards (NIS)
Giza, Egypt

Thalia Silva Valério
Department of Plant Production
College of Agricultural and Technological Sciences
São Paulo State University (Unesp)
São Paulo, Brazil

Usha Rani
School of Biosciences
Swami Rama Himalayan University
Dehradun, India

Victor Hugo Cruz
Department of Plant Production
College of Agricultural and Technological Sciences
São Paulo State University (Unesp)
São Paulo, Brazil

Vijay Kumar
School of Biosciences
Swami Rama Himalayan University
Dehradun, India

Vishal Rajput
School of Biosciences
Swami Rama Himalayan University
Dehradun, India

Vivek Kumar
School of Biosciences
Swami Rama Himalayan University
Dehradun, India

Yanca Araujo Frias
Department of Plant Production
College of Agricultural and Technological Sciences
São Paulo State University (Unesp)
São Paulo, Brazil

Preface

In the 21st century, the increasing population requires a huge food supply, pushing the need to increase agricultural productivity with limited land and water availability. Agriculture faces several other challenges, such as environmental pollution, degradation, and climate change, which threaten sustainable food production. Thus, there is a need for innovation integrating cutting-edge multidisciplinary approaches and scientific advancements with sustainable agricultural practices.

This book explores various strategies, such as biofertilizer, plant growth promoters, microbiome engineering, CRISPR-based genetic modifications, nanotechnology, and phytoremediation, to enhance agricultural productivity while minimizing environmental impact. The role of microorganisms such as bacteria, fungi, and yeast is crucial to soil health, plant growth, protection, and environmental sustainability through the synthesis of essential enzymes and bioactive compounds. These microorganisms are present as epiphytes, endophytes, and rhizosphere and are key indicators of soil health and even useful in managing plant growth under stressed conditions.

One of the key themes of this book is the role of biofertilizers and plant growth-promoting rhizobacteria (PGPR) in improving water productivity, nutrient uptake, and stress resilience under biotic and abiotic conditions. By leveraging microbial applications, we can mitigate plant stress, enhance soil fertility, and improve crop yields. Furthermore, microbiome engineering offers a promising avenue for enhancing plant health and optimizing resource use efficiency.

One of the thematic areas of the book covers the role of biofertilizers and PGPRs in enhancing water productivity, nutrient uptake, and stress resilience under biotic and abiotic circumstances. The application of microbes can reduce plant stress, increase soil fertility, and improve crop yields. Furthermore, microbiome engineering presents a promising approach to improving plant health and optimizing the effectiveness of resource use.

Additionally, this book provides insight into transformative technologies to improve crop resilience and productivity, such as multi-omics approaches and CRISPR/Cas9-mediated modifications. Environmental protection and conservation via phytoremediation, soil carbon sequestration, industrial waste remediation, and wastewater reuse in irrigation are also the focus of this book, covering how environmental sustainability continues to be at the forefront of agricultural advancement. Furthermore, methods based on nanotechnology are investigated for their potential to enhance pest resistance, soil health, and nutrient delivery.

This book can be a useful resource for researchers, academics, and policymakers as it offers vital insights into microbial innovations and sustainable agricultural practices meant to develop a cleaner, more resilient, and ecologically sustainable agroecosystem. This book could be an asset in providing insights to address the issues of resource scarcity and climate change.

Acknowledgments

First of all, we would like to convey our gratitude toward CRC Press and the series editors of the book series *Climate Change Resilience through Biorefinery Technologies* for providing us the opportunity to submit this book titled *Microbial Solutions for Sustainable Agriculture* under the book series.

This book is a result of support from researchers and academicians who have contributed their expertise and provided insights throughout the review and revision process. We extend our sincere gratitude to the contributing authors, whose dedication and knowledge have been instrumental in bringing this work to fruition. We are also indebted to the reviewers who have supported us in peer reviewing the book. Their collective efforts showcase the dynamic aspects of associating the microbial solution in advancing research and its transformative potential in shaping a sustainable future. We also hope that this book serves as a source of inspiration for future innovations, driving the transition toward a sustainable future and maintaining environmental health while meeting global food and energy needs.

Dr. Ghada Abd-Elmonsef Mahmoud and Dr. Bikash Kumar extend their heartfelt gratitude to their respective institutions – the Botany and Microbiology Department, Faculty of Science, Assiut University, Assiut, Egypt, Amity Institute of Biotechnology, Amity University Chhattisgarh, Raipur and the Indian Institute of Technology, Indore, India, respectively– for providing invaluable infrastructural support and a conducive environment for teaching and research. Their academic and research experiences have offered deep insights into the needs of academicians, students, and researchers, which greatly influenced the development of this book.

Dr. Bikash Kumar expresses his sincere appreciation to DST-SERB for awarding him the Postdoctoral Fellowship (NPDF/2022/001781) at IIT Indore, which provided him with the opportunity to engage in advanced research and academic pursuits. The knowledge and insights gained during this fellowship played a crucial role in shaping the perspectives presented in this book. The fellowship not only facilitated access to essential resources but also fostered a research environment that greatly contributed to the book's development. Dr. Kumar is also profoundly grateful to his current NPDF mentor, Prof. Kiran Bala, for her constant motivation and unwavering support in his academic endeavors.

We are deeply thankful to God and our parents for their blessings. Dr. Ghada expresses her sincere appreciation to her parents, whose spirit has been supportive throughout her journey. She also dedicates a special dedication to her husband and son, who have shared this journey with her and supported her along the way. Dr. Kumar extends his heartfelt appreciation to his parents – Arbind Kumar, the late Mamta Devi, and Rajkumari Devi – and his siblings, Vatan, Kanchan, Kamal, Shrishti, and Akash, for their encouragement and support. A special mention goes to his wife, Sheetal, and daughter, Shrinika, whose love and support continue to inspire and strengthen him.

1 Role of Microorganisms and Biofertilizers in Improving Water Productivity for Sustainable Agriculture

Mohamed Hefzy, Reda K. Darwesh, Alaa Z. El-Bably, Sayed A. Abd El Hafez, and Ghada Abd-Elmonsef Mahmoud

1.1 INTRODUCTION

Water is the essential source of life for all species on Earth and ranks as the second most crucial natural resource, behind air, due to its finite quantity, including fresh water, salt water, surface water, and groundwater. Consequently, water requirements and the developmental process are interdependent, as human civilization and advancement are quantified by daily water use. The economics of water utilization and its long-term prospects need the exploration of alternatives and the assessment of current water supplies, as well as potential future sources. Water shortage is the primary constraint hindering the ambitious development of agricultural land necessary to bridge the existing gap between food production and consumption (Swain et al. 2020). In regions like Egypt, situated in an arid or semi-arid zone, the water needs of vegetation constitute a substantial portion of overall water usage. The pressures of population expansion and escalating demands from local and other sectors for water create further problems for agriculture. The Nile River serves as Egypt's primary water supply. According to the 1959 Nile Waters Agreement, Egypt's allocation is 55.5 billion m^3/year, as seen in Table 1.1, which depicts the water sources in Egypt (Abd Ellah 2020) (Figure 1.1).

Agricultural practices in Egypt utilize 80%–85% of water supplies. Effective water governance should be implemented to tackle these difficulties, focusing on minimizing losses and maximizing benefits per unit of water. Enhancing water productivity and plant yield significantly relies on supplying crops with the required amount of water, informed by the interplay of climate, vegetation, and soil conditions. Numerous methods exist to enhance water productivity, including the enhancement of irrigation canals in established lands to improve transport efficiency, the

DOI: 10.1201/9781003714484-1

TABLE 1.1
Microbial Roles of Phosphate-Solubilizing Bacteria in Enhancing Plant Growth, Water Productivity, and Soil Health Under Normal and Stress Conditions

Name	Effects of Inoculating the Bacterial Phosphate Solubilizers on the Soil and Plant's Health	Reference
Bacillus subtilis	Soil inoculation increased exoenzyme in the rhizosphere.	Iqbal et al. (2024)
	Improved common bean growth and physiological and nutritional qualities.	Souza et al. (2023)
	Increased the dwarf tomato size, root length, weight, and phosphorus content.	Jensen et al. (2024)
	Enhanced the plant growth parameters, physiological properties, water relations, and biochemical properties of sunflowers under cadmium stress.	Abeed et al. (2022)
	Inoculation enhanced cotton growth in semi-arid areas. Increased root length and *B. subtilis* also produced plant growth-promoting substances that enhanced cotton growth and health.	Ahmad et al. (2021)
Bacillus megaterium	Soil inoculation increased exoenzyme in the rhizosphere.	Iqbal et al. (2024)
	Plant inoculation improved common bean growth and the nutritional qualities.	Souza et al. (2023)
	Plant inoculation improved maize growth, phosphorus contents, and mineral uptakes.	de Oliveira-Paiva et al. (2024)
Bacillus sp.	Inoculation-induced plant resistance against bacterial canker disease of tomatoes and increased the defense enzymes and the plant growth.	Bakki et al. (2024)
	Bacterial inoculation gives the highest oil yield of *Mentha arvensis* and menthol content.	Prakash and Arora (2019).
Bacillus cerkularice	Enhanced *Pisum sativum* growth, and yield, and increased the antioxidants of the yield and quality.	Mahmoud et al. (2023)
Bacillus aryabhattai, Paenibacillus polymyxa	Bacterial inoculation enhanced the cotton growth in semi-arid areas. Increased shoot and root lengths, phosphorus contents, and dry weights.	Ahmad et al. (2021)
Paenibacillus polymyxa	Improvement of the durum wheat seedling's weight, chlorophyll, phosphorus contents, and growth.	Cherchali et al. (2019)
Bacillus altitudinis	Increased the nitrogen and phosphorus concentrations of corn seedlings by 100% and 47.9%, respectively.	Zhao et al. (2022)
Bacillus licheniformis	Increased the germination rates and seedling growth under saline irrigation. Raised leaf chlorophyll, P, and K uptake.	Mahdi et al. (2020)
Pseudomonas sp.	Increased shoot and root lengths, enhancing leaf chlorophyll, and phosphorus uptake.	Bakki et al. (2024)

Pseudomonas aeruginosa	Higher levels of phosphate solubilization were received after six days of soil inoculation and higher enzymatic activities.	Soares et al. (2023)
Pseudomonas fluorescens	Increases the nutrient uptake, especially available phosphorus of *Italian ryegrass*.	Amri et al. (2022)
Pseudomonas putida	Enhanced the growth parameters of *Brassica juncea*, nutrient uptakes, phosphorus contents, and quality.	Mishra et al. (2020)
Enterobacter asburiae	Increased potassium plant accumulation reduced sodium contents, and enhanced phosphorus uptakes.	Mahdi et al. (2020)
Rhizobium radiobacter	Enhanced loose-leaf lettuce quality, nutrient contents, and yield in saline soil.	Verma et al. (2020)
Rhizobium leguminosarum	Improved available phosphorus in the soil, and Kikuyu grass yield.	Torres-Cuesta et al. (2023)
Pantoea agglomerans	Increased phosphorus availability in the soil, and enhanced the soil beneficial microbiota in semi-arid soil.	Saadouli et al. (2021)
	Enhanced tomato seedling growth during the greenhouse experiment and elevated its phosphorus content and nutrient uptakes.	Suleimanova et al. (2023)
Acinetobacter rhizosphaerae	Improved wheat germination in cold stressed conditions, nutrient uptake, especially phosphorus, and enhanced their physiological properties.	Kour and Yadav (2023)
Streptomyces sp.	Produced siderophores regulate phosphorus mobilization within iron-rich soil. Increased available phosphorus in the soil cultivated with *Camellia oleifera*.	Cui et al. (2022)

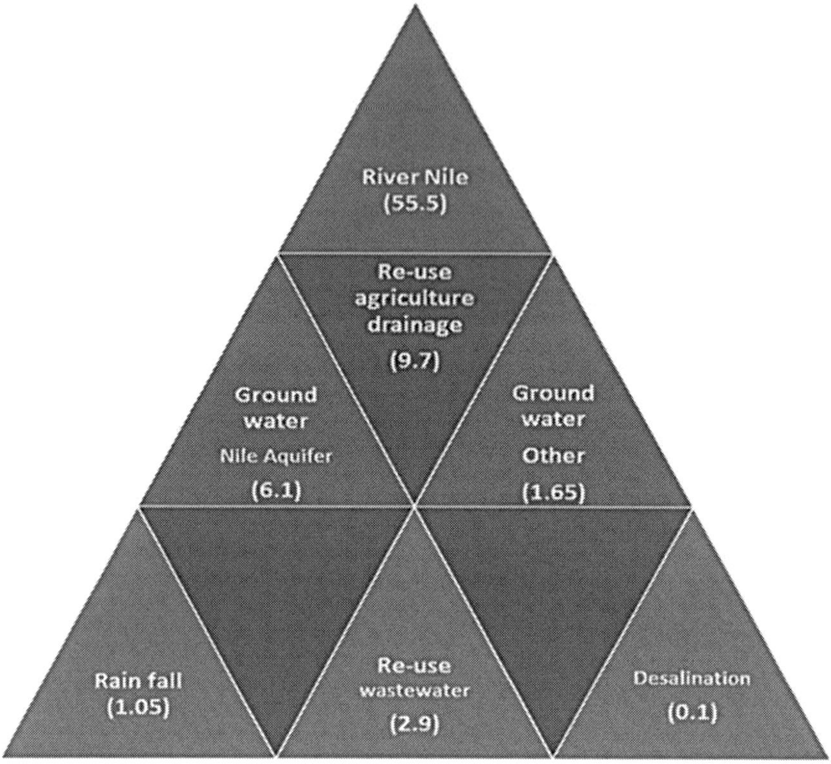

FIGURE 1.1 Egyptian water resources (billion m^3/year).

implementation of pressurized irrigation systems in newly reclaimed areas, and the adoption of on-farm water management techniques such as laser land leveling (Meena et al. 2024). Additional strategies encompass raised bed irrigation, irrigation scheduling, intercropping, crop intensification, and mulching; the application of soil conditioners and organic fertilizers; and the utilization of short-lived, drought-resistant varieties alongside plant growth-promoting microorganisms (PGPMs) to bolster crop health, productivity, and water uptake efficiency (Mahmoud et al. 2024).

Egypt is, among the nations, most severely affected by global and climate change. Challenges encompass population growth, heightened needs for food, water, and energy, along with evolving land use patterns and urbanization. The Mediterranean region of Egypt features a very intricate hydrological system, as it experiences minimal precipitation (with Cairo averaging 30 mm/year) and relies entirely on the flow of the Nile River (Mahmoud and El-Bably 2017). The anticipated expansion of the Egyptian population and economy would result in heightened water demand, with allocation priorities established as follows: first for potable water, followed by industrial use, and any residual resources will be designated for agriculture and ecological needs. Innovations that enhance yields and food security through scientifically grounded methods to address future drought and water shortages, resulting in "more crop per drop," are essential (Wouter et al. 2013).

The supply of water and nutrients (chemical, organic, or biological sources) are the primary determinants of productivity in irrigated agriculture. Enhancing the efficiency of these components is the objective of effective management and is essential in arid and semi-arid regions where water supplies are scarce. Furthermore, in dryland irrigated agriculture, nitrogen emerges as the primary limiting factor for crop output. Under these circumstances, the efficient utilization of both irrigation water and nitrogen is frequently suboptimal and is significantly influenced by the application technique (Huang et al. 2022). Enhancing irrigation schedules and methodologies can augment water efficiency. Methods such as supplementary irrigation, deficit irrigation, precision irrigation techniques, and soil-water conservation practices improve water production. These methods influence water management, soil fertility, insect control, crop selection, and market accessibility. Proper irrigation is crucial for crop development; nevertheless, over-irrigation may result in water loss. Achieving the optimal equilibrium between irrigation levels and plant density is crucial for maximizing water productivity (Cao et al. 2022).

At the same time, plants also require minerals and nutrients for their growth and differentiation besides water. Nitrogen, phosphorus, and potassium are the primary macronutrients essential for plant development. These nutrients have a vital role in cellular growth, morphological attributes, and physiological and biochemical properties (Mahmoud et al. 2023). Chemical fertilizers, while regarded as the primary source of essential nutrients, lead to numerous environmental and health issues due to their excessive application in agriculture. They contain heavy metals, produce significant quantities of nitrite, contribute to air pollution, contaminate groundwater with nitrates, and alter soil pH and structure (Velimirovic et al. 2021). Biofertilizers serve as an environmentally sustainable alternative to chemical fertilizers, addressing prior issues. The combination of biofertilizers and customized irrigation methods will enhance water efficiency and meet the individual requirements of crops.

1.2 WATER PRODUCTIVITY

Water productivity (WP) is the net yield per unit of water used and in agriculture is essential for sustainable water supply, agricultural yield, and climate resilience. It diminishes water wastage, augments soil vitality, and bolsters agricultural sustainability. Effective water utilization also aids in the preservation of natural ecosystems, biodiversity, and environmental protection (Chawla et al. 2023). The objectives of enhancing water productivity are to generate increased food, income, improved livelihoods, and ecosystem services while using less water. However, crop water productivity is defined as the marketable yield of a crop relative to its consumptive water usage. Consumptive water usage pertains not to the extraction of water from surface or groundwater sources, but rather to the fraction of water removed that is depleted, so lost to the system and not recovered or recycled into the groundwater or surface water systems (Molden et al. 2010). This depletion or evapotranspiration (ET) encompasses all evaporation and transpiration, whether advantageous or disadvantageous, within a certain region. Consequently, agricultural water productivity may be determined either as the quantity of product per unit of evapotranspiration (kg/m^3) or as net farmer income per unit of evapotranspiration (\$/m^3). In regions with limited water resources or where overexploitation occurs, genuine water conservation is attained by diminishing ET. Water usage efficiency (WUE) is defined as the ratio of net CO_2 assimilation to water use, and CO_2 assimilation may be assessed by net CO_2 exchange, dry matter accumulation, and economic yield, whilst water utilization may be evaluated by mass or molar unit (Bacon 2004).

1.3 MICROBIAL ROLES IN SUSTAINABLE AGRICULTURE AND WATER PRODUCTIVITY

Microbes provide enhanced flexibility and variety in promoting crop yield and agri-food production by serving as natural nitrogen fixators, mineral solubilizers, nano-mineral synthesizers, and inducers of plant growth regulators (Mahmoud et al. 2024). They actively degrade organic materials and remediate heavy metals and pollutants in soils, serving as soil-water binders. Moreover, microorganisms residing in the plant rhizosphere exude biochemicals that have harmless effects on both the host and the environment. These biochemicals may function as biocides in the management of agricultural pests, infections, and illnesses (Noor-Hassim et al. 2023).

Sustainable agriculture seeks to demonstrate the efficacy of natural biofertilizers in mitigating issues related to water, temperature, and soil. Numerous microorganisms, such as algae, bacteria, and fungi, have been utilized in commercial applications as efficient, environmentally sustainable, and cost-effective biofertilizers to improve crop yield, soil health, and quality. Furthermore, these natural biofertilizers assist plants in mitigating biotic and abiotic stresses by boosting their metabolic activities and defense mechanisms (Mahmoud et al. 2023). Biofertilizer refers to any chemical containing live microorganisms that promotes nutrient availability in soil ecosystems, colonizes the plant rhizosphere, and positively affects plants by enhancing nutrient accessibility (Altabbaa et al. 2023). The efficacy of biofertilizers mostly relies on their successful colonization in the soil, particularly inside the plant rhizosphere (Trivedi et al. 2020). These microorganisms may possess various distinctive features such as atmospheric nitrogen fixation and the solubilization of minerals, including phosphorus, potassium, calcium, and sulfur (Mahmoud et al. 2023; Zhao et al. 2022). These microorganisms might enhance agricultural yields, increase plant tolerance, and preserve soil fertility, hence benefiting future cultivation of the same soil for more crops (Berruto and Demirer 2024).

Plants may directly uptake nutrients in soluble form, but insoluble nutrient forms, particularly phosphate (phosphate rock) and potassium salts (feldspar), are not absorbable from the soil. Phosphate is regarded as the second necessary macronutrient for plant growth, following nitrogen. The soil harbors substantial amounts of insoluble phosphorus resulting from the overuse of chemical fertilizers, constituting around 90% of soil phosphate in an insoluble form (Ghorai and Ghosh 2022). Phosphorus-solubilizing microorganisms address these challenges by secreting phytase, nuclease, and phosphatase enzymes to hydrolyze and convert insoluble organic phosphorus into phosphate (Singh et al. 2020), as well as releasing organic acids (lowering the pH) to solubilize phosphate rock and render insoluble phosphate accessible (Yin et al. 2021). Potassium-solubilizing microorganisms function similarly by secreting protons, enzymes, organic acids, and exo-polysaccharides, which transform insoluble potassium into soluble forms (Muthuraja and Muthukuma 2021).

Plant growth-promoting rhizobacteria are soil bacteria that reside on or near the root surface and contribute to plant growth and development, either directly or indirectly, through the creation and release of numerous regulatory compounds in the rhizosphere. They can regulate plant development and enhance output by creating plant hormones (gibberellin, cytokinin, and indole acetic acid), siderophores, enzymes, antioxidants, peptides, lipids, and polysaccharides (Wang et al. 2022a; Mahmoud et al. 2024). Plant growth-promoting rhizobacteria enhance plant growth directly by aiding in resource acquisition (nitrogen, phosphorus, and other essential minerals) or by modulating plant hormone levels, and indirectly by mitigating the detrimental effects of various pathogens on plant growth and development as biocontrol agents (Ahemad and Kibret 2014).

Crop water productivity serves as an indication of biogeochemical cycling within agricultural ecosystems and may also be affected by the soil microbial community. The majority of pertinent research has concentrated on the impact of root microbial inoculation on agricultural water productivity (Di et al. 2021). Inoculation with root microbes, including mineral-solubilizing microorganisms, arbuscular mycorrhiza, and phytohormone microbial producers, has demonstrated beneficial effects on crop water productivity. This could be attributed to crops in agricultural environments predominantly facing nutritional deficiencies, particularly in nitrogen and phosphorus, primarily due to their extraction during harvest. The microorganism's inoculation can enhance the breakdown of soil organic matter and increase nutrient availability, hence facilitating greater nutrient uptake by crops for their growth and development (Cheng et al. 2023).

1.4 MICROBIAL ROLES IN SOIL AND PLANT HEALTH

Biofertilizers are rich in microbes that enhance the physical, chemical, and biological characteristics of soil while ensuring nutrition availability for plants. The application of microorganisms can enhance nutrient absorption by modifying the physicochemical features of the rhizosphere, including the augmentation of soil cation exchange capacity and other metabolic reactions in root tissues (Adhikary 2012). The enhancement of root nutrient absorption under diverse environmental situations, such as water stress, may be attributed to the fact that plant growth-promoting microbial treatments augment the translocation of photo-assimilation to root tissues, enhancing the activity of enzymes involved in the nitrogen metabolism, hence augmenting nitrogen content under water deficiency conditions (Ansari and Ahmad 2019). Certain bacteria and fungi, when subjected to abiotic stress, can enhance physiological and biochemical processes, including nutrient absorption, photosynthesis, metabolism, osmotic potential, protein functionality, and phytohormone synthesis (Jalal et al. 2023) (Figure 1.2).

The utilization of microbial fertilizers enhances soil microbial diversity, elevates the population of beneficial microorganisms, enhances aeration, augments soil porosity, stabilizes soil aggregates, and promotes soil compaction and soil water permeability (Liu et al. 2022; Wei et al. 2024). Soil microorganism inoculation bolsters plant disease resistance by controlling soil pH, augmenting soil organic carbon content, and fighting pathogens for essential nutrients. *Trichoderma* species have a significant soil-water-holding capacity with their mycelium, which improves soil water repellence

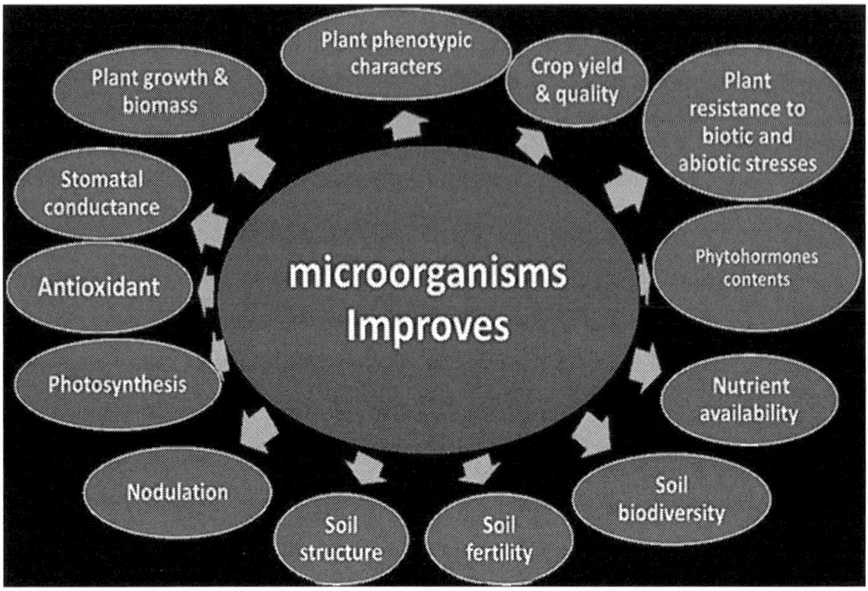

FIGURE 1.2 The role of microorganisms in soil and plant health.

(Scudeletti et al. 2021). *Bacillus* species recovered from arid and desert soils generate lipopolysaccharides that function as biosurfactants and bioemulsifiers, exhibiting polyextreme tolerance in diverse soils for 30 months (Raddad et al. 2018). Biosurfactant- and bioemulsifier-producing microbes enhance the soil hydrological characteristics, alleviate drought stress in arid regions, and enhance plant survival under water constraints (Ramirez-Valdespino and Orrantia-Borunda 2021).

Inoculation of the maize rhizosphere with *Bacillus licheniformis* enhanced root and shoot dry weight, enzymatic activities, and water usage efficiency by up to 46% in drought circumstances (Akhtar et al. 2020). Inoculating grapevines with *Pseudomonas corrugata* and *Enterobacter soli* improved the root-adhering soil/root tissue ratios, soil stability, elevated nitrogen and phosphorus concentrations in both soil and plant, enhanced relative water content, and photosynthetic rate (Duan et al. 2021). Inoculating sugarcane plants with *Trichoderma asperellum* in drought stressed conditions enhanced plant photosynthetic rate, stomatal conductance, and WUE (Scudeletti et al. 2021).

1.5 MICROBIAL ROLES IN THE NUTRIENT'S AVAILABILITY IN SOIL

Plants may directly absorb soluble nutrients like nitrogen, potassium, and phosphorus; however, they cannot absorb their insoluble salts (Mahmoud et al. 2023). Nitrogen, phosphorus, potassium, and other nutrients are absorbed by plants through the rhizosphere, facilitating their participation in the nutrient cycle (Moreau et al. 2019). Nitrogen-fixing bacteria employ the enzyme nitrogenase to transform nitrogen into ammonia or nitrate, which is assimilated by plants (Rees et al. 2005). The majority of phosphorus in soil, encompassing organic phosphorus and inorganic phosphorus, is phosphorus-solubilizing microorganisms (*Bacillus* sp. and *Trichoderma* sp.) capable of secreting phytase and phosphatase to hydrolyze organic phosphorus (insoluble) to absorbable phosphate (Singh et al. 2020). These microbes may release organic acids, which solubilize phosphate by lowering the pH (Yin et al. 2021). The majority of potassium in the soil is encapsulated inside minerals like feldspar and vermiculite, which makes them also unavailable for plants. Potassium-solubilizing microorganisms can excrete substances, including extracellular polysaccharides, organic acids, and enzymes, which facilitate potassium release by lowering pH and degrading the crystal structure of insoluble potassium (Muthuraja and Muthukuma 2021).

1.5.1 Nitrogen-Fixing Bacteria

Plants, like all living beings, require nitrogen, a crucial element of nucleotides and proteins, and a fundamental component of chlorophyll. It encompasses protein, organic acid, DNA, and other intricate organic compounds. Although nitrogen is plentiful, the majority exists as inert nitrogen in the atmosphere, rendering it unusable for plants (Moreau et al. 2019). Soluble nitrogen in soil exhibits significant spatial and temporal variability, and most plants seem to be extremely opportunistic in utilizing transient microscale nitrogen patches. Plants mostly absorb inorganic nitrogen in the forms of nitrate and ammonium. The absorption of nitrate is an active, energy-dependent process since it occurs against an electrochemical gradient (Liu et al. 2024).

The soil nitrogen cycle primarily includes nitrogen fixation, nitrification, denitrification, ammonization after nitrogen fixation, and ammonium reduction, predominantly facilitated by nitrogen-fixing microorganisms and microorganisms involved in the dissimilatory reduction of nitrate to ammonium. The use of nitrogen-fixing microbes remains one of the most efficacious methods to enhance crop output (Xiong et al. 2021). To comprehend the influence of these microbes on nitrogen absorption and transformation in plants, it is essential to ascertain the forms of nitrogen that plants can assimilate, the availability of these compounds in the soil, and the impact of plant characteristics on nitrogen nutrient uptake. Plants often absorb inorganic nitrogen in the form of nitrate and ammonium salts (Liu et al. 2024).

Nitrogen-fixing bacteria are crucial microbial functional flora in plant soil, often found in the plant rhizosphere environment, and play a significant role in controlling nitrogen nutrition in plants. The nitrogen conversion process facilitated by microorganisms is typically characterized as a sequence of sequential chemical steps: a nitrogen molecule is converted into ammonia (nitrogen fixation), subsequently turned into biological organic nitrogen (assimilation and absorption), and ultimately converted into ammonium salt, which is oxidized to nitrate (nitrification) and then reduced to nitrogen gas (denitrification) or anaerobic ammonia (Xiong et al. 2021; Liu et al. 2024). The currently identified nitrogen-fixing microorganisms are classified into aerobic nitrogen-fixing bacteria (*Azotobacter* sp. and *Azotomonas* sp.), microaerobic (*Arthrobacter* sp., *Dexia* sp., *Alcaligenes* sp., and *Methylomonas* sp.), facultative anaerobic (*Enterobacter* sp., *Klebsiella* sp., and *Bacillus* sp.), anaerobic (*Clostridium* species), and photoautotrophic (*Rhodobacter* sp. and *Rhodospirillum* sp.) (Mahmud et al. 2020; Xiong et al. 2021; Liu-Xu et al. 2024). In addition to the free-living nitrogen-fixing bacteria, cyanobacteria like *Nostoc* and *Anabaena* species also play significant roles in nitrogen assimilation (Kaschuck and Hungria 2017).

Azotobacter has a high efficiency of nitrogen fixation, reaching 20 kg N/ha/year (Esmailpour et al. 2013). Inoculating the previous nitrogen regulators, microbes demonstrated their roles in enhancing the plant productivity; inoculating *Azospirillum brasilense* and *Azotobacter chroococcum*, bacteria with *Trichoderma lignorum* fungi in sugarcane effectively supply the plants with 60% of their required nitrogen (Serna-Cock et al. 2011). Inoculating *Azotobacter* in wheat increases the grain yield up to 19.42% more than the control (Mahato and Kafle 2018). Inoculating sugarcane with the bacterial combination of *Pseudomonas* sp., *H. seropedicae*, and *Bacillus megaterium* increases the plant yield by up to 57% (Antunes et al. 2019). Inoculating *Pisum sativum* with *Azotobacter chroococcum* nitrogen-fixing bacteria in combination with *Bacillus megaterium* and *B. cerkularice* raised the crop yield and increased the antioxidant quality of pea (Mahmoud et al. 2023).

1.5.2 Phosphate-Solubilizing Microbes

Phosphorus is the principal macronutrient that impedes plant development, ranking just below nitrogen. Owing to its rapid fixation, it is a crucial element for plants and is generally deficient in the soil. Phosphorus exists in both organic and inorganic forms throughout the soil (Rawat et al. 2021). Approximately 35%–70% of the total phosphorus in soil exists as inorganic phosphorus, predominantly in the form of insoluble mineral complexes (Khan et al. 2019). Plants experiencing phosphorus deficiency secrete more enzymes capable of degrading phosphorus and are significantly impacted by organic acids. Soil phosphatases are required for the mineralization of organic phosphorus. Microbial phosphatases, rather than those from plants, release orthophosphates from organic phosphorus in the soil. A diverse array of phosphoesters was dephosphorylated by soil enzymes, using around 90% of the available organic phosphates. Over 50% of soil phosphorus exists in the form of phytate, an organic phosphorus compound. Nevertheless, due to their intricate association with cations or their adsorption onto various organic constituents of soil, they are not readily assimilated by plants (Iftikhar et al. 2024).

Phosphate solubilizers are prevalent in soil and are essential to the biogeochemical cycle of phosphorus, by being involved in processes like solubilization, weathering, mineralization, and immobilization (Tian et al. 2023). Phosphate-solubilizing bacteria are distinctive for their ability to produce the enzyme 1-aminocyclopropane-1-carboxylic acid deaminase, which reduces ethylene levels in stressed plants, hence aiding plants in enduring or thriving under adverse conditions (Iftikhar et al. 2024). Using these microbial solubilizers can reduce the utilization of phosphate fertilizers by 50% without significantly affecting grain output (Yazdani et al. 2009). For efficient use of phosphate-solubilizing microorganisms, certain parameters like soil pH, moisture, and organic matter concentration need to be adjusted as they will affect on the efficient microbial colonization in the crop's rhizosphere (Timofeeva et al. 2022).

New research and scientific lines have started to focus more on phosphate-solubilizing microorganisms as substitutions for chemical phosphate fertilizers. Nacoon et al. (2020) designed experiments to clarify the interactions between microbial phosphate solubilizers and other microorganisms in the soil, focusing on the role of phosphate solubilizers in soil health and fertility. Liu et al. (2020) developed new approaches for phosphate solubilizer engineering to maximize their solubilizing abilities and enhance soil fertility using CRISPR-Cas9 technology. Yahya et al. (2022) utilized metagenomics to understand the diversity and the function of these microbes in the soil and their relations with enhancing plant growth and also productivity with environmental stress. Song et al. (2022) investigated the other roles of phosphate solubilizers in soil remediation, especially heavy metals, and found their efficient contaminant removal roles in the soil.

A range of microorganisms, including bacteria and fungi, was hypothesized to have the capability to solubilize insoluble phosphorus with 0.1%–0.5% for fungi and 1%–50% for bacteria (Iftikhar et al. 2024). Numerous bacteria have demonstrated a positive role in phosphate when supplied near plant roots, enhancing crop yields, including *Pseudomonas* spp., *Bacillus* spp., *Rhizobium* spp., *Azotobacter* spp., *Micrococcus* spp., *Flavobacterium* spp., *Burkholderia* spp., *Achromobacter* spp., *Erwinia* spp., and *Agrobacterium* spp. (Wang et al. 2023; Iftikhar et al. 2024) as illustrated in Table 1.1. Table 1.1 clearly shows the recently used types of phosphate-solubilizing bacteria and their roles in enhancing plant growth, soil health, water consumption, and yield quality under normal and stressful conditions. Inoculating the rhizosphere area of plants with *Pseudomonas* and *Bacillus* species has significantly improved the soil nutrient levels, excreting plant growth regulators, and inhibiting soil-borne pathogens (Sadiq et al. 2013). Application of two phosphate-solubilizing bacteria (*Alcaligenes faecalis* and *Pseudomonas geniculate*) increased sesame yield by 21.97% and 23.11% compared to untreated plants in the first and second seasons, respectively (Omer and Abd-Elnaby 2017).

Phosphate-solubilizing fungi involve a large group of filamentous fungi and yeast types, including *Aspergillus* sp., *Candida* sp., *Penicillium* sp., *Trichoderma* sp., and *Mucor ramosissimus* (Wang et al. 2023; Iftikhar et al. 2024; Song et al. 2024). *Trichoderma* is a saprotrophic (soil-borne) filamentous fungus that is found worldwide. *Trichoderma* spp. demonstrate a strong ability to take advantage of various ecological circumstances and adapt to any biotic and abiotic stress. Inoculating the soil with *Trichoderma* spp. reduced soil pH, raised alkaline phosphatases, and enhanced maize root and shoot weights and lengths (de Oliveira et al. 2024). Citric, malic, gluconic, oxalic, lactic, tartaric, and propionic acids are all organic acids produced by phosphate solubilizers utilized for adsorption, chelate anions, pH adjustment, and interaction with other metal ions to facilitate the dissolution of inorganic phosphates (Wang et al. 2023; Iftikhar et al. 2024).

Trichoderma spp. are unique mineral solubilizers, especially of phosphorus and potassium; they even have a promising role in the cell regulation of resistance genes, especially in drought stress. *Trichoderma asperellum* was recognized for modulating a unique defense gene that influenced the jasmonic acid and salicylic acid pathways (De Sousa et al. 2020). Under drought conditions, inoculation with *Trichoderma asperellum* enhanced the net photosynthetic rate by 50% in non-stressed plants and by 68% in drought-stressed plants (Scudeletti et al. 2021). Khoshmanzar et al. (2020) stated that co-inoculation of *Trichoderma asperellum* and *Bacillus amyloliquefaciens* in wheat stimulated antioxidant enzymatic activities and proline and enhanced the plant growth and resistance to drought stress.

Phosphate-solubilizing fungi like *Trichoderma harzianum were* found to have enhancing abilities on sunflower growth parameters, physiological properties, water relations, and biochemical properties under cadmium stress (Abeed et al. 2022), enhancing the availability of micronutrients in the soil, especially phosphorus, zinc, and manganese (Ali et al. 2022). *Trichoderma* inoculation to the plant's rhizosphere enhanced the growth properties and regulated the water consumption and the physiological properties of maize using *T. harzianum* (de Oliveira et al. 2024) and tomato (Bader et al. 2020) as shown in Table 1.2, which clarifies the recent roles of *Trichoderma* species with single and co-inoculation in enhancing the growth of various plant types in the form of nutrient uptake (especially phosphorus), water consumption, physiological traits, plant length, weight, and yield quality.

1.5.3 Potassium Solubilization

Potassium is vital for plant growth, development, metabolism, stress resilience, and reproduction; it ranks as the seventh most prevalent element in the Earth's crust, and plant physiologists assert that, regarding the stimulation of plant growth, it ranks second to nitrogen. Although soil potassium is plentiful, just 2% is accessible to plants in a freely soluble state, while the remainder is bound to other soil minerals (Olaniyan et al. 2022; Babar et al. 2024). The production of ATP, sugars, protein, and starch synthesis, movements of water and nutrients, and enzyme activity are all affected by soil potassium levels (Xu et al. 2020). The imbalance of potassium in the soil has emerged as a significant limitation on agricultural output in several developing nations. It improves plant health, growth, and development; enhances disease resistance; elevates crop quality and extends the shelf life of produce; and bolsters plants' resilience to stresses such as drought and disease. It also facilitates photosynthesis, affects carbohydrates, oils, and fat synthesis, and regulates water absorption by roots (Han et al. 2023). Soil has four distinct forms of potassium: exchangeable-ionic K, non-exchangeable-fixed K, accessible-soluble K, and unavailable-mineral K. Exchangeable potassium, or ionic potassium, is readily assimilated by the plant's root system and displaces potassium at the exchange sites, whereas fixed potassium acts as a reserve supply of potassium. Some of these are also present in the organic components of the soil microbial community (Olaniyan et al. 2022).

Microorganisms (potassium-solubilizing microorganisms) play a vital part in the natural recycling of potassium, converting it from insoluble sources into forms accessible to plants. Ahmad et al. (2016) stated that rhizospheric microorganisms, particularly bacteria, have the ability to solubilize potassium from minerals for efficient absorption by plants. A diverse array of microorganisms has been documented to liberate potassium from minerals and provide it in an accessible form to plants in the soil. Soil inoculation with potassium bacteria solubilizes and promotes potassium ion absorption over sodium ions in plants, even in saline-stress situations (Han et al. 2023).

TABLE 1.2
Microbial Roles of Phosphate-Solubilizing Fungi in Enhancing Plant Growth, Water Productivity, and Soil Health Under Normal and Stressed Conditions

Name	Effects of Inoculating the Bacterial Phosphate Solubilizers on the Soil and Plants Health	Reference
Trichoderma harzianum	Inoculation of the soil with *T. harzianum* reduced soil pH, raised alkaline phosphatases, and enhanced maize root and shoot weights, phosphorus uptake, and length.	de Oliveira et al. (2024)
	Co-inoculation of bacteria and *T. harzianum* enhanced the plant growth parameters, physiological properties, water relations, and biochemical properties of sunflowers under cadmium stress.	Abeed et al. (2022)
	Soil inoculation enhanced the availability of micronutrients in the soil, especially phosphorus, zinc, and manganese.	Ali et al. (2022)
	Efficiently controlled wilt disease in tomatoes and enhanced the plant growth, phosphorus contents, and resistance enzymes.	Bader et al. (2020)
	Raised the tolerance abilities of tomato plants in drought conditions, nutrient uptakes, especially phosphorus, tomato growth, and the physiological properties.	Khoshmanzar et al. (2020)
Trichoderma viride	Enhanced phosphorus uptake and growth parameters of *Chloris virgata* under deficient phosphorus conditions	Song et al. (2024)
	Enhanced phosphorus contents in *Melilotus officinalis* seedlings and the growth parameters.	Song et al. (2023)
	The seed-priming technique with *T. viride* reduced up to 20% of the phosphorus fertilizers dose recommended for soybeans.	Paul and Rakshit (2021)
Trichoderma atroviride, *T. citrinoviride*	The combination of the two isolates enhances growth, phosphorus uptake, yield, and quality of *Brassica chinensis*	Chen et al. (2021)
Trichoderma koningiopsis	Soil inoculation increased available phosphorus in the soil under the abiotic stress of drought and alkaline stress.	Tandon et al. (2020)
Trichoderma gamsii	Enhanced photosynthetic pigments, growth, phosphorus contents, and quality of common bean	Bedine et al. (2022)
Trichoderma orientale	Enhanced chickpea germination of the seeds, nutrient uptakes, and seedling growth.	Bekkar and Zaim (2023)
Trichoderma sp.	Increased pepper growth, soluble proteins, phosphorus, mineral uptakes, and sugar contents of pepper.	Duan et al. (2023b)
	Root colonization with *Trichoderma* sp. regulates plant growth, water consumption, and development.	Prasad et al. (2023)
	Enhanced soybean growth from 2.1% to 41.1% and raised phosphorus uptake by 141%	Bononi et al. (2020)

The use of potassium-solubilizing bacteria as biofertilizers is promising method instead of artificial fertilizers. They could enhance soil fertility, cultivate high-quality crops, and provide accessibility for farmers, which present significant potential in addressing the worldwide food problem and the economic exports of many nations (Xu et al. 2020). These comprise bacteria such as *Acidothiobacillus* spp., *Bacillus circulans, Aminobacter* spp., *Bacillus cereus, Arthrobacter* sp., *Burkholderia* spp., *Bacillus amyloliquefaciens, Bacillus coagulans, B. megaterium, Bacillus subtilis, Enterobacter* spp., and *Pseudomonas* spp., as shown in Table 1.3, which shows the recent roles of potassium-solubilizing bacterial species in enhancing various plant types in the form of nutrient uptake (especially potassium), water consumption, physiological traits, and yield quality.

TABLE 1.3
Microbial Roles of Potassium-Solubilizing Bacteria in Enhancing Plant Growth, Water Productivity, and Soil Health Under Normal and Stressful Conditions

Name	Effects of Inoculating the Bacterial Phosphate Solubilizers on the Soil and Plant's Health	References
Burkholderia vietnamiensis, Staphylococcus hominis	Significantly enhanced the seed germination of rice by 95.3% compared to 82% of untreated seeds. The consortium also improved rice growth characteristics under deficient potassium conditions.	Nguyen et al. (2024)
Pseudomonas migulae, P. poae, P. extremaustralis	Enhanced the physiological properties of apples, including sugar, protein, and chlorophyll, and highly increases the potassium content of the apple seedlings.	Jiao et al. (2024)
Pseudomonas sp., *Mammaliiococcus* sp., *Erwinia, Citrobacter* sp., *and Pantoea* sp.	Inoculation of the potassium-solubilizing consortium to rice plants enhanced K-assimilation, plant growth, physiological characteristics, and cytosolic K/Na ratio under salt stress.	Nawaz et al. (2023)
Bacillus sp., *Pseudomonas* sp., *Enterobacter* sp., *Acidithiobacillus* sp.	Co-inoculation enhanced soil bioavailability of potassium through the production of various types of organic acids (lactic acid, citric acid, tartaric acid, and gluconic acid) and enhanced plant growth and yield production.	Babar et al. (2024)
Bacillus subtilis	Enhanced plant growth parameters, physiological properties, water relations, and biochemical properties of sunflower under cadmium stress.	Abeed et al. (2022)
Bacillus cereus	Significantly increased the potato yield by 21% above the control plants. The K-feldspar applications with *Bacillus cereus* inoculation to potato gave 40-ton/ha tuber yield.	Ali et al. (2020)
Burkholderia sp., *Paraburkholderia* sp., *Collimonas* sp., *Pseudomonas, Agrobacterium* sp.	The co-inoculations with various phosphate-solubilizing types enhanced the potassium uptake levels of rice seedlings, dry weight, and seedling length.	Zhang et al. (2024)

(Continued)

TABLE 1.3 (*Continued*)

Name	Effects of Inoculating the Bacterial Phosphate Solubilizers on the Soil and Plant's Health	References
Bacillus sp., *Pseudomonas* sp.	Co-inoculation increased the wheat root length of the seedlings by 16%–40% and increased the biomass two to three times.	Gandhi et al. (2022)
Bacillus aryabhattai	Improved effectively available potassium in *Populus alba* rhizosphere soil and promoted plant growth.	Chen et al. (2020)
Bacillus sp., *Enterobacter* sp., *Pseudomonas* sp., *Achromobacter* sp.	It increases maize potassium content, nutrient uptakes, leaf number, shoot, and root length, and chlorophyll, and enhances the expression of enzymes related to stress.	Saheewala et al. (2023)
Escherichia vulneris, *Bacillus cereus*	Increased the available soil potassium by 28.41 mg/kg and plant height of garlic by 23.53% and also increased garlic leaf width, length, and fresh and dry weight.	Wang et al. (2022b)
Bacillus licheniformis, *B. cenocepacia*	They improved the growth, height, root/shoot ratio, leaf area, root length of tomato plants, potassium uptake, water consumption, and tissue potassium content under greenhouse conditions.	Raji and Thangavelu (2021)
Pseudomonas gessardii	Influenced positively on the barley length, weight, carotenoids, enzymes, nutrient uptake, potassium content, and sugar content.	Kaur et al. (2021)
Streptomyces griseorubens and *Nocardiopsis alba*	They have implementation strategies to enhance the soil fertility, wheat growth, yield, and potassium uptake levels in wheat.	Boubekri et al. (2021)
Burkholderia sp.	Increase potassium uptake in *Mikania micrantha* and stimulate plant growth and quality.	Sun et al. (2020)
Paenibacillus sabinae, *Leptolyngbya* sp.	Co-application had a significant effect on available potassium in the soil; dry weight and root length of wheat increased by 20%–70%. Also, the potassium/sodium ratio in wheat increased by 41%.	Duan et al. (2023a)
Pantoea agglomerans, *Rahnella aquatilis*	The co-inoculation enhanced canola growth, yield, water consumption, nutrient uptake, and potassium concentrations under saline stress.	Beaicknejad Daroonkolaei et al. (2022)
Priestia megaterium	Produced extracellular potassium-binding proteins that solubilize potassium in feldspar.	Wu et al. (2023)
Bacillus circulans	Inoculation of wheat with 1.5% *B. circulans* with foliar spraying significantly increased the wheat grain yield by 16.41% more than the control treatment.	El-Egami et al. (2024)

The potassium-solubilizing mechanism encompasses the creation of various organic acids, polysaccharides, and protons, and the formation of microbial biofilm on mineral surfaces. Insoluble potassium-bearing minerals, including micas, biotite, muscovite, feldspar, and orthoclase. The secretion of these acids leads to lower pH, which solubilizes potassium minerals by the gradual release of exchangeable and easily accessible potassium (Nawaz et al. 2023). Organic acids liberate potassium ions from potassium-bearing minerals by chelating aluminum, ferrous, and calcium ions, which potentially disrupt the lattice structure and release potassium into the soil (Sattar et al. 2019). It was clear that potassium-solubilizing bacteria liberated potassium ions from phlogopite via crystal lattice acid solubilization and aluminum chelation (Ahmad et al. 2016). Moreover, *Bacillus altitudinis* potassium solubilizer generated organic acids that solubilized feldspar and released aluminum and ferrous elements (Etesami et al. 2017). Microorganisms manufacture unique exopolymer layers called biofilms consisting of microbial consortia cells that adhere to the mineral surfaces. The biofilm cells synthesize polysaccharides, proteins, and lipids from the gelatinous layer on mineral or rock surfaces to create regulated optimum microenvironments surrounding their cells for efficient solubilization through both organic acids and secondary metabolites. The mechanism may involve the decrease of rhizospheric pH to enhance the solubilization of potassium-bearing minerals and their availability for plant absorption (Sattar et al. 2019).

1.6 MICROORGANISMS AS SOIL WATER CONDITIONERS

Soil water retention refers to the correlation between the soil water content and potential. Mycorrhizal fungi are found to have the efficiency to alter the plant water relations, which will reflect on the soil structure. Mycorrhizal fungi behaved as a soil conditioner beside or far from the root through enhanced soil water storage in sandy soil (Pauwels et al. 2023). They also have efficient abilities to tolerate high drought conditions and hard climate changes (Begum et al. 2020). The drought tolerance mechanism of these fungi sustained stomatal opening, higher water potentials of plant, activation of aquaporins in the plant root, changed osmolyte contents, and maximized root hydraulic conductivities, especially in low water content soils (Begum et al. 2020; Abdalla and Ahmed 2021).

Nitrogen-fixing, phosphate, and potassium microbial solubilizers have unique properties in enhancing water productivity in various crops through single and co-inoculation techniques. The inoculation of potato plants with *Azotobacter chroococcum* and *Azospirillum brasilense* with nitrogen fixation properties increased water productivity by 0.99%, 11.64%, and 37.4% compared with non-inoculation under irrigation levels of 100%, 80%, and 60%, respectively. Inoculation of potato plants with *Bacillus megaterium* as phosphorus solubilization bacteria increased water productivity by 6.73%, 8.86%, and 36.9% compared with non-inoculation under irrigation levels of 100%, 80%, and 60%, respectively. However, inoculation of potato plants with *Bacillus circulans* as potassium solubilization bacteria increased water productivity by 18.22%, 23.26%, and 29.25% compared with non-inoculation under irrigation levels of 100%, 80%, and 60%, respectively (Zahran et al. 2020).

1.7 CONCLUSION

Microbial mineral solubilizers and nitrogen-fixing bacteria cleared magnificent properties in enhancing plant growth, water production, physiological properties, and yield quality through single and co-inoculation to the soil. It also enhanced the soil quality and mineral availability in normal and stressed conditions. In drought conditions, these microbes efficiently utilize the available water and express tolerance genes through the plant to activate the defense mechanism and tolerance properties. To get the highest benefits from these microbes, some points need to be kept in mind: inoculum preparation represents the most important item to optimize the right microbial concentration, production of these microbes using low-cost optimized fermentation techniques on a large scale, and the most efficient and cost-effective plant treatment technique, including seed treatments, and when selecting these microbes, the pathogenicity test needs to be performed, and the antagonistic activities of the microbes in the soil need to be considered. Moreover, adding natural sources of potassium and phosphorus with low percentages will speed up the solubilization process.

1.8 FUTURE PROSPECTS

In the near future, it is expected that microbial research will keep advancing and become more focused on how to improve these microbial isolates through genetic engineering to maximize the mineral solubilization properties and improve nitrogen fixation efficiency and effectiveness. Utilizing these microbes is more efficient in promoting the plant's growth, especially under water shortage and drought conditions. Hoping to utilize metagenomics to understand the relations between the inoculated microbes and the normal microbiota of the soil and even understand the criteria of saving water in stressed conditions and enhancing the relation between the inoculated microbes and the normal microbiota with positive effects on the plant and the soil. Moreover, new formulation techniques of these beneficial microbes to keep them active for a long time and keep their survival are urgent, especially since we are facing severe climate changes every day threatening the whole agriculture sustainability.

ABBREVIATIONS

PGPMs Plant growth-promoting microorganisms
WP Water productivity
WUE Water usage efficiency

ARTIFICIAL INTELLIGENCE DECLARATION

The authors have not used regenerative artificial intelligence tools while writing the chapter.

REFERENCES

Abd Ellah, R.G. 2020. Water resources in Egypt and their challenges, Lake Nasser case study. *Egypt. J. Aquat. Res.* 46, 1–12.

Abdalla, M., Ahmed, M.A. 2021. Arbuscular mycorrhiza symbiosis enhances water status and soil-plant hydraulic conductance under drought. *Front. Plant Sci.* 12, 722954.

Abeed, A.H.A., Mahdy, R.E., Alshehri, D., Hammami, I., Eissa, M.A., Abdel Latef, A.A.H., Mahmoud, G.A. 2022. Induction of resilience strategies against biochemical deteriorations prompted by severe cadmium stress in sunflower plant when Trichoderma and bacterial inoculation were used as biofertilizers. *Front. Plant Sci.* 13, 1004173. doi: 10.3389/fpls.2022.1004173

Adhikary, S. 2012. Vermicompost, the story of organic gold: a review. *Agric. Sci.* 37, 905–917.

Ahemad, M., Kibret, M. 2014. Mechanisms and applications of plant growth promoting rhizobacteria: current perspective. *J. King Saud Univ. Sci.* 26, 1–20.

Ahmad, I., Ahmad, M., Hussain, A. 2021. Integrated use of phosphate-solubilizing *Bacillus subtilis* strain IA6 and zinc-solubilizing *Bacillus* sp. strain IA16: a promising approach for improving cotton growth. *Folia Microbiol.* 66, 115–125.

Ahmad, M., Nadeem, S.M., Naveed, M., Zahir, Z.A. 2016. Potassium-solubilizing bacteria and their application in agriculture. In: V.S. Meena, B.R. Maurya, J.P. Verma, R.S. Meena (eds) *Potassium Solubilizing Microorganisms for Sustainable Agriculture*. Springer India, 293–313.

Akhtar, S.S., Amby, D.B., Hegelund, J.N., Fimognari, L., Großkinsky, D.K., Westergaard, J.C., Müller, R., Moelbak, L., Liu, F., Roitsch, T. 2020. *Bacillus licheniformis* FMCH001 increases water use efficiency via growth stimulation in both normal and drought conditions. *Front. Plant Sci.* 11, 297.

Ali, A.M., Awad, M.Y.M., Hegab, S.A., Gawad, A.M.A.E., Eissa, M.A. 2020. Effect of potassium solubilizing bacteria (*Bacillus cereus*) on growth and yield of potato. *J. Plant Nutr.* 44(3), 411–420.

Ali, S., Khan, M.J., Anjum, M.M. 2022. *Trichoderma harzianum* modulates phosphate and micronutrient solubilization in the rhizosphere. *Gesunde Pflanzen* 74, 853–862.

Altabbaa, S., Mann, N.A., Chauhan, N., Utkarsh, K., Thakur, N., Mahmoud, G.A.-E. 2023. Era connecting nanotechnology with agricultural sustainability: issues and challenges. *Nanotechnol. Environ. Eng.* 8(2), 481–498.

Amri, M., Mateus, D., Gatrouni, M., Rjeibi, M.R., Asses, N., Abbes, C. 2022. Co-inoculation with phosphate-solubilizing microorganisms of rock phosphate and phosphogypsum and their effect on growth promotion and nutrient uptake by ryegrass. *Appl. Biosci.* 1(2), 179–197.

Ansari, F.A., Ahmad, I., 2019. Alleviating drought stress of crops. Through PGPR: mechanism. In Dhananjaya P. Singh, Vijai Kumar Gupta, Ratna Prabha (eds) *Microbial Interventions in Agriculture and Environment: 2: Rhizosphere, Microbiome and Agro-Ecology*, 341–358. Springer, Singapore. doi: 10.1007/978-981-13-8383-0_11

Antunes, J.E.L., De Freitas, A.D.S., Oliveira, L.M.S., De Lyra, M.D.P., Fonseca, M.A.C., Santos, C.E.R.S., De Oliveira, J.P., De Araújo, A.S.F., Figueiredo, M.V.B. 2019. Sugarcane inoculated with endophytic diazotrophic bacteria: effects on yield, biological nitrogen fixation and industrial characteristics. *An. Acad. Bras. Cienc.* 91, e20180990.

Babar, S., Baloch, A., Qasim, M., Wang, J., Wang, X., Li, Y., Khalid, S., Jiang, C. 2024. Unearthing the soil-bacteria nexus to enhance potassium bioavailability for global sustainable agriculture: a mechanistic preview, *Microbiol. Res.* 288, 127885.

Bacon, M.A. 2004. *Water Use Efficiency in Plant Biology*. Blackwell, Oxford, UK.

Bader, A.N., Salerno, G.L., Covacevich, F., Consolo, V.F. 2020. Native *Trichoderma harzianum* strains from Argentina produce indole-3 acetic acid and phosphorus solubilization, promote growth and control wilt disease on tomato (Solanum lycopersicum L.). *J. King Saud Univ. Sci.* 32, 867–873.

Bakki, M., Banane, B., Marhane, O., Esmaeel, Q., Hatimi, A., Barka, E.A., Azim, K., Bouizgarne, B. 2024. Phosphate solubilizing Pseudomonas and Bacillus combined with rock phosphates promoting tomato growth and reducing bacterial canker disease. *Front. Microbiol.* 15, 1289466.

Beaicknejad Daroonkolaei, S., Amerian, M.R., Pirdashti, H., Bakhshandeh, E., Gholami, A. 2022. Evaluation of the impact of potassium solubilizing bacteria on potassium efficiency and yield of canola under saline and non-saline soil conditions. *Arch. Agron. Soil Sci.* 69(9), 1437–1453.

Bedine, M.A., Iacomi, B., Tchameni, S.N., Sameza, M.L., Fekam, F.B. 2022. Harnessing the phosphate-solubilizing ability of Trichoderma strains to improve plant growth, phosphorus uptake and photosynthetic pigment contents in common bean (Phaseolus vulgaris). *Biocatal. Agric. Biotechnol.* 45, 102510.

Begum, N., Ahanger, M.A., Zhang, L.A.M.F. 2020. Inoculation and phosphorus supplementation alleviates drought induced growth and photosynthetic decline in Nicotiana tabacum by up-regulating antioxidant metabolism and osmolyte accumulation. *Environ. Exp. Bot.* 176, 104088.

Bekkar, A.A., Zaim S. 2023. Phosphate solubilization and the enhancement of chickpea growth by new rhizospheric microorganisms Bacillus tequilensis and Trichoderma orientale. *Arch. Biol. Sci.* 75(4), 419–422.

Berruto, C.A., Demirer, G.S. 2024. Engineering agricultural soil microbiomes and predicting plant phenotypes. *Trends Microbiol.*, S0966-842X(24)00043-X. *Advance online publication.* doi: 10.1016/j.tim.2024.02.003

Bononi, L., Chiaramonte, J.B., Pansa, C.C. et al. 2020. Phosphorus-solubilizing *Trichoderma* spp. from Amazon soils improve soybean plant growth. *Sci. Rep.* 10, 2858.

Boubekri, K., Soumare, A., Mardad, I., Lyamlouli, K., Hafidi, M., Ouhdouch, Y., Kouisni, L. 2021. The screening of potassium- and phosphate-solubilizing actinobacteria and the assessment of their ability to promote wheat growth parameters. *Microorganisms* 9(3), 470. doi: 10.3390/microorganisms9030470

Cao, X., Li, Y., Wu, M. 2022. Irrigation water use and efficiency assessment coupling crop cultivation, commutation and consumption processes. *Agric. Water Manag.* 261, 107370. doi: 10.1016/j.agwat.2021.107370

Chawla, R., Dubey, S., Khose, S.B. 2023. Water productivity in agriculture: a key to sustainable food production. *Agric. Food Paper* 5(12), 326–329.

Chen, D., Hou, Q., Jia, L., Sun, K. 2021. Combined use of two *Trichoderma* strains to promote growth of Pakchoi (*Brassica chinensis* L.). *Agronomy* 11(4), 726.

Chen, Y., Ye, J., Kong, Q. 2020. Potassium-solubilizing activity of Bacillus aryabhattai SK1-7 and its growth-promoting effect on Populus alba L. *Forests* 11(12), 1348. doi: 10.3390/f11121348

Cheng, Y., Narayanan, M., Shi, X., Li, Z., Ma, Y. 2023. Phosphate-solubilizing bacteria: their agroecological function and optimistic application for enhancing agro-productivity. *Sci. Total Environ.* 901(25), 166468. doi: 10.1016/j.scitotenv.2023.166468

Cherchali, A., Boukhelata, N., Kaci, Y., Abrous-Belbachir, O., Djebbar, R. 2019. Isolation and identification of a phosphate-solubilizing Paenibacillus polymyxa strain GOL 0202 from durum wheat (Triticum durum Desf.) rhizosphere and its effect on some seedlings morphophysiological parameters. *Biocat. Agri. Biotechnol.* 19, 101087.

Cui, K., Xu, T., Chen, J., Yang, H., Liu, X., Zhuo, R., Peng, Y., Tang, W., Wang, R., Chen, L., Zhang, X., Zhang, Z., He, Z., Wang, X., Liu, C., Chen, Y., Zhu, Y. 2022. Siderophores, a potential phosphate solubilizer from the endophyte Streptomyces sp. CoT10, improved phosphorus mobilization for host plant growth and rhizosphere modulation. *J. Clean. Prod.* 367, 133110.

de Oliveira, H.P., de Melo, R.O., Cavalcante, V.S. et al. 2024. Phosphate fertilizers coated with phosphate-solubilising *Trichoderma harzianum* increase phosphorus uptake and growth of *Zea mays*. *Plant Soil.* doi: 10.1007/s11104-024-06818-0

de Oliveira-Paiva, C.A., Bini, D., de Sousa, S.M., Ribeiro, V.P., dos Santos, F.C., de Paula Lana, U.G., de Souza, F.F., Gomes, E.A., Marriel, I.E. 2024. Inoculation with Bacillus megaterium CNPMS B119 and Bacillus subtilis CNPMS B2084 improve P-acquisition and maize yield in Brazil. *Front. Microbiol.* 15, 1426166.

De Sousa, T.P., Chaibub, A.A., da Silva, G.B., de Filippi, M.C.C. 2020. Trichoderma asperellum modulates defense genes and potentiates gas exchanges in upland rice plants. *Physiol. Mol. Plant Pathol.* 112, 101561. doi: 10.1016/j.pmpp.2020.101561

Di, W.U., Andales, A.A., Yang, H., Sun, Q., Chen, S., Guo, X., Li, D., Du, T. 2021. Linking crop water productivity to soil physical, chemical and microbial. *Front. Agric. Sci. Eng.* doi: 10.15302/J-FASE-2020349

Duan, B., Li, L., Chen, G., Su-Zhou, C., Li, Y., Merkeryan, H., Liu, W., Liu, X. 2021. ACC deaminase-producing PGPRs improve drought stress tolerance in grapevine (Vitis vinifera L.) *Front. Plant Sci.* 12, 706990. doi: 10.3389/fpls.2021.706990

Duan, H., Liu, W., Zhou, L., Han, B., Huo, S., El-Sheekh, M., Dong, H., Li, X., Xu, T., Elshobary, M. 2023a. Improving saline-alkali soil and promoting wheat growth by co-applying potassium-solubilizing bacteria and cyanobacteria produced from brewery wastewater. *Front. Environ. Sci.* 11, 1170734. doi: 10.3389/fenvs.2023.1170734

Duan, X., Zou, C., Jiang, Y., Yu, X., Ye, X. 2023b. Effects of reduced phosphate fertilizer and increased Trichoderma application on the growth, yield, and quality of pepper. *Plants* 12(16), 2998.

El-Egami, H.M., Hegab, R.H., Montaser, H., El-Hawary, M.M. 2024. Hasanuzzaman M. Impact of potassium-solubilizing microorganisms with potassium sources on the growth, physiology, and productivity of wheat crop under salt-affected soil conditions. *Agronomy* 14(3), 423. doi: 10.3390/agronomy14030423

Esmailpour, A., Hassanzadehdelouei, M., Madani, A. 2013. Impact of livestock manure, nitrogen and biofertilizer (Azotobacter) on yield and yield components wheat (Triticum Aestivum L.). *Cerc. Agron. Mold.* 46(2), 5–15.

Etesami, H., Emami, S., Likhani, H.A. 2017. Potassium solubilizing bacteria (KSB): mechanisms, promotion of plant growth, and future prospects – a review. *J. Soil Sci. Plant Nutr.* 17(4), 897–911.

Gandhi, R., Prittesh, P., Jinal, H.N., Chavan, S.M., Paul, D., Amaresan, N. 2022. Evaluation of the effect of potassium solubilizing bacterial strains on the growth of wheat (Triticum aestivum L.). *J. Plant Nutr.* 46(8), 1479–1490. doi: 10.1080/01904167.2022.2074859

Ghorai, P., Ghosh, D. 2022. Ameliorating the performance of NPK biofertilizers to attain sustainable agriculture with special emphasis on bioengineering. *Bioresour. Technol. Rep.* 19, 101117. doi: 10.1016/j.biteb.2022.101117

Han, T., Li, D., Liu, K., Huang, G., Zhang, L., Liu, S., Shah, A., Liu, L., Feng, G., Zhang, H. 2023. Soil potassium regulation by initial K level and acidification degree when subjected to liming: a meta-analysis and long-term field experiment. *Catena* 232, 107408.

Huang, Y., Zhang, Z., Li, Z., Dai, D., Li, Y. 2022. Evaluation of water use efficiency and optimal irrigation quantity of spring maize in Hetao Irrigation District using the Noah-MP Land Surface Model. *Agric. Water Manag.* 264, 107498. doi: 10.1016/j.agwat.2022.107498

Iftikhar, A., Farooq, R., Akhtar, M., Khalid, H., Hussain, N., Ali, Q., Saif ul Malook, Ali, D. 2024. Ecological and sustainable implications of phosphorous-solubilizing microorganisms in soil. *Discov. Appl. Sci.* 6(2). doi: 10.1007/s42452-024-05683-x

Iqbal, Z., Ahmad, M., Raza, M.A. et al. 2024. Phosphate-solubilizing *Bacillus* sp. modulate soil exoenzyme activities and improve wheat growth. *Microb. Ecol.* 87, 31. doi: 10.1007/s00248-023-02340-5

Jalal, A., da Silva Oliveira, C.E., Galindo, F.S., Rosa, P.A.L., Gato, I.M.B., de Lima, B.H., Teixeira Filho, M.C.M. 2023. Regulatory mechanisms of plant growth-promoting rhizobacteria and plant nutrition against abiotic stresses in brassicaceae family. *Life* 13, 211.

Jensen, C.N.G., Pang, J.K.Y., Gottardi, M. et al. 2024. *Bacillus subtilis* promotes plant phosphorus (P) acquisition through P solubilization and stimulation of root and root hair growth. *Physiol. Plant.* 176(3), e14338.

Jiao, H., Wang, R., Qin, W., Yang, L. 2024. Screening of rhizosphere nitrogen fixing, phosphorus and potassium solubilizing bacteria of Malus sieversii (Ldb.) Roem. and the effect on apple growth. *J. Plant Physiol.* 292, 154142.

Kaschuck, G., Hungria, M. 2017. Diversity and importance of diazotrophic bacteria to agricultural sustainability in the tropics. In *Diversity and Benefits of Microorganisms from the Tropics*; Springer International Publishing, New York, NY, USA, 269–292.

Kaur, T., Devi, R., Kour, D., Yadav, A., Yadav, A.N. 2021. Plant growth promotion of barley (Hordeum vulgare L.) by potassium solubilizing bacteria with multifarious plant growth promoting attributes. *Plant Sci. Today* [Internet] 31(spl), 17–24. https://horizonepublishing.com/journals/index.php/PST/article/view/1377

Khan, M.A., Asaf, S., Khan, A.L., Ullah, I., Ali, S., Kang, S.M., Lee, I.J. 2019. Alleviation of salt stress response in soybean plants with the endophytic bacterial isolate Curtobacterium sp. SAK1. *Ann. Microbiol.* 69, 797–808.

Khoshmanzar, E., Aliasgharzad, N., Neyshabouri, M.R., Khoshru, B., Arzanlou, M., Asgari Lajayer, B. 2020. Effects of Trichoderma isolates on tomato growth and inducing its tolerance to water-deficit stress. *Int. J. Environ. Sci. Technol. 17*, 869–878. doi: 10.1007/s13762-019-02405-4

Kour, D., Yadav, A.N. (2023). Alleviation of cold stress in wheat with psychrotrophic phosphorus solubilizing *Acinetobacter rhizosphaerae* EU-KL44. *Braz. J. Microbiol.* 54, 371–383.

Liu, H., Li, S., Qiang, R., Lu, E., Li, C., Zhang, J., Gao, Q. 2022. Response of soil microbial community structure to phosphate fertilizer reduction and combinations of microbial fertilizer. *Front. Environ. Sci.* 10, 899727.

Liu, T., Duan, H., Lücker, S. 2024. Sustainable wastewater management through nitrogen-cycling microorganisms. *Nat. Water* doi: 10.1038/s44221-024-00307-5

Liu, Y., Wang, Y., Kong, W.H., Liu, W.L., Xie, X.H., Wu, X. 2020. Identification, cloning and expression patterns of the genes related to phosphate solubilization in Burkholderia multivorans WS-FJ9 under different soluble phosphate levels. *AMB Express* 10, 108. doi: 10.1186/s13568-020-01032-4

Liu-Xu, L., González-Hernández, A.I., Camañes, G., Vicedo, B., Scalschi, L., Llorens, E. 2024. Harnessing green helpers: nitrogen-fixing bacteria and other beneficial microorganisms in plant–microbe interactions for sustainable agriculture. *Horticulturae* 10, 621.

Mahato, S., Kafle, A. 2018. Comparative study of Azotobacter with or without other fertilizers on growth and yield of wheat in Western hills of Nepal. *Ann. Agrar. Sci.* 16, 250–256.

Mahdi, I., Fahsi, N., Hafidi, M., Allaoui, A., Biskri, L. 2020. Plant growth enhancement using rhizospheric halotolerant phosphate solubilizing bacterium *Bacillus licheniformis QA1* and *Enterobacter asburiae QF11* isolated from *Chenopodium quinoa* willd. *Microorganisms* 8(6), 948.

Mahmoud, G.A.E., Abeed, A.H.A., Mostafa, H.H.A., Abdel Monsef, O. 2023. responses of pea (*Pisum sativum* l.) to single and consortium bio-fertilizers in clay and newly reclaimed soils. *Plants* 12, 3931. doi: 10.3390/plants12233931

Mahmoud, G.A.E., Hefzy, M., Zahran, M.M.A.A. 2024. Synergistic effects of microbial gibberellic acid and vitamins on onion (*Allium cepa* L.) yield, and quality in low-fertility soil. *J. Soil Sci. Plant Nutr.* 24, 5342–5359. doi: 10.1007/s42729-024-01911-w

Mahmoud, M.M.A., El-Bably, A.Z. 2017. Crop Water Requirements and Irrigation Efficiencies in Egypt. In *The Handbook of Environmental Chemistry*. doi: 10.1007/698_2017_42

Mahmud, K., Makaju, S., Ibrahim, R., Missaoui, A. 2020. Current progress in nitrogen fixing plants and microbiome research. *Plants (Basel, Switz)* 9(1), 97. doi: 10.3390/plants9010097

Meena, R.P., Karnam, V., Sujatha, H.T., Tripathi, S.C., Singh, G. 2024. Practical approaches to enhance water productivity at the farm level in Asia: a review. *Irrig. Drain.* 73(2), 770–793. doi: 10.1002/ird.2891

Mishra, I., Fatima, T., Egamberdieva, D., Arora, N.K. 2020. Novel bioformulations developed from *Pseudomonas putida* BSP9 and its biosurfactant for growth promotion of *Brassica juncea* (L.). *Plants* 9(10), 1349.

Molden, D., Oweis, T., Steduto, P., Bindraban, P., Hanjra, M.A., Kijne, J. 2010. Improving agricultural water productivity: between optimism and caution. *Agric. Water Manag.* 97(4), 528–535. doi: 10.1016/j.agwat.2009.03.023

Moreau, D., Bardgett, R.D., Finlay, R.D., Jones, D.L., Philippot, L. 2019. A plant perspective on nitrogen cycling in the rhizosphere. *Funct. Ecol.* 33(4), 540–552. doi: 10.1111/1365-2435.13303

Muthuraja, R., Muthukuma, T. 2021. Isolation and characterization of potassium solubilizing Aspergillus species isolated from saxum habitats and their effect on maize growth in different soil types. *Geomicrobiol J.* 38 (8), 672–685. doi: 10.1080/01490451.2021.1928800

Nacoon, S., Jogloy, S., Riddech, N., Mongkolthanaruk, W., Kuyper, T.W., Boonlue, S. 2020. Interaction between phosphate solubilizing bacteria and arbuscular mycorrhizal fungi on growth promotion and tuber inulin content of helianthus tuberosus L. *Sci. Rep.* 10, 4916. doi: 10.1038/s41598-020-61846-x

Nawaz, A., Qamar, Z.U., Marghoob, M.U., Imtiaz, M., Imran, A., Mubeen, F. 2023. Contribution of potassium solubilizing bacteria in improved potassium assimilation and cytosolic K+/Na+ ratio in rice (Oryza sativa L.) under saline-sodic conditions. *Front. Microbiol.* 14, 1196024. doi: 10.3389/fmicb.2023.1196024

Nguyen, N.K., Vo, D.T., Le, T.X., Morton, L.W., Tran, H., Robatjazi, J., Lasar, H.J.L., Tecimen, H.B. 2024. Isolation, and selection of indigenous potassium solubilizing bacteria from Vietnam Mekong Delta rhizospheric soils and their effects on diverse cropping systems. *Biocatal. Agric. Biotechnol.* 58, 103200.

Noor-Hassim, M.F., Ng, C.L., Teo, H.M., Wahizatul-Afzan, A.Z.M., Muhamad-Zalan, N.B. Karim, N.B., Ahmad, A. 2023. The utilization of microbes for sustainable food production. *BioTechnologia* 104(2), 209–216.

Olaniyan, F.T., Alori, E.T., Adekiya, A.O., Ayorinde, B.B., Daramola, F.Y., Osemwegie, O.O., Babalola, O.O. 2022. The use of soil microbial Potassium solubilizers in potassium nutrient availability in soil and its dynamics. *Ann. Microbiol.* 72(1), 45. doi: 10.1186/s13213-022-01701-8

Omer, A.M., Abd-Elnaby, A.S. 2017. Effect of phosphate dissolving bacteria on physiological behavior of some sesame cultivars under saline conditions at Sahle Eltina-North Sinai. *Alex. Sci. Exch. J.* 38(4), 687–698.

Paul, S., Rakshit, A. 2021. Effect of seed bio-priming with *Trichoderma viride* strain BHU-2953 for enhancing soil phosphorus solubilization and uptake in soybean (*Glycine max*). *J. Soil Sci. Plant Nutr.* 21, 1041–1052.

Pauwels, R., Graefe, J., Bitterlich, M. 2023. An arbuscular mycorrhizal fungus alters soil water retention and hydraulic conductivity in a soil texture specific way. *Mycorrhiza* 33(3), 165–179. doi: 10.1007/s00572-023-01106-8

Prakash, J., Arora, N.K. 2019. Phosphate-solubilizing *Bacillus* sp. enhances growth, phosphorus uptake and oil yield of *Mentha arvensis* L.. *3 Biotech* 9(4), 126. doi: 10.1007/s13205-019-1660-5

Prasad, A., Dixit, M., Meena, S.K., Kumar, S.A. 2023. Qualitative and quantitative estimation for phosphate solubilizing ability of Trichoderma isolates: a natural soil health enhancer. *Mater Today Proc.* 81, 360–366.

Raddad, N., Giacomucci, L., Marasco, R., Daffonchio, D., Cherif, A., Fava, F. 2018. Bacterial polyextremotolerant bio-emulsifiers from arid soils improve water retention capacity and humidity uptake in sandy soil. *Microbiol. Cell Fact.* 17(83), 1–12. doi: 10.1186/s12934-018-0934-7

Raji, M., Thangavelu, M. 2021. Isolation and screening of potassium solubilizing bacteria from saxicolous habitat and their impact on tomato growth in different soil types. *Arch. Microbiol.* 203, 3147–3161. doi: 10.1007/s00203-021-02284-9

Ramirez-Valdespino, C.O., Orrantia-Borunda, E. 2021. Trichoderma and nanotechnology in sustainable agriculture: a review. *Front. Fungal Biol.* 2, 764675. doi: 10.3389/ffunb.2021.764675

Rawat, P., Das, S., Shankhdhar, D. et al. 2021. Phosphate-solubilizing microorganisms: mechanism and their role in phosphate solubilization and uptake. *J. Soil Sci. Plant Nutr.* 21, 49–68. doi: 10.1007/s42729-020-00342-7

Rees, D.C., Akif Tezcan, F., Haynes, C.A., Walton, M.Y., Andrade, S., Einsle, O., Howard, J.B. 2005. Structural basis of biological nitrogen fixation. *Phil. Trans. R. Soc. A* 363 (1829), 971–984.

Saadouli, I., Mosbah, A., Ferjani, R., Stathopoulou, P., Galiatsatos, I., Asimakis, E., Marasco, R., Daffonchio, D., Tsiamis, G., Ouzari, H.I. 2021. The impact of the inoculation of phosphate-solubilizing bacteria *Pantoea agglomerans* on phosphorus availability and bacterial community dynamics of a semi-arid soil. *Microorganisms* 9(8), 1661. doi: 10.3390/microorganisms9081661

Sadiq, H.M., Jahangir, G.Z., Nasir, I.A., Iqtidar, M., Iqbal, M. 2013. Isolation and characterization of phosphate-solubilizing bacteria from rhizosphere soil. *Biotechnol. Biotechnol. Equip.* 27(6), 4248–4255. doi: 10.5504/BBEQ.2013.0091

Saheewala, H., Sanadhya, S., Upadhyay, S.K., Mohanty, S.R., Jain, D. 2023. Polyphasic characterization of indigenous potassium-solubilizing bacteria and its efficacy studies on maize. *Agronomy* 13(7), 1919. doi: 10.3390/agronomy13071919

Sattar, A., Naveed, M., Ali, M., Zahira, Z.A., Nadeem, S.M., Yaseen, M., Meena, V.S., Farooq, M., Singh, R., Rahman, M., Meena, H.N. 2019. Perspectives of potassium solubilizing microbes in sustainable food production system: a review. *Appl. Soil Ecol.* 133(12), 146–159.

Scudeletti, D., Crusciol, C.A.C., Bossolani, J.W., Moretti, L.G., Momesso, L., Tubaña, B.S., de Castro, S.G.Q., De Oliveira, E.F., Hungria, M. 2021. Trichoderma asperellum inoculation as a tool for attenuating drought stress in sugarcane. *Front. Plant Sci.* 12, 645542. doi: 10.3389/fpls.2021.645542

Serna-Cock, L., Arias-García, C., Valencia Hernandez, L.J. 2011. Effect of biofertilization on the growth of potted sugarcane plants (Saccharum officinarum). *Biotecnol. Sector Agropecuario Agroind.* 9, 85–95.

Singh, B., Boukhris, I., Pragya, Kumar, Yadav, A.N., Farhat-Khemakhem, A., Kumar, A., Singh, D., Blibech, M., Chouayekh, H., Alghamdi, O.A. 2020. Contribution of microbial phytases to the improvement of plant growth and nutrition: a review. *Pedosphere* 30 (3), 295–313. doi: 10.1016/s1002-0160(20)60010-8

Soares, A.S., Nascimento, V.L., de Oliveira, E.E. et al. 2023. *Pseudomonas aeruginosa* and *Bacillus cereus* isolated from Brazilian Cerrado soil act as phosphate-solubilizing bacteria. *Curr. Microbiol.* 80, 146.

Song, M., Lin, X., Wei, X., Zeng, Q., Mu, C., Zhou, X. 2024. *Trichoderma viride* improves phosphorus uptake and the growth of *Chloris virgata* under phosphorus-deficient conditions. *Front. Microbiol.* 15, 1425034.

Song, M., Wang, X., Xu, H. et al. 2023. Effect of *Trichoderma viride* on insoluble phosphorus absorption ability and growth of *Melilotus officinalis*. *Sci. Rep.* 13, 12345.

Song, P., Xu, D., Yue, J., Ma, Y., Dong, S., Feng, J. 2022. Recent advances in soil remediation technology for heavy metal contaminated sites: a critical review. *Sci. Total Environ.* 838, 156417. doi: 10.1016/j.scitotenv.2022.156417

Souza, A.E.S.D., Filla, V.A., Silva, J.P.M.D., Barbosa Júnior, M.R., Oliveira-Paiva, C.A.D., Coelho, A.P., Lemos, L.B. 2023. Application of *Bacillus* spp. phosphate-solubilizing bacteria improves common bean production compared to conventional fertilization. *Plants* 12, 3827. doi: 10.3390/plants12223827

Suleimanova, A., Bulmakova, D., Sokolnikova, L., Egorova, E., Itkina, D., Kuzminova, O., Gizatullina, A., Sharipova, M. 2023. Phosphate solubilization and plant growth promotion by *Pantoea brenneri* soil isolates. *Microorganisms* 11(5), 1136.

Sun, F., Ou, Q., Wang, N., Guo, Z., Ou, Y., Li, N., Peng, C. 2020. Isolation and identification of potassium-solubilizing bacteria from Mikania micrantha rhizospheric soil and their effect on M. micrantha plants. *Glob. Ecol. Conserv.* 23, e01141.

Swain, S.S., Mishra, A., Sahoo, B., Chatterjee, C. 2020. Water scarcity-risk assessment in data-scarce river basins under decadal climate change using a hydrological modelling approach. *J. Hydrol.* 590, 125260. doi: 10.1016/j.jhydrol.2020.125260

Tandon, A., Fatima, T., Shukla, D., Tripathi, P., Srivastava, S., Singh, P.C. 2020. Phosphate solubilization by Trichoderma koningiopsis (NBRI-PR5) under abiotic stress conditions. *J. King Saud Univ. Sci.* 32, 791–798.

Tian, D., Zhang, X., Wang, L., Han, M., Zhang, C., Ye, X. 2023. Lead remediation is promoted by phosphate-solubilizing fungi and apatite via the enhanced production of organic acid. *Front. Bioeng. Biotechnol.* doi: 10.3389/fbioe.2023.1180431

Timofeeva, A., Galyamova, M., Sedykh, S. 2022. Prospects for using phosphate-solubilizing microorganisms as natural fertilizers in agriculture. *Plants (Basel).* 11(16), 2119. doi: 10.3390/plants11162119

Torres-Cuesta, D., Mora-Motta, D., Chavarro-Bermeo, J.P., Olaya-Montes, A., Vargas-Garcia, C., Bonilla, R., Estrada-Bonilla, G. 2023. Phosphate-solubilizing bacteria with low-solubility fertilizer improve soil P availability and yield of Kikuyu grass. *Microorganisms* 11(7), 1748.

Trivedi, P., Leach, J.E., Tringe, S.G., Sa, T., Singh, B.K. 2020. Plant–microbiome interactions: from community assembly to plant health. *Nat. Rev. Microbiol.* 18(11), 607–621. doi: 10.1038/s41579-020-00490-8

Velimirovic, A., Jovovic, Z., Przulj, N. 2021. From neolithic to late modern period: brief history of wheat. *Genetika* 53(1), 407–417.

Verma, M., Singh, A., Dwivedi, D.H. et al. 2020. Zinc and phosphate solubilizing *Rhizobium radiobacter* (LB2) for enhancing quality and yield of loose leaf lettuce in saline soil. *Environ. Sustain.* 3, 209–218.

Wang, C., Pan, G., Lu, X., Qi, W. 2023. Phosphorus solubilizing microorganisms: potential promoters of agricultural and environmental engineering. *Front. Bioeng. Biotechnol.* 11, 1181078. doi: 10.3389/fbioe.2023.1181078

Wang, X., Zhang, J., Wang, X., An, J., You, C., Zhou, B., Hao, Y. 2022b. The growth promoting mechanism of Brevibacillus laterosporus AMCC100017 on apple rootstock Malus robusta. *Hortic. Plant J.* 8(1), 22–34. doi: 10.1016/j.hpj.2021.11.005

Wang, Y., Yan, X., Su, M., Li, J., Man, T., Wang, S., Li, C., Gao, S., Zhang, R., Zhang, M., Wang, P., Jia, X., Ren, L. 2022a. Isolation of potassium solubilizing bacteria in soil and preparation of liquid bacteria fertilizer from food wastewater. *Biochem. Eng. J.* 181, 108378.

Wei, X., Xie, M., Wan, C., Song, R., Zhong, W., Xin, S., Song, K. 2024. Enhancing soil health and plant growth through microbial fertilizers: mechanisms, benefits, and sustainable agricultural practices. *Agronomy* 14(609), 1–16.

Wouter, W., Samia, E.G., El Deen, S. Magdy, R.K., Robert, S., Froebrich, J. 2013. Assessment and management of water resources in Egypt to face drought and water scarcity. *Geophys. Res. Abstr.* 15, EGU2013-10647-1.

Wu, X., Zhao, Z., Zhao, Z., Zhang, Y., Li, M., Yu, Q. 2023. Analysis of the potassium-solubilizing priestia megaterium strain NK851 and its potassium feldspar-binding proteins. *Int. J. Mol. Sci.* 24(18), 14226. doi: 10.3390/ijms241814226

Xiong, Q., Hu, J., Wei, H., Zhang, H., Zhu, J. 2021. Relationship between plant roots, rhizosphere microorganisms, and nitrogen and its special focus on rice. *Agriculture* 11(3), 234.

Xu, X., Du, X., Wang, F., Sha, J., Chen, Q., Tian, G., Zhu, Z., Ge, S., Jiang, Y. 2020. Effects of potassium levels on plant growth, accumulation and distribution of carbon, and nitrate metabolism in apple dwarf rootstock seedlings. *Front. Plant Sci.* 11, 904. doi: 10.3389/fpls.2020.00904

Yahya, M., Rasul, M., Sarwar, Y. et al. 2022. Designing synergistic biostimulants formulation containing autochthonous phosphate-solubilizing bacteria for sustainable wheat production. *Front. Microbiol.* 13, 889073. doi: 10.3389/fmicb.2022.889073

Yazdani, M., Bahmanyar, M.A., Pirdashti, H., Esmaili, M.A. 2009. Effect of phosphate solubilization microorganisms (PSM) and plant growth promoting rhizobacteria (PGPR) on yield and yield components of corn (Zea mays L.). *World Acad. Sci. Eng. Technol.* 49, 90–92.

Yin, J., Sui, Z.M., Huang, J.G. 2021. Mobilization of soil inorganic phosphorus and stimulation of crop phosphorus uptake and growth induced by Ceriporia lacerata HG2011. *Geoderma* 383, 114690. doi: 10.1016/j.geoderma.2020.114690

Zahran, M.M.A.A., Hefzy, M., Mostafa, H.H.A. 2020. Enhancing the productivity of potato crop under drought stress by using some biological treatments. *Int. J. Environ.* 9(2), 83–103.

Zhang, L., Tan, C., Li, W. et al. 2024. Phosphorus-, potassium-, and silicon-solubilizing bacteria from forest soils can mobilize soil minerals to promote the growth of rice (Oryza sativa L.). *Chem. Biol. Technol. Agric.* 11, 103. doi: 10.1186/s40538-024-00622-9

Zhao, D., Ding, Y., Cui, Y. et al. 2022. Isolation and genome sequence of a novel phosphate-solubilizing rhizobacterium *Bacillus altitudinis* GQYP101 and its effects on rhizosphere microbial community structure and functional traits of corn seedling. *Curr. Microbiol.* 79, 249.

2 Plant Growth Promoter Mechanisms Under Biotic and Abiotic Stresses and Nutrient Uptake

Aya A.M. Abdellatif, Ibrahim M.K. Elmaghraby, Maissara M.K. Elmaghraby, Mostafa A.S. Abdelgalil, and Hassan H.A. Mostafa

2.1 INTRODUCTION

Due to the continuously growing world population, great pressure is placed on biological resources, water, and arable land to meet the demand for food while preserving the ecosystems' integrity. In the last 50 years, under the modern agriculture system, the implementation of chemical fertilizers has increased aggressively (Mahmoud et al. 2023). Moreover, the negative impact of biotic and abiotic stresses causes a significant loss in plant productivity and poses a major challenge for maintaining world food demands and food industrial processes. Therefore, considering the environmental damage, increasing costs, and the risk of increasing soil salinity regarding the use of current fertilization practices, the necessity of new farming technologies is highly required.

The plant biostimulants, as microbial or non-microbial (i.e., humic and fulvic acids, seaweed extracts, protein hydrolysates, amino acids, silicon, and chitosan), are a new generation of available products on the agricultural market used to achieve agricultural sustainability (Rouphael and Colla 2020; Castiglione et al. 2021). Biostimulants are any natural substances that have beneficial effects on plant development and growth, productivity, quality, and stress tolerance. Jalal et al. (2023) reported that using microbes to produce agricultural products and lessen the adverse impacts of stress is thought to be a novel sustainable technology. A microbial biostimulant includes a microorganism or a consortium of microorganisms, which consists of four diverse genera (*Azotobacter* spp., *Azospirillum* spp., mycorrhizal fungi, and *Rhizobium* spp.). The types and classification of biostimulants are shown in Figure 2.1. Recently, the conducted research linked to the positive impact of microorganisms is continuously rising. Numerous microorganisms play a pivotal function

 DOI: 10.1201/9781003714484-2

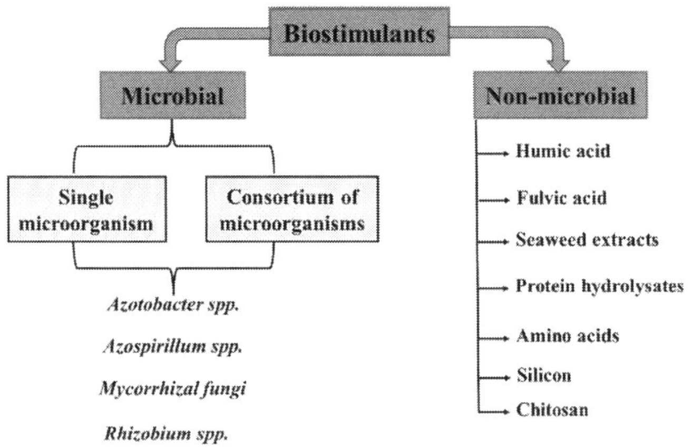

FIGURE 2.1 Classification of biostimulants.

in atmospheric nitrogen fixation, organic wastes, plant disease mitigation, detoxification of pesticides, and bioactive compound production, like hormones, vitamins, and enzymes (Munir et al. 2022).

The plant growth-promoting rhizobacteria (PGPR) improved the productivity and growth of various agriculture plants; however, it varied in different crops and in various varieties of the same crop. Therefore, here we focused on how PGPR contributes to plant productivity, growth, and health of various agriculture crops and nutrient uptake under normal and stressful conditions.

2.1.1 PGPR Contributions to Plant Growth and Development

In the era of climate change, there is a growing requirement for crop production with a considerable decrease in the application of synthetic fertilizers. Numerous plant growth-promoting rhizobacteria (PGPR) play a crucial role in the sustainable agriculture business (Vejan et al. 2016). Soil application with PGPR showed a strong capability to boost plant yield and growth through multiple mechanisms, including phosphate solubilization, nitrogen fixation, biosynthesis of phytohormones, antimicrobial compounds, and siderophore productions (Table 2.1).

Biofertilizers are natural products containing living beneficial microorganisms that are applied to plant surfaces, seeds, or soil to colonize the rhizosphere or interior parts of the plant to promote plant growth and participate in soil fertilization by biofixing of atmospheric nitrogen and bio-solubilization of insoluble nutrients like potassium, phosphorus, and some micronutrients (Aloo et al. 2020). There are numerous types of biofertilizers categorized based on their functions in plant rhizospheres. Markedly, a single biofertilizer consists of one strain of PGPR with solo or multiple PGP traits. Furthermore, some commercial biofertilizers are based on advanced formulations of microbial groups with diverse PGP traits. Consequently, PGPRs are principally recognized as important biological agents in integrated soil nutrient management for sustainable agricultural productivity, as mentioned in Elmaghraby et al. (2023).

TABLE 2.1
The Role of PGPR in Plant Nutrient Dynamics and Growth Promotion

PGPR	PGP Trait	Host Plant	Reference
Azotobacter chroococcum, *Azospirillum brasilense*	Biological Nitrogen Fixation and Phosphorus Solubilization	*Cassia angustifolia* Vhal (Senna) and *Cyamopsis tetragonoloba* (L.)	Junia et al. (2021)
Pseudomonas koreensis CY4 and *P. entomophila* CN11	Biological Nitrogen Fixation	Sugarcane (*Saccharum officinarum* L)	Singh et al. (2023)
Bacillus siamensis Gxun-6	Siderophore Production	Banana	Shen et al. (2022)
Bacillus pumilus strain *SH-9*	Regulation of phytohormones, Phosphate Solubilization, Siderophore, and Exopolysaccharides Production.	Soybean (*Glycine max*)	Shaffique et al. (2023)
Pseudomonas sp. Strain WS32	Phosphate Mobilization	Wheat (*Triticum aestivum*)	Ou et al. (2023)
Serratia marcescens strain NPKC3_2_21	Phosphate Mobilization	Rice (*Oryza sativa*)	Sutio et al. (2023)
Bacillus safensis TS3	Phosphatase and IAA production	Oat (*Avena sativa* L.)	Chebotar et al. (2024)
Pantoea vagans ZHS-1	Potassium Mobilization and IAA production	Rice (*Oryza sativa*)	Tian et al. (2024)
Pseudomonas aeruginosa A10	Potassium Solubilization and Organic Acids Production	Cotton (*Gossypium herbaceum*)	Zhao et al. (2024)

2.1.2 Biological Nitrogen Fixation

Nitrogen (N) is an important limiting factor and essential nutrient for plants, as it is the key building block of amino acids, proteins, and adenosine triphosphate, and it is the main component of chlorophyll, the most important pigment required for the photosynthetic process. N is accessible to the plant in two distinguished forms, ammonium (NH_4^+) or nitrate (NO_3^-). There are two mechanisms to regulate nitrogen use efficiency: nitrogen uptake and utilization efficiency. Additionally, regarding nitrogen uptake efficiency, there are two transport strategies engaged in nitrate uptake inside the plant: one at a low nitrogen concentration called the high-affinity transport system (HATS), and the other involved at a higher nitrate concentration called the low-affinity transport system (LATS). The nitrate transporter 1 (NRT1) gene family is involved in HATS, while the NRT2 gene family is characterized as an LATS. However, for nitrogen utilization efficiency, there are several enzymes that participate in nitrogen intake after transporting nitrate into the plant cell: nitrite reductase (NiR), reductase (NR), glutamine synthetase (GS), glutamate 2-oxoglutarate amino transferase (GOGAT), glutamic-oxaloacetic transaminase (GOT), and glutamic-pyruvic

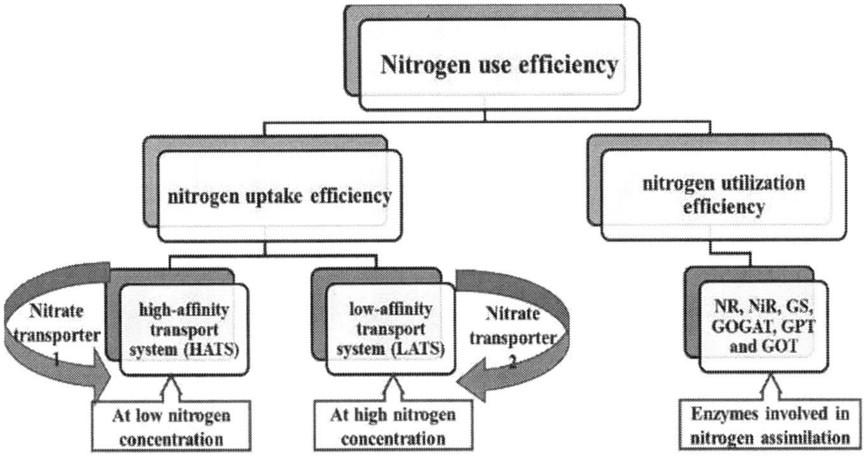

FIGURE 2.2 Summarize the nitrogen use efficiency mechanism.

transaminase (GPT) (Figure 2.2), Mostafa et al. (2021). In most agricultural systems, N deficiency results in significant limitations in plant growth (Sanow et al. 2022). Nitrogen-fixing biofertilizers contain PGPR that performs biological nitrogen fixation (BNF), thereby enhancing soil nitrogen supply to crops. BNF fixation is limited to bacteria that have an enzymatic complex (nitrogenase), which catalyzes the decrease of atmospheric nitrogen into ammonia. Nitrogen-fixing prokaryotes include both free-living rhizospheric bacterial species such as *Azotobacter* (Sumbul et al. 2020) and *Azospirillum* (Fukami et al. 2018), in addition to symbiotic rhizospheric bacteria, including *Rhizobium*, *Bradyrhizobium*, *Azorhizobium*, *Sinorhizobium*, *Allorhizobium*, and *Mesorhizobium*, that could fix nitrogen only when they become in symbiotic association with specific plants (Suzaki et al. 2015).

The members of *Azotobacter* are free-living, aerobic, gram-negative motile bacteria that are embedded in a thick-walled cyst. *Azotobacter* can be inoculated in soil for its high capability for converting nitrogen into ammonia, which is later on absorbed by the plants. Some species of *Azotobacter* could fix approximately 20 kg N per hectare annually (Sumbul et al. 2020). In this context, it was found that inoculation of the tomato rhizosphere with *A. chroococcum* promoted several growth parameters under reduced nitrogen conditions and salt stresses (Van Oosten et al. 2018). Additionally, the field application of *A. chroococcum*, *Bacillus megaterium*, *B. circulans*, *B. subtilis*, and *Pseudomonas fluorescens* recorded positive impacts on nutrient uptake, plant defense enzymes, and numerous growth parameters of onion under organic agricultural practices (Abdelrahman et al. 2021). Several strains of *Klebsiella* are salt-tolerant and could fix the atmospheric nitrogen. *Klebsiella pneumoniae* JCK-2201 is PGPR that has nitrogen-fixing activity and showed biocontrol activity against bacterial wilt in tomatoes (Kim et al. 2022). Additionally, Li et al. (2017) studied the plant-growth-promoting traits and nitrogenase activity of *Pseudomonas* species isolated from sugarcane rhizosphere. The results revealed that out of all the investigated species, *P. koreensis* and *P. entomophila* displayed the most notable plant growth-promoting traits and root colonization ability, as well as nitrogenase activity.

Recently, certain members of *Bacillus* have been reported for their potentiality to fix nitrogen to a wide range of host plants. Karagoz and Dursun (2019) stated that the application of a bacterial mixture consisting of nitrogen-fixing and phosphate-solubilizing bacteria *B. megaterium* and *P. polymyxa* increased the number of bulbs and later improved the satisfactory quality of tulip cultivars.

Rhizobia is a genus of Gram-negative bacteria that perform symbiotic nitrogen fixation in the roots of nodule plants, and such a process is mostly controlled by *nod*, *nif*, and *fix* genes. However, *Frankia* and *Cyanobacteria* are two other groups of bacteria that have the capability to fix atmospheric nitrogen in symbiotic association with plants (Lindström and Mousavi 2019). Recently, the application of *Rhizobia* in combination with other effective PGPR maximizes its modulation activity. Zhang et al. (2023) revealed that soil inoculation with PGPR, *Enterobacter* sp. Z1, and *Klebsiella* sp. Z2 enhanced the symbiosis between rhizobia and soybeans and obviously boosted plant productivity through the effective production of indole acetic acid and taurine, which increased nodule production, chlorophyll content, and nitrogenase activity in soybeans. In another study, Win et al. (2023) clarified that the dual application of ACC deaminase-producing *Pseudomonas* and *Bradyrhizobium* mixtures is a promising approach to enhancing the symbiotic efficiency of nodulation and mitigating salinity stress in soybeans. Also, Sibponkrung et al. (2020) revealed that co-inoculation of soil with *Bradyrhizobium diazoefficiens* and *Bacillus velezensis* resulted in enhancement in nitrogen-fixing efficiency and nodulation in soybean plants, which in turn increased the biomass yield.

2.1.3 Phosphorus and Potassium Solubilization

Phosphorus (P) is a vital macronutrient in agriculture, essential for plant development in agriculture due to its crucial roles in plant structure, energy transfer, biomass buildup, and respiration chain reactions. Regrettably, soluble inorganic phosphate in such chemical fertilizers is rapidly immobilized and remains fixed in soil as phosphates of iron, aluminum, and calcium. The application of phosphate-solubilizing bacteria (PSB) as a single strain or in combination can be considered a promising and sustainable approach to increasing phosphorus availability and increasing crop yields (Elmaghraby et al. 2023). PSB transforms phosphates into soluble forms by secreting enzymes like phosphatases and/or organic acids.

Numerous *Bacillus* species are effective in bioleaching phosphorus into available forms for plants. Different mechanisms could be performed by *Bacillus* spp., such as synthesizing organic acids that facilitate the solubilization of inorganic phosphate compounds (dicalcium phosphate, tricalcium phosphate, rock phosphate, and hydroxyapatite) and inducing activity of acid phosphatases that cause mineralization of organic phosphorous (Saeid et al. 2018). A recent study showed the highest capability of two PGPR strains named *Bacillus megaterium* and *B. subtilis* to produce natural organic acids in Pikovskaya's broth culture medium, which helps in the solubilization of phosphate, reduction of chemical fertilization usage, and promotion of wheat productivity (Iqbal et al., 2024). Another study stated that the soil application of *Pseudomonas* and *Bacillus* biofertilizers was able to solubilize rock phosphate by organic acid production and induce systemic resistance to the causal agent of canker disease in tomatoes. The inoculated tomato recorded high root and shoot lengths and weights compared with the control (Bakki et al. 2024). Co-inoculation with *B. subtilis* and *Serratia marcescens* was effective in phosphate solubilization due to organic acid secretion even in soils contaminated with common pesticides such as oxamyl, thiophanate methyl, and captan (Mohamed et al. 2018). Furthermore, *S. marcescens* isolated from the tea rhizosphere recorded various fertilization functions such as solubilizing insoluble phosphate, enhancing the growth of maize seedlings, and secreting volatile organic compounds such as dimethyl disulfide, which has broad antifungal activity (Gong et al. 2022). Lactic acid bacteria (LAB) can be used as eco-friendly biofertilizers in sustainable agriculture for their efficacy in producing a wide range of natural organic acids that play a vital role in the bioleaching of raw phosphate. *Limosilactobacillus* sp. LF-17 was effective as a phosphate solubilizer in calcareous soil and promoted the growth of maize seedlings (Li et al. 2024).

Potassium (K) is a vital macronutrient implicated in plant photosynthesis; activity affects plant metabolisms and stimulates their resilience to different abiotic and biotic stressors. Only a small amount of K is available to plants in free-soluble form because more than 90% of potassium exists in soils as an insoluble rock. Potassium-solubilizing biofertilizers have the capability to free potassium from minerals and supply K in an easily reached form to plants through the secretion of natural organic acids such as tartaric acid, acetic acid, oxalic acid, and citric acid to release K from its minerals (Olaniyan et al. 2022).

A consortium of five strains, identified as *Citrobacter braakii*, *Enterobacter hormaechei*, *Pseudomonas putida*, *Pantoea agglomerans*, and *Erwinia iniecta*, was used as soil inoculant to investigate its contribution to cytosolic K^+/Na^+ ratio, K intake, and subsequent yield enhancement in rice grown under saline-sodic conditions. The results revealed that both root and shoot K^+/Na^+ ratios were improved with a remarkable increase in the biomass of rice (*Oryza sativa* L.) under field conditions (Nawaz et al. 2023). The biofertilization of potatoes with potassium-solubilizing *Bacillus cereus* significantly increased K uptake by potato tubers by 62% compared to control plants when potassium feldspar was applied at a rate of 240 kg K_2O/ha (Ali et al. 2020).

2.1.4 Phytohormone Production

By definition, phytohormones are low-molecular organic molecules that are synthesized in certain plant organs. Phytohormone-producing PGPR has a direct impact on plant growth by improving mineral nutrition and inducing resistance to abiotic stress and plant pathogens. Plant hormone-mediated promotion in root growth is able to enhance water relations and mineral nutrition. Moreover, phytohormone-induced stimulation in plant antioxidant systems plays a central role in plant protection against oxidative stress that accompanies most dramatic environmental conditions (Ali et al. 2020).

Several biofertilizers encourage the growth of roots by satisfactory production of plant growth regulators (PGRs) such as indole-3-acetic acid (IAA) and auxin. Bacterial IAA production is strain-dependent, and it is a key selection as a desirable trait in bacterial strains during the design of commercial bioformulation (Pantoja-Guerra et al. 2023).

In this regard, Novel *Ignatzschineria* sp. is a promising phytohormone producer that recorded high production of IAA while grown in Luria-Bertani broth media supplemented with a precursor such as tryptophan. The apple root mass and length were highly stimulated by the application of *Ignatzschineria* sp. culture supernatant (Park et al. 2021). Similarly, *Bacillus velezensis*, when introduced into barley plants, acts as a rich source of IAA in the soil, enhancing both root length and weight (Abo Elsoud et al. 2023). Additionally, the endophytic *Bacillus safensis* has proven its ability to produce phytohormones that increase root surface area, promote cell proliferation and development, and mitigate salt stress, ultimately improving plant yield (Chebotar et al. 2024). Gibberellins are a class of tetracyclic diterpenoid carboxylic acids that are known for their capacity to elongate plant shoots and regulate shifts in the phases between growth and development of plants, including processes like elongation and cell division (Camara et al. 2018). Baliyan et al. (2021) optimized the production of gibberellic acid in a new endophytic strain, *Bacillus cereus* MEN8, to be applied as a plant growth regulator in Chickpea under field conditions. The results recorded that biofertilization of *B. cereus* MEN8 is a promising approach to boosting plant growth, seed germination, and crop productivity. Another recent investigation revealed that four promising rhizospheric bacterial species, i.e., *Rhizobium* sp., *Agrobacterium* sp., *Bacillus* sp., and *Phyllobacterium* sp., were isolated from *Acacia cyanophylla* root nodules and proved for their high potentiality to mobilize inorganic phosphate and produce IAA (Lebrazi et al. 2020).

Indole-3-acetic acid IAA is a common natural auxin phytohormone that is produced efficiently by plant growth-promoting Actinobacteria. Optimization of fermentation conditions is a key step for maximizing the phytohormones and antibiotics synthesized during large-scale production. The utmost production of IAA was achieved by *Streptomyces fradiae* NKZ-259 when grown on an induction medium enriched with 2 g/L tryptophan via tryptophan-dependent pathways (Myo et al. 2019). Mahmoud and Mostafa (2017) produced IAA by culturing the *Fusarium oxysporum*, and the maximum value was 300.4 mg/L when the fermentation conditions were as follows: 25°C temperature, pH 7, incubation period of five days, shaking rate of 150 rpm, and inoculum size of 2%.

2.2 ROLE OF PGPR IN PLANT HEALTH ENHANCEMENT

2.2.1 Support Plant Tolerance Under Abiotic Stressed Conditions

Plants are susceptible to a variety of environmental stresses, including drought, high temperature, and high salt levels, that have an impact on growth and productivity and can disrupt normal cellular activity (Bhat et al. 2023). Recently, agricultural fields have become more salinized as a result of climate change and a decline in the supply of clean water. This increase in salinity represents an important abiotic stress that negatively impacts gene expression and plant physiology. Consequently, vital processes like growth development, seed germination, and yield are dramatically affected (Acharya et al. 2024). Additionally, drought is the most harmful abiotic stress, resulting in massive losses of crop yield around the globe. Soil salinity has appeared as a significant danger to agricultural ecosystems around the globe, as it affects many continents across the world, reducing crop productivity significantly (Mishra et al. 2021). The presence of excessive water-soluble salts in soils has negative effects on different plant physiological processes, i.e., photosynthesis, seed germination, antioxidants, membrane transport, and ethylene production (Chang et al. 2014). Recent promising approaches, such as the application of genetically modified organisms, are effective in eliminating the effects of high salt and drought stresses (Raza et al. 2023). Organic amendments, including biochar, biofertilizer, vermiwash, and vermicompost, were proved as candidate approaches that could improve the drought tolerance and salinity of agricultural crops, leading to enhanced crop productivity (Hoque et al. 2022).

Plant growth-promoting rhizobacteria showed great ability to decrease the effects of drought stress by regulating stress-responsive genes and altering root morphology and biofilm production and by producing phytohormones, siderophores, osmolytes, exopolysaccharides, and volatile organic compounds (Poudel et al. 2021). Phytohormones such as indole acetic acid (IAA), abscisic acid (ABA), gibberellins, ethylene, and cytokinin play a crucial part in controlling the physiological and biochemical processes that promote yield and plant development under both optimum and stressful circumstances (Nahakpam et al. 2021). It has been shown that PGPR soil inoculations stimulate plant growth during drought conditions by modifying plant hormones. For example, abscisic acid (ABA) is essential for the growth of plants and has the ability to reorganize a variety of biochemical and physiological signal transduction pathways in order to help plants adapt to environmental stressors, including drought. Abscisic acid-mediated stomatal conductance reduces transpiration water loss during drought stress (Aslam et al. 2022). Gowtham et al. (2021) found that the rhizospheric *Bacillus marisflavi* was effective in stimulating the resistance against drought stress in mustard seedlings by the production of an ABA analogue. Under severe drought stress, plants maximize their formation of stress-generating ethylene (ET), which inhibits the growth of plants, especially seed germination and root development. Certain PGPR species secrete effectively. Aminocyclopropane-1-Carboxylate deaminase (ACCd), which could cleave the plant ethylene precursor ACC to ammonia and α-ketobutyrate and so enhance plant growth even under drought stress (Poudel et al. 2021). In this context, Saleem et al. (2018) stated that the application of ACCd-producing *Bacillus* sp. and *Enterobacter* sp. could alleviate drought stress in velvet beans (*Mucuna pruriens*). Bacterial biofilm is another mechanism for plant drought stress tolerance. PGPR biofilm components called exopolysaccharides enhance plant cell protection against desiccation during drought stress by improving water retention and by organizing the diffusion of organic carbon sources (Chandra et al. 2021). Several strains of *Pseudomonas* proved successful in reducing the effects of drought stress by expressing stress-responsive genes. *In vitro* investigation of these strains revealed ACCd activity and abscisic acid, gibberellic acid, and indole acetic acid, as well as exopolysaccharide production (Yasmin et al. 2021). Yang et al. (2023) reported that inoculation of *Arabidopsis* with bioproducts based on Streptomyces effectively alleviates salinity and drought stress by producing spiroketal polyketide pteridic acid H and its isomer F. Transcriptomics profiles, after pteridic acid treatment, revealed a remarkable enhancement in the expression of different stress-responsive genes in *Arabidopsis* seedlings. Additionally, inoculation of a drought-sensitive variety of wheat by *Streptomyces pactum* significantly promoted different growth parameters, photosynthetic efficiency, chlorophyll content, and antioxidant enzyme activity such as peroxidase (POD) and superoxide dismutase (SOD) when compared to untreated control in drought field environments (Yang et al. 2024).

Exposure to salinity is known to increase the severity of many plant pathogens due to poor defense mechanisms of plants caused by hormonal imbalance, osmotic stress, and repressed expression of defense response genes and primary metabolism-related genes (Gowtham et al. 2024). Shahzad et al. (2017) found that an endophytic bacterium, *Bacillus amyloliquefaciens*, was able to produce ABA, which has the ability to enhance plant resistance to salinity stress. The application of such endophytic bacterium boosted various growth parameters of rice plants as compared to control plants (non-inoculated) under salinity stress. The well-known salt-tolerant bacteria *Streptomyces* is able to survive in different hard conditions, including saline soils. Gao et al. (2022) demonstrated that the application of salt-tolerant *Streptomyces paradoxus* effectively enhanced soybean tolerance to saline-alkali stress and promoted seedling growth and crop productivity. Moreover, *Klebsiella variicola* was able to grow luxuriously under high salinity stressed conditions and secreted numerous active metabolites with principal salinity ameliorating traits, such as IAA, ACCd, and exopolysaccharides, as well as osmolytes, including proteins, sugars, proline, amino acids, and organic acids. Soil application of halotolerant *K. variicola* alleviated salinity stress and boosted both nutrients and growth in maize and wheat (Kusale et al. 2021).

Generally, the promotion of plant growth parameters in the presence of abiotic stresses is attributed to the ability of PGPR to secrete copious active compounds such as IAA, ACCD, EPS, and osmoprotectants that help in plant growth and relief from the destructive impacts of various environmental stresses (Table 2.2).

2.2.2 The Role of PGPR in Suppression of Plant Disease and Pests

Phytopathogens reduce both crop quality and yield, resulting in significant losses in agricultural productivity. Plant fungal diseases result in an annual estimated loss of approximately 10%–15% of the world's economic crops. Some plant diseases are induced by multiple pathogens, which are known as complex infection. Several factors are responsible for the pathogeny of fungi, including plant toxins, cell wall-degrading enzymes, growth regulators, and their analogs (Peng et al. 2021).

Traditional agriculture relies largely on chemical pesticides that are harmful to the environment and induce significant depletion of natural resources. Chemical-based agriculture poses environmental challenges, emphasizing the critical need for sustainable solutions in the face of climate change (Tripathi et al. 2024). Numerous studies have discussed the role of PGPR in promoting disease resistance and plant growth in crops. Many plant pathogens, mainly fungi, bacteria, and nematodes, cause serious plant diseases, which are considered an important barrier to sustainable crop yield and health. Biocontrol of plant pathogens is a promising, sustainable, and safe approach that reduces the riskiness of various crop diseases (Ayaz et al. 2023). The potentiality of PGPR as biocontrol agents and for replacing fungicides and chemical

TABLE 2.2
Role of PGPR in Mitigation of Abiotic Stresses

PGPR	Abiotic Stress	Mitigation Mechanism	Host Crop	Reference
Bacillus amyloliquefaciens S-134	Drought	Regulation of indole-3-acetic acid (IAA), indole-3-lactic acid (ILA), and indole-3-carboxylic acid (ICA) production	Wheat (*Triticum aestivum*)	Raheem et al. (2018)
Enterobacter cloacae	Drought	Production of ACC deaminase that catabolizes the ethylene produced in response to drought stress	*Maize (Zea mays L.)*	Danish et al. (2020)
Bacillus subtilis-FAB1 & *Pseudomonas azotoformans*-FAP3	Drought	Biofilm formation	Wheat (*Triticum aestivum)*	Ansari et al. (2023)
Bacillus subtilis IB22	Salinity	Accumulation of Abscisic Acid	Barley	Akhtyamova et al. (2021)
Pseudomonas putida	Drought	Indole acetic acid, and catalase production.	Canola (*Brassica napus L.)*	Tanveer et al. (2023)
Streptomyces Paradoxus D2-8	Soda Saline-Alkali Stress	IAA, ACCd, siderophore, and organic acids production	Soybean (*Glycine max*)	Gao et al. (2022)
Serratia liquefaciens KM4	Salinity	Regulating ion homeostasis, leaf gas exchange, redox potential, and stress-related genes expression.	Maize (*Zea mays*)	El-Esawi et al. (2018)
Lactobacillus planturum YQM48	Saline-Alkali Soil	Production of natural organic acids such as lactic acid, acetic acid, and butyric acid	Alfalfa (Medicago *sativa*)	Liu et al. (2023)
Bacillus pumilus, Bacillus licheniformis, Bacillus firmus, Bacillus safensis, and *Bacillus cereus*	Salinity	Production of ACCd, siderophores, and IAA.	Tomato (Lycopersicon esculentum Mill.)	Patani et al. (2023)
Bacillus safensis	High-temperature	Preserving functional and structural integrity of important cellular organelles like mitochondria and chloroplast via elimination of excess ROS.	Wheat (*Triticum aestivum*)	Sarkar et al. (2021)

pesticides may be related to multiple mechanisms of action, including direct competition for nutrients and niches, signal interference, enzyme lysis, antibiosis, and indirect induction of host resistance. Additionally, beneficial rhizosphere microorganisms can form biofilms and secrete secondary metabolites, including iturin, surfactin, fengycin, and bacillomycin, which inhibit the phytopathogen population by establishing robust plant–microbial interactions in the rhizosphere area (Ayaz et al. 2022).

Rhizosphere competence is known as the potentiality of any soil organism to grow fast and compete with others for vital nutritious compounds and exudates produced by plant roots, with their capability to colonize the root surface of host plants in rhizosphere soil (Igiehon et al. 2024). Physical sites and nutrients are important factors for the establishment of populations of both pathogenic and nonpathogenic epiphytic microorganisms. Generally, biocontrol agents have the ability to effectively limit the growth of phytopathogens through competitive colonization of space and successful survival in root zones. Numerous species belonging to PGPR could act as biocontrol agents by outcompeting the plant pathogens either for binding sites or for nutrients in the host plant root area. Such competition helps in constraining plant pathogens from binding to the plants, therefore hindering their proliferation and suppressing plant infection (Etesami and Maheshwari 2018). For example, *Pseudomonas putida* showed significant biocontrol activity via the mechanism of competition for niches and nutrients. This biocontrol effect of *P. syringae* may be because of siderophores in the gene clusters of this bacterial strain, which chelate essential trace metal ions (Afordoanyi et al. 2022). Moreover, *Bacillus amyloliquefaciens* produce *surfactins*, which play a crucial role in bacterial biofilm formation and cell spreading. Such characteristics help *B. amyloliquefaciens* to compete in the living space and colonize plant roots persistently, reducing the colonization of phytopathogens in the rhizosphere area (Luo et al. 2022).

Fusarium is one of the most destructive fungal genera, severely affecting the yield and quality of plants globally. The effectiveness of *Bacillus* members in the biocontrol of *Fusarium* spp. is related to their persistence in highly competitive niches in the rhizosphere as well as the synthesizing of various volatile compounds and soluble bioactive secondary metabolites (Boulahouat et al. 2023).

Alternaria is a well-known fungal infection that causes early blight and a dramatic drop in the yield of many crops worldwide. The infection starts on the lower leaves and then spreads upwards, where little brown spots form on the leaves (Attia et al. 2022). PGPR *B. amyloliquefaciens* is an endophytic bacterium that showed significant rapid growth and antifungal activity against *A. alternata* in pepper. Furthermore, this bacterial strain produces antioxidant enzymes and solubilizes nutrients that enhance host development (Soliman et al. 2023).

Recently, the comparative genomic analysis revealed that the growth-promoting characteristics of PGPR, including nitrogen and carbon acquisition as well as phytohormone production, are deterministic for their successful colonization and population establishment in the rhizosphere (de Souza et al. 2019).

Antibiosis was defined as the production of a heterogeneous community of low-molecular-weight active compounds (antibiotics) by beneficial microorganisms, which play an important role in the management of plant diseases (Köhl et al. 2019). PGPR plays a significant role in the manufacture of a wide range of antibiotics that have negative impacts on plant pathogens. *Bacillus* species and *Pseudomonas* are the most common PGPR that aid in pathogen elimination by producing inhibitory, antagonistic toxins as part of their defensive mechanism against hazardous microbial strains (Wang et al. 2021). Mainly, the antibiotics could be categorized into volatile and nonvolatile compounds. The volatile antibiotics include aldehydes, alcohols, hydrogen cyanide, sulfides, and ketones, while heterocyclic nitrogenous compounds come under the category of nonvolatile antibiotics (Gouda et al. 2018).

Several *Bacillus* species can produce a large spectrum of antimicrobial peptide antibiotics, making them potential candidates for biotechnological applications and precursors in the industry for novel inhibitory products. Cyclic lipopeptides are a significant group of antibiotics that are produced effectively by *Bacillus* species. These peptides are well known for their antagonistic activity against a vast range of phytopathogens by interacting with the cell membrane of the target pathogens, inducing changes in its permeability and structure through solubilization, disruption, or formation of ion-conducting pores, which help in and help colonization and persistence of *Bacillus* in the root zone, and trigger plant defense mechanisms (Miljaković et al. 2020). Surfactins, fengycins, and iturins are important families of antibiotics that are produced by numerous strains of *Bacillus* spp. The characterization of partially purified metabolites produced by *B. amyloliquefaciens* JFL21 revealed the presence of multiple lipopeptide mixtures belonging to fengycin, which plays a central role in the antifungal and antibacterial activity of JFL21 (Lin et al. 2020). Iturin A is an antifungal lipopeptide biosurfactant produced by certain strains of *B. subtilis*, which recorded high antagonistic potentiality against *Phytophthora infestans*, the causative agent of potato late blight (Wang et al. 2020). A scanning electron microscopy study revealed that Iturin A-induced cell membrane malformations as well as disruption in internal cell structure. Stoll et al. (2021) found that surfactin produced by *B. velezensis* has a role in biofilm formation and stable colonization on tomato root and leaf surfaces, which in turn triggered its systematic resistance against *Botrytis cinerea*. Additionally, six active compounds, identified as 2,3-butanedione, acetic acid, 2-isopropoxy ethylamine, 3-methyl-butyric acid, 2-methyl-butyric acid, and octanoic acid, were detected in *Bacillus altitudinis* AMCC 1040 secondary metabolites and proved for their strong lethal activity against root-knot nematodes (Zhou et al. 2022).

Several bacteria strains belonging to *Pseudomonas* species are important plant growth-promoting bacteria that proved high efficacy against a vast range of soil-borne diseases. For instance, *P. rhizobacterium* exhibited a wide spectrum against fungal pathogens and plant parasitic nematodes. Four nematicidal volatile organic compounds, octanoic acid ethyl ester, 2-methyl-1-butanol, isobutyl acetate, and ethyl acetate, were characterized and identified in the crude filtrate of this bacterial strain (Zhang et al. 2020). *Klebsiella pneumoniae* was successful in reducing the bacterial wilt incidence in tomatoes, and its disease control efficiency was demonstrated to be mainly due to the production of Meso-2,3-Butanediol in its fermentation broth (Kim et al. 2022).

Bacteriocins are heat-stable, ribosomally synthesized antimicrobial peptides produced effectively by lactic acid bacteria (LAB). In this context, *Lactiplantibacillus plantarum*, *Lacticaseibacillus paracasei*, and *Lentilactobacillus buchneri* showed antifungal ability against *Rhizoctonia solani, Colletotrichum gleosporoides, Alternaria alternata*, and *Sclerotinia sclerotiorum* and promoted both productivity and growth of wheat under field conditions (Gwiazdowski et al. 2023). Also, lactic acid, a potent disinfectant, is produced efficiently by this plant growth-promoting LAB and is capable of suppressing *Fusarium solani*, which is the causative agent of *Fusarium* wilt in a wide range of economic crops (Atwa et al. 2021). Furthermore, the natural organic acids such as succinic, maleic, lactic, and acetic acid synthesized by *Pediococcus pentosaceus* have been proven to be promising nematicidal compounds (Ibrahim et al. 2022).

The field application of *Streptomyces* spp. is known to stimulate the availability of minerals and nutrients, enhance plant growth regulators, and promote yield productivity (Boukhatem et al. 2022). Moreover, *Streptomyces* spp. generates secondary metabolites with diverse biological activities and functions, such as insecticidal, antibacterial, and antifungal activities (Katarzyna et al. 2018). *S. rochei, S. griseus, and S. sampsonii* recorded potent antifungal activity and inhibited the mycelial growth of *Botrytis cinerea, Macrophomina phaseolina, Rhizoctonia solani, and Sclerotinia sclerotiorum* fungi *in vitro*. GC–MS analysis showed the presence of antifungal active compounds that were categorized as fatty acids, volatile organic compounds, and plant growth regulators (Ghanem et al. 2022). Also, *Streptomyces* sp. JCK-6131 recorded broad-spectrum biocontrol activity against numerous plant pathogenic fungi and bacteria. ESI-MS/MS and RP-HPLC analyses revealed that JCK-6131 can synthesize various antimicrobial compounds, three of which were characterized as streptothricin D, streptothricin E acid, and 12-carbamoyl streptothricin D (Le et al. 2021). Three novel Streptomyces strains exhibited antiviral activity by producing behenic alcohol (docosanol), which provided clues to potential antiviral mechanisms able to down-regulate P1 gene expression responsible for zucchini yellow mosaic virus movement from cell to cell and replication to induce systemic infection (Ghanem et al. 2023).

Chitinases are hydrolytic enzymes that catalyze the degradation of chitin, the main structural component of fungal cell walls and nematode eggshells. These enzymes are secreted effectively by diverse bacteria, fungi, actinobacteria, insects, and plants (Poria et al. 2021). Chitinolytic PGPR is gaining interest as promising candidates for controlling plant diseases, including those caused by plant pathogenic fungi (Loc et al. 2019). Extracellular crude chitinases, produced by soil microorganisms, were reported to destroy nematode larvae, damage the eggshell, and reduce egg hatching as chitin is the structural component of the nematode eggshell (Soliman et al. 2019). Four soil chitinolytic *Streptomyces* isolates were identified as *S. fumosus, S. rimosus, S. flavogriseus, and S. albus* and proved for their potent antifungal activity against *Rhizoctonia solani*, *Magnaporthe oryzae, Botrytis cinerea*, and *Fusarium graminearum* (Ekundayo et al. 2022). Kurniawan et al. (2022) optimized the medium compositions and cultural conditions to induce chitinase production by plant growth-promoting *Bacillus paramycoides*. The bacterial crude chitinase showed great potential to suppress the growth of *Fusarium solani and Penicillium chrysogenum* up to 80% and induce significant distribution in cell wall composition. Moreover, the chitinolytic activity of five plant growth-prompting bacteria, *Bacillus toyonensis*, *Chryseobacterium daecheongense*, *Pseudomonas lini*, *Klebsiella oxytoca*, and

Lactobacillus helveticus, was investigated by Abdellatif Aya et al. (2023). The results revealed that induction of chitinolytic bacterial activity maximized their nematicidal efficacy and stimulated both yield productivity and growth parameters of pepper plants under field conditions.

The development of systemic resistance is indirect but a crucial tactic to suppress different biotic stresses in plants. PGPRs are diverse groups of plant-associated microorganisms, which can reduce the incidence or severity of plant disease by influencing induced systemic resistance (ISR) to elicit a defense response in host plants. PGPR strains have stimulated systemic resistance against different crop-damaging diseases. When several PGPRs are applied as seed treatments, the cell wall structural integrity is improved. Both physiological and biochemical alterations result in the production of proteins, peptides, and other compounds involved in plant defense mechanisms (Meena et al. 2020).

In addition, the insecticidal activity of some PGPR has been confirmed by some mechanisms, such as the production of biosurfactants that cause histopathological changes in the midgut of the treated insects. Also, some strains showed the ability to secrete special toxins, such as Cry toxins, which involve the loosening of the columnar cells, microvilli damage, cytoplasmic vacuolization, and the release of cytoplasm into the lumen (Ruiu 2020). *Bacillus thuringiensis* is a successful bioinsecticidal agent that is effective in the production of toxic proteins and other virulence factors. Recently, some strains of *B. thuringiensis* have been reported as root endophytes or fast root surface colonizers and showed distinct biofertilization traits such as solubilization of nutrients and production of growth regulators (Gomis-Cebolla and Berry 2023). The potent chitinolytic activity of *Serratia marcescens* enables it to prove promising virulence against insect pests, as chitin constitutes the major structural component of the insects' exoskeleton. The cultural filtrate of novel chitinase-producing S. *marcescens* strain TC-1 could induce a lethal effect on second- and third-instar larvae of *Spodoptera exigua* (Tao et al. 2022).

Generally, plant growth-promoting rhizobacteria naturally inhabit the rhizosphere and are widely involved in the biocontrol of plant pathogens by competition for space and nutrients, lytic enzyme production, antibiotic production, volatile compounds synthesizing, and induction of systemic resistance in the host, which makes them a perfect candidate to be commercialized for field application to enhance plant health as safe and sustainable alternatives to harmful synthetic pesticides.

2.3 PGPR AS A BIOSTIMULATOR

The plant biostimulants that enhance plant growth, productivity, and health are divided into two types: substances and microorganisms. Calvo et al. (2014) mentioned the definition and concept of plant biostimulants according to the European Biostimulants Industry Council as follows: "Plant biostimulants are substances and/or microorganisms that, when applied to plants or the rhizosphere, work to promote natural processes that improve crop quality, tolerance to abiotic stress, nutrient uptake, and nutrient efficiency" (European Biostimulants Industry 2012). The agriculture market related to PGPR products due to the beneficial utilization of microorganisms is growing at an exponential rate as a result of the new technologies.

A microbial plant biostimulant contains a microorganism or a consortium of microorganisms. The PGPR consists of four various genera (*Rhizobium* spp., *Azotobacter* spp., mycorrhizal fungi, and *Azospirillum* spp.). It is well documented that under nutritional conditions and abiotic stress or during the restoration of polluted soils, different plant growth-promoting rhizobacteria (PGPRs) are able to exert a beneficial effect on plant growth (Castiglione et al. 2021). PGPRs mitigated abiotic stresses through different mechanisms, boosting different biochemical and physiological processes, i.e., photosynthesis, source-sink relationships, and nutrient uptake; phytohormone production, i.e., indole-3-acetic acid and 1-aminocyclopropane-1-carboxylic acid deaminase; metabolism and the regulation of homeostasis; protein function; osmotic potential; nutrient solubilization; and enzymatic activity (Dogra et al. 2019; Jalal et al. 2023).

Under different soil types (clay and newly reclaimed soil), PGPRs enhanced the growth and productivity of pea plants with single and consortium inoculation (*Azotobacter chroococcum*, *Bacillus megaterium*, and *Bacillus cerkularice*). However, bacterial inoculation in the consortium was more effective than single bacterial inoculation (Mahmoud et al. 2023). The same trend was observed when treating potato plants with the bacterial consortium under calcareous soil conditions. In addition, microbial inoculation improved the growth and productivity of potato plants under various irrigation systems (Zahran et al. 2020). The positive impact of PGPR could be attributed to enhancing the efficiency of nutrient and water absorption. Bottini et al. (2004) reported that *Azospirillum spp.* and *Azotobacter spp.* stimulate plant growth and enhance plant productivity by producing some phytohormones such as cytokinin, auxin, abscisic acid, and gibberellins. Treated soil with PGPR could stimulate plant growth by providing better-growing conditions in the soil for the plant through improved soil structure and becoming more friable, less compact, and better drained. Moreover, *Azotobacter* spp., as a result of its ability to produce antifungal compounds, can support the plant in defending against numerous plant diseases and enhance plant growth and productivity (Jen-Hshuan 2006).

2.4 FUTURE PROSPECT

PGPR offers a natural approach to control plant diseases and pests by secreting antimicrobial and insecticidal compounds or by inducing systemic resistance in plants, which in turn reduces the consumption of harmful synthetic pesticides. Thus, PGPR has a bright opportunity in organic farming to be a promising strategy that increases the bioavailability of nutrients, enhances nutrient acquisition, and maximizes plant productivity, which helps farmers to meet the demand of consumers for organic products. Furthermore, there is promising interest in the formulation of microbial mixtures that could synergistically interact and cover a wider range of plant needs, from the solubilization of nutrients to the management of plant pathogens and pests. Microencapsulation, nano-biofertilizers, and other innovative formulation techniques will allow better control of PGPR release into soil, improving their survivability in soil even under environmental stressors. Development and application of well-designed biological programs will help in creating a healthy rhizosphere rich with a variety of microbial communities with unique physiological characteristics that can stimulate crop productivity even under various environmental challenges. Effective formulation of PGPR will contribute to a significant decrease in the application of synthetic pesticides and fertilizers. The cost-effective application of PGPR will make biocontrol approaches more advantageous than harmful chemical pesticides. Developing novel approaches for crop improvement would require the use of biotechnological techniques for the identification and characterization of genes related to lytic enzymes, antibiosis, and other antagonistic features. Furthermore, the application of bioproducts based on novel PGPR in crop breeding may result in cultivars that will be more resilient to different biotic and abiotic stressors.

2.5 CONCLUSION

This chapter emphasizes the vital role of PGPR-dependent instruments in addressing the issues related to global food production. One of the largest challenges facing the agriculture sector is the high growth in the world's population, particularly in light of the numerous intricate issues originating from global climate change. Abiotic and biotic stresses are two categories of environmental challenges that affect agricultural quality and productivity. Based on the updated literature, this chapter highlighted the significant role of PGPR in the enhancement of soil fertility via nitrogen fixation, phosphate solubilization, and production of growth-promoting substances that support the long-term health of the soil and maximize plant productivity to ensure food and nutritional security. Additionally, PGPR harnesses a wide range of mechanisms, or a combination of them, to combat different abiotic and biotic stresses such as salinity, drought, and phytopathogens. These mechanisms include phytohormone production, ROS enzyme activities, signaling pathways, and gene regulation and expression, as well as the production of important secondary metabolites such as volatile organic compounds, organic acids, bioactive compounds, and lytic enzymes. Generally, PGPR has a key role to play in the future of sustainable and regenerative agriculture, especially with the current global focus on food security and environmental conservation.

ARTIFICIAL INTELLIGENCE DECLARATION

The authors have not used regenerative artificial intelligence tools while writing the chapter.

REFERENCES

Abdellatif Aya, A. M., Abd-Elrahman, T. M. A., Sayed, M. A. E., Ragab, A. A. E. H., & Ibrahim, D. S. S. (2023). Isolation, identification and biocontrol activity of novel chitinolytic bacteria against Meloidogyne incognita infecting capsicum annuum l. *Caspian Journal of Environmental Sciences*, 21(1), 49–68.

Abdelrahman, H., Zaghloul, R., Abou-Aly, H., Ragab, A., & Elmaghraby, M. K. (2021). Application of some organic farming methods to enhancement the growth and production of green onion. *Journal of Agricultural Chemistry and Biotechnology*, 12(4), 79–89. https://doi.org/10.21608/jacb.2021.162616

Abo Elsoud, M. M., Hasan, S. F., & Elhateir, M. M. (2023). Optimization of indole-3-acetic acid production by *Bacillus velezensis* isolated from pyrus rhizosphere and its effect on plant growth. *Biocatalysis and Agricultural Biotechnology*, 50, 102714. https://doi.org/10.1016/j.bcab.2023.102714

Acharya, B. R., Gill, S. P., Kaundal, A., & Sandhu, D. (2024). Strategies for combating plant salinity stress: The potential of plant growth-promoting microorganisms. *Frontiers in Plant Science*, 15. https://doi.org/10.3389/fpls.2024.1406913

Afordoanyi, D. M., Diabankana, R. G., Miftakhov, A. K., Kuchaev, E. S., & Validov, S. Z. (2022). Genomic features of pseudomonas putida PCL1760: A biocontrol agent acting via competition for nutrient and niche. *Applied Microbiology*, 2(4), 749–765. https://doi.org/10.3390/applmicrobiol2040057

Akhtyamova, Z., Arkhipova, T., Martynenko, E., Nuzhnaya, T., Kuzmina, L., Kudoyarova, G., & Veselov, D. (2021). Growth-promoting effect of Rhizobacterium (Bacillus subtilis IB22) in salt-stressed Barley depends on Abscisic acid accumulation in the roots. *International Journal of Molecular Sciences*, 22(19), 10680. https://doi.org/10.3390/ijms221910680

Ali, A. M., Awad, M. Y., Hegab, S. A., Gawad, A. M., & Eissa, M. A. (2020). Effect of potassium solubilizing bacteria (*Bacillus cereus*) on growth and yield of potato. *Journal of Plant Nutrition*, 44(3), 411–420. https://doi.org/10.1080/01904167.2020.1822399

Aloo, B. N., Makumba, B. A., & Mbega, E. R. (2020). Plant growth promoting rhizobacterial biofertilizers for sustainable crop production: The past, present, and future. https://doi.org/10.20944/preprints202009.0650.v1

Ansari, A. F., Ahmad, I., & Pichtel, J. (2023). Synergistic effects of biofilm-producing PGPR strains on wheat plant colonization, growth and soil resilience under drought stress. *Saudi Journal of Biological Sciences*, 30(6), 103664. https://doi.org/10.1016/j.sjbs.2023.103664

Aslam, M. M., Waseem, M., Jakada, B. H., Okal, E. J., Lei, Z., Saqib, H. S. A., Yuan, W., Xu, W., & Zhang, Q. (2022). Mechanisms of abscisic acid-mediated drought stress responses in plants. *International Journal of Molecular Sciences*, 23(3), 1084. https://doi.org/10.3390/ijms23031084

Attia, M. S., Hashem, A. H., Badawy, A. A., & Abdelaziz, A. M. (2022). Biocontrol of early blight disease of eggplant using endophytic aspergillus terreus: Improving plant immunological, physiological and antifungal activities. *Botanical Studies*, 63(1). https://doi.org/10.1186/s40529-022-00357-6

Atwa, N., Maany, D., El-Diwany, A., Daba, G., Elmaghraby, I., Elmaghraby, M., & Amer, M. (2021). Studies on the antimicrobial activity of a locally isolated lactic acid bacterium and its application on Fusarium solani-infected plants. *Egyptian Pharmaceutical Journal* 20(3), 207. https://doi.org/10.4103/epj.epj_14_21

Ayaz, M., Ali, Q., Jiang, Q., Wang, R., Wang, Z., Mu, G., Khan, S. A., Khan, A. R., Manghwar, H., Wu, H., Gao, X., & Gu, Q. (2022). Salt tolerant bacillus strains improve plant growth traits and regulation of phytohormones in wheat under salinity stress. *Plants*, 11(20), 2769. https://doi.org/10.3390/plants11202769

Ayaz, M., Li, C., Ali, Q., Zhao, W., Chi, Y., Shafiq, M., Ali, F., Yu, X., Yu, Q., Zhao, J., Yu, J., Qi, R., & Huang, W. (2023). Bacterial and fungal biocontrol agents for plant disease protection: Journey from lab to Field, current status, challenges, and global perspectives. *Molecules*, 28(18), 6735. https://doi.org/10.3390/molecules28186735

Bakki, M., Banane, B., Marhane, O., Esmaeel, Q., Hatimi, A., Barka, E. A., Azim, K., & Bouizgarne, B. (2024). Phosphate solubilizing pseudomonas and bacillus combined with rock phosphates promoting tomato growth and reducing bacterial canker disease. *Frontiers in Microbiology*, 15. https://doi.org/10.3389/fmicb.2024.1289466

Baliyan, N., Dhiman, S., Dheeman, S., Kumar, S., Arora, N. K., & Maheshwari, D. K. (2021). Optimization of gibberellic acid production in endophytic *Bacillus cereus* using response surface methodology and its use as plant growth regulator in chickpea. *Journal of Plant Growth Regulation*, 41(7), 3019–3029. https://doi.org/10.1007/s00344-021-10492-2

Bhat, M. A., Mishra, A. K., Jan, S., Bhat, M. A., Kamal, M. A., Rahman, S., Shah, A. A., & Jan, A. T. (2023). Plant growth promoting Rhizobacteria in plant health: A perspective study of the underground interaction. *Plants*, 12(3), 629. https://doi.org/10.3390/plants12030629

Bottini, R., Cassan, F., & Piccoli, P. (2004). Gibberellin production by bacteria and its involvement in plant growth promotion and yield increase. *Applied Microbiology and Biotechnology*, 65(5). https://doi.org/10.1007/s00253-004-1696-1

Boukhatem, Z. F., Merabet, C., & Tsaki, H. (2022). Plant growth promoting Actinobacteria, the most promising candidates as Bioinoculants? *Frontiers in Agronomy*, 4. https://doi.org/10.3389/fagro.2022.849911

Boulahouat, S., Cherif-Silini, H., Silini, A., Bouket, A. C., Luptakova, L., Alenezi, F. N., & Belbahri, L. (2023). Biocontrol efficiency of Rhizospheric bacillus against the plant pathogen Fusarium oxysporum: A promising approach for sustainable agriculture. *Microbiology Research*, 14(3), 892–908. https://doi.org/10.3390/microbiolres14030062

Calvo, P., Nelson, L., & Kloepper, J. W. (2014). Agricultural uses of plant biostimulants. *Plant and Soil*, 383(1–2), 3–41. https://doi.org/10.1007/s11104-014-2131-8

Camara, M. C., Vandenberghe, L. P., Rodrigues, C., De Oliveira, J., Faulds, C., Bertrand, E., & Soccol, C. R. (2018). Current advances in gibberellic acid (GA3) production, patented technologies and potential applications. *Planta*, 248(5), 1049–1062. https://doi.org/10.1007/s00425-018-2959-x

Castiglione, A. M., Mannino, G., Contartese, V., Bertea, C. M., & Ertani, A. (2021). Microbial biostimulants as response to modern agriculture needs: Composition, role and application of these innovative products. *Plants*, 10(8), 1533. https://doi.org/10.3390/plants10081533

Chandra, P., Wunnava, A., Verma, P., Chandra, A., & Sharma, R. K. (2021). Strategies to mitigate the adverse effect of drought stress on crop plants – influences of soil bacteria: A review. *Pedosphere*, 31(3), 496–509. https://doi.org/10.1016/s1002-0160(20)60092-3

Chang, P., Gerhardt, K. E., Huang, X., Yu, X., Glick, B. R., Gerwing, P. D., & Greenberg, B. M. (2014). Plant growth-promoting bacteria facilitate the growth of Barley and oats in salt-impacted soil: Implications for Phytoremediation of saline soils. *International Journal of Phytoremediation*, 16(11), 1133–1147. https://doi.org/10.1080/15226514.2013.821447

Chebotar, V. K., Zaplatkin, A. N., Chizhevskaya, E. P., Gancheva, M. S., Voshol, G. P., Malfanova, N. V., Baganova, M. E., Khomyakov, Y. V., & Pishchik, V. N. (2024). Phytohormone production by the endophyte *Bacillus safensis* TS3 increases plant yield and alleviates salt stress. *Plants*, 13(1), 75. https://doi.org/10.3390/plants13010075

Danish, S., Zafar-ul-Hye, M., Fahad, S., Saud, S., Brtnicky, M., Hammerschmiedt, T., & Datta, R. (2020). Drought stress alleviation by ACC Deaminase producing Achromobacter xylosoxidans and Enterobacter cloacae, with and without timber waste Biochar in maize. *Sustainability*, 12(15), 6286. https://doi.org/10.3390/su12156286

de Souza, R. S., Armanhi, J. S., Damasceno, N. D., Imperial, J., & Arruda, P. (2019). Genome sequences of a plant beneficial synthetic bacterial community reveal genetic features for successful plant colonization. *Frontiers in Microbiology*, 10. https://doi.org/10.3389/fmicb.2019.01779

Dogra, N., Yadav, R., Kaur, M., Adhikary, A., Kumar, S., & Ramakrishna, W. (2019). Nutrient enhancement of chickpea grown with plant growth promoting bacteria in local soil of Bathinda, northwestern India. *Physiology and Molecular Biology of Plants*, 25(5), 1251–1259. https://doi.org/10.1007/s12298-019-00661-9

Ekundayo, F. O., Folorunsho, A. E., Ibisanmi, T. A., & Olabanji, O. B. (2022). Antifungal activity of chitinase produced by Streptomyces species isolated from grassland soils in FUTA area, Akure. *Bulletin of the National Research Centre*, 46(1). https://doi.org/10.1186/s42269-022-00782-4

El-Esawi, M. A., Alaraidh, I. A., Alsahli, A. A., Alzahrani, S. M., Ali, H. M., Alayafi, A. A., & Ahmad, M. (2018). Serratia liquefaciens KM4 improves salt stress tolerance in maize by regulating redox potential, ion homeostasis, leaf gas exchange and stress-related gene expression. *International Journal of Molecular Sciences*, 19(11), 3310. https://doi.org/10.3390/ijms19113310

Elmaghraby, M. M., Abdellatif, A. A., Amer, M. N., & Sahu, P. K. (2023). Enhancing productivity through multiple microbial Inoculants. In: Kaur, S., Dwibedi, V., Sahu, P. K., Kocher, G. S. (eds) *Metabolomics, Proteomes and Gene Editing Approaches in Biofertilizer Industry*, pp. 117–137, SpringerNature, Singapore. https://doi.org/10.1007/978-981-99-3561-1_7

Etesami, H., & Maheshwari, D. K. (2018). Use of plant growth promoting rhizobacteria (PGPRs) with multiple plant growth promoting traits in stress agriculture: Action mechanisms and future prospects. *Ecotoxicology and Environmental Safety*, 156, 225–246. https://doi.org/10.1016/j.ecoenv.2018.03.013

European Biostimulants Industry Council (2012). EBIC and biostimulants in brief. http://www.biostimulants.eu/

Fukami, J., Cerezini, P., & Hungria, M. (2018). Azospirillum: Benefits that go far beyond biological nitrogen fixation. *AMB Express*, 8(1). https://doi.org/10.1186/s13568-018-0608-1

Gao, Y., Han, Y., Li, X., Li, M., Wang, C., Li, Z., Wang, Y., & Wang, W. (2022). A salt-tolerant streptomyces paradoxus D2-8 from rhizosphere soil of phragmites communis augments soybean tolerance to soda saline-alkali stress. *Polish Journal of Microbiology*, 71(1), 43–53. https://doi.org/10.33073/pjm-2022-006

Ghanem, G. A., Gebily, D. A., Ragab, M. M., Ali, A. M., Soliman, N. E., & El-Moity, T. H. (2022). Efficacy of antifungal substances of three Streptomyces spp. against different plant pathogenic fungi. *Egyptian Journal of Biological Pest Control*, 32(1). https://doi.org/10.1186/s41938-022-00612-9

Ghanem, G. A., Mahmoud, A. M., Kheder, A. A., Gebily, D. A., & Osamy, A. (2023). Antiviral activities of three Streptomyces spp. against zucchini yellow mosaic virus (ZYMV) infecting squash (Cucurbita pepo L.) plants. *Egyptian Journal of Biological Pest Control*, 33(1). https://doi.org/10.1186/s41938-023-00750-8

Gomis-Cebolla, J., & Berry, C. (2023). *Bacillus thuringiensis* as a biofertilizer in crops and their implications in the control of phytopathogens and insect pests. *Pest Management Science*, 79(9) 2992–3001. https://doi.org/10.20944/preprints202210.0254.v1

Gong, A., Wang, G., Sun, Y., Song, M., Dimuna, C., Gao, Z., Wang, H., & Yang, P. (2022). Dual activity of Serratia marcescens pt-3 in phosphate-solubilizing and production of antifungal volatiles. *BMC Microbiology*, 22(1). https://doi.org/10.1186/s12866-021-02434-5

Gouda, S., Kerry, R. G., Das, G., Paramithiotis, S., Shin, H., & Patra, J. K. (2018). Revitalization of plant growth promoting rhizobacteria for sustainable development in agriculture. *Microbiological Research*, 206, 131–140. https://doi.org/10.1016/j.micres.2017.08.016

Gowtham, H., Duraivadivel, P., Ayusman, S., Sayani, D., Gholap, S., Niranjana, S., & Hariprasad, P. (2021). ABA analogue produced by bacillus marisflavi modulates the physiological response of brassica juncea L. under drought stress. *Applied Soil Ecology*, 159, 103845. https://doi.org/10.1016/j.apsoil.2020.103845

Gowtham, H., Murali, M., Shilpa, N., Amruthesh, K., Gafur, A., Antonius, S., & Sayyed, R. (2024). Harnessing abiotic elicitors to bolster plant's resistance against bacterial pathogens. *Plant Stress*, 11, 100371. https://doi.org/10.1016/j.stress.2024.100371

Gwiazdowski, R., Kubiak, K., Juś, K., Marchwińska, K., & Gwiazdowska, D. (2023). The biocontrol of plant pathogenic fungi by selected lactic acid bacteria: From laboratory to field study. *Agriculture*, 14(1), 61. https://doi.org/10.3390/agriculture14010061

Hoque, M. N., Imran, S., Hannan, A., Paul, N. C., Mahamud, M. A., Chakrobortty, J., Sarker, P., Irin, I. J., Brestic, M., & Rhaman, M. S. (2022). Organic amendments for mitigation of salinity stress in plants: A review. *Life*, 12(10), 1632. https://doi.org/10.3390/life12101632

Ibrahim, D. S., Amer, M. N., El-Shishtawy, H. M., Elmaghraby, M. M. K., Abdellatif, A. A. M., & El-Deriny, M. M. (2022). Nematicidal activity of lactic acid bacteria against root-knot nematodes. *Open Access Journal of Microbiology & Biotechnology*, 7 (1), 000216. https://doi.org/10.23880/oajmb-16000216

Igiehon, B. C., Babalola, O. O., & Hassen, A. I. (2024). Rhizosphere competence and applications of plant growth-promoting rhizobacteria in food production – A review. *Scientific African*, 23, e02081. https://doi.org/10.1016/j.sciaf.2024.e02081

Iqbal, Z., Ahmad, M., Raza, M. A., Hilger, T., & Rasche, F. (2024). Phosphate-solubilizing bacillus Sp. Modulate soil exoenzyme activities and improve wheat growth. *Microbial Ecology*, 87(1). https://doi.org/10.1007/s00248-023-02340-5

Jalal, A., Da Silva Oliveira, C. E., Galindo, F. S., Rosa, P. A., Gato, I. M., De Lima, B. H., & Teixeira Filho, M. C. (2023). Regulatory mechanisms of plant growth-promoting Rhizobacteria and plant nutrition against abiotic stresses in Brassicaceae family. *Life*, 13(1), 211. https://doi.org/10.3390/life13010211

Jen-Hshuan, C. (2006). The combined use of chemical and organic fertilizers and/or biofertilizer for crop growth and soil fertility. *International Workshop on Sustained Management of the Soil Rhizosphere System for Efficient Crop Production and Fertilizer Use, Bangkok*, 16–20, 1–10.

Junia, R., Jain, N., Sohal, J., Sharma, D., Khare, N., & Aseri, G. (2021). Comparative impact of bioinoculants on nutrient uptake, enzyme activities and growth of *Cassia angustifolia* Vhal (Senna) and *Cyamopsis tetragonoloba* (L.) (Guar) in feldspar mine spoil. *South African Journal of Botany*, 140, 434–443. https://doi.org/10.1016/j.sajb.2020.12.006

Karagoz, F. P., & Dursun, A. (2019). Effects of nitrogen fixing and phosphate solubilizing bacteria on growth and bulbs production of tulip cultivars. *Journal of Agriculture faculty of Ege University*, 56(2), 241–248. https://doi.org/10.20289/zfdergi.477345

Katarzyna, J. K., Aleksandra, R. M., Adam, G., Joannam, Z., & Jolanta, S. (2018). Secondary metabolites of actinomycetes and their antibacterial, Antifungal and antiviral properties. *Polish Journal of Microbiology*, 67(3), 259–272. https://doi.org/10.21307/pjm-2018-048

Kim, B., Park, A. R., Song, C. W., Song, H., & Kim, J. (2022). Biological control efficacy and action mechanism of *klebsiella pneumoniae* JCK-2201 producing meso-2,3-Butanediol against tomato bacterial wilt. *Frontiers in Microbiology*, 13. https://doi.org/10.3389/fmicb.2022.914589

Köhl, J., Kolnaar, R., & Ravensberg, W. J. (2019). Mode of action of microbial biological control agents against plant diseases: Relevance beyond efficacy. *Frontiers in Plant Science*, 10. https://doi.org/10.3389/fpls.2019.00845

Kurniawan, E., Leelakriangsak, M., & Panphon, S. (2022). Antifungal Chitinase production by bacillus paramycoides B26 using squid pen powder as a carbon source. *Journal of Pure and Applied Microbiology*, 16(4), 2496–2506. https://doi.org/10.22207/jpam.16.4.09

Kusale, S. P., Attar, Y. C., Sayyed, R. Z., El Enshasy, H., Hanapi, S. Z., Ilyas, N., Elgorban, A. M., Bahkali, A. H., & Marraiki, N. (2021). Inoculation of klebsiella variicola alleviated salt stress and improved growth and nutrients in wheat and maize. *Agronomy*, 11(5), 927. https://doi.org/10.3390/agronomy11050927

Le, K. D., Kim, J., Nguyen, H. T., Yu, N. H., Park, A. R., Lee, C. W., & Kim, J. (2021). Streptomyces Sp. JCK-6131 protects plants against bacterial and fungal diseases via two mechanisms. *Frontiers in Plant Science*, 12. https://doi.org/10.3389/fpls.2021.726266

Lebrazi, S., Niehaus, K., Bednarz, H., Fadil, M., Chraibi, M., & Fikri-Benbrahim, K. (2020). Screening and optimization of indole-3-acetic acid production and phosphate solubilization by rhizobacterial strains isolated from acacia cyanophylla root nodules and their effects on its plant growth. *Journal of Genetic Engineering and Biotechnology*, 18(1), 71. https://doi.org/10.1186/s43141-020-00090-2

Li, H., Chen, S., Wang, M., Shi, S., Zhao, W., Xiong, G., Zhou, J., & Qu, J. (2024). Phosphate solubilization and plant growth properties are promoted by a lactic acid bacterium in calcareous soil. *Applied Microbiology and Biotechnology*, 108(1). https://doi.org/10.1007/s00253-023-12850-4

Li, H., Singh, R. K., Singh, P., Song, Q., Xing, Y., Yang, L., & Li, Y. (2017). Genetic diversity of nitrogen-fixing and plant growth promoting pseudomonas species isolated from sugarcane rhizosphere. *Frontiers in Microbiology*, 8. https://doi.org/10.3389/fmicb.2017.01268

Lin, L., Zheng, Q., Wei, T., Zhang, Z., Zhao, C., Zhong, H., Xu, Q., Lin, J., & Guo, L. (2020). Isolation and characterization of Fengycins produced by bacillus amyloliquefaciens JFL21 and its broad-spectrum antimicrobial potential against multidrug-resistant foodborne pathogens. *Frontiers in Microbiology*, 11. https://doi.org/10.3389/fmicb.2020.579621

Lindström, K., & Mousavi, S. A. (2019). Effectiveness of nitrogen fixation in rhizobia. *Microbial Biotechnology*, 13(5), 1314–1335. https://doi.org/10.1111/1751-7915.13517

Liu, Y., Wang, Y., Zhang, L., Liu, L., Cai, T., Chang, C., Sa, D., Yin, Q., Jiang, X., Li, Y., & Lu, Q. (2023). The effect of lactobacillus planturum YQM48 inoculation on the quality and microbial community structure of alfalfa silage cultured in saline-alkali soil. *Fermentation*, 9(6), 511. https://doi.org/10.3390/fermentation9060511

Loc, N. H., Huy, N. D., Quang, H. T., Lan, T. T., & Thu Ha, T. T. (2019). Characterisation and antifungal activity of extracellular chitinase from a biocontrol fungus, Trichoderma asperellum PQ34. *Mycology*, 11(1), 38–48. https://doi.org/10.1080/21501203.2019.1703839

Luo, L., Zhao, C., Wang, E., Raza, A., & Yin, C. (2022). Bacillus amyloliquefaciens as an excellent agent for biofertilizer and biocontrol in agriculture: An overview for its mechanisms. *Microbiological Research*, 259, 127016. https://doi.org/10.1016/j.micres.2022.127016

Mahmoud, G. A. & Mostafa, H. H. A. (2017). Statistical optimization as a powerful tool for indole acetic acid production by Fusarium oxysporum. *European Journal of Biological Research*, 7 (4), 315–323. https://doi.org/10.5281/zenodo.1012348

Mahmoud, G. A.-E., Abeed, A. H. A., Mostafa, H. H. A., & Abdel Monsef, O. (2023). Responses of pea (Pisum sativum L.) to single and consortium bio-fertilizers in clay and newly reclaimed soils. *Plants*, 12, 3931. https://doi.org/10.3390/plants12233931

Meena, M., Swapnil, P., Divyanshu, K., Kumar, S., Harish, Tripathi, Y. N., Zehra, A., Marwal, A., & Upadhyay, R. S. (2020). PGPR-mediated induction of systemic resistance and physiochemical alterations in plants against the pathogens: Current perspectives. *Journal of Basic Microbiology*, 60(10), 828–861. https://doi.org/10.1002/jobm.202000370

Miljaković, D., Marinković, J., & Balešević-Tubić, S. (2020). The significance of bacillus spp. in disease suppression and growth promotion of Field and vegetable crops. *Microorganisms*, 8(7), 1037. https://doi.org/10.3390/microorganisms8071037

Mishra, P., Mishra, J., & Arora, N. K. (2021). Plant growth promoting bacteria for combating salinity stress in plants – Recent developments and prospects: A review. *Microbiological Research*, 252, 126861. https://doi.org/10.1016/j.micres.2021.126861

Mohamed, E. A., Farag, A. G., & Youssef, S. A. (2018). Phosphate solubilization by Bacillus subtilis and Serratia marcescens isolated from tomato plant rhizosphere. *Journal of Environmental Protection*, 09(3), 266–277. https://doi.org/10.4236/jep.2018.93018

Mostafa, H. H., Li, B., Zhu, X., & Song, C. (2021). Nitrogen assimilation under osmotic stress in maize (Zea mays L.) seedlings. *Plant Growth Regulation*, 94(1), 87–99. https://doi.org/10.1007/s10725-021-00698-5

Munir, N., Hanif, M., Abideen, Z., Sohail, M., El-Keblawy, A., Radicetti, E., Mancinelli, R., & Haider, G. (2022). Mechanisms and strategies of plant microbiome interactions to mitigate abiotic stresses. *Agronomy*, 12(9), 2069. https://doi.org/10.3390/agronomy12092069

Myo, E. M., Ge, B., Ma, J., Cui, H., Liu, B., Shi, L., Jiang, M., & Zhang, K. (2019). Indole-3-acetic acid production by *Streptomyces fradiae* NKZ-259 and its formulation to enhance plant growth. *BMC Microbiology*, 19(1). https://doi.org/10.1186/s12866-019-1528-1

Nahakpam, S., Shah, K., Kundu, M., & Heikham, R. S. (2021). Role of phytohormones as master regulators during the abiotic stress. *Stress Tolerance in Horticultural Crops*, 347–369. https://doi.org/10.1016/b978-0-12-822849-4.00002-4

Nawaz, A., Qamar, Z. U., Marghoob, M. U., Imtiaz, M., Imran, A., & Mubeen, F. (2023). Contribution of potassium solubilizing bacteria in improved potassium assimilation and cytosolic K+/Na+ ratio in rice (Oryza sativa L.) under saline-sodic conditions. *Frontiers in Microbiology*, 14. https://doi.org/10.3389/fmicb.2023.1196024

Olaniyan, F. T., Alori, E. T., Adekiya, A. O., Ayorinde, B. B., Daramola, F. Y., Osemwegie, O. O., & Babalola, O. O. (2022). The use of soil microbial potassium solubilizers in potassium nutrient availability in soil and its dynamics. *Annals of Microbiology*, 72(1). https://doi.org/10.1186/s13213-022-01701-8

Ou, K., He, X., Cai, K., Zhao, W., Jiang, X., Ai, W., Ding, Y., & Cao, Y. (2023). Phosphate-solubilizing *Pseudomonas* Sp. Strain WS32 rhizosphere colonization-induced expression changes in wheat roots. *Frontiers in Microbiology*, 13. https://doi.org/10.3389/fmicb.2022.927889

Pantoja-Guerra, M., Valero-Valero, N., & Ramírez, C. A. (2023). Total auxin level in the soil–plant system as a modulating factor for the effectiveness of PGPR inocula: A review. *Chemical and Biological Technologies in Agriculture*, 10(1). https://doi.org/10.1186/s40538-022-00370-8

Park, S., Kim, A., Hong, Y., Shin, J., & Joo, S. (2021). A highly efficient auxin-producing bacterial strain and its effect on plant growth. *Journal of Genetic Engineering and Biotechnology*, 19(1), 179. https://doi.org/10.1186/s43141-021-00252-w

Patani, A., Prajapati, D., Ali, D., Kalasariya, H., Yadav, V. K., Tank, J., Bagatharia, S., Joshi, M., & Patel, A. (2023). Evaluation of the growth-inducing efficacy of various bacillus species on the salt-stressed tomato (Lycopersicon esculentum mill.). *Frontiers in Plant Science*, 14. https://doi.org/10.3389/fpls.2023.1168155

Peng, Y., Li, S. J., Yan, J., Tang, Y., Cheng, J. P., Gao, A. J., Yao, X., Ruan, J. J., & Xu, B. L. (2021). Research progress on Phytopathogenic fungi and their role as biocontrol agents. *Frontiers in Microbiology*, 12. https://doi.org/10.3389/fmicb.2021.670135

Poria, V., Rana, A., Kumari, A., Grewal, J., Pranaw, K., & Singh, S. (2021). Current perspectives on chitinolytic enzymes and their agro-industrial applications. *Biology*, 10(12), 1319. https://doi.org/10.3390/biology10121319

Poudel, M., Mendes, R., Costa, L. A., Bueno, C. G., Meng, Y., Folimonova, S. Y., Garrett, K. A., & Martins, S. J. (2021). The role of plant-associated bacteria, fungi, and viruses in drought stress mitigation. *Frontiers in Microbiology*, 12. https://doi.org/10.3389/fmicb.2021.743512

Raheem, A., Shaposhnikov, A., Belimov, A. A., Dodd, I. C., & Ali, B. (2018). Auxin production by rhizobacteria was associated with improved yield of wheat (Triticum aestivum L.) under drought stress. *Archives of Agronomy and Soil Science* 64, 574–587. https://doi.org/10.1080/03650340.2017.1362105

Raza, A., Tabassum, J., Fakhar, A. Z., Sharif, R., Chen, H., Zhang, C., Ju, L., Fotopoulos, V., Siddique, K. H., Singh, R. K., Zhuang, W., & Varshney, R. K. (2023). Smart reprograming of plants against salinity stress using modern biotechnological tools. *Critical Reviews in Biotechnology*, 43(7), 1035–1062. https://doi.org/10.1080/07388551.2022.2093695

Rouphael, Y., & Colla, G. (2020). Editorial: Biostimulants in agriculture. *Frontiers in Plant Science*, 11. https://doi.org/10.3389/fpls.2020.00040

Ruiu, L. (2020). Plant-growth-promoting bacteria (PGPB) against insects and other agricultural pests. *Agronomy*, 10(6), 861. https://doi.org/10.3390/agronomy10060861

Saeid, A., Prochownik, E., & Dobrowolska-Iwanek, J. (2018). Phosphorus solubilization by *Bacillus* species. *Molecules*, 23(11), 2897. https://doi.org/10.3390/molecules23112897

Saleem, A. R., Brunetti, C., Khalid, A., Della Rocca, G., Raio, A., Emiliani, G., De Carlo, A., Mahmood, T., & Centritto, M. (2018). Drought response of Mucuna pruriens (L.) DC. inoculated with ACC deaminase and IAA producing rhizobacteria. *PLoS One*, 13(2), e0191218. https://doi.org/10.1371/journal.pone.0191218

Sanow, S., Kuang, W., Schaaf, G., Huesgen, P., Schurr, U., Roessner, U., Watt, M., & Arsova, B. (2022). Molecular mechanisms of pseudomonas-assisted plant nitrogen uptake: Opportunities for modern agriculture. *Molecular Plant-Microbe Interactions*, 36(9), 536–548. https://doi.org/10.1094/mpmi-10-22-0223-cr

Sarkar, J., Chakraborty, U., & Chakraborty, B. (2021). High-temperature resilience in bacillus safensis primed wheat plants: A study of dynamic response associated with modulation of antioxidant machinery, differential expression of HSPs and osmolyte biosynthesis. *Environmental and Experimental Botany*, 182, 104315. https://doi.org/10.1016/j.envexpbot.2020.104315

Shaffique, S., Imran, M., Adhikari, A., Aaqil Khan, M., Rahim, W., Alomrani, S. O., Yun, B. W., Kang, S., & Lee, I. (2023). A newly isolated bacillus pumilus strain SH-9 modulates response to drought stress in soybean via endogenous phytohormones and gene expression (Daegu, South Korea). *Plant Stress*, 10, 100279. https://doi.org/10.1016/j.stress.2023.100279

Shahzad, R., Khan, A. L., Bilal, S., Waqas, M., Kang, S., & Lee, I. (2017). Inoculation of abscisic acid-producing endophytic bacteria enhances salinity stress tolerance in oryza sativa. *Environmental and Experimental Botany*, 136, 68–77. https://doi.org/10.1016/j.envexpbot.2017.01.010

Shen, N., Li, S., Li, S., Zhang, H., & Jiang, M. (2022). The siderophore-producing bacterium, bacillus siamensis gxun-6, has an antifungal activity against Fusarium oxysporum and promotes the growth of banana. *Egyptian Journal of Biological Pest Control*, 32(1). https://doi.org/10.1186/s41938-022-00533-7

Sibponkrung, S., Kondo, T., Tanaka, K., Tittabutr, P., Boonkerd, N., Yoshida, K., & Teaumroong, N. (2020). Co-inoculation of *Bacillus velezensis* strain S141 and Bradyrhizobium strains promotes nodule growth and nitrogen fixation. *Microorganisms*, 8(5), 678. https://doi.org/10.3390/microorganisms8050678

Singh, P., Singh, R. K., Li, H., Guo, D., Sharma, A., Verma, K. K., Solanki, M. K., Upadhyay, S. K., Lakshmanan, P., Yang, L., & Li, Y. (2023). Nitrogen fixation and phytohormone stimulation of sugarcane plant through plant growth promoting diazotrophic Pseudomonas. *Biotechnology and Genetic Engineering Reviews*, 1–21. https://doi.org/10.1080/02648725.2023.2177814

Soliman, G. M., Ameen, H. H., Abdel-Aziz, S. M., & El-Sayed, G. M. (2019). In vitro evaluation of some isolated bacteria against the plant parasite nematode Meloidogyne incognita. *Bulletin of the National Research Centre*, 43(1). https://doi.org/10.1186/s42269-019-0200-0

Soliman, S. A., Abdelhameed, R. E., & Metwally, R. A. (2023). In vivo and in vitro evaluation of the antifungal activity of the PGPR bacillus amyloliquefaciens RaSh1 (MZ945930) against Alternaria alternata with growth promotion influences on capsicum annuum L. plants. *Microbial Cell Factories*, 22(1). https://doi.org/10.1186/s12934-023-02080-8

Stoll, A., Salvatierra-Martínez, R., González, M., & Araya, M. (2021). The role of Surfactin production by bacillus velezensis on colonization, Biofilm formation on tomato root and leaf surfaces and subsequent protection (ISR) against botrytis cinerea. *Microorganisms*, 9(11), 2251. https://doi.org/10.3390/microorganisms9112251

Sumbul, A., Ansari, R. A., Rizvi, R., & Mahmood, I. (2020). Azotobacter: A potential biofertilizer for soil and plant health management. *Saudi Journal of Biological Sciences*, 27(12), 3634–3640. https://doi.org/10.1016/j.sjbs.2020.08.004

Sutio, G., Iskandar, I., Indriyati, L. T., & Djajakirana, G. (2023). Endophytic testing of *Serratia marcescens* strain NPKC3_2_21 against INPARA 3 rice variety. *Biovalentia: Biological Research Journal*, 9(1), 48–55. https://doi.org/10.24233/biov.9.1.2023.372

Suzaki, T., Yoro, E., & Kawaguchi, M. (2015). Leguminous plants: Inventors of root nodules to accommodate symbiotic bacteria. *International Review of Cell and Molecular Biology*, 111–158. https://doi.org/10.1016/bs.ircmb.2015.01.004

Tanveer, S., Akhtar, N., Ilyas, N., Sayyed, R., Fitriatin, B. N., Perveen, K., & Bukhari, N. A. (2023). Interactive effects of pseudomonas putida and salicylic acid for mitigating drought tolerance in Canola (*Brassica napus* L.). *Heliyon*, 9(3), e14193. https://doi.org/10.1016/j.heliyon.2023.e14193

Tao, A., Wang, T., Pang, F., Zheng, X., Ayra-Pardo, C., Huang, S., Xu, R., Liu, F., Li, J., Wei, Y., Wang, Z., Niu, Q., & Li, D. (2022). Characterization of a novel chitinolytic Serratia marcescens strain TC-1 with broad insecticidal spectrum. *AMB Express*, 12(1). https://doi.org/10.1186/s13568-022-01442-6

Tian, S., Xu, Y., Zhong, Y., Qiao, Y., Wang, D., Wu, L., Yang, X., Yang, M., & Wu, Z. (2024). Exploring the organic acid secretion pathway and potassium solubilization ability of *Pantoea vagans* ZHS-1 for enhanced rice growth. *Plants*, 13(14), 1945. https://doi.org/10.3390/plants13141945

Tripathi, A., Pandey, V. K., Jain, D., Singh, G., Brar, N. S., Taufeeq, A., Pandey, I., Dash, K. K., Samrot, A. V., & Rustagi, S. (2024). An updated review on significance of PGPR-induced plant signalling and stress management in advancing sustainable agriculture. *Journal of Agriculture and Food Research*, 16, 101169. https://doi.org/10.1016/j.jafr.2024.101169

Van Oosten, M. J., Di Stasio, E., Cirillo, V., Silletti, S., Ventorino, V., Pepe, O., Raimondi, G., & Maggio, A. (2018). Root inoculation with azotobacter chroococcum 76a enhances tomato plants adaptation to salt stress under low N conditions. *BMC Plant Biology*, 18(1). https://doi.org/10.1186/s12870-018-1411-5

Vejan, P., Abdullah, R., Khadiran, T., Ismail, S., & Nasrulhaq Boyce, A. (2016). Role of plant growth promoting Rhizobacteria in agricultural sustainability – A review. *Molecules*, 21(5), 573. https://doi.org/10.3390/molecules21050573

Wang, H., Liu, R., You, M. P., Barbetti, M. J., & Chen, Y. (2021). Pathogen biocontrol using plant growth-promoting bacteria (PGPR): Role of bacterial diversity. *Microorganisms*, 9(9), 1988. https://doi.org/10.3390/microorganisms9091988

Wang, Y., Zhang, C., Liang, J., Wu, L., Gao, W., & Jiang, J. (2020). Iturin a extracted from bacillus subtilis WL-2 affects Phytophthora infestans via cell structure disruption, oxidative stress, and energy supply dysfunction. *Frontiers in Microbiology*, 11. https://doi.org/10.3389/fmicb.2020.536083

Win, K. T., Wasai-Hara, S., Tanaka, F., Oo, A. Z., Minamisawa, K., Shimoda, Y., & Imaizumi-Anraku, H. (2023). Synergistic N2-fixation and salt stress mitigation in soybean through dual inoculation of ACC deaminase-producing pseudomonas and Bradyrhizobium. *Scientific Reports*, 13(1). https://doi.org/10.1038/s41598-023-43891-4

Yang, B., Wen, H., Wang, S., Zhang, J., Wang, Y., Zhang, T., Yuan, K., Lu, L., Liu, Y., Xue, Q., & Shan, H. (2024). Enhancing drought resistance and yield of wheat through inoculation with Streptomyces pactum Act12 in drought Field environments. *Agronomy*, 14(4), 692. https://doi.org/10.3390/agronomy14040692

Yang, Z., Qiao, Y., Konakalla, N. C., Strøbech, E., Harris, P., Peschel, G., Agler-Rosenbaum, M., Weber, T., Andreasson, E., & Ding, L. (2023). Streptomyces alleviate abiotic stress in plant by producing pteridic acids. *Nature Communications*, 14(1). https://doi.org/10.1038/s41467-023-43177-3

Yasmin, H., Bano, A., Wilson, N. L., Nosheen, A., Naz, R., Hassan, M. N., Ilyas, N., Saleem, M. H., Noureldeen, A., Ahmad, P., & Kennedy, I. (2021). Drought-tolerant Pseudomonas sp. showed differential expression of stress-responsive genes and induced drought tolerance in Arabidopsis thaliana. *Physiologia Plantarum*, 174(1). https://doi.org/10.1111/ppl.13497

Zahran, M. M. A. A., Hefzy, M., & Mostafa, H. H. A. (2020). Enhancing the productivity of potato crop under drought stress by using some biological treatments. *International Journal of Environment*. https://doi.org/10.36632/ije/2020.9.2.6

Zhang, C., Xu, M., Gong, Y., Li, X., Huang, J., Zhou, S., Xing, K., & Qin, S. (2020). Identification and characterization of nematicidal activity of organic volatiles from a pseudomonad rhizobacterium. *Rhizosphere*, 16, 100244. https://doi.org/10.1016/j.rhisph.2020.100244

Zhang, Y., Xu, Q., Wang, G., & Shi, K. (2023). Mixed Enterobacter and klebsiella bacteria enhance soybean biological nitrogen fixation ability when combined with rhizobia inoculation. *Soil Biology and Biochemistry*, 184, 109100. https://doi.org/10.1016/j.soilbio.2023.109100

Zhao, Y., Liang, H., Zhang, J., Chen, Y., Dhital, Y. P., Zhao, T., & Wang, Z. (2024). Isolation and characterization of potassium-solubilizing Rhizobacteria (KSR) promoting cotton growth in saline–sodic regions. *Microorganisms*, 12(7), 1474. https://doi.org/10.3390/microorganisms12071474

Zhou, B., Ye, L., Wang, J., Liu, X., Guan, Q., Dou, N., Li, J., Zhang, Q., Gao, Y., Wang, M., & Li, J. (2022). Nematicidal activity of volatile organic compounds produced by bacillus altitudinis AMCC 1040 against Meloidogyne incognita. *Archives of Microbiology*. https://doi.org/10.21203/rs.3.rs-1500056/v1

3 Microbiome Engineering
A Pathway to Sustainable Agricultural Productivity

Mona Shaban E. M. Badawy, Lamiaa Ibrahim Ahmed, and Omnia Karem M. Riad

3.1 INTRODUCTION

In the natural environment, a variety of biotic and abiotic stressors limit the growth and production of plants. It will be extremely difficult for scientists worldwide to provide food for the world's growing population of 7.6 billion, which is expected to grow to 9.5–10 billion by 2050. Increasing crop yields and production will be very sustainably challenging with less chemical fertilizer and agrochemical use (Kumar and Dubey, 2020).

Plants have contact with trillions of microorganisms in the environment; these cells settle and occupy rhizospheres, rhizoplanes, endospheres, and photospheres of plants. For this reason, plant's secondary genomes are believed to be present (Dubey et al., 2019). To combat biotic and abiotic stress, beneficial plant microbiomes and their concoctions are applied; it is a rapidly developing and fascinating field of study (Kumar and Dubey, 2020).

Through the examination of the plant microbiome, we have additional information on plants and environments that support a diverse range of microbial communities. The microbial diversity found in the phyllosphere, rhizosphere, and endosphere across different regions and organs has been extensively documented by researchers (Fendrihan and Pop, 2021; Gupta et al., 2021). Usually, plants employ several strategies to encourage and facilitate microbial colonization, such as the generation of secondary metabolites or the presence of specialized structures like trichomes or hairs. Numerous labs worldwide are still working to develop a thorough method for interpreting plant–microbe interaction (Gupta et al., 2021). The diversity of the microbial population is affected by a number of variables, including host species, selection pressure, environment, developmental stage, and agricultural practices (Gupta et al., 2022).

It is now possible to modify plant-beneficial microorganisms because of the development of these high-throughput technologies. In this case, understanding, modifying, and creating corresponding technologies for cultivating microbial populations, which are essential to plant health and productivity, might be accomplished through microbiome engineering (Nadarajah and Abdul Rahman, 2023).

DOI: 10.1201/9781003714484-3

Therefore, this chapter discusses the role of beneficial microbes in sustainable agriculture, microbiome engineering, and field applications.

3.2 PLANT MICROBIOME

The environment surrounding plants that are home to soil-dwelling microbes is complex, supporting a wide variety of adaptable and diverse microbes and animals that account for at least 25% of Earth's biodiversity (Orgiazzi et al., 2016). It is possible to find hundreds to thousands of different microbial species in a single gram of soil, including viruses, bacteria, fungi, oomycetes, and archaea, according to the recent report by Afridi et al. (2022).

Up to 99.9% of soil microorganisms remain uncultivated and require special handling to grow in laboratory conditions (Savary et al., 2019). Under its broad heading, "plant microbiome," a multitude of related elements that are subterranean (rhizosphere), internal (endosphere), and aboveground (phyllosphere) are included (Afridi et al., 2022). Beneficial microorganisms are those that help plants absorb nutrients, shield them from toxic ones, and give them the support they need to withstand biotic and abiotic stressors (Figure 3.1). On the other hand, plant parasitic fungi and nematodes are dangerous microorganisms since they pose the greatest risk to plant health by causing economically significant illnesses in important crop plants that jeopardize food security (Afridi et al., 2022).

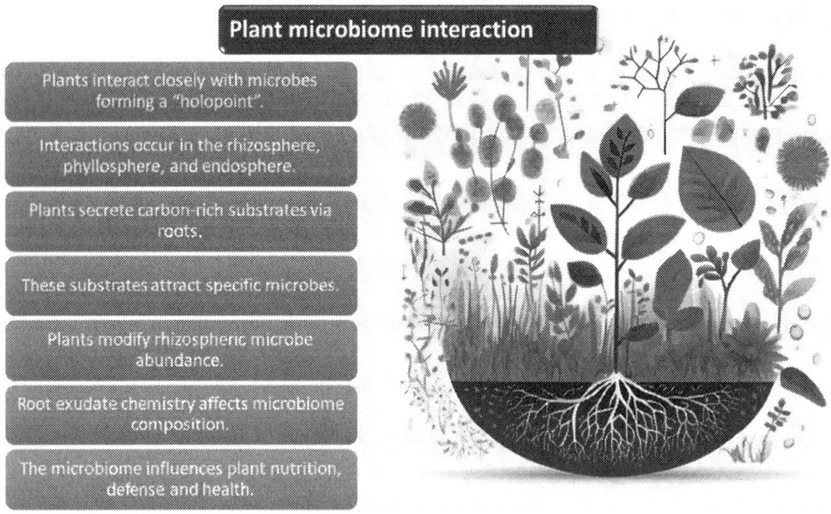

FIGURE 3.1 Plant microbiome interaction.

3.3 ENHANCING AGRICULTURAL PRODUCTIVITY THROUGH MICROBIOME-BASED RHIZOSPHERE ENGINEERING

The rhizosphere is home to a multitude of microbial communities that have a wide range of functions and impacts on plant growth. They can act as plant pathogens, aid in the cycling of nutrients, and offer defense against biotic and abiotic stresses. These rhizosphere-based microbial activities affect the microbial component and change the kind, quantity, and makeup of root exudates released by plants (Hakim et al., 2021).

3.3.1 Plant–Microbe Interaction in the Rhizosphere

Rhizosphere-associated microorganisms can sense and decipher signals from plants, other bacteria, and themselves. They can also affect their plant host by releasing signaling chemicals. The main signaling mechanisms in the rhizosphere can be divided into three groups based on the participants and the communication route (Pantigoso et al., 2022) as follows.

3.3.1.1 Microbe-to-Microbe

Quorum sensing (QS), a mechanism that is also known as autoinducer synthesis and detection, is the main way that microbial intraspecies and interspecies signaling are carried out. These autoinducers either activate or inhibit the transcription of several QS-regulated genes, including those involved in chemotaxis, biofilm formation, and virulence factors. Thus, microorganisms in microbial communities are able to monitor cell density and collectively track behavioral changes (Venturi and Keel, 2016). *N*-acyl homoserine lactones (AHL) autoinducer is a better-studied family of QS compounds that affect Gram-negative bacteria, including well-known soil bacterial taxa, including *Pseudomonas* spp., *Burkholderia* spp., and *Serratia* spp. Gram-positive rhizobacteria also use peptides as one of their extra means of communication, both within and between species (Monnet et al., 2016).

Numerous substances, such as volatile organic compounds, oxalic acid, trehalose, glucose, and thiamine, have been shown to act as signaling molecules in addition to QS inducer chemicals (Scherlach and Hertweck, 2018). According to Hassani et al. (2018), these tactics can aid bacteria in surviving in nutrient-poor conditions, gaining access to difficult-to-break-down compounds, getting rid of hazardous metabolites, and exchanging electrons.

3.3.1.2 Plant-to-Microbe

Most plants produce a variety of root exudates, which have been identified as signals to soil microorganisms. Specific receptors for pattern recognition (PRRs) are used by plants to identify related rhizosphere microbes (Venturi and Keel, 2016). It has been demonstrated that plant-secreted molecules play a role in signaling from plants to microorganisms and several beneficial interactions with plants (Mhlongo et al., 2018).

3.3.1.3 Microbe-to-Plant

According to Venturi and Keel (2016), signaling molecules that affect gene expression, development, and immunological and stress responses are released by microbes and are detected by plants. Through microbial elicitors known as microbe-associated molecular patterns (MAMPs), which include chitin, flagellin, lipopolysaccharide, and peptidoglycans, microorganisms in the rhizosphere trigger plant responses (Millet et al., 2010). The process by which MAMPs stimulate and initiate systemic defensive responses in rhizosphere-beneficial microorganisms is known as induced systemic resistance (ISR). MAMPs can also produce systemic acquired resistance (SAR), which is primarily caused by infections (Conrath et al., 2015).

3.3.2 Rhizosphere Microbiome

The rhizosphere, a small area of soil near plant roots that acts as a barrier between soil and roots, is home to a wide variety of microorganisms and small animals that inhabit the soil (Dubey et al., 2019). The ectorhizosphere and the endorhizosphere are the two distinct compartments that make up the rhizosphere. Nonetheless, a greater number of habitats are inhabited by diverse microorganisms, and numerous researchers have described their interactions with roots (Dubey et al., 2019). Only a small percentage of the diversity of microbes connected to plants is found to be harmful, while the majority have beneficial interactions that increase plant fitness and survival (Egamberdieva et al., 2017).

Due to their potential to fix nitrogen, endophytes have long been underappreciated but are currently receiving a lot of attention (Gallego-Giraldo et al., 2014). Endophytes are found in non-nodulating strains of endophytes such as *Microbacterium trichothecenolyticumn*, *Brevibacillus choshinensis*, *Endobacter medicaginis*, and *Micromonospora* spp., as well as inside the root nodules of various crop plants such as *Rhizobium* spp. (Benidire et al., 2017). The physiology of the plant-associated microbial population helps to improve a number of diseases and boosts plant resistance to stress by dispersing and delivering a range of nutrients (Kumar and Dubey, 2020).

3.3.3 Rhizosphere Engineering

Plant's benefit from the environment should be optimized through microbial engineering. Improving nutrient cycling, nutrient sequestration, resistance to heavy metals and salinity, disease and pest resistance, and water-holding capacity would be the obvious course of action. Many products on the market today are derived from the plant-growth-promoting bacteria (PGPB) and fungal microbial consortium. Examples include Bactophospin, which uses *Bacillus mucilaginous* (a Russian microorganism), and Mamezo, which uses *Rhizobium* sp. (a Japanese microorganism) (Mishra and Arora, 2016).

Additionally, PGPBs have been found in laboratory and greenhouse trials; however, despite their great potential, they did not perform well in the field (Khan et al., 2019b). The incapacity of these strains to colonize the environment was a major factor in their subpar performance. Therefore, it is crucial to keep in mind that to guarantee long-term placement in the community and structure within that ecosystem, the bioinoculant must be able to establish strong and durable relationships in the soil and with the plant (Wang and Coleman, 2019).

Now the environment and plants could be intentionally altered thanks to the omics tools available today (Moreira et al., 2023). To replicate the symbiotic interactions that naturally exist between the soil, microbes, and plants, the mechanism that changes the environment and establishes identical conditions in the field must be understood. Once this is finished, the bioinoculant will flourish in the planted habitat (Nadarajah and Abdul Rahman, 2023).

3.4 SHAPING THE RHIZOSPHERE'S MICROBIOME

Shaping the rhizosphere's microbiome can be achieved through the cultivation of microorganisms to confer specific functions and properties or by using plants with special characteristics through plant engineering or plant breeding (Yue et al., 2023).

3.4.1 Microbe-Mediated Practices to Modify the Rhizosphere Microbiota

Numerous rhizosphere engineering techniques necessitate the cultivation of microorganisms to enhance their cultivability within the rhizosphere. The microbes that can be grown exhibit specific functional abilities; however, their behavior in varying environmental conditions remains uncertain (Kumar and Dubey, 2020).

Data regarding the PGPR employed as a possible biofertilizer that coexists in symbiotic association with their host plants should be collected and put into a database in order to apply these bacterial formulations in the field later.

Different types of phosphate-solubilizing bacteria (PSB), siderophore-producing bacteria, and arbuscular mycorrhizal fungi (AMF) boost access to different nutrients, such as iron, phosphorus, zinc, copper, and cadmium (Hashem et al., 2017). These rhizobacteria are recognized as possible biocontrol agents in addition to generating antibiotic substances such as oligomycin, phenazine, HCN, and bacteriocins (Nisin). Additionally, they generate antifungal substances such as pyoluteorin, phenazines, and phoroglucinols (Khan et al., 2019b). Furthermore, by producing a variety of defense compounds, the inoculation of plants with consortiums of PGPR and AMF helps to alleviate various abiotic and biotic stresses (Kumar and Dubey, 2020).

The growth and development of plants depend on phytohormones, which are believed to be an essential part of interactions between plants and microbes (Gupta et al., 2018). Numerous phytohormones, including auxins or indole-3-acetic acid (IAA), gibberellins (GA), and cytokinin, were discovered to be produced by microbial communities. Examining how three compounds – ethylene, salicylic acid, and jasmonic acid – mediate crosstalk and impact plants' responses to activate systemic acquired resistance (SAR) and induce systemic resistance (ISR) is crucial. Inoculating plants with non-pathogenic bacteria can help them acquire resistance against a wide range of pathogenic germs in both above-ground and below-ground parts (Kumar and Dubey, 2020).

Plants that are prepared to respond to the pathogen assault more swiftly and forcefully. ISR has been found for several microorganisms as well as their cellular derivative determinants, or MAMPs, which include flagella, cell envelope elements, and siderophores (Vandenkoornhuyse et al., 2015). Remarkably, certain

PGPR causes an ISR reaction and stimulates plant growth by releasing volatile organic compounds (VOCs) (Chowdhury et al., 2018). Several species of *Pseudomonas*, *Bacillus*, *Serratia*, and *Trichoderma harzianum* are well-characterized ISR-inducing microbes.

The ethylene and jasmonic acid signaling pathways are the main sources of support for this ISR. As a result, plants are ready to react to the pathogen attack more quickly and powerfully. Microorganisms and their cellular derivative determinants, such as flagella, cell envelope elements, and siderophores, have been found to have ISRs (Vandenkoornhuyse et al., 2015). Surprisingly, some PGPRs release volatile organic compounds (VOCs) that trigger an ISR reaction and promote plant growth (Chowdhury et al., 2018).

Other substances known to have hormonal activity include brassinosteroids and strigolactones. *Miscanthus* seedlings are injected with the temperate grass endophyte *Herbaspirillum frisingense* (GSF30T) to stimulate root and shoot growth. According to transcriptome analyses, the ethylene and jasmonic acid signaling pathways are regulated, suggesting that phytohormone activity either promotes or modifies plant growth (Sheibani-Tezerji et al., 2015).

From sweet potatoes, a distinct class of bacterial endophytes was cultivated. Endophytic bacterial strains that yield auxin and indole acetic acid (IAA) were introduced into the cuttings through inoculation. Compared to non-inoculated cuttings, those cuttings sprout roots more quickly. It has been shown that GSF30T *Herbaspirillum frisingense* also produces IAA in the culture, and it has been deduced that auxin production is the reason why wheat plants grow more seedlings when inoculated with *B. subtilis* (Kumar and Dubey, 2020).

It is well known that *Azospirillum* species fix nitrogen and produce auxin, which both promote plant growth. *Azospirillum* sp. strain B510, isolated from surface-sterilized rice stems, increases the yield of paddy fields and rice plants by re-inoculating seedlings. In contrast, three strains of *Pseudomonas* improved the growth and spike length of wheat plants in both field and laboratory conditions (Jiménez-Vicente et al., 2015). The biocontrol activity of these microbes has been extensively studied in both laboratory and field settings, leading to the development of a wide range of commercial products.

By producing antimicrobial compounds, endophytic bacteria block pathogenic quorum sensing and, in turn, prevent virulence, biofilm formation, and communication without halting bacterial growth. By generating cell-free lysates, endophytic bacteria inhibit the formation of biofilm in *P. aeruginosa* PAO1 and are also able to break down quorum-sensing molecules (Kumar and Dubey, 2020).

3.4.2 Plant-Mediated Practices to Modify the Rhizosphere Microbiome

Genetic engineering and plant breeding are two methods used to manipulate plant characteristics through plant-mediated strategies. A microbial community can be selected using plant breeding, which aims to increase crop yield by providing resistance to a variety of stresses (Ryan et al., 2009). Hence, microbiome selection was integrated into plant breeding programs to target very important taxa and functions. For instance, Neal et al. (1973) employed chromosome swapping between two wheat lines in their study to increase resistance to root rot disease and maintain the rhizosphere's beneficial bacterial populations.

According to Koyama et al. (2000), transgenic plants grow more readily on phosphate-limited soil due to their increased capacity to secrete citrate from their roots. This finding implies that crop plants with improved utilization of a phosphate evolved a greater capacity to grow in acidic soils and a tolerance to aluminum. As a result, the process underlying the inherent "suppressiveness" of soil against soil-borne illnesses has not been fully covered. In his instance, *Nicotiana tabacum* and *Arabidopsis* transgenic lines are used to control the pH of the rhizosphere. Several phenotypes, including increased H^+ efflux from the plant's roots, a more acidic rhizosphere, enhanced resistance to salinity and drought stress (AVP1) and enhanced resistance to salinity stress in the tobacco line, were caused by these plants being modified to overexpress H^+ ATPase protein (AVP1 phosphatase in *Arabidopsis* and PMA4 in tobacco) (Kumar and Dubey, 2020).

According to Ellouze et al. (2013), certain chickpea cultivars draw a more beneficial microbiota that aids in shaping durum wheat plants in North America's semi-arid grasslands. The goal of many studies conducted by different researchers is to manipulate plants by changing the production of vital exudates that direct the development of certain interactions between the plant and the microbiome. In recent years, site-directed modification techniques have been developed using CRISPR/CRISPR-associated protein 9 (Cas9), finger nucleases, meganucleases, zinc transcription activator-like effector nucleases (TALENs), and clustered, repeatedly interspaced short palindrome repeats (CRISPR). Because of its increased versatility, low cost, ease of design and implementation, and higher success rate than other genome editing techniques, CRISPR/Cas9 has recently become quite popular (Ricroch et al., 2017).

The assembly and metabolic processes of the rhizosphere microbiome are significantly influenced by plant domestication. This also showed that host plants can utilize the distinct ecological function of root-associated keystone microbiomes by releasing root exudates, which in turn support underground multi-nutrient cycles and plant growth (Yue et al., 2023).

3.5 ALTERED MICROBIOMES

3.5.1 Genetically Modified Microbes

Using genome-based methods, several genes in microorganisms that mediate interactions between plants and bacteria have been identified (Levy et al., 2018). Microbial genes must be altered to promote beneficial traits in order to produce better and safer microbial inoculants. The symbiotic bee gut bacterium *Snodgrassella alvi* has recently been genetically modified to synthesize double-stranded RNA, causing extensive RNA interference (RNAi) in bee hosts. By activating RNAi, the engineered bacteria shield honeybees from mite and virus infections (Leonard et al., 2020). Similar microbiome upgrade techniques can also be used for endophytic engineering, which triggers a plant's defense system against pests and diseases. The possibilities for genome editing have been greatly increased by the very efficient clustered regularly interspaced short palindromic repeats (CRISPR)/CRISPR associated protein

CRISPR/Cas systems (site-directed nucleases) (Uniyal et al., 2019). Additionally, it has significantly increased the potential for genome and gene editing to enhance the advantageous traits of bacteria associated with plants. These technologies can be used to develop inducible regulatory circuits that control the induction of beneficial traits in response to signals produced by the signaling process between the plant host and the target organisms (in the case of biocontrol). Considerable effort is being made to restructure microbial gene networks for efficient agricultural usage using synthetic biology technologies and biosensors (Goold et al., 2018). Banana Guard, a product with a genetically altered *Pseudomonas putida* strain, was developed to prevent the *Fusarium oxysporum* infection that causes Panama disease in bananas (Kemal et al., 2016).

Endophytic or epiphytic bacteria found in cereal crops have recently been altered to enhance nitrogen fixation. In reaction to substances in the rhizosphere and seeds that have been demonstrated to change plant microbiomes, the engineered nitrogen-fixing strains were further altered to express N_2-fixing genes (Ryu et al., 2020).

Through contemporary targeted genome editing tools, it is now possible to create unique designer circuits for genes in microbes that react to plant signals in order to maximize microbe interactions. Numerous efforts are being made to improve crop-associated microbes' ability to fix nitrogen; if successful, this will lower fertilizer use and runoff, which harms rivers (Goold et al., 2018).

3.5.2 Targeted Plant Microbiome Engineering

The most significant element affecting a plant's growth and development is its root microbiome. It promotes tolerance to biotic stresses, microbiota regulation, and nutrient uptake (Arif et al., 2020). In the rhizosphere, a number of microorganisms can interact with one another through functional consortium (Dubey et al., 2019). For instance, PGPR and arbuscular mycorrhizal (AM) fungi worked in tandem in a model grassland by restricting nutrients and fulfilling distinct ecological functions (Vyas et al., 2018). Lastly, the modification can be integrated into the rhizosphere microbiome by activating particular substrates and molecules in the signaling pathway or by conditioning the soil with some conventional/traditional soil additives to captivate and entice the desired microbiota (Pascale et al., 2019).

It is also possible to introduce particular microbial strains into soil to change and modify the makeup of different microbial communities (del Carmen Orozco-Mosqueda et al., 2018). Examples include the following:

1. Even in the face of heavy metal and salt stress, sunflower plants developed profusely when a powerful bacterial and fungal consortium (*Pseudomonas libanensis* TR1 (bacterial part) + [*Arbuscular mycorrhiza*] AM fungus, *Claroideoglomus claroideum* BEG210 [fungal part]) was applied to the rhizosphere section (Ma et al., 2019).
2. *Acinetobacter sp.* (UQO2), *Bacillus amyloliquefaciens*, and *Bacillus velezensis* bacterial cultures applied to the roots of developing chilli plants enhanced plant growth and, consequently, the plants' resistance to soil-borne *Phytophthora capsicum* infection (Syed-Ab-Rahman et al., 2019).

3. *Agrobacterium* sp. 10C2-grown beans yielded more nodules and plant biomass in their harvested pods, along with increased levels of phosphorus, flavonoids, and antioxidants (Chihaoui et al., 2015). Many PGPR (plant growth-promoting rhizobacteria) species, including *Actinomycetes* and *Brevibacterium*, started to colonize the soil surrounding the plant after 15 days of inoculation. It has been demonstrated that the seed microbiome, which is inherited from parents, enhanced germination, plant survival, and performance (Acuña et al., 2023). When *Pseudomonas stutzeri* and *Stenotrophomonas maltophilia* were co-inoculated into tomato seedling rhizospheres, the plants grew more quickly, and both organisms produced compounds that were effective against Botrytis cinerea leaf disease (dimethyl disulfide, for instance). Similar to how root exudates boost beneficial microbes in the rhizosphere, roots serve as essential gatekeepers, choosing advantageous microorganisms that may reach the plant as endophytes. Additionally, roots have the ability to directly eat related bacteria and use them as a source of nitrogen (Mendes et al., 2019). Panke-Buisse et al. (2017) demonstrated that the rhizosphere's microbial biomass had an unidentified role in plant growth. To find out if plants prefer to eat particular microorganisms, more investigation is needed. More resilient crops that can resist pest and disease attacks are the goal of recent plant breeding efforts. However, plants usually cannot fight both biotic and abiotic stress at the same time since the related pathways are often antagonistic (He et al., 2010).

3.6 STRATEGIES TO MODIFY MICROBES FOR BENEFICIAL ACTIVITIES

The sustainability of crop production and the production of food globally are seriously threatened by several biotic and abiotic stressors, including a shortage of water, a shortage of land, global warming, pests, and diseases, which lead to a high rate of crop failure globally (Cerda et al., 2017). There is an urgent need to develop better agricultural practices that will increase crop output, combat climate change with minimal degradation of the ecosystem and soil, and preserve human health to secure a sustainable food supply (Cerda et al., 2017). The rhizosphere is crucial for several ecosystem functions (Arif et al., 2020). Most of the bacteria in the rhizosphere microbiome are plant growth-promoting rhizobacteria (PGPR). The majority of them interact favorably with plants and promote their development and survival (Kumar and Dubey, 2020). They include a range of bacteria that are frequently employed as legume inoculants, including *Rhizobium, Sinorhizobium, Bradyrhizobium, Azorhizobium, Allorhizobium, Ochrobactrum*, and *Mesorhizobium* (Hakim et al., 2018, 2020a). *Arthrobacter, Curtobacterium, Micromonospora, Microbacterium, Mycobacterium, Acinetobacter, Agrobacterium, Blastobacter, Bosea, Devosia, Enterobacter, Herbaspirillum, Pantoea, Pseudomonas, Ralstonia, Stenotrophomonas, Bacillus, Brevibacillus, Paenibacillus, Chryseobacterium, and Sphingobacterium* are among the non-rhizobia that colonize the interior of root nodules (Hakim et al., 2020b). Extracellular PGPR (ePGPR), known as free living rhizobacteria, exist in soil but do not live in plant tissues; *Arthrobacter*, *Bacillus*, *Azospirillum*, *Azotobacter*, *Micrococcus*, *Pseudomonas*, and *Serratia* are a few examples (Adesemoye and Egamberdieva, 2013). Actinomycetes, which exhibit numerous beneficial traits for improving plant productivity, also form a major group of microbes in the rhizosphere (Muleta and Assefa, 2018). However, the overall functioning of the ecosystem and rhizosphere biology could be impacted by elevated temperatures and unpredictable climate change through both direct and indirect mechanisms. Therefore, the engineering of plant microbiomes is an alternative but unexplored strategy (Kumar and Dubey, 2020).

Table 3.1 shows the advantages and possible disadvantages of the various approaches taken to mold the rhizosphere microbiome.

3.7 METHODS FOR MODIFYING THE RHIZOSPHERE MICROBIOME MEDIATED BY MICROBIOMES

To improve the cultivability of the bacteria present in the rhizosphere, the culturing of microbes is a necessary step in many rhizosphere engineering procedures. These cultivable microorganisms showed some functional potential, but further research is required to assess how they behave in various environmental settings before they can be used in the field (Dubey et al., 2019). As a result, the biocontrol activity associated with these bacteria has been thoroughly investigated in both laboratory and field settings, leading to the development of multiple commercial treatments, as shown in Table 3.2.

3.8 NEW APPROACHES TO IN SITU MODIFICATION OF THE PLANT MICROBIOME

Numerous techniques are being developed and tried to regulate the plant's complete microbiome without the need for culture. Several approaches are being studied and tested to mobilize the whole microbiome without requiring its corresponding culture, including the use of transgenic or conventional means. Unfortunately, because of how the public views transgenic plants, applying this idea to crops is still somewhat restricted. Alternative non-transgenic approaches to manipulate and control the microbiome in situ are being explored. Here, a few of these methods are examined: Utilizing chemicals involved in plant–microbe signal transduction to take advantage of plant–microbe communication: Different signaling molecules are produced by plants and microbes to communicate their needs to their partners. For example, signal particles released by phosphorus-starved plants trigger an upregulation of phosphorus-assembling genes in rhizosphere organisms (Zhang et al., 2019). In addition, plants release these signaling molecules when they are threatened by pests or diseases or when they want to communicate with other plants through volatile organic compounds (VOCs) to promote the colonization of beneficial microbes.

These compounds may prove useful for cultivation, detection, and identification, but they are difficult to cultivate, detect, and identify because the amounts are so small. Utilizing blended concoctions (cocktails) of microorganisms that do not

TABLE 3.1
Advantages and Possible Disadvantages of the Various Approaches Taken to Mold the Rhizosphere Microbiome

Strategy		Benefits/ Potential Drawbacks	Reference
Microbiome-mediated methods	PGPR, AMF, endophytes, rhizobia, and ecto mycorrhiza	• Improve plant health and disease-fighting biocontrol. • Phytohormone Production – Increases SAR and ISR in the plant. • Increase the soil's fertility. Aids in nodulation and nitrogen fixation. • Disadvantages: Competition with local microorganisms for resources.	Hashem et al. (2019)
	Strains of recombinant microbes.	• Certain genes are transferred by horizontal gene transfer (HGT), which promotes the expression of advantageous functionalities. • Disadvantages include loss of the target gene over time and the undesirable outcomes associated with HGT.	Kumar and Dubey (2020)
Plant-based methods	Breeding plants and modifying their genetic makeup Increased exudate production attracts particular advantageous microbial functional groups.	• It does not require any modifications to the field sites' management or infrastructure. • Impacts microbial diversity by promoting the growth of specific rhizosphere-dwelling microorganisms. • Disadvantages: Plant exudates may be decomposed rapidly in the soil. • Evaluations of plant lines for interactions with the soil microbiota are lacking	Dubey et al. (2019)
	Selection of mutants with improved capacity for mutual symbiosis development.	• A better ability to get nutrients • Has a negative impact when exposed to excessive nutrient levels.	Kumar and Dubey (2020)
	Modification of genetics Modification in the quantity of organic exudates, residues, and signaling molecules that seep into the soil.	• Produce useful functional features such as siderophores, antibiotics, antimicrobials, and other biocontrol agents. • Enhancing resilience to unfavorable environmental circumstances. • Cons: Genes are passed from one species to another. The compounds may be inactivated in the soil and quickly disintegrated once the target gene was successfully inserted into the plant, or the rate of exudation may be too sluggish to have an impact on the rhizosphere.	Imam et al. (2016)

(*Continued*)

TABLE 3.1 *(Continued)*

Strategy		Benefits/ Potential Drawbacks	Reference
Meta-organism based	Controlling and choosing plants and microbiomes that complement each other	• Suppressive soils are introduced to control soil diversity. • Enhancing the physical-chemical properties of the soil. • Increase in the amount of organic carbon and the cycling of nutrients. • Effective advancement in disease prevention and agricultural techniques. • Drawbacks: It's unclear how the mechanisms work.	Kumar et al. (2015)
	Bacteria that can use opine are co-inoculated with plants that synthesize opine	• Creating a direct line of communication between the two interaction partners.	Kumar and Dubey (2020)
	Application of mineral fertilizers such as ammonium nitrate, urea, phosphates, and sulfates.	• Increases crop residue, system productivity, and soil organic matter • Drawbacks: The use of P affects AMF root colonization, while N reduces soil pH and encourages soil acidification.	Hashem et al. (2019)
	Utilizing organic fertilizers such as animal manure, composts, and biosolids	• Boost the soil's biological activity. • Drawbacks: Biosolids: hazardous materials that can damage soil microbiota may be present. Unable to accurately duplicate the composition of compost.	Kumar and Dubey (2020)
Biochemical pathway	Create native microbiomes to produce volatile organic compounds (VOCs) and other naturally occurring chemicals that are produced by microorganisms and plants.	• Improved adaptation against stress, suppression of pathogens, and microbial quorum sensing through targeted chemical release. • Cons: Copying quorum sensing could cause the beneficial bacterial communities to stop communicating with one another.	Schulz-Bohm et al. (2018)
Molecular Strategies	Plasmids introduce desired genes and metabolic processes, like the breakdown of toluene.	• New crop varieties can be raised such as the toluene degradation pathway into the endophytic bacterium *Burkholderia cepacia*, or the engineered pTOM toluene-degradation plasmid into B. cepacian to inoculate lupine seeds. • Disadvantages: Difficult to be implemented in a dynamic environment like the soil.	Kishan et al. (2021)
Agronomic Approach	Crop rotation, organic amendments, and inorganic fertilizers	• Increased soil organic carbon content, better physico-chemical properties, and higher soil biological activity. • Drawbacks: acidity of the soil and potential buildup of hazardous substances in the soil microbiota.	Kishan et al. (2021)

TABLE 3.2
Studies Used Different Strategies to Modify Microbes for Beneficial Activities: In Situ

Strategy	Involved Microbes	Technology Involved	Reference
Microbe-mediated strategies	• Strains of AMF • siderophore-producing bacteria • PSB (phosphate-solubilizing bacteria)	Enhancing the availability of various nutrients (phosphorus, iron, copper, zinc, and cadmium).	Hashem et al. (2017)
	Rhizobacteria (*Streptomyces*, *Bacillus*, *Pseudomonads*)	• Producing antifungal substances (pyoluteorin, phoroglucinols, and phenazines). • Creating antibiotic chemicals (oligomycin, HCN, DAPG, bacteriocines [Nisin]).	Khan et al. (2019a)
	• *Bacillus* spp. • *Pseudomonas* spp. • *Serratia* species • other well-characterized ISR-inducing microorganisms	• Release of volatile organic compounds • Certain PGPR induced an ISR response and stimulated plant development.	Chowdhury et al. (2018)
	Strain B510 of *Azospirillum* sp. isolated from surface-sterilized rice stems	Enhanced the growth and spiked length of wheat plants in both field and laboratory conditions.	Jiménez-Vicente et al. (2015)
	Soil microbiomes	Engineered to create resistance to the aboveground herbivorous insects	Pineda et al. (2020)
	Acinetobacter sp. (UQO2), *Bacillus amyloliquefaciens*, and *Bacillus velezensis* on the roots of chilli plants	Enhanced plant development and stability to Phytophthora capsicum infection	Syed-Ab-Rahman et al. (2019)
	BHUJPCS-15 and BHUJPVCRS-1 isolated from chickpea rhizosphere soil and chickpea seed	Raised soil enzymes, microbial populations, and NPK output.	Mukherjee et al. (2022)
	PGPR	Introduced resistance mechanisms in *Arachis hypogaea* against illnesses in the early stages	Syed et al. (2020)

(Continued)

TABLE 3.2 *(Continued)*

Strategy	Involved Microbes	Technology Involved	Reference
Plant-mediated strategies	*Rhizosphere*	a chromosomal swap between two wheat lines enhanced resistance to root rot disease and maintained the beneficial bacterial populations	Kumar and Dubey (2020)
	Streptomyces griseoflavus and Bradyrhizobium	Enhanced Glycine max's nodulation, nitrogen fixation, and nutrient uptake	Htwe et al. (2018)
CRISPR for crop improvement	Genetically modified strain of *Pseudomonas putida*	Fighted *Fusarium oxysporum*, the culprit behind Panama disease in bananas.	Kemal et al. (2016)
	CRISPR-Cas9 technique	Inactivated and modified mutants were unable to produce fumonisins (mycotoxins).	Ferrara et al. (2019)
	Trichorderma atroviride	Increased the synthesis of genes involved in polyketide biosynthesis. Control specific diseases caused by soil pathogens.	Fang and Chen (2018)
	Indica root microbiome	Increased the efficiency of nitrogen consumption in Japanese rice.	Xin et al. (2021)

directly assist plants can boost the native microflora's positive motility. In essence, the microbes in these mixed potions have high concentrations of signaling chemicals.

1. Understanding the role of crop species microbial hubs in microbiome construction and activity is essential when changing the overall microbiome in situ.
2. Synthetic biology, a subfield of industrial science, can also be used to produce crop probiotics with unique and predictable qualities, which, when combined with a plant, can increase the beneficial organisms' activity.
3. The genome of the in situ microbiome can be directly engineered and ultimately altered or modified using designing techniques such as in situ genome designing (Sheth et al., 2016). It is possible to introduce directional and desired capabilities to native microflora by introducing transportable hereditary components like plasmids.
4. In plants, microRNA (miRNA) controls the rhizosphere microbiome, plant–microbe collaboration, structural makeup, and functional range. It has been used to restore digestive health (Liu and Weiner, 2016), and it has the potential to be a very useful tool for identifying recipient organisms to increase agricultural yields.
5. By offering dynamic interaction imaging, the application of microfluidics-based technology provides a unique perspective into the intriguing root microenvironment for monitoring the interactions between the plant and the microbiome. With the use of this technology, the current understanding of the microbiome and its practical management techniques will be further advanced (Bashiardes et al., 2016).
6. Combining ecological technologies and techniques with the upgraded technologies mentioned above has the potential to change host microbiomes, reduce plant and soil diseases, and ultimately increase sustainable productivity (incorporating crop rotation and non-tillage practices with microbiome-specific breeding techniques). But since these technologies are still in their infancy, more studies are necessary before they can be used effectively and without causing unfavorable effects (Nitika et al., 2023) (Table 3.3).

3.9 PLANT ENGINEERING THROUGH THE UTILIZATION OF OPTIMIZED MICROBIOME

To enhance microbial-mediated functions or carry out specific microbiome assembly, plants can be genetically engineered or changed. For instance, cinnamoyl-CoA reductase (CCR) activity has been inhibited in poplar plants through genetic modification. The root endophytic population's structure and function are not affected by this change in the microbiome of the root. Host genetics influence the composition of the root microbiome. Plant breeding can benefit immensely from the use of microbiome-based features (Beckers et al., 2016).

TABLE 3.3
The Impact of PGPR-Assisted Rhizosphere Engineering on Plant Development and Stressor Tolerance

Plant	Microbes Used	Beneficial effects	References
Growth under normal conditions in the rhizosphere			
Lens culinaris var. PL-	*Chryseobacterium* sp. PSR10	Phosphate solubilization	Singh et al. (2018)
Wheat	*Enterobacter* sp., *A. chlorophenolicus*, *S. marcescens*, *B. megaterium*	P-solubilization, IAA, HCN, N-fixation, Gibberellin, Siderophores	Kumar et al. (2015)
Potato	*Azospirillum* sp.	P N-fixation, IAA production	Naqqash et al. (2016)
Soybean	*Bacillus cereus*	Sideophore, IAA, P-solubilization, EPS	Arif et al. (2020)
Arabidopsis thaliana	*Bacillus amyloliquefaciens*	Phytohormone production, Lipopeptide	Asari et al. (2017)
Sarracenia species	Endophytic diazotrophic bacteria	Nitrogen fixation	Sexton et al. (2019)
Potato	*Brevundimonas* spp.	P-solubilization, N-fixation	Naqqash et al. (2020)
Rice	*Acinetobacter soli*, *Bacillus* sp., *Pseudomonas putida*, *Pseudomonas mosselii*, *Arthrobacter woluwensis*	P solubilization, ACC deaminase activity, Siderophores production	Aw et al. (2019); Khan et al. (2019a)
Cotton	*Bacillus amyloliquefaciens*, *Bacillus* spp.	Production of phytohormones, N-fixation, P solubilization, and antibiotic activity	Fahimi et al. (2014)
Cabbage	*B. subtilis*	production Gibberellins	Kang et al. (2019)
Plant	**Microbes Used**	**Pathogen Control**	**References**
Disease infested soils			
Tomato	*P. fluorescens*	*Ralstonia solanacearum*	Hakim et al. (2021)
	Bacillus velezensis	Antibiosis activity, Lipopeptides production	
Rice	*Pseudomonas* sp.	*Xanthomonas oryzae*	
		Peroxidase activity, polyphenol-oxidase & phenylalanine-ammonia lyase	
Potato	*Lysinibacillus* sp., *Pseudomonas fluorescens*, *Bacillus subtilis*,	*Ralstonia solanacearum*	
		Antibiosis	

Cajanus cajan and Festuca	*Azotobacter chroococcum, Bacillus megaterium*, and *P. fluorescens*	Impaired growth of Listeria monocytogenes	Sharma et al. (2020)
Black pepper	*B. subtilis, B. siamensis, Brevibacillus gelatini, P. geniculata, P. beteli, Burkholderia ubonensis, and Burkholderia territorii*	Antagonistic to soil-borne Fusarium solani	Lau et al. (2020)
Peanut	*P. fluorescens*	*Sclerotium rolfsii* Antimicrobial compound	Hakim et al. (2021)
Cucumber	*Bacillus amyloliquefaciens*	*Fusarium oxysporum* Secondary metabolites, ISR, Phytohormones	
Tomato	*Paenibacillus lentimorbus* *Bacillus cereus*	*Sclerotium rolfsii, Alternaria solani, Fusarium oxysporum* Expression of defense-related & autophagy-related genes IAA, Siderophore, Ammonia, ACC deaminase, Catalase, N-fixation	
Chili	*Talaromyces funiculosus*	*Colletotrichum capsica* ISR, Callose & Lignin deposition	
Black Mustard	*Bacillus subtilis, Pseudomonas fluorescens*	*Turnip mosaic virus* Catalase activity, ISR, Phenol production	
Periwinkle	*B. pumilus, B. subtilis*	*Cucumber mosaic virus* Phenol, alkaloid, and flavonoid production	
Chili	*Bacillus amyloliquefaciens*	*Groundnut-bud necrosis virus* Flagellin, ISR, Phytohormone	

(Continued)

TABLE 3.3 *(Continued)*

Plant	PGPR	PGPR characteristics	References
Medick	*Sinorhizobium medicae*	Nodulation, Nutrient uptake	Staudinger et al. (2016)
Maize	*Azospirillum* sp.	Siderophore, N-fixation, IAA, P-solubilization	García et al. (2017)
Chickpea	*Bacillus* sp	IAA, HCN, EPS, Ammonia production	Khan et al. (2019a)
Mint	*Azospirillum brasilense*, *Azotobacter chroococcum*	Osmolyte accumulation, Phenol production, Antioxidant activity, ABA	Asghari et al. (2020)
Salt stress			
Wheat	*Planococcus rifietoensis*	ACC-deaminase, P solubilization, IAA production	Rajput et al. (2018)
	Aeromonas spp.	ACC-deaminase, Zn-/P solubilization, IAA production	
Okra	*Enterobacter* sp.	ACC deaminase, Antioxidant activity	Habib et al. (2016)
Sweet and chili pepper	*Rhizobium massiliae*, *Brevibacterium iodinum*, *Microbacterium oleivorans*	Proline, Antioxidant enzyme, Sugar	Hahm et al. (2017)
False Flax	*P. migulae*	ACC activity, Phytohormone production, ABA	Heydarian et al. (2018)
Phaseolus vulgaris	*Aneurinibacillus aneurinilyticus* and *Paenibacillus* sp.	1-Aminocyclopropane-1-carboxylic acid (ACC) deaminase activity alleviating salinity stress	Gupta and Pandey (2020)
Rice	*Glutamicibacter* sp.	ACC deaminase activity, IAA production	Ji et al. (2020)
Wheat	*P. putida*	Antioxidant activity, Proline, Total proteins, Sugars, Amino acids	Ali et al. (2020)
	Azospirillum brasilense, *Bacillus amyloliquefaciens*	Secondary metabolites, Antioxidant enzymes	Abd El-Daim et al. (2014)
	B. safensis	Antioxidant enzymes, osmolyte accumulation	Sarkar et al. (2018)
Soybean	*B. aryabhattai*	JA, IAA, Antioxidant enzymes	Park et al. (2017)
Tomato	*Bacillus cereus*	ACC Deaminase Activity, EPS	Mukhtar et al. (2020)
Avicennia marina	*Halomonas* sp.	Heavy metal stress	Mukherjee et al. (2019)

The taxa *Amycolatopsis* sp., *Serratia proteamaculans*, *Pedobacter* sp., *Arthrobacter* sp., *Stenotrophomonas* sp., *Fusarium merismoides*, and *Fusicolla* sp. improved canola yields in field studies (Lay et al., 2018). Despite advances in our understanding of how the microbiome influences host performance, the successful implementation of these strategies in breeding programs necessitates a more in-depth understanding of genotype interactions. Future research should focus on identifying plant genes that influence the composition of the microbiome. To obtain the most reliable results, these investigations must be conducted in natural fields managed by best practices, with multiple locations and multiyear field trials (Wille et al., 2019).

3.10 MICROBIOME OPTIMIZATION USING A VARIETY OF MANAGEMENT TECHNIQUES

Mulching, crop rotation, cover crops, and applying organic manure are all efficient management techniques that encourage soil growth that resists disease. Mixing small amounts of conducive and suppressive soils allows the disease's suppressive qualities to be readily transmitted and transferred to suppressive soils (Raaijmakers and Mazzola, 2016). Agroecosystems can grow healthier as a result of the transferable positive microbial effects of specific management techniques. Plant physiology, especially the generation of plant hormones, is influenced by gene expression, which is influenced by the legacy effect of management techniques on soil microbiome assembly (Blair et al., 2018).

It has been demonstrated that soil nutrient cycling and soil microbiome structure and activity were enhanced by organic farming and related holistic techniques (Mendes et al., 2011). However, a number of factors impact and influence management choices, including the choice of variety and location, soil health, the use of manure and fertilizer, nutrient applications, yield and profitability, the use of machinery, and the measurement of benefit-to-cost ratios (Trivedi et al., 2020).

3.11 MULTI-OMICS

A genetic viewpoint to learn more about the interactions, activities, and composition of microbiomes. Integrated multi-omics techniques are revealing different facets of microbiomes.

These integrated methods offer a comprehensive understanding of how particular microbial communities and bacteria foster synergistic interactions between the plant and the microbiome, guaranteeing plant health and resilience to environmental stressors.

3.12 REVOLUTIONARY GENOME MECHANISM: A RELIABLE GENE EDITING TECHNIQUE

A major advance in gene editing techniques is represented by the clustered regularly interspaced short palindromic repeats (CRISPR)/CRISPR-associated protein (Cas). In a variety of economically significant crops, the CRISPR technique is emerging as a potent tool for introducing desired traits (disease resistance) and genes of interest.

Additionally, numerous studies showed that CRISPR-mediated created resistance can successfully integrate gene characteristics, producing a broad-spectrum effect against a wide range of pathogen groups (Oliva et al., 2019). Additionally, the commercialization of CRISPR crop varieties is not limited to protected cultivation circumstances or laboratory experiments and is supported by research and literature. In the US, the marketing and commercialization of *C. sativa* (CRISPR-Crop) with enriched omega-3 oil were setting new market records (Waltz, 2018). In addition to giving the crop disease resistance, CRISPR technology aids in the de novo induction of meristem, which eliminates the need for laborious and time-consuming tissue culture procedures to produce dicotyledonous plants (Maher et al., 2020). Additionally, CRISPR/Cas genome editing is being utilized to improve food storage, post-harvest quality, and the development of resistance to a variety of environmental challenges (Zhang et al., 2021). Accordingly, the CRISPR-edited tomato demonstrated enhanced storage capacity and extended shelf life without influencing other agronomic factors like plant size and fruit firmness (Yu et al., 2017).

3.13 GOALS FOR SUSTAINABLE DEVELOPMENT: THE GENETIC MARCH TO END HUNGER

Aiming to eradicate hunger by 2030, the Sustainable Development Goals (SDGs) are being thwarted by global economic downturns brought on by the COVID-19 pandemic and climate change (Veillet et al., 2019). Even while transgenic and conventionally cross-bred crop varieties have greatly decreased the usage of pesticides, fungicides, and herbicides, they still need to be closely watched to stop the development of weeds and insect pests through natural selection (Qin et al., 2020). We must properly boost food production and guarantee equitable distribution if we hope to end poverty. Maintaining yield while remaining sustainable, enhancing grain quality and nutritional content, and building resilience to biotic and abiotic stresses are some of the obstacles that agricultural production must overcome (Cai et al., 2020). By 2030, scientists will have a big role to play in advancing sustainable agriculture, enhancing nutrition, and ending hunger. One way to guarantee food security is to create sustainable crops that can adjust to shifting environmental conditions. Furthermore, the potential application of transgenic crops is limited by issues with their pricing, effectiveness, bioethics, regulation, and public acceptance. Ending world hunger by 2030 is the second sustainable development goal (SDG2) set forth by the UN in 2015 (Xing et al., 2021). Increased and more consistent agricultural output is required to meet the nutritional needs of the world's population and eradicate hunger by 2030 using the previously mentioned advanced techniques.

3.14 CONCLUSION AND FUTURE PERSPECTIVES

In conclusion, the pressing challenges of food security, climate change, and environmental degradation necessitate innovative solutions for sustainable agricultural productivity. This chapter highlighted the transformative potential of microbiome engineering, particularly through the targeted manipulation of the rhizosphere microbiome and the application of genetic engineering techniques. By optimizing the microbial communities associated with plant roots, we can significantly enhance crop nutrient acquisition, water use efficiency, disease resistance, and overall resilience to stressors. Microbiome engineering can play a pivotal role in transforming agricultural practices, contributing to sustainable food production, and promoting environmental stewardship. As we embrace these innovative approaches, we can foster a resilient agricultural system capable of meeting the demands of a growing global population while safeguarding the planet's resources.

Several key areas warrant further exploration:

Holistic Understanding of Microbial Ecosystems: Continued research is essential to unravel the complexities of soil and plant-associated microbial communities. This knowledge will guide the effective selection and application of microbial consortia tailored to specific crops and environments.

Integration with Traditional Practices: Developing strategies to seamlessly integrate microbiome engineering with conventional agricultural methods will be crucial for widespread adoption. This includes education and collaboration with farmers to ensure that these innovative techniques complement established practices.

Regulatory Frameworks and Public Acceptance: As microbiome engineering technologies advance, establishing clear regulatory frameworks will be vital. Engaging with the public to address concerns and enhance acceptance of these technologies will facilitate their adoption in mainstream agriculture.

Scalability and Economic Viability: Research should focus on the scalability of microbiome engineering techniques, ensuring they are economically viable for farmers. This includes exploring cost-effective production methods for microbial inoculants and assessing the long-term benefits on crop yields and soil health.

Exploration of Untapped Microbial Diversity: Investigating underutilized microbial species and their potential benefits can lead to novel applications in agriculture. Bioprospecting for beneficial microorganisms in diverse ecosystems may uncover new solutions for enhancing agricultural productivity.

AI DECLARATION

I declare that we didn't use artificial intelligence in our chapter.

BIBLIOGRAPHY

Abd El-Daim IA, Bejai S, and Meijer J. (2014). Improved heat stress tolerance of wheat seedlings by bacterial seed treatment. *Plant Soil* 379, 337–350. https://doi.org/10.1007/s11104-014-2063-3

Acuña JJ, Hu J, Inostroza NG, Valenzuela T, Perez P, Epstein S, Sessitsch A, Zhang Q, and Jorquera MA. (2023). Endophytic bacterial communities in ungerminated and germinated seeds of commercial vegetables. *Sci. Rep.* 13(1), 19829. https://doi.org/10.1038/s41598-023-47099-4

Adesemoye AO, and Egamberdieva D. (2013). Beneficial effects of plant growth-promoting rhizobacteria on improved crop production: Prospects for developing economies. *In Bacteria in Agrobiology: Crop Productivity* (pp. 45–63). Berlin, Heidelberg: Springer Berlin Heidelberg.

Afridi MS, Javed MA, Ali S, De Medeiros FHV, Ali B, Salam A, Sumaira, Marc RA, Alkhalifah DHM, Selim S, and Santoyo G. (2022). New opportunities in plant microbiome engineering for increasing agricultural sustainability under stressful conditions. *Front. Plant Sci.* 13, 899464. https://doi.org/10.3389/fpls.2022.899464

Akinsanya, M. A., Goh, J. K., Lim, S. P., & Ting, A. S. Y. (2015). Metagenomics study of endophytic bacteria in Aloe vera using next-generation technology. *Genom. Data* 6, 159–163.

Ali S, Hameed S, Shahid M, Iqbal M, Lazarovits G, and Imran A. (2020). Functional characterization of potential PGPR exhibiting broad-spectrum antifungal activity. *Microbiol. Res.* 232, 126389. https://doi.org/10.1016/j.micres.2019.126389

Aliche EB, Screpanti C, De Mesmaeker A, Munnik T, and Bouwmeester HJ. (2020). Science and application of strigolactones. *New Phytol.*, 227(4), 1001–1011.

Alqarawi AA, Hashem A, Kumar A, Al-Arjani ABF, Abd Allah EF, Dar BA, et al. (2018). Allelopathic effects of the aqueous extract of Rhazya stricta on growth and metabolism of Salsola villosa. *Plant Biosyst.* 152, 1263–1273. https://doi.org/10.1080/11263504.2018.1439117

Arias-Sánchez, FI, Vessman B, and Mitri S. (2019). Artificially selecting microbial communities: If we can breed dogs, why not microbiomes? *PLoS Biol.* 17(8), 1–8. https://doi.org/10.1371/journal.pbio.3000356

Arif MS, Muhammad R, Shahzad SM, Yasmeen T, Shafaqat A, and Akhtar MJ. (2020). Phosphorus-mobilizing rhizobacterial strain Bacillus cereus GS6 improves symbiotic efficiency of soybean on an Aridisol amended with phosphorus-enriched compost. *Pedosphere* 27, 1049–1061. https://doi.org/10.1016/S1002-0160(17)60366-7

Asari S, Tarkowská D, Rolčík J, Novák O, Velázquez D. (2017). Analysis of plant growth-promoting properties of Bacillus amyloliquefaciens UCMB5113 using Arabidopsis thaliana as host plant. *Planta* 245, 15–30. https://doi.org/10.1007/s00425-016-2580-9

Asghari B, Khademian R, and Sedaghati B. (2020). Plant growth promoting rhizobacteria (PGPR) confer drought resistance and stimulate biosynthesis of secondary metabolites in pennyroyal (Mentha pulegium L.) under water shortage condition. *Sci. Hortic.* 263, 109132. https://doi.org/10.1016/j.scienta.2019.109132

Aw X, Li Z, Wc L, and Zh Y. (2019). The effect of plant growthpromoting rhizobacteria (PGPR) on arsenic accumulation and the growth of rice plants (Oryza sativa L.). *Chemosphere* 242, 125136. https://doi.org/10.1016/j.chemosphere.2019.125136

Bainard LD, Navarro-Borrell A, Hamel C, Braun K, Hanson K, Gan Y. (2017). Increasing the frequency of pulses in crop rotations reduces soil fungal diversity and increases the proportion of fungal pathotrophs in a semiarid agroecosystem. *Agric. Ecosyst. Environ.* 240, 206–214. https://doi.org/10.1016/j.agee.2017.02.020

Bashiardes S, Zilberman-Schapira G, Elinav E. (2016). Use of metatranscriptomics in microbiome research. *Bioinform. Biol. Insights* 10, 4610.

Beckers, B, Op De Beeck M, Weyens N, Van Acker R, Van Montagu M, Boerjan W, and Vangronsveld J. (2016). Lignin engineering in field-grown poplar trees affects the endosphere bacterial microbiome. *Proc. Natl. Acad. Sci.* 113(8), 2312–2317. https://doi.org/10.1073/pnas.1523264113

Bell-Dereske L, Takacs-Vesbach C, and Kivlin SN. (2017). Leaf endophytic fungus interacts with precipitation to alter belowground microbial communities in primary successional dunes. *FEMS Microbiol. Ecol.* 93. https://doi.org/10.1093/femsec/fix036

Benidire L, Lahrouni M, Khalloufi FEL, Göttfert M, and Oufdou K. (2017). Effects of Rhizobium leguminosarum inoculation on growth, nitrogen uptake and mineral assimilation in Vicia faba plants under salinity stress. *J. Agric. Sci. Technol.* 19, 889–901.

Berg G, Grube M, Schloter M, and Smalla K. (2014). Unraveling the plant microbiome: Looking back and future perspectives. *Front. Microbiol.* 5, 148. https://doi.org/10.3389/fmicb.2014.00148

Blair PM, Land ML, Piatek MJ, Jacobson D, Lu T-Y, Doktycz MJ, and Pelletier DA. (2018). Exploration of the biosynthetic potential of the *populus* microbiome. *MSystems* 3(5). https://doi.org/10.1128/msystems.00045-18

Caccamo M. (2023). New precision-breeding law unlocks gene editing in England. *Nat. Biotechnol.*, 41, 1–2.

Cai Y, Chen L, Zhang Y, Yuan S, Su Q, Sun S, Wu C, Yao W, Han T, and Hou W. (2020). Target base editing in soybean using a modified CRISPR/Cas9 system. *Plant Biotechnol. J.* 18(10), 1996–1998. https://doi.org/10.1111/pbi.13386

Cerda A, Rodrigo-Comino J, Gimenez-Morera A, Keesstra SD. (2017). An economic, perception and biophysical approach to the use of oat straw as mulch in Mediterranean rainfed agriculture land, *Ecol. Eng.* 108(2017) 162–171.

Chaithanya BH. (2016). Biocontrol potentiality and plant growth promoting activity of pseudomonas fluorescens—A review. *Int. J. Agric. Sci. Res.* 6, 169–176.

Cheng X, Wang M, Yuan MM, Li J, and Xiong W (2023). Editorial: Rhizosphere microbiome engineering for crop cultivation. *Front. Bioeng. Biotechnol.* 11, 1267442. https://doi.org/10.3389/fbioe.2023.1267442

Chihaoui SA, Trabelsi D, Jdey A, Mhadhbi H, Mhamdi R (2015). Inoculation of *Phaseolus vulgaris* with the nodule-endophyte *Agrobacterium* sp. 10C2 affects the richness and structure of rhizosphere bacterial communities and enhances nodulation and growth. *Arch. Microbiol.* 197, 805–813.

Chowdhury FT, Islam MR, Islam MR, Khan H. (2018). Diversity of plant endophytic volatile organic compound (VOC) and their potential. *Endophytes Secondary Metabol.* 2018, 1–27. https://doi.org/10.1007/978-3-319-76900-4_10-1

Conrath U, Beckers GJ, Langenbach CJ, and Jaskiewicz MR. (2015). Priming for enhanced defense. *Annu. Rev. Phytopathol.* 53, 97–119. https://doi.org/10.1146/annurev-phyto-080614-120132

Cui Y, Jiang N, Xu Z, and Xu Q. (2020). Heterotrimeric G protein are involved in the regulation of multiple agronomic traits and stress tolerance in rice. *BMC Plant Biol.* 20(1), 90. https://doi.org/10.1186/s12870-020-2289-6

Dong N, Yin W, Liu D, Zhang X, Yu Z, Huang W, Liu J, Yang Y, Meng W, Niu M, and Tong H. (2020). Regulation of brassinosteroid signaling and salt resistance by SERK2 and potential utilization for crop improvement in rice. *Front. Plant Sci.* 11, 621859. https://doi.org/10.3389/fpls.2020.621859

Dubey A, Kumar A, Abd Allah EF, Hashem A, Khan ML. (2019). Growing more with less: Breeding and developing drought resilient soybean to improve food security. *Ecol. Indic.* 105, 425–437. https://doi.org/10.1016/j.ecolind.2018.03.003

Egamberdieva D, Wirth S, Behrendt U, Ahmad P, and Berg G. (2017). Antimicrobial activity of medicinal plants correlates with the proportion of antagonistic endophytes. *Front. Microbiol.* https://doi.org/10.3389/fmicb.2017.00199

Ellouze W, Hamel C, Vujanovic V, Gan Y, Bouzid S, and St-Arnaud M. (2013). Chickpea genotypes shape the soil microbiome and affect the establishment of the subsequent durum wheat crop in the semiarid North American Great Plains. *Soil Biol. Biochem.* 63, 129–141. https://doi.org/10.1016/j.soilbio.2013.04.001

Fahimi A, Ashouri A, Ahmadzadeh M, Hoseini Naveh V, Asgharzadeh A, Maleki F, et al. (2014). Effect of PGPR on population growth parameters of cotton aphid. *Arch. Phytopathol. Plant Protect.* 47, 1274–1285. https://doi.org/10.1080/03235408.2013.840099

Fang C, and Chen X. (2018). Potential biocontrol efficacy of Trichoderma atroviride with cellulase expression regulator ace1 gene knock-out. *3 Biotech* 8, 302.

Fang Y, and Tyler BM. (2016). Efficient disruption and replacement of an effector gene in the oomycete Phytophthora sojae using CRISPR/Cas9. *Mol. Plant Pathol.* 17, 127–139. https://doi.org/10.1111/mpp.12318

Fendrihan S, and Pop C-E. (2021). Biotechnological potential of plant associated microorganisms. *Rom. Biotechnol. Lett.* 26, 2700–2706.

Ferrara, M., Haidukowski, M., Logrieco, A. F., Leslie, J. F., & Mulè, G. (2019). A CRISPR-Cas9 system for genome editing of Fusarium proliferatum. *Sci. Rep.*, *9*(1), 19836

Gallego-Giraldo C, Hu J, Urbez C, Gomez MD, Sun TP, and Perez-Amador MA. (2014). Role of the gibberellin receptors GID1 during fruit-set in Arabidopsis. *Plant J.*, 79(6), 1020–1032. https://doi.org/10.1111/tpj.12603

Gan Y, Hamel C, O'Donovan JT, Cutforth H, Zentner RP, Campbell CA, et al. (2015). Diversifying crop rotations with pulses enhances system productivity. *Sci. Rep.* 5. https://doi.org/10.1038/srep14625

García JE, Maroniche G, Creus C, Suárez-Rodríguez R, Ramirez-Trujillo JA, and Groppa MD. (2017). In vitro PGPR properties and osmotic tolerance of different Azospirillum native strains and their effects on growth of maize under drought stress. *Microbiol. Res.* 202, 21–29. https://doi.org/10.1016/j.micres.2017.04.007

Goold HD, Wright P, and Hailstones D. (2018). Emerging opportunities for synthetic biology in agriculture. *Gene* 9, 341.

Gu X, Liu L, and Zhang H. (2021). Transgene-free genome editing in plants. *Front. Genome Edit.* 3, 805317.

Gu Y, Dong K, Geisen S, Yang W, Yan Y, Gu D, Liu N, Borisjuk N, Luo Y, and Friman V-P. (2020). The effect of microbial inoculant origin on the rhizosphere bacterial community composition and plant growth-promotion. *Plant Soil* 452(1), 105–117. https://doi.org/10.1007/s11104-020-04545-w

Gupta A, Sharma A, Pathak R, and Kumar A, Sharma S. (2018). Solid state fermentation of non-edible oil seed cakes for production of proteases and cellulases and degradation of anti-nutritional factors abstract. *J. Food Biotechnol. Res.* 2(1), 3–8.

Gupta A, Singh UB, Sahu PK, Paul S, Kumar A, Malviya D, Singh S, et al. (2022). Linking soil microbial diversity to modern agriculture practices: A review. *Int. J. Environ. Res. Public Health* 19, 3141.

Gupta R, Anand G, Gaur R, and Yadav D. Plant–microbiome interactions for sustainable agriculture: A review. *Physiol. Mol. Biol. Plants* 2021, 27, 165–179.

Gupta S, and Pandey S. (2020). Enhanced salinity tolerance in the common bean (Phaseolus vulgaris) plants using twin ACC deaminase producing rhizobacterial inoculation. *Rhizosphere* 16, 100241. https://doi.org/10.1016/j.rhisph.2020.100241

Habib SH, Kausar H, and Saud HM. (2016). Plant growth-promoting rhizobacteria enhance salinity stress tolerance in okra through ROS-scavenging enzymes. *Biomed. Res. Int.* 2016, 6284547. https://doi.org/10.1155/2016/6284547

Hahm M-S, Son J-S, Hwang Y-J, Kwon D-K, and Ghim S-Y. (2017). Alleviation of salt stress in pepper (Capsicum annum L.) plants by plant growth-promoting rhizobacteria. *J. Microbiol. Biotechnol.* 27, 1790–1797. https://doi.org/10.4014/jmb.1609.09042

Hakim S, Imran A, and Mirza MS. (2020a). Phylogenetic diversity analysis reveals Bradyrhizobium yuanmingense and Ensifer aridi as major symbionts of mung bean (Vigna radiata L.) in Pakistan. *Braz. J. Microbiol.*, 1–14. https://doi.org/10.1007/s42770-020-00397-9

Hakim S, Mirza BS, Imran A, Zaheer A, Yasmin S, Mubeen F, et al. (2020b). Illumina sequencing of 16S rRNA tag shows disparity in rhizobial and non-rhizobial diversity associated with root nodules of mung bean (Vignaradiata L.) growing in different habitats in Pakistan. *Microbiol. Res.* 231, 126356. https://doi.org/10.1016/j.micres.2019.126356

Hakim S, Mirza BS, Zaheer A, Mclean JE, Imran A, Yasmin S, et al. (2018). Retrieved 16S rRNA and nifH sequences reveal co-dominance of Bradyrhizobium and Ensifer (Sinorhizobium) strains in field-collected root nodules of the promiscuous host Vigna radiata (L.) R. *Wilczek. Appl. Microbiol. Biotechnol.* 102, 485–497. https://doi.org/10.1007/s00253-017-8609-6

Hakim S, Naqqash T, Nawaz MS, Laraib I, Siddique MJ, Zia R, Mirza MS, and Imran A (2021). Rhizosphere engineering with plant growth-promoting microorganisms for agriculture and ecological sustainability. *Front. Sustain. Food Syst.* 5, 617157. https://doi.org/10.3389/fsufs.2021.617157

Hamel C, Gan Y, Sokolski S, Bainard LD. (2018). High frequency cropping of pulses modifies soil nitrogen level and the rhizosphere bacterial microbiome in 4-year rotation systems of the semiarid prairie. *Appl. Soil Ecol.* 2018. https://doi.org/10.1016/j.apsoil.2018.01.003

Hartmann M, Frey B, Mayer J, Mäder P, and Widmer F. (2014). Distinct soil microbial diversity under long-term organic and conventional farming. *ISME J.* 9(5), 1177–1194. https://doi.org/10.1038/ismej.2014.210

Hashem A, Abd Allah EF, Alqarawi AA, Radhakrishnan R, and Kumar A. (2017). Plant defense approach of Bacillus subtilis (BERA 71) against Macrophomina phaseolina (Tassi) Goid in mung bean. *J. Plant Interact.* 12, 390–401. https://doi.org/10.1080/17429145.2017.1373871

Hashem A, Kumar A, Al-Dbass AM, Alqarawi AA, Al-Arjani AF, Singh G, Farooq M, and Abd Allah EF. (2019). Arbuscular mycorrhizal fungi and biochar improves drought tolerance in chickpea. *Saudi J. Biol. Sci.* 26(3), 614–624. https://doi.org/10.1016/j.sjbs.2018.11.005

Hassani MA, Durán P, and Hacquard S. (2018). Microbial interactions within the plant holobiont. *Microbiome* 6(1), 1–17.

He F, Yang Y, Yang G, and Yu L. (2010). Studies on antibacterial activity and antibacterial mechanism of a novel polysaccharide from streptomyces virginia H03. *Food Control* 21(September), 1257–1262. https://doi.org/10.1016/j.foodcont.2010.02.013

Heydarian Z, Gruber M, Glick BR, and Hegedus DD. (2018). Gene expression patterns in roots of Camelina sativa with enhanced salinity tolerance arising from inoculation of soil with plant growth promoting bacteria producing 1-aminocyclopropane-1-carboxylate deaminase or expression the corresponding acds gene. *Front. Microbiol.* 9, 1297. https://doi.org/10.3389/fmicb.2018.01297

Htwe AZ, Moh SM, Moe K, and Yamakawa T. (2018). Effects of co-inoculation of Bradyrhizobium japonicum SAY3-7 and Streptomyces griseoflavus P4 on plant growth, nodulation, nitrogen fixation, nutrient uptake, and yield of soybean in a field condition. *Soil Sci. Plant Nutr.* 64, 222–229. https://doi.org/10.1080/00380768.2017.1421436

Imam J, Singh PK, and Shukla P. (2016). Plant microbe interactions in post genomic era: Perspectives and applications. *Front. Microbiol.* 7. https://doi.org/10.3389/fmicb.2016.01488

Imran A, Mirza MS, Shah TM, Malik KA, and Hafeez FY. (2015). Differential response of kabuli and desi chickpea genotypes toward inoculation with PGPR in different soils. *Front. Microbiol.* 6, 859. https://doi.org/10.3389/fmicb.2015.00859

Ji J, Yuan D, Jin C, Wang G, Li X, and Guan C. (2020). Enhancement of growth and salt tolerance of rice seedlings (Oryza sativa L.) by regulating ethylene production with a novel halotolerant PGPR strain Glutamicibacter sp. YD01 containing ACC deaminase activity. *Acta Physiol. Plant.* 42, 1–17. https://doi.org/10.1007/s11738-020-3034-3

Jiménez-Vicente E, Hernandez JA, Echavarri-Erasun C, and Rubio LM. (2015). Biosynthesis of the iron-molybdenum cofactor of nitrogenase. *Biological Nitrogen Fixation* 1–2, 75–86. https://doi.org/10.1002/9781119053095.ch7

Jochum, MD, McWilliams KL, Pierson EA, and Jo Y-K. (2019). Host-mediated microbiome engineering (HMME) of drought tolerance in the wheat rhizosphere. *PLoS One* 14(12), 1–15. https://doi.org/10.1371/journal.pone.0225933

Kang S-M, Hamayun M, Khan MA, Iqbal A, and Lee I-J (2019). Bacillus subtilis JW1 enhances plant growth and nutrient uptake of Chinese cabbage through gibberellins secretion. *J. Appl. Bot. Food Qual.* 92, 172–178. https://doi.org/10.5073/JABFQ.2019.092.023

Kazan K, and Lyons R. (2015). The link between flowering time and stress tolerance. *J. Exp. Bot.* 67(1), 47–60. https://doi.org/10.1093/jxb/erv441

Kemal, R. A., Islamiah, P. W., Sa'adah, M., & Lusiany, T. (2016, August). Synthetic biology for bio-control: a mini-review. In *Proceeding International Conference on Global Resource Conservation* (Vol. 6, No. 1).

Khan MA, Ullah I, Waqas M, Hamayun M, Khan AL, Asaf S, et al. (2019a). Halo-tolerant rhizospheric Arthrobacter woluwensis AK1 mitigates salt stress and induces physio-hormonal changes and expression of GmST1 and GmLAX3 in soybean. *Symbiosis* 77, 9–21. https://doi.org/10.1007/s13199-018-0562-3

Khan N, Bano A, Rahman MA, Guo J, Kang Z, and Babar MA. Comparative physiological and metabolic analysis reveals a complex mechanism involved in drought tolerance in chickpea (Cicer arietinum L.) induced by PGPR and PGRs. *Sci. Rep.* 2019b, 9. https://doi.org/10.1038/s41598-019-38702-8

Kishan M, Missaoui A, Lee K, Ghimire B, Presley HW, and Makaju S. (2021). Rhizosphere microbiome manipulation for sustainable crop production. *Curr. Plant Biol.* 27(2021), 100210. https://doi.org/10.1016/j.cpb.2021.100210

Koyama H, Kawamura A, Kihara T, Hara T, Takita E, and Shibata D. (2000). Overexpression of mitochondrial citrate synthase in Arabidopsis thaliana improved growth on a phosphorus-limited soil. *Plant Cell Physiol.* 41, 1030–1037. https://doi.org/10.1093/pcp/pcd029

Kumar A, and Dubey A. (2020). Rhizosphere microbiome: Engineering bacterial competitiveness for enhancing crop production. *J. Adv. Res.* 24, 337–352. https://doi.org/10.1016/j.jare.2020.04.014

Kumar A, Maurya BR, and Raghuwanshi R. (2015). Characterization of bacterial strains and their impact on plant growth promotion and yield of wheat and microbial populations of soil. *Afr. J. Agric. Res.* 10, 1367–1375. https://doi.org/10.5897/AJAR2014.8894

Lau ET, Tani A, Khew CY, Chua YQ, and San Hwang S. (2020). Plant growth-promoting bacteria as potential bio-inoculants and biocontrol agents to promote black pepper plant cultivation. *Microbiol. Res.* 240, 126549. https://doi.org/10.1016/j.micres.2020.126549

Lay C-Y, Bell TH, Chantal Hamel K, Harker N, Mohr R, Greer CW, Yergeau É, and St-Arnaud M. (2018). Canola root–associated microbiomes in the Canadian prairies. *Front. Microbiol.* 9. https://doi.org/10.3389/fmicb.2018.01188

Leite J, Fischer D, Rouws LF, Fernandes-Júnior PI, Hofmann A, Kublik S, et al. (2017). Cowpea nodules harbor non-rhizobial bacterial communities that are shaped by soil type rather than plant genotype. *Front. Plant Sci.* 7, 2064. https://doi.org/10.3389/fpls.2016.02064

Leonard SP, Powell JE, Perutka J, Geng P, Heckmann LC, Horak RD, Davies BW, Ellington AD, Barrick JE, Moran NA (2020). Engineered symbionts activate honey bee immunity and limit pathogens. *Science* 367, 573–576.

Levy A, Conway JM, Dangl JL, and Woyke T. (2018). Review elucidating bacterial gene functions in the plant microbiome. *Cell Host Microbe* 24(4), 475–485. https://doi.org/10.1016/j.chom.2018.09.005

Liang Z, Zhang K, Chen K, and Gao C. (2014). Targeted mutagenesis in Zea mays using TALENs and the CRISPR/Cas system. *J. Genet. Genom.* 41, 63–68. https://doi.org/10.1016/j.jgg.2013.12.001

Liu S, and Weiner HL. (2016). Control of the gut microbiome by fecal microRNA. *Microbial Cell* 3(4), 176.

Liu XH, Lyu YS, Yang W, Yang ZT, Lu SJ, and Liu JX. (2020). A membrane-associated NAC transcription factor OsNTL3 is involved in thermotolerance in rice. *Plant Biotechnol. J.* 18(5), 1317–1329. https://doi.org/10.1111/pbi.13297

Ma Y, Rajkumar M, Oliveira RS, Zhang C, and Freitas H. (2019). Potential of plant beneficial bacteria and arbuscular mycorrhizal fungi in phytoremediation of metal-contaminated saline soils. *J. Hazard. Mater.* 379, 120813.

Maher MF, Nasti RA, Vollbrecht M, Starker CG, Clark MD, and Voytas DF. (2020). Plant gene editing through de novo induction of meristems. *Nat. Biotechnol.* 38(1), 84–89. https://doi.org/10.1038/s41587-019-0337-2

Marihal A, and Jagadeesh K. (2013). Plant-microbe interaction: A potential tool for enhanced bioremediation. *Plant Microbe Symb. Fund. Adv.* 2013, 396–410. https://doi.org/10.1007/978-81-322-1287-4_15

Mazzola M. (2002). Mechanisms of natural soil suppressiveness to soilborne diseases. Antonie van Leeuwenhoek. *Int. J. General Mol. Microbiol.* 81, 557–564. https://doi.org/10.1023/A:1020557523557

Mendes LW, Gonçalves de Chaves M, de Cassia Fonseca M, Mendes R, Raaijmakers JM, and Tsai SM. (2019). Resistance breeding of common bean shapes the physiology of the rhizosphere microbiome. *Front. Microbiol.* 10. https://doi.org/10.3389/fmicb.2019.02252

Mendes LW, Raaijmakers JM, De Hollander M, Mendes R, and Tsai SM. Influence of resistance breeding in common bean on rhizosphere microbiome composition and function. *ISME J.* 2018. https://doi.org/10.1038/ismej.2017.158

Mendes R, Garbeva P, and Raaijmakers JM. (2013). The rhizosphere microbiome: Significance of plant beneficial, plant pathogenic, and human pathogenic microorganisms. *FEMS Microbiol. Rev.* 37, 634–663. https://doi.org/10.1111/1574-6976.12028

Mendes R, Kruijt M, de Bruijn I, Dekkers E, van der Voort M, Schneider JHM, et al. (2011). Deciphering the rhizosphere microbiome for disease-suppressive bacteria. *Science* 332(6033), 1097–1100. https://doi.org/10.1126/science.1203980

Mhlongo MI, Piater LA, Madala NE, Labuschagne N, and Dubery IA. (2018). The chemistry of plant–microbe interactions in the rhizosphere and the potential for metabolomics to reveal signaling related to defense priming and induced systemic resistance. *Front. Plant Sci.* 9, 112.

Millet YA, Danna CH, Clay NK, Songnuan W, Simon MD, Werck-Reichhart D, et al. (2010). Innate immune responses activated in Arabidopsis roots by microbe-associated molecular patterns. *Plant Cell* 22(3), 973–990.

Mishra J, and Arora NK (2016). Bioformulations for plant growth promotion and combating phytopathogens: A sustainable approach. In Arora N, Mehnaz S and Balestrini R (eds) *Bioformulations: For Sustainable Agriculture* (pp. 3–33). New Delhi, India: Springer.

Mohapatra S, Mishra SS, Bhalla P, and Thatoi H. Engineering grass biomass for sustainable and enhanced bioethanol production. *Planta* 2019, 250, 395–412.

Monnet V, Juillard V, and Gardan R. (2016). Peptide conversations in Gram-positive bacteria. *Crit. Rev. Microbiol.* 42(3), 339–351.

Moreira ZPM, Chen MY, Ortuno DLY, and Haney CH. (2023). Engineering plant microbiomes by integrating eco-evolutionary principles into current strategies. *Curr. Opin. Plant Biol.* 71, 102316.

Mukherjee A, Singh S, Gaurav AK, Chouhan GK, Jaiswal DK, de Araujo Pereira AP, et al. (2022). Harnessing of phytomicrobiome for developing potential biostimulant consortium for enhancing the productivity of chickpea and soil health under sustainable agriculture. *Sci. Total Environ.* 836, 155550. https://doi.org/10.1016/j.scitotenv.2022.155550

Mukherjee P, Mitra A, and Roy M. (2019). Halomonas rhizobacteria of Avicennia marina of Indian Sundarbans promote rice growth under saline and heavy metal stresses through exopolysaccharide production. *Front. Microbiol.* 10, 1207. https://doi.org/10.3389/fmicb.2019.01207

Mukhtar T, Smith D, Sultan T, Seleiman MF, Alsadon AA, Ali S, et al. (2020). Mitigation of heat stress in Solanum lycopersicum l. by acc-deaminase and exopolysaccharide producing Bacillus cereus: Effects on biochemical profiling. *Sustainability* 12, 2159. https://doi.org/10.3390/su12062159

Muleta A, and Assefa F. (2018). Isolation and screening of antibiotic producing actinomycetes from rhizosphere and agricultural soils. *Afr. J. Biotechnol.* 17, 700–715. https://doi.org/10.5897/AJB2017.16080

Nadarajah K, and Abdul Rahman NSN. (2023). The microbial connection to sustainable agriculture. *Plants* 2023(12), 2307. https://doi.org/10.3390/plants12122307

Naqqash T, Hameed S, Imran A, Hanif MK, Majeed A, and Van Elsas JD. (2016). Differential response of potato toward inoculation with taxonomically diverse plant growth promoting rhizobacteria. *Front. Plant Sci.* 7, 144. https://doi.org/10.3389/fpls.2016.00144

Naqqash T, Imran A, Hameed S, Shahid M, Majeed A, Iqbal J, et al. (2020). First report of diazotrophic Brevundimonas spp. as growth enhancer and root colonizer of potato. *Sci. Rep.* 10, 1–14. https://doi.org/10.1038/s41598-020-69782-6

National Academies of Sciences, Medicine, Division of Behavioral, Social Sciences, Board on Environmental Change, Medicine Division, … & Committee on Science Breakthroughs 2030: A Strategy for Food and Agricultural Research. (2019). *Science Breakthroughs to Advance Food and Agricultural Research by 2030.* National Academies Press.

Neal JL, Larson RI, and Atkinson TG. (1973). Changes in rhizosphere populations of selected physiological groups of bacteria related to substitution of specific pairs of chromosomes in spring wheat. *Plant Soil* 39, 209–212. https://doi.org/10.1007/BF00018061

Nitika T, Nigam M, Ann Mann N, Gupta Chaudhery S, Hussain M, Shukla SK, Ali Shah A, Casini R, Elansary HO, and Aslam Khan S. (2023). Host-mediated gene engineering and microbiome-based technology optimization for sustainable agriculture and environment. *Funct. Integr. Genomics* 23, 57. https://doi.org/10.1007/s10142-023-00982-9

Oliva R, Ji C, Atienza-Grande G, Huguet-Tapia JC, Perez-Quintero A, Li T, et al. (2019). Broad-spectrum resistance to bacterial blight in rice using genome editing. *Nat. Biotechnol.* 37(11), 1344–1350. https://doi.org/10.1038/s41587-019-0267-z

Orgiazzi A, Bardgett RD, and Barrios E. (2016). *Global Soil Biodiversity Atlas.* Brussels, Belgium: European Commission.

Orozco-Mosqueda MDC, Rocha-Granados MDC, Glick BR, and Santoyo G. (2018). Microbiome engineering to improve biocontrol and plant growth-promoting mechanisms. *Microbiol. Res.* 208, 25–31. https://doi.org/10.1016/j.micres.2018.01.005

Panke-Buisse K, Lee S, and Kao-Kniffin J. (2017). Cultivated sub-populations of soil microbiomes retain early flowering plant trait. *Microb. Ecol.* 73(2), 394–403. https://doi.org/10.1007/s00248-016-0846-1

Pantigoso HA, Newberger D, and Vivanco JM. (2022). The rhizosphere microbiome: Plant-microbial interactions for resource acquisition. *J. Appl. Microbiol.* 133(5), 2864–2876. https://doi.org/10.1111/jam.15686. PMID: 36648151; PMCID: PMC9796772.

Park YG, Mun BG, Kang SM, Hussain A, Shahzad R, Seo CW, et al. (2017). Bacillus aryabhattai SRB02 tolerates oxidative and nitrosative stress and promotes the growth of soybean by modulating the production of phytohormones. *PLoS One* 12, e0173203. https://doi.org/10.1371/journal.pone.0173203

Pascale A, Proietti S, Pantelides IS, and Stringlis IA. (2019). Modulation of the root microbiome by plant molecules: The basis for targeted disease suppression and plant growth promotion. *Front. Plant Sci.* 10, 1741. https://doi.org/10.3389/fpls.2019.01741

Pineda A, Kaplan I, Hannula SE, Ghanem W, and Bezemer TM (2020). Conditioning the soil microbiome through plant–soil feedbacks suppress an aboveground insect pest. *New Phytol.* 226, 595–608.

Qin J, Jun Li M, Wang P, Zhang MQ, Wang J. (2011). ChIP-seq/chip and microarray gene expression data to discover direct/indirect targets of a transcription factor. *Nucleic Acids Res.* 39, 430–436. https://doi.org/10.1093/nar/gkr332

Qin L, Li J, Wang Q, Zhongping X, Sun L, Alariqi M, Manghwar H, et al. (2020). High-efficient and precise base editing of C•G to T•A in the allotetraploid cotton (Gossypium hirsutum) genome using a modified CRISPR/Cas9 system. *Plant Biotechnol. J.* 18(1), 45–56. https://doi.org/10.1111/pbi.13168

Raaijmakers JM, and Mazzola M. (2016). Soil immune responses. *Science* 352(6292), 1392–1393. https://doi.org/10.1126/science.aaf3252

Rajput L, Imran A, Mubeen F, and Hafeez FY. (2018). Wheat (Triticum aestivum L.) growth promotion by halo-tolerant PGPR-consortium. *Soil Environ.* 37, 178–189.

Rasul M, Yasmin S, Suleman M, Zaheer A, Reitz T, Tarkka MT, et al. (2019). Glucose dehydrogenase gene containing phosphobacteria for biofortification of phosphorus with growth promotion of rice. *Microbiol. Res.* 223, 1–12. https://doi.org/10.1016/j.micres.2019.03.004

Ricroch A, Clairand P, and Harwood W. (2017). Use of CRISPR systems in plant genome editing: Toward new opportunities in agriculture. *Emerg. Top. Life Sci* 2017. https://doi.org/10.1042/etls20170085

Rodríguez M, Torres M, Blanco L, Béjar V, Sampedro I, and Llamas I. (2020). Plant growth-promoting activity and quorum quenching-mediated biocontrol of bacterial phytopathogens by Pseudomonas segetis strain P6. *Sci. Rep.* 10, 4121. https://doi.org/10.1038/s41598-020-61084-1

Ryan PR, Dessaux Y, Thomashow LS, Weller DM. (2009). Rhizosphere engineering and management for sustainable agriculture. *Plant Soil Biol.* 321, 363–383. https://doi.org/10.1007/s11104-009-0001-6

Ryu MH, Zhang J, Toth T, Khokhani D, Geddes BA, Mus F, et al. (2020). Control of nitrogen fixation in bacteria that associate with cereals. *Nat. Microbiol.* 5, 314–330.

Sarkar J, Chakraborty B, and Chakraborty U. (2018). Plant growth promoting rhizobacteria protect wheat plants against temperature stress through antioxidant signalling and reducing chloroplast and membrane injury. *J. Plant Growth Regul.* 37, 1396–1412. https://doi.org/10.1007/s00344-018-9789-8

Savary S, Willocquet L, Pethybridge SJ, Esker P, McRoberts N, and Nelson A. (2019). The global burden of pathogens and pests on major food crops. *Nat. Ecol. Evol.* 3, 430–439.

Savka MA, Dessaux Y, Oger P, and Rossbach S. (2002). Engineering bacterial competitiveness and persistence in the phytosphere. *Mol. Plant Microbe Interact.* 15, 866–874. https://doi.org/10.1094/MPMI.2002.15.9.866

Schaeffer SM, and Nakata PA. (2015). CRISPR/Cas9-mediated genome editing and gene replacement in plants: Transitioning from lab to field, *Plant Sci.* 240(2015), 130–142.

Scherlach K, and Hertweck C. (2018). Mediators of mutualistic microbe–microbe interactions. *Nat. Prod. Rep.* 35(4), 303–308. https://doi.org/10.1039/c7np00035a

Schmidt SM, Belisle M, and Frommer WB. (2020). The evolving landscape around genome editing in agriculture: Many countries have exempted or move to exempt forms of genome editing from gmo regulation of crop plants. *EMBO Rep.* 21(6), e50680.

Schulz-Bohm K, Gerards S, Hundscheid M, Melenhorst J, de Boer W, and Garbeva P. (2018). Calling from distance: Attraction of soil bacteria by plant root volatiles. *ISME J.* 12(2018), 1252–1262.

Sexton WK, Fidero M, Spain JC, Jiang L, Bucalo K, Cruse-Sanders JM, et al. (2019). Characterization of endophytic bacterial communitie within greenhouse and field-grown rhizomes of three rare pitcher plant species (Sarracenia oreophila S. leucophylla, and S. purpurea spp. venosa) with an emphasis on nitrogen-fixing bacteria. *Plant Soil* 447, 257–279. https://doi.org/10.1007/s11104-019 04372-8

Sharma R, Gal L, Garmyn D, Bisaria VS, Sharma S, and Piveteau P. (2020). Evidence of biocontrol activity of bioinoculants against a human pathogen, Listeria monocytogenes. *Front. Microbiol.* 11, 350. https://doi.org/10.3389/fmicb.2020.00350

Sheibani-Tezerji R, Rattei T, Sessitsch A, Trognitz F, and Mitter B. (2015). Transcriptome profiling of the endophyte burkholderia phytofirmans psjn indicates sensing of the plant environment and drought stress. *MBio* 6, 1–11. https://doi.org/10.1128/mBio.00621-15

Sheth RU, Cabral V, Chen SP, and Wang HH (2016). Manipulating bacterial communities by in situ microbiome engineering. *Trends Genet.* 32, 189–200.

Singh AV, Prasad B, and Goel R. (2018). Plant growth promoting efficiency of phosphate solubilizing Chryseobacterium sp. PSR 10 with different doses of N and P fertilizers on lentil (Lens culinaris var. PL-5) growth and yield. *Int. J. Curr. Microbiol. App. Sci.* 7, 2280–2289. https://doi.org/10.20546/ijcmas.2018.705.265

Staudinger C, Mehmeti-Tershani V, Gil-Quintana E, Gonzalez EM, Hofhansl F, Bachmann G, et al. (2016). Evidence for a rhizobia-induced drought stress response strategy in Medicago truncatula. *J. Proteome* 136, 202–213. https://doi.org/10.1016/j.jprot.2016.01.006

Straub T, Zabel A, Gilfillan GD, Feller C, and Becker PB. (2013). Different chromatin interfaces of the Drosophila dosage compensation complex revealed by high shear ChIP-seq. *Genome Res.* 23, 473–485. https://doi.org/10.1101/gr.146407.112

Syed S, Tollamadugu NP, and Lian B. (2020). Aspergillus and Fusarium control in the early stages of Arachis hypogaea (groundnut crop) by plant growthpromoting rhizobacteria (PGPR) consortium. *Microbiol. Res.* 240, 126562. https://doi.org/10.1016/j.micres.2020.126562

Syed-Ab-Rahman SF, Xiao Y, Carvalhais LC, Ferguson BJ, and Schenk PM. (2019). Suppression of Phytophthora capsici infection and promotion of tomato growth by soil bacteria. *Rhizosphere* 9, 72–75.

Thakur N, Nigam M, Mann NA, Gupta S, Hussain CM, Shukla SK, Shah AA, Casini R, Elansary HO, and Khan SA. (2023). Host-mediated gene engineering and microbiome-based technology optimization for sustainable agriculture and environment. *Funct. Integr. Genomics* 23(1), 57. https://doi.org/10.1007/s10142-023-00982-9

Trivedi P, Leach JE, Tringe SG, Sa T, and Singh BK. (2020). Plant–microbiome interactions: From community assembly to plant health. *Nat. Rev. Microbiol.* 18(11), 607–621. https://doi.org/10.1038/s41579-020-0412-1

Uniyal AP, Mansotra K, Yadav SK, and Kumar V. (2019). An overview of designing and selection of SgRNAs for precise genome editing by the CRISPR-Cas9 system in plants. *3 Biotech* 9(6), 223. https://doi.org/10.1007/s13205-019-1760-2

van Elsas JD, Chiurazzi M, Mallon CA, Elhottova D, Kristufek V, and Salles JF. (2012). Microbial diversity determines the invasion of soil by a bacterial pathogen. *Proc. Natl. Acad. Sci.* 109, 1159–1164. https://doi.org/10.1073/pnas.1109326109

Vandenkoornhuyse P, Quaiser A, Duhamel M, Le Van A, and Dufresne A. (2015). The importance of the microbiome of the plant holobiont. *New Phytol.* 206, 1196–1206. https://doi.org/10.1111/nph.13312

Veillet F, Perrot L, Chauvin L, Kermarrec MP, Guyon-Debast A, Chauvin J-E, Nogué F, and Mazier M. (2019). Transgene-free genome editing in tomato and potato plants using agrobacterium-mediated delivery of a CRISPR/Cas9 cytidine base editor. *Int. J. Mol. Sci.* 20(2). https://doi.org/10.3390/ijms20020402

Venturi V, and Keel C. (2016). Signaling in the rhizosphere. *Trends Plant Sci.* 21(3), 187–198.

Vyas P, Kumar D, Dubey A, and Kumar A. (2018). Screening and characterization of Achromobacter Xylosoxidans isolated from rhizosphere of Jatropha Curcas L. (energy crop) for Plant-growth-promoting traits. *J. Adv. Res. Biotechnol.* 3. https://doi.org/10.15226/2475-4714/3/1/00134

Waltz E. (2018). With a free pass, CRISPR-edited plants reach market in record time. *Nat. Biotechnol.* 36(1), 6–7. https://doi.org/10.1038/nbt0118-6b

Wang Q, and Coleman JJ. (2019). CRISPR/Cas9-mediated endogenous gene tagging in Fusarium oxysporum. *Fungal Genet. Biol.* 126, 17–24.

Wille, L., Messmer, M. M., Studer, B., & Hohmann, P. (2019). Insights to plant–microbe interactions provide opportunities to improve resistance breeding against root diseases in grain legumes. *Plant Cell Environ* 42(1), 20–40.

Williamson KE, Fuhrmann JJ, Eric Wommack K, and Radosevich M. (2017). Viruses in soil ecosystems: An unknown quantity within an unexplored territory. *Ann. Rev. Virol.* 4(1), 201–219. https://doi.org/10.1146/annurev-virology-101416-041639

Xin W, Wang J, Li J, Zhao H, Liu H, Zheng H, et al. (2021). Candidate gene analysis for nitrogen absorption and utilization in Japonica rice at the seedling stage based on a genome-wide association study. *Front. Plant Sci.* 12, 670861.

Xing Z, Li L, and Wang W. (2021). Evaluation for education of the sustainable development of engineering college students base on NLP method. In *2021 2nd International Conference on Information Science and Education (ICISE-IE)*, 705–709. https://doi.org/10.1109/ICISE-IE53922.2021.00165

Yu K, Liu Z, Gui H, Geng L, Wei J, Liang D, Lv J, Xu J, and Chen X. (2021). Highly efficient generation of bacterial leaf blight-resistant and transgene-free rice using a genome editing and multiplexed selection system. *BMC Plant Biol.* 21(1), 197. https://doi.org/10.1186/s12870-021-02979-7

Yu Q-h, Wang B, Li N, Tang Y, Yang S, Yang T, et al. (2017). CRISPR/Cas9-induced targeted mutagenesis and gene replacement to generate long-shelf life tomato lines. *Sci. Rep.* 7(1), 11874. https://doi.org/10.1038/s41598-017-12262-1

Yue H, Yue W, Jiao S, Kim H, Lee Y, Wei G, Song W, and Shu D. (2023). Plant domestication shapes rhizosphere microbiome assembly and metabolic functions. *Microbiome* 11, 70. https://doi.org/10.1186/s40168-023-01513-1

Zaheer A, Mirza BS, Mclean JE, Yasmin S, Shah TM, Malik KA, et al. (2016). Association of plant growth-promoting Serratia spp. with the root nodules of chickpea. *Res. Microbiol.* 167, 510–520. https://doi.org/10.1016/j.resmic.2016.04.001

Zhang D, Zhang Z, Unver T, and Zhang B. (2021). CRISPR/Cas: A powerful tool for gene function study and crop improvement. *J. Adv. Res.* 29(March), 207–221. https://doi.org/10.1016/j.jare.2020.10.003

Zhang J, Liu Y-X, Zhang N, Hu B, Jin T, Xu H, et al. (2019). NRT1.1B is associated with root microbiota composition and nitrogen use in field-grown rice. *Nat. Biotechnol.* 37, 676–684. https://doi.org/10.1038/s41587-019-0104-4

Zhang Y, Yeo WS, Ng KH, and Yap HK. (2015). CRISPR/Cas9 for genome engineering: The next genomic revolution. *J. Biochem. Mol. Biol. Res.* 1, 112–117.

4 Advancing Agricultural Productivity

Integrating Multi-Omics Approaches

Hanaa Metwally Barakat

4.1 INTRODUCTION

The importance of genomics can be fully appreciated only when one considers its far-reaching implications and potential benefits. Genomics aids in the creation of a global database of information that is easily accessible, providing ample opportunity for data mining (Alonge et al., 2020). In an agricultural sense, genomes provide a detailed blueprint for constructing species traits and, given modern biofortification techniques, their nutritional value (Medina-Lozano & Díaz, 2022). This blueprint illustrates all possible targets for crop improvement. For instance, genetic maps can pinpoint the specific genes responsible for a specific disease resistance (Huang et al., 2015).

The past few decades showed rapid advances in the development of high-throughput technologies that resulted in the generation of a group of biological data (Schuster, 2008). The availability of a full sequence of data, in combination with currently developed high-throughput technologies for scientific analysis, has made it possible for the first time to attempt. This embraces the comprehensive analysis of the entire set of genes and/or the gene expression for a species (Bayer et al., 2020). With genomics, integrating information from other Omics fields, as in Figure 4.1, is essential for deepening our understanding of plant biology.

4.1.1 Overview of Omics-Based Approaches

Omics techniques are a viable tool for agricultural gene modification and can provide a comprehensive understanding of an organism at the molecular level. Genetically modified (GM) crops can be thoroughly analyzed using the Omics (Chen et al., 2024a). The process can be observed in the production of phytase transgenic maize. To identify any unintended effects in the GM line, proteomic analyses were employed to compare the transgenic line to an unmodified one (Tan et al., 2017). Thus, the safety assessments of transgenic crops can be much improved by Omics technology's capacity to identify unintended effects (Ladics et al., 2015).

 DOI: 10.1201/9781003714484-4

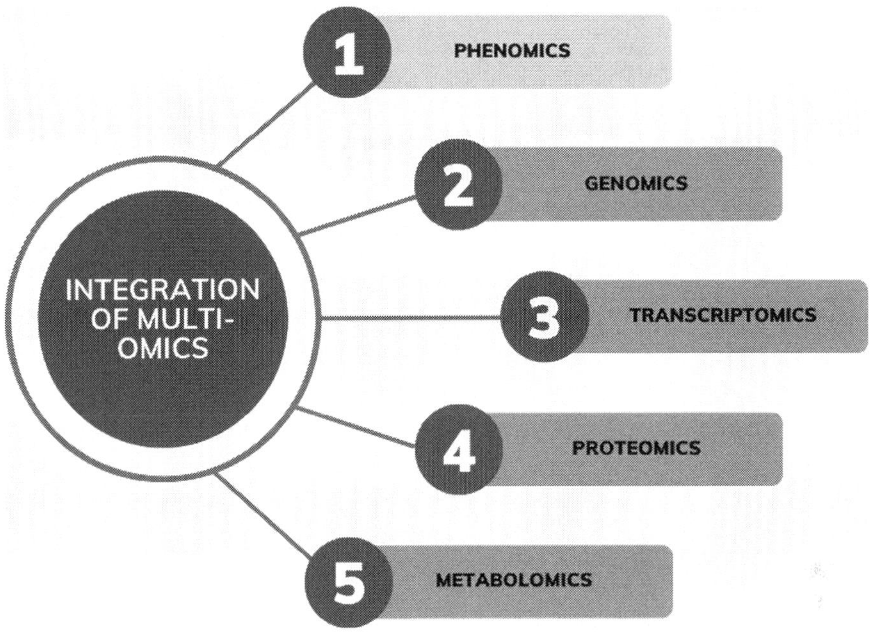

FIGURE 4.1 Integrating various types of Omics technology.

Due to the massive data generated in Omics studies, the development of tools like bioinformatics is essential. Bioinformatics can manage and analyze biological data through computer science, resulting in the development of databases, data storage and retrieval systems, and data analysis algorithms (O'Connor et al., 2023). This made bioinformatics a vital biological instrument in the era of Omics (Shi et al., 2021). Recent studies have examined transcriptome data from many platforms, including GEO, using related analysis methods. Such R package and other tools have displayed potential data, like visualizing the gene networks. Furthermore, combining various datasets notably improves the annotation of unidentified genes (Zainal-Abidin et al., 2022). For instance, it has been confirmed that the genes that exhibit interactions often have similar functions (Ovens et al., 2021). In agricultural applications such as metabolic engineering and transcriptomics, the use of gene network tools like PlaNet and RECoN (Krishnan et al., 2017; Proost & Mutwil, 2017) facilitates species comparison and helps uncover gene functions. Metabolic engineering is a technology that is used to modify the metabolic process of an organism. Omics approach enables definitive identification of genes, transcripts, proteins, and metabolites, thereby accelerating the targeting of specific metabolic pathways (Jansson & Northen, 2010). By targeting specific pathways within the cell, we can enhance compound production (Mipeshwaree Devi et al., 2023). Omics also shed light on the intricacies of gene function and regulation. This can be investigated by studying the whole gene expression using microarrays. Studies have demonstrated that an unknown gene function can be identified by examining the patterns of gene expression in transcriptomic data (De Evsikova et al., 2019).

Omics technology provides a thorough understanding of several aspects of an organism by combining transcriptomics, proteomics, metabolomics, and genomics (Micheel et al., 2012). These methods are becoming increasingly important in both biology and agriculture (Yang et al., 2021). This field involves studying the molecules of an organism through cutting-edge techniques like DNA sequencing (Dai & Shen, 2022). Since the advent of Omics, biology has shifted from focusing on individual molecules to a more holistic one (Kuleshov et al., 2020).

4.1.2 Importance of OMICS in Agriculture

The future of biological experiments will be improved by the use of Omics in agriculture (Li & Yan, 2020). Crop plant improvement is the central goal of agricultural research, and results from applied genetic experiments eventually go into field production. Crop science has two high-level objectives. One is to generate an increased understanding of crop biology in such a way as to significantly increase the rate of future crop improvement (Ray et al., 2013). The second is to efficiently solve industry problems by applying knowledge from basic biology to the development of new crop products (Tester & Langridge, 2010). By streamlining the experimental process, Omics significantly impact both objectives (Arai-Kichise et al., 2011). Omics studies can first generate testable hypotheses to be validated in crop plants (Ma et al., 2017). As theories are validated and new crop phenotypes are developed, new questions will continue to arise. This will lead to a cycle of more basic research employing Omics to generate a further understanding of crop biology and more application of this knowledge toward future crop products (Ma et al., 2017). The result will be a self-generating system of research hypothesis and testing that is directly targeted at understanding and improving crop biology (Wang et al., 2003). Self-perpetuating research will have a crucial impact on the first objective of crop science (Ray et al., 2013). The ability to generate a widespread understanding of crop biology will have a major effect on the rate of crop improvement (Tester & Langridge, 2010). Currently, much crop experimentation is performed through relatively simple genetic manipulation (Ahmad & Mukhtar, 2017). While this is effective for changing crop phenotypes, it often does so in a way that fails to understand the underlying biology (Della Coletta et al., 2021). High-level crop improvement can best be achieved by applying knowledge of underlying biology to develop crops with very specific phenotypes (Ronald, 2014). The second objective of applying knowledge from basic biology to solve industry problems will be directly impacted by the nature of the testing cycle (Valizadeh & Bijani, 2020). As new crop products are developed, they can be quickly tested for efficacy with Omics studies. Ultimately, all agricultural research is done with the goal of improving crop products (Boden & Østergaard, 2019).

4.2 GENOMICS IN AGRICULTURE

The role of genomics in agriculture has been established in recent years (Marudamuthu et al., 2023). Such technology will change the current landscape of agriculture (Pourkheirandish et al., 2020). Genetic modification of plants has provided considerable information on gene function, far more quickly than reverse genetics strategies based on the identification of mutations caused by chemical or radiation treatments (Siddique et al., 2023). Generation of a DNA sequence and comparisons between species can provide information about gene function on which genes are fundamental to plant or crop biology and which are specific to individual species (Della Coletta et al., 2021; Soltis & Soltis, 2003). This knowledge will be essential in enhancing new crops or traits important for food security and bioenergy in response to climate change, increasing world population, and changes in land use patterns (Yadav et al., 2019). In all cases, knowledge analysis will be pivotal in determining which genes should be modified to achieve a desired effect on the plant and crop (Chen et al., 2024b).

A major benefit of genomics approaches is the ability to simultaneously analyze large numbers of genes or whole genomes, and there are now an array of powerful and high-throughput technologies that are providing data to increase our understanding of crop plants (Bagger et al., 2024; Gill et al., 2022). Genomics fundamentally relies on phenotype data, as it utilizes this data to enhance breeding processes (Kaur et al., 2021). According to Arti Singh et al. (2016), phenotypes associated with traits such as resistance and environmental stress responses can be identified using phenomics methods (Singh et al., 2016). However, further research investigation is necessary to broaden the use of these tools beyond specific crops and to gather information from extensive breeding programs. Several studies have used transcriptome analysis to identify genes that are differently expressed in response to agricultural diseases and stressors (Du et al., 2017; Pan et al., 2020). For example, it has been reported to use RNA-seq data to examine the alterations in gene expression in response to various environmental stresses. Three main analytical methods were employed by the researchers: integrated network modeling, Gene Co-expression Network (GCN) design, and Differentially Expressed Gene (DEG) analysis. While GCN construction laid out the interactions between these genes and identified important regulatory players, DEG analysis identified important genes associated with stress situations. With a focus on transcription factors and NLR genes essential for plant stress responses, important regulatory modules and interacting genes were found by combining abiotic and biotic stress networks using the integrated network technique. This analysis thoroughly identified resistance-related genes in this investigation, indicating their importance in deciphering complex regulatory mechanisms of plant stress responses (Goche et al., 2020). Comparing transcriptome data across species on a large scale yields useful information for agricultural improvement (Kumar et al., 2022).

4.2.1 Genomics Technology in Crop Improvement

By thoroughly analyzing an organism's genetic molecules, genomics can pinpoint the genes that give rise to particular traits (Della Coletta et al., 2021). The development of novel molecular methods and informatics is anticipated to have the biggest influence on this field (Walter et al., 2015). Traditional breeding, according to Fasoula et al., depends on the application of germplasm collections to modify plant features (Fasoula et al., 2020). The most advantageous genotypes have been chosen to produce the next crop generation after these collections were evaluated for desirable features (Franco-Duran et al., 2019). This process is constrained by the genes governing the traits, the presence of genetic variety in germplasm collections, and the interaction of the new genotype with the environment (Singh & Behera, 2022). During the development of agronomically superior crop varieties, significant traits have often been lost due to a lack of understanding of genetic diversity at the molecular level (Franco-Duran et al., 2019). In order to eliminate the need for germplasm collections and breeding to improve crops, the focus is now on identifying and assessing gene-controlled traits thanks to the recent sequencing of model plant species and key crops. Breeding and, eventually, genetic manipulation can be used to accomplish this (Cooper et al., 2021; Marudamuthu et al., 2023).

4.2.2 Genomic Analysis in Plant Breeding/Disease Area

Plant breeding can be greatly accelerated by breakthroughs in genome analysis (Crossa et al., 2017). Identifying DNA markers that are strongly associated with quantitative trait loci (QTL) of desired qualities has long been the foundation of marker-assisted selection (Kushanov et al., 2021). However, more potent options for the selection of superior genotypes have been made available by recent advancements in genome resequencing and single-nucleotide polymorphism (SNP) genotyping (Li et al., 2023). These methods are referred to as genomic selection. Using a large amount of genome data, the objective is to select the best parents and estimate the value of individuals for target traits (Pourkheirandish et al., 2020). This approach makes it simple to incorporate genetic merit estimations into cultivar development and compare them across all traits (Clark et al., 2013).

Researchers are working to develop pest-resistant crops as an efficient and eco-friendly substitute for chemical pesticides due to the extensive use of both organic and inorganic chemical pesticides, their high cost, their negative effects on the environment, and their effects on rural development (Lefebvre et al., 2020). According to Pourkheirandish et al. (2020), this plant is a cost-effective, durable, and environmentally beneficial crop protection solution (Pourkheirandish et al., 2020). This method may be faster and less expensive than phenotypic screening for disease reaction and resistance levels in several hybrid groups. By introducing resistance genes into the genome of a vulnerable hybrid or variety, transgenesis can be used to create resistant cultivars (Flavell, 2009). Comparing transgenic and non-transgenic isogenic lines for the expression of disease on the host in a controlled setting and on several occasions is required to demonstrate the idea of the biological effect of the transgene (Raveendran & Balasubramanian, 2009). It is now feasible to generate host-induced mutants of the pathogen to function as a virulent gene differential to the wild type and validate the identity of the resistance gene in the host thanks to developments in gene silencing and alteration of pathogen virulence genes (Rato et al., 2021). According to Longdon et al. (2014), these techniques produce high-quality biological material appropriate for genetic study of specific disease resistance (Longdon et al., 2014). Intricate information about the functional biology of the host and pathogen as well as the alterations in gene expression in response to the disease can be gathered by integrating Omics platforms into these studies (Zand Karimi & Innes, 2022). In addition to offering diagnostic tools for breeding resistant types, this will greatly advance our understanding of the intricate genetics of resistance and susceptibility (Luo et al., 2023).

4.3 PROTEOMICS IN AGRICULTURE

Proteomics has emerged as a transformative tool in agricultural research, enabling the detailed analysis of crop proteomes to elucidate the molecular mechanisms underlying development, growth, and stress responses. The integration of mass spectrometry with bioinformatics has facilitated significant breakthroughs in plant biology (Yan et al., 2022). These advancements are critical for driving progress in areas such as improving crop yields, enhancing resistance to diseases, and developing stress-tolerant varieties (Komatsu et al., 2013). For instance, proteomic studies have identified key proteins associated with drought tolerance, offering new avenues for crop improvement and enabling early disease detection for more effective agricultural management (Alharbi, 2020; Hu et al., 2015).

In the context of sustainable agriculture, proteomics plays a pivotal role in optimizing resource utilization and minimizing environmental impacts, as outlined in Table 4.1. By analyzing protein expression profiles in plants subjected to varying environmental conditions, researchers can develop strategies to improve nutrient use efficiency and reduce reliance on chemical fertilizers (Ahmad et al., 2016). Additionally, proteomics is instrumental in studying plant–microbe interactions, which can promote beneficial symbiotic relationships, enhance soil health, and support crop productivity (Gupta et al., 2021). Furthermore, the integration of proteomics with other Omics technologies, such as genomics and metabolomics, provides a comprehensive understanding of plant systems biology, paving the way for innovative agricultural solutions (Yan et al., 2022). As the field continues to

TABLE 4.1
Overview of Proteomics Applications in Agriculture

Application	Description	Key Benefits	References
Crop stress response	Analysis of proteins under abiotic and biotic stressed conditions	Identifies stress-inducible proteins; enhances stress tolerance	Cavanagh et al. (2023); Yang et al. (2024)
Nutrient use efficiency	Examination of protein expression profiles for nutrient optimization	Improves fertilizer use; reduces environmental impact	Ahmad et al. (2016)
Plant–microbe interactions	Investigation of interactions between plants and beneficial microbes	Enhances plant growth; improves soil quality	Gupta et al. (2021); Yurgel et al. (2021)
Disease resistance	Identification of proteins related to disease resistance	Aids in early disease detection; develops resistant varieties	Hu et al. (2015); Alharbi (2020)
Integrated Omics approaches	Combining proteomics with genomics and metabolomics	Provides a holistic understanding of plant system biology	Yan et al. (2022)

evolve, proteomics will remain a cornerstone of agricultural research, driving advancements in crop enhancement and sustainable farming practices (Komatsu et al., 2013).

4.3.1 Proteomic Profiling for Identifying Stress-Inducible Proteins

The life cycle of crop plants encompasses a range of environmental conditions. The effect of these stress factors may be critical and can significantly reduce crop yield (Waqas et al., 2019). Understanding how crop plants respond to stress is crucial for developing stress-resistant varieties (Zhu et al., 2021). In this context, proteome analysis has already generated important insights (Ge et al., 2013; Wang et al., 2014). As genotype or gene can be affected by the environment, it results in alterations in its physiology, proteome, or metabolome, making the study of these changes under unstable global climate conditions both timely and valuable (Zhu, 2016). The responses to abiotic and biotic stresses are regulated by the expression of numerous genes (Baillo et al., 2019). These gene-encoded proteins orchestrate a cascading bunch of events, leading to observable alterations in plant physiology (Zhu, 2016). The study by *Cavanagh et al.* showed the role of Rubisco during various abiotic stresses in Arabidopsis. This showed that the change in the expression of a single gene can prompt the synthesis of distinct protein forms, altering plant physiology and necessitating the identification of specific proteins under various stressed conditions (Cavanagh et al., 2023). Given the multifaceted nature of stress factors and the intricate mechanisms underlying stress responses, alterations in gene expression and protein abundance are inevitable, leading to observable changes in the proteome (Punzo et al., 2020). This underscores the utility of comparative proteomics as a powerful approach for elucidating plant stress responses. By generating detailed protein profiles, this methodology reveals which proteins are activated and quantifies the extent of their involvement in stress-related processes (Yang et al., 2024). Such insights are critical for advancing our understanding of how plants adapt to and cope with stressed conditions.

4.3.2 Proteomics for Plant–Microbe Interactions

Plant–microbe interactions have garnered significant attention within the context of sustainable agricultural systems (Bakker et al., 2018). The challenges associated with the widespread use of chemical inputs in agriculture have prompted a shift toward soil and crop management practices that minimize reliance on such inputs, offering both economic and environmental benefits (Glick, 2020). Among these practices, the use of biological control agents, such as plant growth-promoting rhizobacteria (PGPR), has emerged as a promising agronomic strategy to enhance plant growth (Vejan et al., 2016). The development of effective and targeted microbiological tools for improving plant health and soil management hinges on a thorough understanding of the interactions between plants and their symbiotic or biocontrol partners (Smith & Smith, 2011). However, this area has thus far been an undeveloped area in plant biology due to the complexity and hidden nature of these interactions. Holistic approaches are needed to unravel the mechanisms by which microbes can affect plant physiological processes and development (Bakker et al., 2018; Glick, 2020). Although functional genomics studies employing microarrays have illustrated transcriptome alterations in response to microbes (Das et al., 2019), a paucity of research has uncovered the precise functions of the implicated genes and transcript variants (Stearns et al., 2012). Investigating protein expression and its functions is pivotal to understanding or engineering a biological process (Singh et al., 2015). An advantage of proteomics is that, proteomics allows for the examination of plant–microbe interactions in both in vitro and in vivo studies (Kosolapova et al., 2022). Studies have revealed that certain pathogens alter root exudation patterns (Bakker et al., 2018). This work can be further extended to look at changes in the proteome of the microorganism interacting with plants (Figure 4.2). A recent study compared the use of proteomics to identify specific bacterial proteins that are responsible for changes in plant metabolism. In the free-living and symbiotic states, the study revealed unique protein expressions, characterized by elevated nitrogenase and stress response proteins in bacteroids, suggesting specialized metabolic adaptations for symbiosis (Yurgel et al., 2021). Another interesting point is that evidence has shown how proteomics can also be used to identify the impact of biocontrol agents on pathogens that are antagonistic to plants. This study concludes that the bacterial biocontrol agents, like *B. velezensis*, inhibit the growth of *L. fungicola* in vitro, implying its suitability as a new disease biocontrol agent for mushrooms (Clarke et al., 2022). Studying the plant–microorganism relationship from both perspectives can greatly expand our understanding of their expression.

4.4 METABOLOMICS IN AGRICULTURE

The significant contribution of metabolomics and its linked approaches, metabolite profiling, metabolomics, metabolic fingerprinting, and metabolite target analysis, to agriculture, has opened a new era to enhance crop productivity and quality (Litvinov et al., 2021). Metabolomics primarily deals with qualitative and quantitative assessment of the wide array of total endogenous and exogenous metabolites present in the biological system (Patti et al., 2012). The information generated by an in-depth study of metabolome using statistical and bioinformatics tools could lead to better information on the physiological and biochemical basis for gene expression and/or specific characteristics in plants (Schrimpe-Rutledge et al., 2016). The availability of a large number of specific analytical methodologies for metabolome analysis now allows for the application of many and varied problems in plant biology (Alseekh & Fernie, 2018). The use of metabolic profiling techniques like Gas Chromatography-Mass Spectrometry (GC-MS), Liquid Chromatography-Mass Spectrometry (LC-MS), and NMR provides wider coverage of different metabolite pools (Roessner & Bowne, 2009). In turn, metabolomics, steps shown in Table 4.2, enable the identification of the metabolic changes occurring in plants due to various factors and therefore act as a bridge between the information-rich genome and the biochemically dynamic phenome (Yu et al., 2024).

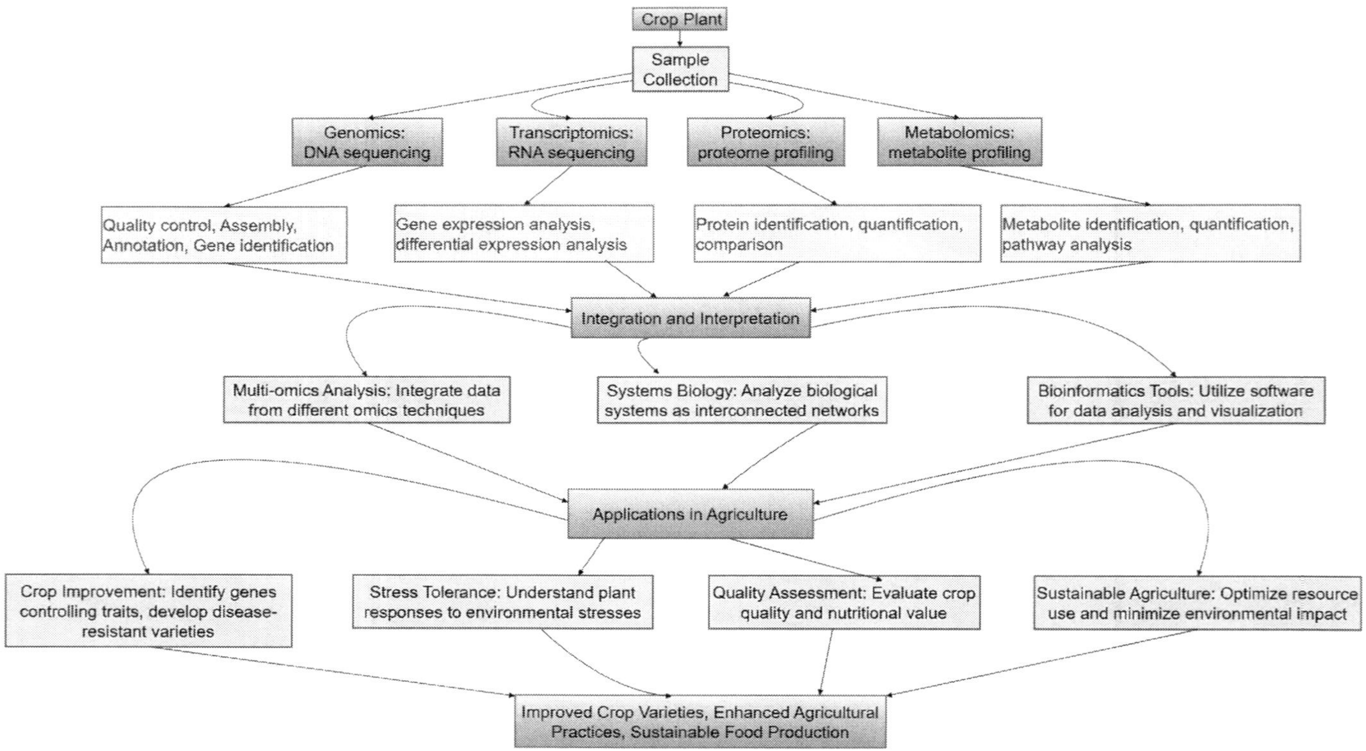

FIGURE 4.2 A memic chart illustrating the process from crop plant sample collection to the utilization of Omics technologies, data analysis, and their applications in agriculture.

TABLE 4.2
Metabolomics in Agriculture: Steps and Applications

Steps	Description	Tools/Technology	Application
Metabolite Profiling	Qualitative and quantitative assessment of metabolites under different environmental conditions (Roessner & Bowne, 2009).	• Gas Chromatography-Mass Spectrometry (GC-MS) • Liquid Chromatography-Mass Spectrometry (LC-MS) • NMR	• Identifies metabolites involved in plant development, stress response, and productivity.
Data Analysis and Integration	Analyzing large metabolomic datasets and integrating them with other Omics data (Johnson et al., 2016).	• Bioinformatics software • Statistical analysis tools	• Creates a holistic view of plant metabolism, linking metabolomics with genomics and proteomics.
Metabolomic Markers Identification	Identification of key metabolites related to crop productivity, quality, and stress response, acting as markers for crop improvement (Yu et al., 2024).	- Metabolomic profiling - Marker discovery techniques	• Enables the selection of traits for breeding programs to improve yield and quality.
Stress Response Analysis	Examination of changes in the plant metabolome under abiotic and biotic stressed conditions (e.g., drought, pests).	• Comparative metabolomics	• Helps in developing stress-resistant varieties
Metabolic Engineering	Manipulation of metabolic pathways by modifying the expression of key enzymes to increase the production of desired metabolites (Heinig et al., 2013).	• Genetic engineering • Enzyme regulation methods	• Improves specific traits such as nutrient content, disease resistance, and environmental resilience.
Crop Improvement and Breeding	Use of metabolomic data to guide crop breeding, focusing on traits such as enhanced productivity and quality (Razzaq et al., 2022).	• Plant breeding techniques • Metabolomic markers	• Results in the development of high-yielding, high-quality, and stress-tolerant crops tailored for both farmers and consumers.

4.4.1 Profiling and Analyzing Plant Metabolites

Metabolites are small molecules that are found as intermediates and products of cellular metabolism (Patti et al., 2012). The plant metabolome is highly dynamic and is vastly affected by endogenous and exogenous factors. These include abiotic and biotic stresses, plant development, and environmental factors (Alseekh & Fernie, 2018). The metabolome is closely tied to the genome, transcriptome, and proteome and often closer represents the phenotype due to its high responsiveness to change (Bujak et al., 2015). Due to the complexity of the metabolome, analytical methods must be highly sensitive and represent a wide dynamic range to detect all metabolites and provide a representation of global changes in the metabolome (Kim et al., 2022). Metabolite profiling needs vary greatly depending on the scientific questions and must be tailored to fit the hypotheses being tested or the goal of the experiment (Trivedi et al., 2017). As with all Omics methods, metabolomics generates a vast amount of data that requires bioinformatics and statistics for interpretation and must be integrated with other levels of biological information for maximal understanding (Johnson et al., 2016).

4.4.2 Metabolic Pathways in Crop Development and Stress Response

Metabolic pathways involve a series of chemical reactions occurring within a cell. These pathways are an integrated response to changes in the cell environment or the cellular signals (Bhambhani et al., 2021). The pathway can stop or start according to the cell's needs (Bobik & Burch-Smith, 2015). Metabolism is defined as the total of all chemical reactions involved in the maintenance and production of the energy required for the cell's existence (Judge & Dodd, 2020). The chemical compounds taking place in these pathways have to be identified for pathway manipulation (Singh et al., 2023). Otherwise, what should be better than an understanding of chemical components is their Omics fingerprint. A metabolic fingerprint is studied by profiling and databasing, taking into account intracellular and external metabolites (Guendel et al., 2021). An intracellular metabolite would be contained within the cell and classified as a metabolite taking part in one of the biochemical pathways. An extracellular metabolite would be a chemical compound involving organic metabolism or classified as a substrate (Elshafie et al., 2023). Data mining is subsequently conducted on the compound, which involves identifying structurally similar compounds, reviewing existing knowledge about the compound, and examining relevant regulatory frameworks associated with it. This process offers valuable insights into the potential applications of the compound and its potential effects on cellular mechanisms (Shang & Huang, 2019).

4.4.3 Metabolomic Markers for Crop Productivity and Quality

Metabolic engineering has become a pivotal strategy in the field of plant biotechnology (Mipeshwaree Devi et al., 2023). Traditional crop improvement methods, such as plant breeding, have historically depended on generating genetic diversity through the crossbreeding of closely related species and selecting desirable traits over successive generations (Shahiba et al., 2023). In contrast, metabolic engineering offers the potential to improve crop productivity and quality by introducing genes from a wide range of organisms that encode enzymes with specific functions. This method enables researchers to precisely influence the metabolome, allowing for targeted enhancements in crop traits (Heinig et al., 2013). A critical aspect of effective metabolic engineering lies in identifying key metabolites that contribute to crop productivity and quality, as well as elucidating the metabolic pathways responsible for their synthesis and degradation (Manickam et al., 2023). Targeted metabolic engineering, which focuses on increasing the production of a specific metabolite, becomes viable when the biosynthetic pathway is well understood and a rate-limiting step has been identified (Zhu et al., 2020). This is typically achieved by amplifying the activity of the rate-limiting enzyme. Alternatively, indirect metabolic engineering aims to modify the overall flux within a metabolic pathway, which can have more extensive and complex effects on the levels of multiple metabolites (Zhu et al., 2020).

The profiling of plant metabolites under varying conditions has enabled the identification of those linked to crop productivity and quality, which can serve as metabolomic markers (Salam et al., 2023). These significant metabolites can be utilized as selection traits in plant breeding programs to develop new crop varieties. This strategy is anticipated to accelerate the creation of high-yielding and high-quality varieties, offering benefits to both producers and consumers (Razzaq et al., 2022). To fully leverage the potential of metabolomics in agriculture, there is a growing need for analytical tools and applications that facilitate the interpretation of the metabolome (Oh et al., 2023).

4.5 INTEGRATION OF OMICS APPROACHES IN AGRICULTURE

The integration of genomics and post-genomics technologies into agricultural research has produced substantial progress, providing important insights to the scientific community (Golicz et al., 2015). These technologies are progressively utilized in plant breeding initiatives, providing practical solutions that augment their efficacy (Varshney et al., 2017). For instance, genomic selection has been effectively employed in tree breeding programs, exhibiting the capacity to double the rate of genetic improvement for numerous economically significant traits (Grattapaglia & Resende, 2011). As for genomics, transcriptomics is also providing valuable information to aid breeding programs (Qiao et al., 2024). In a previous study on stripe rust of wheat, scientists used integrated data to analyze genomic and transcription profiles. By identifying genes and transcripts that are causative of the plant response to disease/resistance, new approaches to plant breeding can be opened (Qiao et al., 2024). In a study relating to seed development, a mixture of proteomics and transcriptomics was utilized to visualize changes in metabolism at different stages of seed maturity. Such advancements have facilitated the identification of changes in flavonoid and fatty acid synthesis, as well as maturation processes (Ye et al., 2021). Moreover, the application of Omics methods across different species data comparison has facilitated the identification of gene functions via co-expression networks, especially within flavonoid biosynthesis pathways (Zainal-Abidin et al., 2022). Alternative methodologies have also proven effective in pinpointing genes associated with responses to environmental stressors (Baldoni et al., 2021). The ability to identify specific genes and proteins involved in crop development offers opportunities to utilize them as markers for improving plant breeding outcomes.

Systems biology, recognized as a powerful framework for studying complex biological systems (Sheth & Thaker, 2014), provides a holistic perspective on the interactions between molecular networks and environmental factors. In agriculture, these approaches have significantly advanced our understanding of crop physiology, metabolic processes, and stress responses (Hong et al., 2016). By integrating diverse Omics datasets with computational modeling, researchers can address the intricacies of plant biology at a systems level (Ali & Georgiev, 2023). This integrative methodology enables the identification of key regulatory networks, signaling pathways, and metabolic mechanisms that control critical agronomic traits (Chao et al., 2023). Furthermore, systems biology allows for the prediction of plant phenotypic responses to varying environmental conditions, supporting the development of crop varieties with improved yield, resilience, and nutritional quality (Pazhamala et al., 2021). Despite its promise, the application of systems biology in agriculture is still in its developmental stages and faces challenges related to traditional practices (Chen et al., 2024b). However, ongoing advancements in Omics technologies, computational tools, and interdisciplinary collaborations are addressing these limitations, unlocking the full potential of systems biology for agricultural innovation (Pazhamala et al., 2021).

For instance, integrated platforms have been applied to study Sorghum, a model crop known for its drought and waterlogging tolerance. Researchers have examined traits, phenotypes, resistance mechanisms, proteins, and metabolites in Sorghum to gain insights into stress tolerance. By analyzing marker expressions identified through these approaches and cross-referencing them with existing databases, scientists can better understand the markers associated with enhanced plant performance under stress (Kumar et al., 2022). A notable example involves the proteomic comparison of two Sorghum varieties – one drought-sensitive and the other drought-tolerant. The drought-tolerant variety demonstrated a unique adaptation mechanism, closing its leaf stomata to conserve water under stressed conditions. This response, regulated by specific protein networks, highlights the plant's ability to maintain water balance during drought, offering valuable insights into the physiological basis of drought resistance (Goche et al., 2020). Collectively, these studies underscore the importance of integrating data from multiple Omics approaches to achieve a comprehensive understanding of plant biology, environmental interactions, and molecular networks. Such integrated insights are instrumental in refining crop improvement strategies, as illustrated in Figure 4.2.

The flowchart in Figure 4.2 depicts the systematic workflow of applying multi-Omics technologies (genomics, transcriptomics, proteomics, and metabolomics) to plant samples, beginning with sample collection. Integrated datasets are analyzed via computational bioinformatics to derive actionable insights for developing improved crop varieties and advancing sustainable agricultural practices.

4.6 CHALLENGES AND FUTURE PROSPECTS

While systems biology presents much promise for enhancing agricultural methods, numerous hurdles must be addressed to effectively harness its advantages. A significant problem is the integration of multi-scale Omics data, often derived from several platforms and experimental circumstances. To extract valuable insights from these intricate datasets, trustworthy computational techniques for data integration, standardization, and analysis must be developed (Chen et al., 2020). Moreover, there are computational and logistical challenges when applying system biology techniques to large-scale agricultural systems. Scalable modeling frameworks and high-performance computing infrastructure are urgently needed to manage the enormous amounts of data created by these systems as they become increasingly intricate and interconnected (Pazhamala et al., 2021). Furthermore, overcoming the gap between basic research and interdisciplinary collaboration and knowledge sharing is crucial (Aubry, 2019). Facilitating collaboration among biologists, agronomists, computational scientists, and industry stakeholders, the translation of systems biology insights into practical, sustainable agricultural solutions can be expedited. Despite the existing challenges, the future of systems biology in agriculture continues to hold considerable promise, with the potential to significantly advance crop improvement, optimize resource management, and enhance environmental sustainability on a global scale (Sahu et al., 2023).

Although a multitude of databases dedicated to crop information are currently available, there is a pressing need to develop additional repositories that encompass a broader range of crop species. The expansion of such resources would facilitate access to more comprehensive datasets across a wider diversity of species, thereby enhancing the efficacy of research efforts and agricultural innovation (Zainal-Abidin et al., 2022). Furthermore, when employing diverse methodologies to generate data, it is imperative to interpret the resulting findings through multiple analytical frameworks. A critical challenge lies in identifying the most appropriate tools for data analysis, as the optimal strategy for addressing this issue remains unclear. This underscores the necessity for further research to address increasingly complex agricultural questions (Lim et al., 2022).

AI DECLARATION

AI tools have been used to improve the readability and language of a few sections in the chapter.

BIBLIOGRAPHY

Ahmad, N., & Mukhtar, Z. (2017). Genetic manipulations in crops: Challenges and opportunities. *Genomics*, 109(5–6). https://doi.org/10.1016/j.ygeno.2017.07.007

Ahmad, P., Abdel Latef, A. A. H., Rasool, S., Akram, N. A., Ashraf, M., & Gucel, S. (2016). Role of proteomics in crop stress tolerance. *Frontiers in Plant Science*, 7(September). https://doi.org/10.3389/fpls.2016.01336

Ainsworth, E. A., & Long, S. P. (2005). What have we learned from 15 years of free-air CO_2 enrichment (FACE)? A meta-analytic review of the responses of photosynthesis, canopy properties and plant production to rising CO_2. *New Phytologist*, 165(2), 351–371.

Alharbi, R. A. (2020). Proteomics approach and techniques in identification of reliable biomarkers for diseases. *Saudi Journal of Biological Sciences*, 27(3). https://doi.org/10.1016/j.sjbs.2020.01.020

Ali, K., & Georgiev, M. I. (2023). Editorial: Omics and its integration: A systems biology approach to understanding plant physiology. *Frontiers in Plant Science*, 14. https://doi.org/10.3389/fpls.2023.1324901

Alonge, M., Wang, X., Benoit, M., Soyk, S., Pereira, L., Zhang, L., Suresh, H., Ramakrishnan, S., Maumus, F., Ciren, D., Levy, Y., Harel, T. H., Shalev-Schlosser, G., Amsellem, Z., Razifard, H., Caicedo, A. L., Tieman, D. M., Klee, H., Kirsche, M., … Lippman, Z. B. (2020). Major impacts of widespread Structural variation on gene expression and crop improvement in Tomato. *Cell*, 182(1). https://doi.org/10.1016/j.cell.2020.05.021

Alseekh, S., & Fernie, A. R. (2018). Metabolomics 20 years on: What have we learned and what hurdles remain? *Plant Journal*, 94(6). https://doi.org/10.1111/tpj.13950

Arai-Kichise, Y., Shiwa, Y., Nagasaki, H., Ebana, K., Yoshikawa, H., Yano, M., & Wakasa, K. (2011). Discovery of genome-wide DNA polymorphisms in a landrace cultivar of japonica rice by whole-genome sequencing. *Plant and Cell Physiology*, 52(2). https://doi.org/10.1093/pcp/pcr003

Aubry, S. (2019). The future of digital sequence information for plant genetic resources for food and agriculture. *Frontiers in Plant Science*, 10. https://doi.org/10.3389/fpls.2019.01046

Bagger, F. O., Borgwardt, L., Jespersen, A. S., Hansen, A. R., Bertelsen, B., Kodama, M., & Nielsen, F. C. (2024). Whole genome sequencing in clinical practice. *BMC Medical Genomics*, 17(1). https://doi.org/10.1186/s12920-024-01795-w

Baillo, E. H., Kimotho, R. N., Zhang, Z., & Xu, P. (2019). Transcription factors associated with abiotic and biotic stress tolerance and their potential for crops improvement. *Genes*, 10(10). https://doi.org/10.3390/genes10100771

Bakker, P. A. H. M., Pieterse, C. M. J., de Jonge, R., & Berendsen, R. L. (2018). The soil-borne legacy. *Cell*, 172(6). https://doi.org/10.1016/j.cell.2018.02.024

Baldoni, E., Frugis, G., Martinelli, F., Benny, J., Paffetti, D., & Buti, M. (2021). A comparative transcriptomic meta-analysis revealed conserved key genes and regulatory networks involved in drought tolerance in cereal crops. *International Journal of Molecular Sciences*, 22(23). https://doi.org/10.3390/ijms222313062

Bayer, P. E., Golicz, A. A., Scheben, A., Batley, J., & Edwards, D. (2020). Plant pan-genomes are the new reference. *Nature Plants*, 6(8). https://doi.org/10.1038/s41477-020-0733-0

Bhambhani, S., Kondhare, K. R., & Giri, A. P. (2021). Diversity in chemical structures and biological properties of plant alkaloids. *Molecules*, 26(11). https://doi.org/10.3390/molecules26113374

Bobik, K., & Burch-Smith, T. M. (2015). Chloroplast signaling within, between and beyond cells. *Frontiers in Plant Science*, 6(October). https://doi.org/10.3389/fpls.2015.00781

Boden, S. A., & Østergaard, L. (2019). How can developmental biology help feed a growing population? *Development (Cambridge)*, 146(3). https://doi.org/10.1242/dev.172965

Bujak, R., Struck-Lewicka, W., Markuszewski, M. J., & Kaliszan, R. (2015). Metabolomics for laboratory diagnostics. *Journal of Pharmaceutical and Biomedical Analysis*, 113. https://doi.org/10.1016/j.jpba.2014.12.017

Cavanagh, A. P., Slattery, R., & Kubien, D. S. (2023). Temperature-induced changes in Arabidopsis Rubisco activity and isoform expression. *Journal of Experimental Botany*, 74(2). https://doi.org/10.1093/jxb/erac379

Chao, H., Zhang, S., Hu, Y., Ni, Q., Xin, S., Zhao, L., Ivanisenko, V. A., Orlov, Y. L., & Chen, M. (2023). Integrating omics databases for enhanced crop breeding. *Journal of Integrative Bioinformatics*, 20(4). https://doi.org/10.1515/jib-2023-0012

Chen, C., Chen, H., Zhang, Y., Thomas, H. R., Frank, M. H., He, Y., & Xia, R. (2020). TBtools: An integrative toolkit developed for interactive analyses of big biological data. *Molecular Plant*, 13(8). https://doi.org/10.1016/j.molp.2020.06.009

Chen, J. Y., Sang, H., Chilvers, M. I., Wu, C. H., & Chang, H. X. (2024a). Characterization of soybean chitinase genes induced by rhizobacteria involved in the defense against Fusarium oxysporum. *Frontiers in Plant Science*, 15. https://doi.org/10.3389/fpls.2024.1341181

Chen, L., Liu, G., & Zhang, T. (2024b). Integrating machine learning and genome editing for crop improvement. *aBIOTECH*. https://doi.org/10.1007/s42994-023-00133-5

Clark, S. A., Kinghorn, B. P., Hickey, J. M., & Van Der Werf, J. H. J. (2013). The effect of genomic information on optimal contribution selection in livestock breeding programs. *Genetics Selection Evolution*, 45(1). https://doi.org/10.1186/1297-9686-45-44

Clarke, J., Grogan, H., Fitzpatrick, D., & Kavanagh, K. (2022). Characterising the proteomic response of mushroom pathogen Lecanicillium fungicola to Bacillus velezensis QST 713 and Kos biocontrol agents. *European Journal of Plant Pathology*, 163(2). https://doi.org/10.1007/s10658-022-02482-1

Cooper, M., Voss-Fels, K. P., Messina, C. D., Tang, T., & Hammer, G. L. (2021). Tackling G × E × M interactions to close on-farm yield-gaps: Creating novel pathways for crop improvement by predicting contributions of genetics and management to crop productivity. *Theoretical and Applied Genetics*, 134(6). https://doi.org/10.1007/s00122-021-03812-3

Crossa, J., Pérez-Rodríguez, P., Cuevas, J., Montesinos-López, O., Jarquín, D., de los Campos, G., Burgueño, J., González-Camacho, J. M., Pérez-Elizalde, S., Beyene, Y., Dreisigacker, S., Singh, R., Zhang, X., Gowda, M., Roorkiwal, M., Rutkoski, J., & Varshney, R. K. (2017). Genomic selection in plant breeding: Methods, models, and perspectives. *Trends in Plant Science*, 22(11). https://doi.org/10.1016/j.tplants.2017.08.011

Dai, X., & Shen, L. (2022). Advances and trends in omics technology development. *Frontiers in Medicine*, 9. https://doi.org/10.3389/fmed.2022.911861

Das, S., Gwon, H. S., Khan, M. I., Van Nostrand, J. D., Alam, M. A., & Kim, P. J. (2019). Taxonomic and functional responses of soil microbial communities to slag-based fertilizer amendment in rice cropping systems. *Environment International*, 127. https://doi.org/10.1016/j.envint.2019.04.012

De Evsikova, C. M., Raplee, I. D., Lockhart, J., Jaimes, G., & Evsikov, A. V. (2019). The transcriptomic toolbox: Resources for interpreting large gene expression data within a precision medicine context for metabolic disease atherosclerosis. *Journal of Personalized Medicine*, 9(2). https://doi.org/10.3390/jpm9020021

Della Coletta, R., Qiu, Y., Ou, S., Hufford, M. B., & Hirsch, C. N. (2021). How the pangenome is changing crop genomics and improvement. *Genome Biology*, 22(1). https://doi.org/10.1186/s13059-020-02224-8

Díaz-Muñoz, S. L., Osorio, A., Bonilla Ramirez, L., Latorre Moratalla, M. L., & García Heredia, J. M. (2018). Big data analysis in bioinformatics. *Advances in Experimental Medicine and Biology*, 1094, 123–141.

Du, H., Zhu, J., Su, H., Huang, M., Wang, H., Ding, S., Zhang, B., Luo, A., Wei, S., Tian, X., & Xu, Y. (2017). Bulked segregant RNA-seq reveals differential expression and SNPs of candidate genes associated with waterlogging tolerance in maize. *Frontiers in Plant Science*, 8. https://doi.org/10.3389/fpls.2017.01022

Elshafie, H. S., Camele, I., & Mohamed, A. A. (2023). A comprehensive review on the biological, agricultural and pharmaceutical properties of secondary metabolites based-plant origin. *International Journal of Molecular Sciences*, 24(4). https://doi.org/10.3390/ijms24043266

Emmert-Streib, F., & Dehmer, M. (2018). Networks for systems biology: Conceptual connection of data and function. *IET Systems Biology*, 12(3), 61–69.

Fang, Y., & Luo, J. (2019). Rethinking plant metabolic engineering in the age of big data. *Plant Communications*, 1(1), 100002.

Fasoula, D. A., Ioannides, I. M., & Omirou, M. (2020). Phenotyping and plant breeding: Overcoming the barriers. *Frontiers in Plant Science*, 10. https://doi.org/10.3389/fpls.2019.01713

Fernie, A. R., & Schauer, N. (2009). Metabolomics-assisted breeding: A viable option for crop improvement? *Trends in Genetics*, 25(1), 39–48.

Flavell, R. (2009). Role of model plant species. *Methods in Molecular Biology*, 513. https://doi.org/10.1007/978-1-59745-427-8_1

Franco-Duran, J., Crossa, J., Chen, J., & Hearne, S. J. (2019). The impact of sample selection strategies on genetic diversity and representativeness in germplasm bank collections. *BMC Plant Biology*, 19(1). https://doi.org/10.1186/s12870-019-2142-y

Ge, P., Hao, P., Cao, M., Guo, G., Lv, D., Subburaj, S., Li, X., Yan, X., Xiao, J., Ma, W., & Yan, Y. (2013). iTRAQ-based quantitative proteomic analysis reveals new metabolic pathways of wheat seedling growth under hydrogen peroxide stress. *Proteomics*, 13(20). https://doi.org/10.1002/pmic.201300042

Gill, T., Gill, S. K., Saini, D. K., Chopra, Y., de Koff, J. P., & Sandhu, K. S. (2022). A comprehensive review of high throughput phenotyping and machine learning for plant stress phenotyping. *Phenomics*, 2(3). https://doi.org/10.1007/s43657-022-00048-z

Glick, B. R. (2020). Beneficial plant-bacterial interactions. *Beneficial Plant-Bacterial Interactions*. https://doi.org/10.1007/978-3-030-44368-9

Goche, T., Shargie, N. G., Cummins, I., Brown, A. P., Chivasa, S., & Ngara, R. (2020). Comparative physiological and root proteome analyses of two sorghum varieties responding to water limitation. *Scientific Reports*, 10(1). https://doi.org/10.1038/s41598-020-68735-3

Golicz, A. A., Bayer, P. E., & Edwards, D. (2015). Skim-based genotyping by sequencing. *Methods in Molecular Biology*, 1245. https://doi.org/10.1007/978-1-4939-1966-6_19

Gomez-Cabrero, D., Abugessaisa, I., Maier, D., & Teschendorff, A. E. (2014). Data integration in the era of omics: Current and future challenges. *BMC Systems Biology*, 8(Suppl 2), I1.

Grattapaglia, D., & Resende, M. D. V. (2011). Genomic selection in forest tree breeding. *Tree Genetics & Genomes*, 7(2). https://doi.org/10.1007/s11295-010-0328-4

Guendel, A., Hilo, A., Rolletschek, H., & Borisjuk, L. (2021). Probing the metabolic landscape of plant vascular bundles by infrared fingerprint analysis, imaging and mass spectrometry. *Biomolecules*, 11(11). https://doi.org/10.3390/biom11111717

Gupta, R., Anand, G., Gaur, R., & Yadav, D. (2021). Plant–microbiome interactions for sustainable agriculture: A review. *Physiology and Molecular Biology of Plants*, 27(1). https://doi.org/10.1007/s12298-021-00927-1

Gutiérrez-Salazar, M. P., Balderas-Martínez, Y. I., & Juárez-Colunga, S. (2018). Challenges and opportunities for synthetic biology in agricultural plants. *Frontiers in Plant Science*, 9, 1324.

Hanahan, D., & Weinberg, R. A. (2011). Hallmarks of cancer: The next generation. *Cell*, 144(5), 646–674.

Heinig, U., Gutensohn, M., Dudareva, N., & Aharoni, A. (2013). The challenges of cellular compartmentalization in plant metabolic engineering. *Current Opinion in Biotechnology*, 24(2). https://doi.org/10.1016/j.copbio.2012.11.006

Hong, J., Yang, L., Zhang, D., & Shi, J. (2016). Plant metabolomics: An indispensable system biology tool for plant science. *International Journal of Molecular Sciences*, 17(6). https://doi.org/10.3390/ijms17060767

Houle, D., Govindaraju, D. R., & Omholt, S. (2010). Phenomics: The next challenge. *Nature Reviews Genetics*, 11(12), 855–866.

Hu, J., Rampitsch, C., & Bykova, N. V. (2015). Advances in plant proteomics toward improvement of crop productivity and stress resistance. *Frontiers in Plant Science*, 6(Apr). https://doi.org/10.3389/fpls.2015.00209

Huang, X., Yang, S., Gong, J., Zhao, Y., Feng, Q., Gong, H., Li, W., Zhan, Q., Cheng, B., Xia, J., Chen, N., Hao, Z., Liu, K., Zhu, C., Huang, T., Zhao, Q., Zhang, L., Fan, D., Zhou, C., … Han, B. (2015). Genomic analysis of hybrid rice varieties reveals numerous superior alleles that contribute to heterosis. *Nature Communications*, 6. https://doi.org/10.1038/ncomms7258

Jansson, C., & Northen, T. (2010). Calcifying cyanobacteria-the potential of biomineralization for carbon capture and storage. *Current Opinion in Biotechnology*, 21(3). https://doi.org/10.1016/j.copbio.2010.03.017

Johnson, C. H., Ivanisevic, J., & Siuzdak, G. (2016). Metabolomics: Beyond biomarkers and towards mechanisms. *Nature Reviews Molecular Cell Biology*, 17(7). https://doi.org/10.1038/nrm.2016.25

Judge, A., & Dodd, M. S. (2020). Metabolism. *Essays in Biochemistry*, 64(4), 607–647. https://doi.org/10.1042/EBC20190041

Kaur, B., Sandhu, K. S., Kamal, R., Kaur, K., Singh, J., Röder, M. S., & Muqaddasi, Q. H. (2021). Omics for the improvement of abiotic, biotic, and agronomic traits in major cereal crops: Applications, challenges, and prospects. *Plants*, 10(10). https://doi.org/10.3390/plants10101989

Kim, Y. X., Son, S. Y., Lee, S., Lee, Y., Sung, J., & Lee, C. H. (2022). Effects of limited water supply on metabolite composition in tomato fruits (Solanum lycopersicum L.) in two soils with different nutrient conditions. *Frontiers in Plant Science*, 13. https://doi.org/10.3389/fpls.2022.983725

Komatsu, S., Mock, H. P., Yang, P., & Svensson, B. (2013). Application of proteomics for improving crop protection/artificial regulation. *Frontiers in Plant Science*, 4(Dec). https://doi.org/10.3389/fpls.2013.00522

Kosolapova, A. O., Belousov, M. V., Sulatsky, M. I., Tsyganova, A. V., Sulatskaya, A. I., Bobylev, A. G., Shtark, O. Y., Tsyganov, V. E., Volkov, K. V., Zhukov, V. A., Tikhonovich, I. A., & Nizhnikov, A. A. (2022). RopB protein of Rhizobium leguminosarum bv. viciae adopts amyloid state during symbiotic interactions with pea (Pisum sativum L.). *Frontiers in Plant Science*, 13. https://doi.org/10.3389/fpls.2022.1014699

Krishnan, A., Gupta, C., Ambavaram, M. M. R., & Pereira, A. (2017). RECoN: Rice environment coexpression network for systems level analysis of abiotic-stress response. *Frontiers in Plant Science*, 8. https://doi.org/10.3389/fpls.2017.01640

Kuleshov, M. V., Stein, D. J., Clarke, D. J. B., Kropiwnicki, E., Jagodnik, K. M., Bartal, A., Evangelista, J. E., Hom, J., Cheng, M., Bailey, A., Zhou, A., Ferguson, L. B., Lachmann, A., & Ma'ayan, A. (2020). The COVID-19 Drug and Gene Set Library. *Patterns*, 1(6). https://doi.org/10.1016/j.patter.2020.100090

Kumar, P., Singh, J., Kaur, G., Adunola, P. M., Biswas, A., Bazzer, S., Kaur, H., Kaur, I., Kaur, H., Sandhu, K. S., Vemula, S., Kaur, B., Singh, V., & Tseng, T. M. (2022). OMICS in fodder crops: Applications, challenges, and prospects. *Current Issues in Molecular Biology*, 44(11). https://doi.org/10.3390/cimb44110369

Kushanov, F. N., Turaev, O. S., Ernazarova, D. K., Gapparov, B. M., Oripova, B. B., Kudratova, M. K., Rafieva, F. U., Khalikov, K. K., Erjigitov, D. S., Khidirov, M. T., Kholova, M. D., Khusenov, N. N., Amanboyeva, R. S., Saha, S., Yu, J. Z., & Abdurakhmonov, I. Y. (2021). Genetic diversity, QTL mapping, and marker-assisted selection technology in cotton (Gossypium spp.). *Frontiers in Plant Science*, 12. https://doi.org/10.3389/fpls.2021.779386

Ladics, G. S., Bartholomaeus, A., Bregitzer, P., Doerrer, N. G., Gray, A., Holzhauser, T., Jordan, M., Keese, P., Kok, E., Macdonald, P., Parrott, W., Privalle, L., Raybould, A., Rhee, S. Y., Rice, E., Romeis, J., Vaughn, J., Wal, J. M., & Glenn, K. (2015). Genetic basis and detection of unintended effects in genetically modified crop plants. *Transgenic Research*, 24(4). https://doi.org/10.1007/s11248-015-9867-7

Lefebvre, V., Boissot, N., & Gallois, J.-L. (2020). Host plant resistance to pests and pathogens, the genetic leverage in integrated pest and disease management. *Integrated Pest and Disease Management in Greenhouse Crops*. https://doi.org/10.1007/978-3-030-22304-5_9

Li, Q., & Yan, J. (2020). Sustainable agriculture in the era of omics: Knowledge-driven crop breeding. *Genome Biology*, 21(1). https://doi.org/10.1186/s13059-020-02073-5

Li, Y., Luo, X., Peng, X., Jin, Y., Tan, H., Wu, L., Li, J., Pei, Y., Xu, X., & Zhang, W. (2023). Development of SNP and InDel markers by genome resequencing and transcriptome sequencing in radish (Raphanus sativus L.). *BMC Genomics*, 24(1). https://doi.org/10.1186/s12864-023-09528-6

Lim, P. K., Zheng, X., Goh, J. C., & Mutwil, M. (2022). Exploiting plant transcriptomic databases: Resources, tools, and approaches. *Plant Communications*, 3(4). https://doi.org/10.1016/j.xplc.2022.100323

Litvinov, D. Y., Karlov, G. I., & Divashuk, M. G. (2021). Metabolomics for crop breeding: General considerations. *Genes*, 12(10). https://doi.org/10.3390/genes12101602

Longdon, B., Brockhurst, M. A., Russell, C. A., Welch, J. J., & Jiggins, F. M. (2014). The evolution and genetics of virus host shifts. *PLoS Pathogens*, 10(11). https://doi.org/10.1371/journal.ppat.1004395

Luo, K., He, D., Guo, J., Li, G., Li, B., & Chen, X. (2023). Molecular advances in breeding for durable resistance against pests and diseases in wheat: Opportunities and challenges. *Agronomy*, 13(3). https://doi.org/10.3390/agronomy13030628

Luo, Z., Xiong, J., Xia, H., Ma, X., Gao, M., Wang, L., Liu, G., Yu, X., & Luo, L. (2020). Transcriptomic divergence between upland and lowland ecotypes contributes to rice adaptation to a drought-prone agroecosystem. *Evolutionary Applications*, 13(9). https://doi.org/10.1111/eva.13054

Ma, J., Li, R., Wang, H., Li, D., Wang, X., Zhang, Y., Zhen, W., Duan, H., Yan, G., & Li, Y. (2017). Transcriptomics analyses reveal wheat responses to drought stress during reproductive stages under field conditions. *Frontiers in Plant Science*, 8. https://doi.org/10.3389/fpls.2017.00592

Manickam, S., Rajagopalan, V. R., Kambale, R., Rajasekaran, R., Kanagarajan, S., & Muthurajan, R. (2023). Plant metabolomics: Current initiatives and future prospects. *Current Issues in Molecular Biology*, 45(11). https://doi.org/10.3390/cimb45110558

Marudamuthu, B., Sharma, T., Purru, S., Soam, S. K., & Rao, C. S. (2023). Next-generation sequencing technology: A boon to agriculture. *Genetic Resources and Crop Evolution*, 70(2). https://doi.org/10.1007/s10722-022-01512-5

Medina-Lozano, I., & Díaz, A. (2022). Applications of genomic tools in plant breeding: Crop biofortification. *International Journal of Molecular Sciences*, 23(6). https://doi.org/10.3390/ijms23063086

Micheel, C. M., Nass, S. J., Omenn, G. S., & Policy, H. S. (2012). Evolution of translational omics Lessons learned and the path forward. *Evolution of Translational Omics Lessons Learned and the Path Forward*, NBK202168, 13297. https://doi.org/10.17226/13297

Mipeshwaree Devi, A., Khedashwori Devi, K., Premi Devi, P., Lakshmipriyari Devi, M., & Das, S. (2023). Metabolic engineering of plant secondary metabolites: Prospects and its technological challenges. *Frontiers in Plant Science*, 14. https://doi.org/10.3389/fpls.2023.1171154

Nagamine, T., & Matsui, H. (2010). Quantitative models of Arabidopsis floral regulatory network. *Bioinformatics*, 26(12), 159–165.

Nie, Y., Wu, G., Yu, J., & Cao, H. (2015). Challenges in integrative approaches to modelling the biology of inflammation and cancer. *Briefings in Bioinformatics*, 16(2), 242–251.

O'Connor, L. M., O'Connor, B. A., Lim, S. Bin, Zeng, J., & Lo, C. H. (2023). Integrative multi-omics and systems bioinformatics in translational neuroscience: A data mining perspective. *Journal of Pharmaceutical Analysis*, 13(8). https://doi.org/10.1016/j.jpha.2023.06.011

Oh, S. W., Imran, M., Kim, E. H., Park, S. Y., Lee, S. G., Park, H. M., Jung, J. W., & Ryu, T. H. (2023). Approach strategies and application of metabolomics to biotechnology in plants. *Frontiers in Plant Science*, 14. https://doi.org/10.3389/fpls.2023.1192235

Ovens, K., Eames, B. F., & McQuillan, I. (2021). Comparative analyses of gene co-expression networks: Implementations and applications in the study of evolution. *Frontiers in Genetics*, 12. https://doi.org/10.3389/fgene.2021.695399

Pan, Y., Zhao, S. W., Tang, X. L., Wang, S., Wang, X., Zhang, X. X., Zhou, J. J., & Xi, J. H. (2020). Transcriptome analysis of maize reveals potential key genes involved in the response to belowground herbivore holotrichia parallela larvae feeding. *Genome*, 63(1). https://doi.org/10.1139/gen-2019-0043

Patti, G. J., Yanes, O., & Siuzdak, G. (2012). Innovation: Metabolomics: The apogee of the omics trilogy. *Nature Reviews Molecular Cell Biology*, 13(4). https://doi.org/10.1038/nrm3314

Pazhamala, L. T., Kudapa, H., Weckwerth, W., Millar, A. H., & Varshney, R. K. (2021). Systems biology for crop improvement. *Plant Genome*, 14(2). https://doi.org/10.1002/tpg2.20098

Pourkheirandish, M., Golicz, A. A., Bhalla, P. L., & Singh, M. B. (2020). Global role of crop genomics in the face of climate change. *Frontiers in Plant Science*, 11. https://doi.org/10.3389/fpls.2020.00922

Proost, S., & Mutwil, M. (2017). Planet: Comparative co-expression network analyses for plants. *Methods in Molecular Biology*, 1533. https://doi.org/10.1007/978-1-4939-6658-5_12

Punzo, P., Grillo, S., & Batelli, G. (2020). Alternative splicing in plant abiotic stress responses. *Biochemical Society Transactions*, 48(5). https://doi.org/10.1042/BST20200281

Qiao, L., Luo, J., Wang, H., Kong, Y., Du, T., Qin, P., & Yang, B. (2024). Integrated genome-wide association and transcriptomic analysis to identify receptor kinase genes to stripe rust resistance in wheat germplasm from southwestern China. *BMC Plant Biology*, 24(1), 319. https://doi.org/10.1186/s12870-024-05020-9

Rato, C., Carvalho, M. F., Azevedo, C., & Oblessuc, P. R. (2021). Genome editing for resistance against plant pests and pathogens. *Transgenic Research*, 30(4). https://doi.org/10.1007/s11248-021-00262-x

Raveendran, M., & Balasubramanian, P. (2009). Pyramiding genes for enhancing tolerance to abiotic and biotic stresses. In *Molecular Techniques in Crop Improvement: 2nd Edition*. https://doi.org/10.1007/978-90-481-2967-6_7

Ray, D. K., Mueller, N. D., West, P. C., & Foley, J. A. (2013). Yield trends are insufficient to double global crop production by 2050. *PLoS One*, 8(6). https://doi.org/10.1371/journal.pone.0066428

Razzaq, A., Wishart, D. S., Wani, S. H., Hameed, M. K., Mubin, M., & Saleem, F. (2022). Advances in metabolomics-driven diagnostic breeding and crop improvement. *Metabolites*, 12(6). https://doi.org/10.3390/metabo12060511

Roessner, U., & Bowne, J. (2009). What is metabolomics all about? *BioTechniques*, 46(5 Spec. issue). https://doi.org/10.2144/000113133

Ronald, P. C. (2014). Lab to farm: Applying research on plant genetics and genomics to crop improvement. *PLoS Biology*, 12(6). https://doi.org/10.1371/journal.pbio.1001878

Sahu, S. K., Waseem, M., & Aslam, M. M. (2023). Editorial: Bioinformatics, big data and agriculture: A challenge for the future. *Frontiers in Plant Science*, 14. https://doi.org/10.3389/fpls.2023.1271305

Salam, U., Ullah, S., Tang, Z. H., Elateeq, A. A., Khan, Y., Khan, J., Khan, A., & Ali, S. (2023). Plant metabolomics: An overview of the role of primary and secondary metabolites against different environmental stress factors. *Life*, 13(3). https://doi.org/10.3390/life13030706

Schrimpe-Rutledge, A. C., Codreanu, S. G., Sherrod, S. D., & McLean, J. A. (2016). Untargeted metabolomics strategies—Challenges and emerging directions. *Journal of the American Society for Mass Spectrometry*, 27(12). https://doi.org/10.1007/s13361-016-1469-y

Schuster, S. C. (2008). Next-generation sequencing transforms today's biology. *Nature Methods*, 5(1). https://doi.org/10.1038/nmeth1156

Shahiba, A. M., Jayalekshmy, V. G., & Chacko, A. (2023). Traditional plant breeding techniques—Exploring crossbreeding and selection methods In *Modern Approaches in Plant Breeding Enhancing Crop Genetics* (pp. 1–14). https://doi.org/10.5281/zenodo.8406082

Shang, Y., & Huang, S. (2019). Multi-omics data-driven investigations of metabolic diversity of plant triterpenoids. *Plant Journal*, 97(1). https://doi.org/10.1111/tpj.14132

Sheth, B. P., & Thaker, V. S. (2014). Plant systems biology: Insights, advances and challenges. *Planta*, 240(1). https://doi.org/10.1007/s00425-014-2059-5

Shi, F., He, Y., Chen, Y., Yin, X., Sha, X., & Wang, Y. (2021). Comparative analysis of multiple neurodegenerative diseases based on advanced epigenetic aging brain. *Frontiers in Genetics*, 12. https://doi.org/10.3389/fgene.2021.657636

Siddique, M. I., Younis, A., Gururani, M. A., & Venkatesh, J. (2023). Application of tilling as a reverse genetics tool to discover mutation in plants genomes for crop improvement. *Mutation Breeding for Sustainable Food Production and Climate Resilience*. https://doi.org/10.1007/978-981-16-9720-3_9

Singh, A., & Behera, C. (2022). Strategies, opportunities, and challenges in crop genetic diversity conservation: A plant breeder's perspective. *Molecular Genetics and Genomics Tools in Biodiversity Conservation*. https://doi.org/10.1007/978-981-16-6005-4_7

Singh, A., Ganapathysubramanian, B., Singh, A. K., & Sarkar, S. (2016). Machine learning for high-throughput stress phenotyping in plants. *Trends in Plant Science*, 21(2). https://doi.org/10.1016/j.tplants.2015.10.015

Singh, G., Agrawal, H., & Bednarek, P. (2023). Specialized metabolites as versatile tools in shaping plant–microbe associations. *Molecular Plant*, 16(1). https://doi.org/10.1016/j.molp.2022.12.006

Singh, R. P., Shelke, G. M., Kumar, A., & Jha, P. N. (2015). Biochemistry and genetics of ACC deaminase: A weapon to "stress ethylene" produced in plants. *Frontiers in Microbiology*, 6(Sep). https://doi.org/10.3389/fmicb.2015.00937

Smith, S. E., & Smith, F. A. (2011). Roles of arbuscular mycorrhizas in plant nutrition and growth: New paradigms from cellular to ecosystem scales. *Annual Review of Plant Biology*, 62. https://doi.org/10.1146/annurev-arplant-042110-103846

Soltis, D. E., & Soltis, P. S. (2003). The role of phylogenetics in comparative genetics. *Plant Physiology*, 132(4). https://doi.org/10.1104/pp.103.022509

Stearns, J. C., Woody, O. Z., McConkey, B. J., & Glick, B. R. (2012). Effects of bacterial ACC deaminase on Brassica napus gene expression. *Molecular Plant-Microbe Interactions*, 25(5). https://doi.org/10.1094/MPMI-08-11-0213

Tan, Y., Tong, Z., Yang, Q., Sun, Y., Jin, X., Peng, C., Guo, A., & Wang, X. (2017). Proteomic analysis of phytase transgenic and non-transgenic maize seeds. *Scientific Reports*, 7(1). https://doi.org/10.1038/s41598-017-09557-8

Tester, M., & Langridge, P. (2010). Breeding technologies to increase crop production in a changing world. *Science*, 327(5967). https://doi.org/10.1126/science.1183700

Trivedi, D. K., Hollywood, K. A., & Goodacre, R. (2017). Metabolomics for the masses: The future of metabolomics in a personalized world. *New Horizons in Translational Medicine*, 3(6).

Valizadeh, N., & Bijani, M. (2020). *Agricultural Research: Applications and Future Orientations*. https://doi.org/10.1007/978-3-319-95675-6_5

Varshney, R. K., Shi, C., Thudi, M., Mariac, C., Wallace, J., Qi, P., Zhang, H., Zhao, Y., Wang, X., Rathore, A., Srivastava, R. K., Chitikineni, A., Fan, G., Bajaj, P., Punnuri, S., Gupta, S. K., Wang, H., Jiang, Y., Couderc, M., … Xu, X. (2017). Pearl millet genome sequence provides a resource to improve agronomic traits in arid environments. *Nature Biotechnology*, 35(10). https://doi.org/10.1038/nbt.3943

Vejan, P., Abdullah, R., Khadiran, T., Ismail, S., & Nasrulhaq Boyce, A. (2016). Role of plant growth promoting rhizobacteria in agricultural sustainability-A review. *Molecules*, 21(5). https://doi.org/10.3390/molecules21050573

Walter, A., Liebisch, F., & Hund, A. (2015). Plant phenotyping: From bean weighing to image analysis. *Plant Methods*, 11(1). https://doi.org/10.1186/s13007-015-0056-8

Wang, W., Vinocur, B., & Altman, A. (2003). Plant responses to drought, salinity and extreme temperatures: Towards genetic engineering for stress tolerance. *Planta*, 218(1). https://doi.org/10.1007/s00425-003-1105-5

Wang, Z. Q., Xu, X. Y., Gong, Q. Q., Xie, C., Fan, W., Yang, J. L., Lin, Q. S., & Zheng, S. J. (2014). Root proteome of rice studied by iTRAQ provides integrated insight into aluminum stress tolerance mechanisms in plants. *Journal of Proteomics*, 98. https://doi.org/10.1016/j.jprot.2013.12.023

Waqas, M. A., Kaya, C., Riaz, A., Farooq, M., Nawaz, I., Wilkes, A., & Li, Y. (2019). Potential mechanisms of abiotic stress tolerance in crop plants induced by thiourea. *Frontiers in Plant Science*, 10. https://doi.org/10.3389/fpls.2019.01336

Yadav, P., Priyanka, P., Kumar, D., Yadav, A., & Yadav, K. (2019). *Bioenergy Crops: Recent Advances and Future Outlook*. https://doi.org/10.1007/978-3-030-14463-0_12

Yan, S., Bhawal, R., Yin, Z., Thannhauser, T. W., & Zhang, S. (2022). Recent advances in proteomics and metabolomics in plants. *Molecular Horticulture*, 2(1). https://doi.org/10.1186/s43897-022-00038-9

Yang, Y., Cheng, Y., Lu, Z., Ye, H., Du, G., & Li, Z. (2024). Comparative proteomic and metabolomic analyses reveal stress responses of hemp to salinity. *Plant Cell Reports*, 43(6), 154. https://doi.org/10.1007/s00299-024-03237-4

Yang, Y., Saand, M. A., Huang, L., Abdelaal, W. B., Zhang, J., Wu, Y., Li, J., Sirohi, M. H., & Wang, F. (2021). Applications of multi-omics technologies for crop improvement. *Frontiers in Plant Science*, 12. https://doi.org/10.3389/fpls.2021.563953

Ye, Z., Yu, J., Yan, W., Zhang, J., Yang, D., Yao, G., Liu, Z., Wu, Y., & Hou, X. (2021). Integrative iTRAQ-based proteomic and transcriptomic analysis reveals the accumulation patterns of key metabolites associated with oil quality during seed ripening of Camellia oleifera. *Horticulture Research*, 8(1). https://doi.org/10.1038/s41438-021-00591-2

Yu, J., Han, T., Hou, Y., Zhao, J., Zhang, H., Wang, X., & Ge, S. (2024). Integrated transcriptomic, proteomic and metabolomic analysis provides new insights into tetracycline stress tolerance in pumpkin. *Environmental Pollution*, 340. https://doi.org/10.1016/j.envpol.2023.122777

Yurgel, S. N., Qu, Y., Rice, J. T., Ajeethan, N., Zink, E. M., Brown, J. M., Purvine, S., Lipton, M. S., & Kahn, M. L. (2021). Specialization in a nitrogen-fixing symbiosis: Proteome differences between Sinorhizobium medicae bacteria and bacteroids. *Molecular Plant-Microbe Interactions*, 34(12). https://doi.org/10.1094/MPMI-07-21-0180-R

Zainal-Abidin, R. A., Harun, S., Vengatharajuloo, V., Tamizi, A. A., & Samsulrizal, N. H. (2022). Gene co expression network tools and databases for crop improvement. *Plants*, 11(13). https://doi.org/10.3390/plants11131625

Zand Karimi, H., & Innes, R. W. (2022). Molecular mechanisms underlying host-induced gene silencing. *Plant Cell*, 34(9). https://doi.org/10.1093/plcell/koac165

Zhang, Y., Massel, K., Godwin, I. D., & Gao, C. (2019). Correction to: Applications and potential of genome editing in crop improvement. *Genome Biology*, 20(1). https://doi.org/10.1186/s13059-019-1622-6

Zhu, J. K. (2016). Abiotic stress signaling and responses in plants. *Cell*, 167(2). https://doi.org/10.1016/j.cell.2016.08.029

Zhu, Q., Wang, B., Tan, J., Liu, T., Li, L., & Liu, Y. G. (2020). Plant synthetic metabolic engineering for enhancing crop nutritional quality. *Plant Communications*, 1(1). https://doi.org/10.1016/j.xplc.2019.100017

Zhu, T., Fonseca De Lima, C. F., & De Smet, I. (2021). The heat is on: How crop growth, development, and yield respond to high temperature. *Journal of Experimental Botany*, 72(21). https://doi.org/10.1093/jxb/erab308

5 CRISPR/Cas9-Meditated Modifications in Sustainable Agriculture

Mahesh Pattabhiramaiah, Shanthala Mallikarjunaiah, and Manjunatha Reddy G. R.

5.1 INTRODUCTION

Agriculture, biodiversity, society, and the environment are all affected by climate change. Global agricultural output is expected to decrease under various scenarios due to more severe climate variations, including global warming. To achieve sustainable food production, it's essential to enhance crop production on a universal scale. CRISPR (Clustered Regularly Interspaced Short Palindromic Repeats)-Cas9 (CRISPR-associated protein 9) technology, the utmost defined genome editing molecular tools today essential for biological research (Dey, 2021), derived from a bacterial virus defense system, would revolutionize agricultural innovation by improving crop quality, inducing disease resistance, stress, and herbicide tolerance, and enabling rapid and precise germplasm creation and improvement of breeding technique (Wang et al., 2018a).

CRISPR-Cas9 is sequence-specific due to its limited core, nucleic acid sites, and bilobed structure (Song et al., 2016). CRISPR/Cas systems are divided into two classes (Class 1 and Class 2), comprising six kinds and numerous subtypes. Initiated from *Streptococcus pyogenes* (SpCas9), the Type 2 CRISPR-Cas9 system is the utmost extensively researched and utilized CRISPR/Cas systems. Nucleic acid sites form a helix bridge between the nuclease (NUC) and recognition (REC) lobes (Doudna and Charpentier, 2014).

Gene editing technologies embrace Zinc Finger Nucleases (ZFNs) and Transcription Activator-Like Effector Nucleases (TALENs). They generate double-strand breaks in DNA at specific sites that are then restored to introduce desired genetic changes. Random mutagenesis methods like zinc finger nucleases and transcription activator-like effector nucleases are less accurate & productive than CRISPR-Cas9 (Table 5.1). Compared to Zinc Finger or TALENs, CRISPR-Cas9 is easier to create, more efficient, and could mark multiple genes consecutively (Cong et al., 2013; Mali et al., 2013) and boost food production despite biosafety concerns. CRISPR/Cas9 is an active, discriminating genome editing process of great crop breeding and plant biology potential.

 DOI: 10.1201/9781003714484-5

TABLE 5.1
Evaluation of the ZFN, TALEN, and CRISPR Technologies

Feature	ZFN	TALEN	CRISPR	Reference
Mechanism	Targets particular DNA sequences and produces double-strand breaks by using zinc finger proteins.	Utilizes nucleases and effectors like transcription activators to create double-strand break at definite locations.	Creates double-strand breaks by targeting a particular DNA sequence with the Cas9 nuclease using guide RNA.	Carroll (2011)
Design Complexity	High; requires custom design for each target site.	Moderate; involves designing TAL effectors for each target site.	Low; guide RNA can be designed relatively easily for various target sites.	Doyon and Livak (2009)
Target Range	Limited to specific DNA sequences due to design constraints.	Flexible; can target a wide range of sequences with proper design.	Highly versatile; can target virtually any sequence with the appropriate guide RNA.	Jinek et al. (2012)
Efficiency	Often lower due to difficulties in achieving high specificity and efficiency.	Generally high, but can be affected by the efficiency of the TALEN assembly.	Usually high because designing guide RNAs is straightforward and the Cas9 enzyme is highly efficient.	Miller et al. (2011)
Off-Target Effects	Higher potential for off-target effects due to less precise targeting.	Moderate; off-target effects can advance but are generally lower than ZFNs.	Generally low; improved versions and techniques reduce off-target effects.	Doyon and Livak (2009)

5.2 CRISPR/CAS9 DELIVERY TO PLANTS

The most common plant transformation methods, like gene transfer and tissue culture, are laborious, time-consuming, and cause random somatic mutations, making them ineffective (Gao, 2018). Many crops reject tissue culture-based regeneration. To edit plant genomes efficiently, CRISPR/Cas9 is transferred directly to plant apical meristems or pollen grains, which can modify plants deprived of tissue culture (Jung and Seo, 2017).

CRISPR/Cas9 works through a guide RNA (gRNA) and points the Cas9 endonuclease to a definite genomic location. The enzyme then creates a double-strand breakdown that could be restored via inserting, deleting, or replacing genetic material (Jinek et al., 2012). This method can enhance factors, viz., immunity against illness, stress tolerance, and nutritional superiority of various crops (Li et al., 2018).

For pathogen recognition, researchers have employed CRISPR/Cas9 to generate disease-resistant crop diversities, viz., rice and wheat (Wang et al., 2014). The CRISPR/Cas9 usage in agriculture is to successfully deliver the genes to suitable plant cells to grow viable plants. The CRISPR/Cas9 technology improved drought stress tolerance and enhanced the nutritional profiles of crops by modifying genes (Zhao et al., 2016).

CRISPR/Cas9 is an extremely proficient gene-editing practice that modifies genomic DNA with great precision, mimicking the natural defense mechanism used by prokaryotes against viruses and foreign DNA. The leading apparatuses of CRISPR are Cas9 and guide RNA (gRNA). CRISPR and trans-activating RNA (crRNA) constitute the gRNA. The crRNA has a complementary ~20 nucleotide spacer to target gene sequences, succeeding a protospacer adjacent motif (PAM). Cas9, guided by the gRNA, causes double-strand breaks (DSBs) three base pairs upstream of the PAM motif. CRISPR-Cas9 uses a specific Cas9 nuclease and a synthetic sgRNA to direct Cas9 to specific DNA sequences in plants.

These breaks are repaired by non-homologous end joining (NHEJ) or homology-directed repair (HDR), allowing for precise nucleotide insertions, deletions, substitutions, or site-specific mutations. The design of sgRNA, which adheres to Watson-Crick base pairing rules, has established CRISPR/Cas9 as a potent and adaptable device for gene editing.

According to Liu et al. (2019), nonhomologous end joining (NHEJ)-repaired cleavage conventionally deletes genes or reduces protein activity. Instead, homology-directed DNA repair (HDR) would activate to introduce an external DNA restoration template in a specific genomic area (Sun et al., 2016). The plant CRISPR/Cas9 system its byproducts knock genes in, out, and down and stimulate their expression, including multiple gene editing (Figure 5.1).

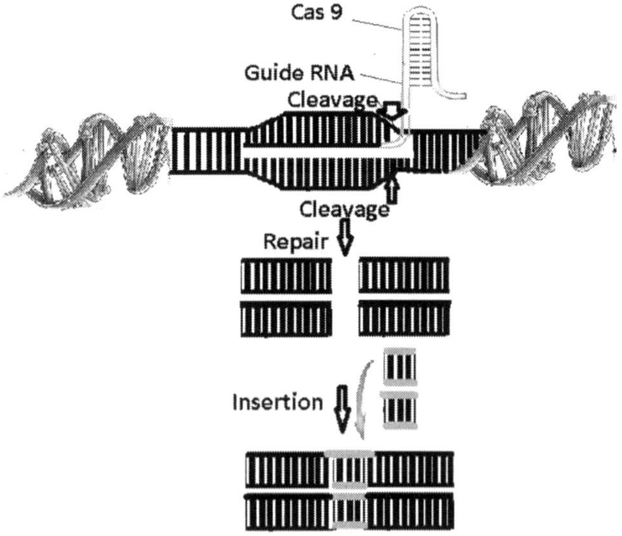

FIGURE 5.1 Plant CRISPR/Cas9 delivery.

There are several gene transfer methods similar to agrobacterium-mediated transformation, biolistic (particle bombardment), electroporation, virus-mediated transformation, and protoplast fusion have revolutionized agricultural practices and research by enabling precise genetic modifications that were not possible with traditional breeding techniques (Su et al., 2023).

Agrobacterium-mediated transfer often transforms many plant species (Tumefaciens or rhizogenes) (Ran et al., 2013; Mikami et al., 2015; Budiani et al., 2018; Hooghvorst et al., 2019; Mao et al., 2019; Pompili et al., 2020). Biolistic delivery replaces *Agrobacterium*-mediated delivery (Carter and Shieh, 2015). Heavy metal particles like tungsten or gold-enveloped DNA are transferred forthrightly into plant cells via "biolistic delivery" (Baltes et al., 2017). When DNA-enveloped elements enter plant cells, they can separate from metal particles and integrate inside the host genome (Baltes et al., 2017). CRISPR/Cas9 utilizing viral vectors has revealed success in plant genome manipulation in wheat (*Triticum aestivum* L.) (Hamada et al., 2018).

5.3 CROP IMPROVEMENT USING CRISPR/CAS9

The CRISPR/Cas9 system is a prevailing tool for development, quality, yield, and stress tolerance in crops for quantitative trait loci (QTLs). The grain size (GS) and grain length (GL) QTLs' effects on *Oryza sativa* (rice) grain size and number were investigated utilizing CRISPR-mediated QTL editing (Shen et al., 2018). A rising sum of monocot & dicot plant species use CRISPR-Cas9 technology to improve their nutritive value, quality, tolerance to biotic and abiotic challenges and productivity (Angon and Habiba, 2023).

5.3.1 CRISPR/Cas9 Gene Editing to Induce Agricultural Plants' Abiotic Stress Tolerance

Salinity, temperature, famine, and cold ominously upset crop growth & harvest (Gong et al., 2020; Liu et al., 2022). Understanding how different crop diversities flourish under varying conditions is crucial, as these stress factors can reduce crop yields. Traditional breeding improves productivity but reduces genetic diversity and takes time, making sustainability and desired traits difficult (Gao, 2018; Wolter et al., 2019), whereas CRISPR/Cas9 gene editing may improve plants' stress tolerance (Figure 5.2).

According to research, due to extreme drought stress, genomic adaptation is necessary for drought resistance. Genome editing by CRISPR/Cas9 can improve famine resistance by targeting drought-susceptible genes or stress response negative regulators. Stress-related ring finger protein 1 *Oryza sativa* Small RNA-Associated F-box Protein 1-(OsSRFP1), famine-induced SINA protein 1-*Oryza sativa* Drought-Induced Stress 1(OsDIS1), and dry and salt-tolerant protein 1, *Oryza sativa*, Drought Stress Tolerance-(OsDST)-silenced-increases antioxidant enzyme levels, lowers H_2O_2 levels, and improves rice drought tolerance (Kumar et al., 2020).

Unique alleles were introduced into the *Arabidopsis* OPEN STOMATA 2 (OST2) gene, which translates a plasma membrane H+ ATPase vital for stomatal activity.

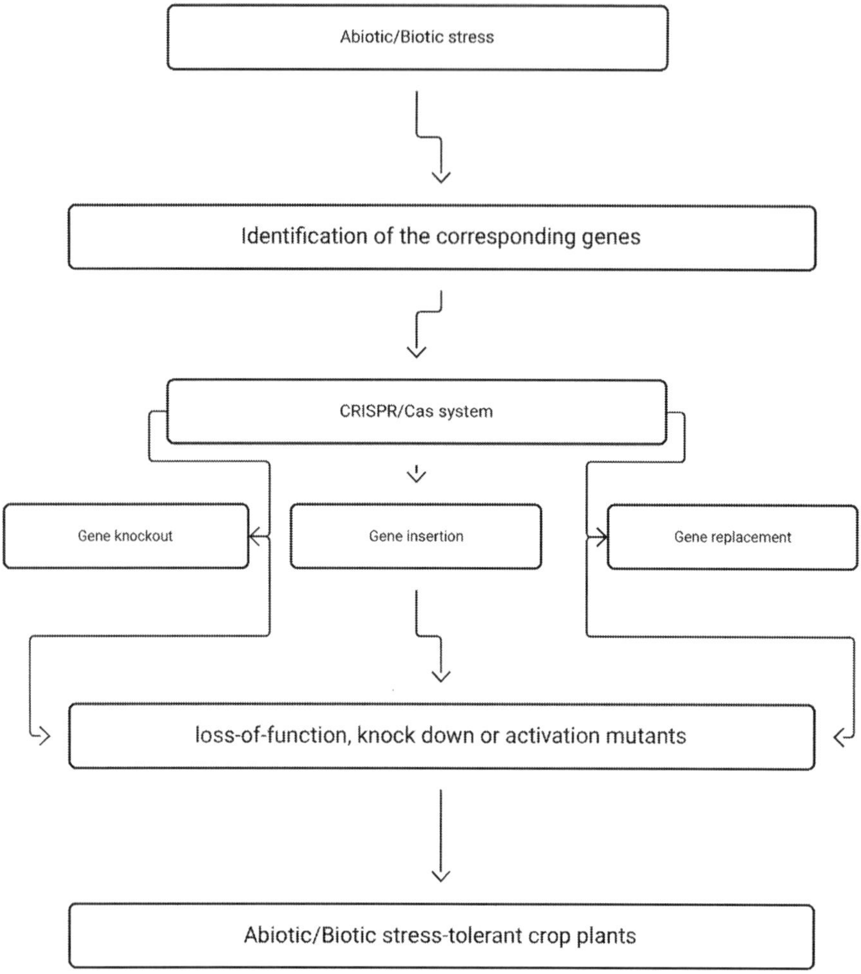

FIGURE 5.2 CRISPR/Cas-induced abiotic/biotic stress tolerance in crop plants.

This ATPase creates a proton gradient for stomatal conductance. Dehydration causes abscisic acid (ABA) to bind to the proton pump's C-terminus, inhibiting H+ ATPase and closing stomata. Importantly, Ost[2] locus mutations completely impair ABA-responsive stomata, causing necrotic lesions and continuous proton pump activity. A highly effective CRISPR–Cas system using Cas9 and a trimmed sgRNA (tru-sgRNA) has identified mutations in transgenic crops without affecting off-target sites (>32%). In comparison to the wild-type, the ost2-CRISPR mutant closed stomata faster and lost less transcriptional water under ABA-mediated conditions.

Thus, CRISPR-Cas9-mediated OST2 gene modifications improved drought tolerance by increasing stomatal responsiveness. CRISPR-Cas9 utilized to generate mutant lines of the non-expresser tomato of pathogenesis-related 1 (NPR1) to test the gene's famine tolerance (Li et al., 2019b). NPR1, a key regulator of plant defense mechanisms, shows a minor part in abiotic stress response. In famine-responsive apple plants, MdNPR1 is downregulated (Bhat et al., 2016).

AtNPR1 overexpression made rice hypersensitive to famine stress (Deng et al., 2018). In uncultivated tomatoes of CRISPR-Cas9 inserted loss-of-function slnpr1 mutations, antioxidant enzyme levels declined, stomatal opening enhanced, MDA content augmented, electrolytic leakage amplified, and famine stress tolerance. Down regulation of drought-responsive genes like SlDREB, SlDHN, and SlGST increased SlNPR1 mutants' drought sensitivity. SlNPR1 regulates stress response in tomatoes and other crops, wherein genome editing can create variants with broad-spectrum scarcity conflict (Li et al., 2019a).

Early drought stress symptoms include reduced height, leaf wilting, flowering, and budding disorder (Zafar et al., 2020). Plant nutrient uptake is repeatedly damaged due to persistent famine. In tomatoes, CRISPR-Cas9 was used to knock down SlMAPK3 to study its role in drought tolerance. The SlMAPK[3]mutants exposed to famine become more susceptible to induce plant wilt and curl in stems than that of wild-type plants and confirmed that SlMAPK3 is indeed engaged in drought tolerance (Wang et al., 2017; Abdelrahman et al., 2018). By the use of CRISPR-Cas9, SlAGL6 for heat stress was knocked down, and creating parthenocarpic fruit in the tomato plant (Klap et al., 2017). A loss-of-function mutant of SAPK2 rice was created by utilizing CRISPR-Cas9. Maize yielded by the ARGOS8 mutant increased grain production under normal conditions in which, ARGOS8-v2 and v1 showed significantly higher transcript levels than wild type (Shi et al., 017). The method CRISPR-Cas9 appears to be operative for the making of allelic variations that permit crop varieties tolerant to drought. A high-efficiency CRISPR-Cas9 system using tru-gRNAs and the AtEF1 tissue-specific promoter to drive Cas9 expression, which caused mutations of OST2/AHA1 with very limited off-target effects (Martignago et al., 2020). This new development is set to allow CRISPR-Cas9 to change crop yield and stress resistance (Abdelrahman et al., 2018).

Global conservation of food is lowered by drought stresses become a challenge to, produce crops on a large scale (Joshi et al., 2020). To increase maize yield during drought, CRISPR/Cas was utilized to substitute the native ARGOS8 promoter with the GOS2 promoter sequence (Shi et al., 2017). The editing of OsERA1 through CRISPR/Cas9 increased the rice drought tolerance (Ogata et al., 2020) and, the editing of TaDREB2 and TaERF3 improved the drought stress tolerance in wheat grain (Kim et al., 2018). By the gene manipulation of OsDST in the indica mega rice cultivar MTU1010, CRISPR/Cas9 was used and it was observed that the transgenic plants grown with a bigger leaf size, dense stomatal density to resist drought stress (Santosh Kumar et al., 2020).

According to Usman et al. (2021), the CRISPR/Cas9-construct OSPY19 mutant showed higher yields in rice with more tolerance to drought stress. By applying CRISPR/Cas9, semi-rolled leaf 1 (SRL1), and semi-rolled leaf 2 (SRL2) mutants of rice conferred, curled leaves increase the tolerance to famines by altering the levels of proteins and reactive oxygen species. CRISPR/Cas9 modified Gibberellin-insensitive dwarf1 (GID1) and SlLBD40 genes in tomatoes enhanced yield with higher leaf liquid content and drought tolerance (Liu et al., 2020a).

According to Al Murad et al. (2020), salinization aggravates with time on Earth's terrestrial (7%) and cultivable land (20%). Osmotic and ion stress induced by salt stress minimize crop superiority and harvest (Siddiqui et al., 2017). The SlHyPRP1 negative-response domain was deleted to improve tomato salty tolerance during sprouting and vegetative stages (Tran et al., 2021). For rice salt tolerance, Alam et al. (2022) utilized CRISPR/Cas9 to hit the OsbHLH024 gene and increase the expression of ion transporter genes like OsHKT1;3, OsHAK7, and OsS. Rice's salt lenience increased without affecting other traits with OsRR22 gene CRISPR/Cas9 mutations (Zhang et al., 2020; Han et al., 2022). The mutant rice with improved saline stress viability was created by modifying OsRAV2 with CRISPR/Cas9 (Liu et al., 2020b). Deleting genes like OsDST, SlARF4, HvITPK1, and OsNAC041 in rice can progress saline hassle lenience in crops (Wang et al., 2016b; Santosh Kumar et al., 2020; Vlčko & Ohnoutková, 2020; Yue et al., 2020).

Increased temperatures affect crop growth, especially blossoming and gametogenesis (Jagadish et al., 2021). Osmotic and oxidative stress from cold stress limits plant growth, distribution, and yields below 20°C and 0°C. CRISPR/Cas9-generated ospin5b, gs3, and osmyb30 mutants augmented spike length, grain size, and cold tolerance (Zeng et al., 2019). CRISPR/Cas9-mediated slmapk3 mutants augmented tomato heat tolerance by regulating antioxidant enzymes and heat shock proteins. Miao et al. (2018) found that CRISPR/Cas9-edited pyl1/4/6 triple hit rice had higher yield, heat tolerance, and condensed pre-harvest germination than wild type.

Heavy metal toxicity, foremost abiotic stress, hinders plant development and yield, making sustainable agriculture difficult. Osmotic stress, oxidative stress, ion imbalance, cytotoxicity, and result in membrane & metabolic disruption (Naeem et al., 2020). Human health is threatened by plant heavy metal accumulation (Kaur et al., 2020). CRISPR/Cas9 tool has the capability to reduce crop heavy metal accumulation. For example, Cadmium levels in rice were reduced by knocking down OsNramp5 and OsLCT1 (Songmei et al., 2019), while eliminating OsNRAMP1 reduced Cd and Pb (Chu et al., 2022). In addition, CRISPR/Cas9 removal of OsARM1, an R2R3 MYB transcription factor, reduced rice arsenic uptake (Wang et al., 2017), and deactivation of the K+ transporter OsHAK1 reduced cesium (Cs) in rice seedling.

5.3.2 CRISPR/Cas9 Gene Editing to Enhance Agricultural Plant's Biotic Stress Tolerance

Fungal, bacterial, and viral diseases cause 20–40% of universal agronomic yield fatalities (Walker, 1984). Increasing host plant pathogen resistance can reduce disease impact and food shortages. CRISPR/Cas9 has created disease-resistant plants (Cheng et al., 2019) (Figure 5.2 and Table 5.2). Powdery mildew, a harmful fungus, reduces crop yields. CRISPR/Cas9 knockout of all wheat TaMLO alleles upgraded ailment confrontation (Wang et al., 2014). Grape and tomato plants were more impervious to powdery mildew after CRISPR/Cas9-mediated SlMLO and VvMOL3 reduction (Wang et al., 2018b). Enhanced tomato resistances were achieved through a CRISPR/Cas9-induced mutation in SlPMR4, but not complete immunity.

TABLE 5.2
Use of CRISPR-Centered Genome Editing in Plants for Abiotic and Biotic Stresses

Plant Species	Target Gene	Type of Stress	CRISPR Strategy	Impact	Reference
Rice	OsWRKY45	Biotic (disease)	Knockout	Enhanced tolerance to bacterial blight	Zhang et al. (2021)
Arabidopsis	AtDREB1A	Abiotic (drought)	Overexpression	Increased drought tolerance and improved growth under water stress	Liu et al. (2020)
Wheat	TaPIP2;5	Abiotic (drought)	Mutagenesis	Improved drought resistance and better water use efficiency	Zhang et al. (2022)
Tomato	SlACS2	Abiotic (salt)	Knockout	Enhanced salt tolerance and better fruit yield	Li et al. (2023)
Soybean	GmNAC1	Abiotic (flood)	Knockout	Increased resistance to flooding and improved root development	Zhang et al. (2020)
Maize	ZmNAC111	Abiotic (cold)	Overexpression	Enhanced cold tolerance and improved seedling growth	Wang et al. (2021)
Barley	HvP5CS	Abiotic (drought)	Gene Editing	Increased drought tolerance and better stress adaptation	Xie et al. (2022)
Cotton	GhHsfA2	Abiotic (heat)	Knockout	Improved heat stress tolerance and enhanced fiber quality	Chen et al. (2023)

CRISPR/Cas9 disruption of OsERF922 and OsSEC3A genes increased rice blast resistance (Wang et al., 2016a). CRISPR/Cas9 mutations in rice Bsr-d1 and Pi21 provided partial rice blast resistance, but not as much as OsERF922 disruption (Nawaz et al., 2020; Zhou et al., 2022).

Tomato late blight (*Phytophthora infestans*) greatly reduces yields. Knocking out miR482b and miR482c together increased fungal confrontation, revealing a new miRNA-mediated resistance mechanism (Chang et al., 2016). Gray mold infection in tomatoes was suggestively decreased by CRISPR/Cas9-mediated PL gene knockout (Silva et al., 2021).

Numerous plant-pathogenic bacteria can cause many diseases in crops (Schloss and Handelsman, 2004). CRISPR/Cas9 genome editing improves crop bacterial resistance. For instance, knocking down the OsSWEET13 gene that translates a sucrose transporter concerned with pathogen relations protected rice from bacterial blight. *Xanthomonas oryzae*'s PthXo2 effector protein increases OsSWEET13 expression and susceptibility (Zhou et al., 2015).

Xanthomonas citri causes citrus bacterial canker (CBC), the utmost common bacterial infection in *Citrus* plants. Jia et al. (2017) improved the CsLOB1 gene's promoter to create CBC-resistant Duncan grapefruit mutants. Peng et al. (2017) found that the CRISPR/Cas9-targeted mutation of the CsLOB1 promoter EBEPthA4 made Wanjincheng orange (*Citrus sinensis*) more resistant to citrus canker.

Coronatine, which opens plant stomata and aids *Pseudomonas syringae* bacterial leaf spot disease, worsens plant survival. CRISPR/Cas9 edited SlJAZ2 to remove the JAZ domain, making it unaffected by bacterial leaf spot disease. This protein is a jasmonate-ZIM domain co-receptor (Ortigosa et al., 2019). CRISPR/Cas9 is utilized in generating mutants of the ACET1a and ACET1b genes to increase confrontation with bacteria and fungi (Jeon et al., 2020). Mutations in the tomato vulnerability gene SlDMR6-1 also protected against bacteria, oomycetes, and fungi (Thomazella et al., 2021). CRISPR/Cas9-mediated osnramp1 mutants showed broad-spectrum resistance, reduced catalase (CAT) activity but increased H_2O_2 and SOD activity (Chu et al., 2022).

Roossinck et al. (2015) classified plant viruses into six groups based on genomic characteristics: reverse-transcribing, positive- and negative-sense single-stranded RNA, ssDNA, dsDNA, and viruses without plant-specific classification. These pathogens cause a risk to critical crops. SsDNA Gemini viruses have been utilized to improve virus confrontation in CRISPR-edited plants (Ali et al., 2015; Baltes et al., 2015; Ji et al., 2015). Gemini viruses reason worldwide yield fatalities in *Euphorbiaceae*, *Cucurbitaceae*, *Malvaceae*, *Solanaceae*, and *Fabaceae* (Zaidi et al., 2016).

These viruses replicate via rolling-circle amplification or recombination-mediated dsDNA replication (Hanley-Bowdoin et al., 2013). The Begomovirus genus infects dicotyledonous plants via sweet potato white fly (*Bemisia tabaci*) and plant phloem (Gilbertson et al., 2015). Begomovirus genomes have two segments (A and B) or one segment (A) with a 220-bp region. CRISPR technology showed resistance to Gemini viruses like BSCTV and BeYDV in *Nicotiana benthamiana* and *Arabidopsis* (Baltes et al., 2015; Ji et al., 2015).

Ji et al. (2015) found 43 BSCTV genome coding and non-coding sgRNA/Cas9 targets, Begomovirus. Multiple sgRNA/Cas9 combinations reduced viral levels in injected leaves, with increased Cas9 and sgRNA expression correlated with viral resistance. Baltes et al. (2015) utilized 11 sgRNAs to target different BeYDV genome regions with similar results. CRISPR/Cas9 increased begomovirus resistance (Ali et al., 2015).

CRISPR/Cas9 targeted and cut viral DNA in the host cell nucleus. A tobacco rattle virus (TRV) vector was utilized to introduce sgRNAs into *Nicotiana benthamiana* plants expressing Cas9. These sgRNAs targeted several regions of the tomato yellow leaf curl virus (TYLCV), together with capsid protein (CP), intergenic region (IR), and replication protein (Rep) motifs. Targeting stem-loop invariant sites in the IR reduced viral replication and accretion significantly. Additionally, the CRISPR/Cas9 system is considered to intend well-maintained stem-loop sequences in both monopartite Beet Curly Top Virus (BCTV) and bipartite *Merremia mosaic virus* (MeMV). This displayed targeting conserved sequences across virus strains can afford broad-spectrum resistance. MeMV, CLCuKoV, and TYLCV strains' coding and non-coding sequences were targeted with CRISPR/Cas9 tools.

Modifying plant genes for viral resistance, isolating CRISPR/Cas9 tools, and emancipating non-transgenic mutants into the field are other viral resistance strategies. RNA viruses live on plant host proteins like eIF4E, eIF(iso)4E, and eIF4G (Sanfaçon, 2015). Chandrasekaran et al. (2016) modified two regions of the host vulnerability gene eIF4E with CRISPR/Cas9 to make cucumbers potyvirus-resistant. This made it resistant to Potyviridae family viruses like PRSV-W, CVYV, and ZYMV. No resistance was seen in heterozygous knockout and non-mutant plants. Macovei et al. (2018) converted rice eIF4G alleles to resist rice tungro spherical virus (RTSV) using CRISPR/Cas9.

After RTSV inoculation, modified rice plants had greater plant height and grain yield than wild-type. CRISPR/Cas9 modified tomato plants' highest-expressed SlDCL2b among the four DCL2 subfamilies (SLDCL2a-SLDCL2d), increasing ToMV resistance (Wang et al., 2018a). Wang et al. (2018a) found that simultaneous modification of SlDCL2a and SlDCL2b improved TMV and PVX resistance.

Pests harm crops and transmit plant diseases, resulting in 20%–40% of universal crop yield reductions (Douglas, 2018). Rising temperatures have increased pest-related agricultural losses (Deutsch et al., 2018). The environment can suffer from excessive pesticide use. OsCYP71A1 disruption in rice stopped serotonin production, increasing salicylic acid levels and plant hopper and stem borer resistance. With Hap3 knocked out via CRISPR/Cas9, the gmcdpk38 mutant was highly resistant to cutworm.

CRISPR/Cas9 has improved tomato plant architecture, resulting in larger fruits and shapes suitable for municipal areas (Lemmon et al., 2018). Eliminating the SlCLV3 promoter, which controls floral stem cells and plant structure, resulted in larger fruit and alterations in floral design. Efforts to domesticate wild tomatoes using CRISPR/Cas9 have also been successful. Researchers were able to enhance fruit size and quantity by up to ten times, as well as improve harvest, nutritional value, and resistance to diseases and environmental stresses, by altering six genes linked to crucial domestication characteristics (Zsögön et al., 2018).

5.4 NUTRITIONAL IMPROVEMENT (BENEFICIAL METABOLITES) USING CRISPR/CAS9

Extreme weather events are disrupting food chains and raising food values, especially in tropical and subtropical provinces with low crop yields. Dry spells make it harder to absorb water-soluble nutrients like silica, calcium, magnesium, nitrate, and sulfate. Uneven rainfall increases soil nitrate leaching. Rising CO_2 levels reduce crop nutrition of wheat, rice, legumes, and vegetables lose protein, zinc, and iron (Soares et al., 2019).

According to climate change modeling, fruit and vegetable crop loss could triple malnutrition-related deaths by 2050 due to decreased availability and access (Springmann et al., 2016). If current CO_2 emission trends continue, 1.9% more people will have zinc deficiency, 1.3% protein deficiency, and 57% of kids& women of childbearing age would thrive in perilous iron deficiency regions by 2050 (Smith and Myers, 2018).

Anthocyanins, lycopene, and carotenoids are plant pigments that contain antioxidant vitamins, which improve nutrition and prevent disease. Beta-carotene boosts immunity and vision, while anthocyanins reduce inflammation and oxidative damage (Tanaka and Ohmiya, 2008).

CRISPR/Cas9 information has been exploited to insert carotenoid biosynthesis genes into rice, increasing carotenoid levels and advancing "Golden Rice." This could benefit poor women and children in rice-growing regions. Similar CRISPR/Cas9 initiatives have increased rice and other crop proanthocyanidins and anthocyanins (Zafar et al., 2020). Gene editing has also produced pink, yellow, and purple tomatoes (Cermák et al., 2015; Filler Hayut et al., 2017; Deng et al., 2018).

Color and nutrition have been improved in genetically modified tomatoes. Inserting a robust promoter upstream of the Myb transcription factor-encoding ANT1 gene produced deeply purple tomatoes with high anthocyanin levels (Cermák et al., 2015). Genome editing of the phytoene synthase gene (Psy1) and carotenoid-related genes increased tomato lycopene content (Li et al., 2018). Gene editing also increased lycopene in domesticated wild tomatoes by 50% (Zsögön et al., 2018) and beta-carotene in bananas (Kaur et al., 2020). These changes boost consumer appeal and pigment intake.

Although gene editing has mostly focused on commodity crops, it could improve non-starchy fruits and vegetables. Gene editing improves the nutritional value of *Physalis* species like groundcherry and goldenberry (Lemmon et al., 2018). These fruits contain antioxidants and essential nutrients (Shenstone et al., 2020). Early research suggests that CRISPR/Cas9 can delete agronomic genes to improve the commercial viability of nutrient-dense berries like ground cherries (Lemmon et al., 2018). The Fatty Acid Desaturase 2 (FAD2) gene in oil crops has been edited, changing monounsaturated fat levels.

Camelina sativa, an emerging oilseed crop, has its oleic acid content increased by over 50% in some cases due to targeted CRISPR/Cas9 editing, while linoleic and linolenic acids have decreased. *Brassica napus* also has higher oleic acid levels (Okuzaki et al., 2018). These changes may improve oil extract nutrition and shelf life (Jiang et al., 2017; Morineau et al., 2017).

5.4.1 Macronutrient Modification

In some crops, gene editing has increased beneficial metabolites and macronutrients. In rice, knocking out the SBEIIb gene, which codes less amylopectin, and more amylose (Sun et al., 2017). However, editing cassava amylose genes reduced starch content (Bull et al., 2018) In potatoes and Strawberries deleting the granule-bound starch synthase (GBSS) gene transformed starch levels (Andersson et al., 2017). Strawberry gene suppression also transformed sugar content. These developments reveal that genome editing can markedly modify macronutrient profiles, enhancing nutritional quality.

5.4.2 Reducing Anti-Nutritional Compounds

Various plant species have anti-nutritional factors reduced using CRISPR-Cas9 gene editing. Editing maize has reduced phytic acid levels, which can hinder nutritional absorption in monogastric animals (Liang et al., 2014). CRISPR/Cas9 and ZFNs can target phytic acid pathway genes to improve iron, zinc, and calcium absorption, reducing anti-nutritional effects (Liang et al., 2014). The genes that produce kafirin, which lowers protein digestibility, were edited in sorghum, a vital crop in drought-prone regions (Li et al., 2018). Using CRISPR/Cas9, rice varieties without cadmium in grains have been created (Tang et al., 2017). One of the most significant challenges in using CRISPR gene editing is off-target DNA cleavage, which occurs when the molecular scissor alters the wrong section of the host genome, resulting in a non-specific, unwanted/unexpected, and even adverse alteration to the genome (Kotagama et al., 2019).

5.5 OFF-TARGET ISSUES RELATED TO CRISPR-CAS GENE EDITING TECHNOLOGY

Even though CRISPR/Cas systems exhibit incredible perspective in agriculture, off-target effects remain a foremost challenge (Fu et al., 2013; Pacesa et al., 2022). Off-target genome editing is a type of unintentional and non-specific genetic modification that occurs when an engineered nuclease is utilized. Possible off-target cleavage activity could still occur on DNA sequences for even three to five base pair mismatches in the PAM-distal part of the sgRNA-guiding sequence, but the occurrence, nature, and location of this genetic modification were not overtly confirmed (Zhang et al., 2015).

The strategy for the identification of potential off-target genome edits is to distinguish between the "unbiased" and "biased" methods. Unbiased methods are based on the genome-wide screening for small DNA modifications or DSBs, which can be consistently connected to the nuclease activity. Young et al. (2019) described a targeted approach to identify Cas9-induced off-target mutations, including the novel biochemical method CLEAVE-Seq. Off-target mutations in the Cas9-edited maize plants in which a gRNA is used that was intentionally selected for its potential of inducing off-target edits. They also analyzed the natural variation in maize control plants at the candidate off-target sites and showed single nucleotide polymorphisms (SNPs) were generated at these sites as a result of either inherent or spontaneous variation. They concluded that, with appropriately designed guide RNAs, off-target mutations are likely to be negligible in the background of existing natural variation. Off-target modifications were more often located in protein-coding regions than non-coding regions. In many cases, these protein-coding regions were homologs of the target gene, as shown, for example, barley (Lawrenson et al., 2015) cauliflower, rice (Hua et al., 2020), and wheat (Ren et al., 2019). As plants might have many alleles/ gene homologues with the same function that often show little sequence diversity.

Biased methods involve targeted analysis of possible DNA mutations at selected genome sites that were predicted to be potential off-target sites by in silico methods. The majority of studies (72%, 77/107) made use of specific bioinformatic tools such as CRISPR-P (http://crispr.hzau.edu.cn/CRISPR2/) and Cas-OFFinder (http://www.rgenome.net/cas-offinder/) for the design of gRNAs and prediction of potential off-target sites (Liu et al., 2017). Alternative methods for the prediction of potential off-target sites are based on, for instance, the use of basic local alignment search tools (BLAST) or selection and analysis of close homologs of the on-target site containing the gene.

In 92% (98/107) of the selected studies, predicted off-target sites were amplified by PCR, and the amplicons were further analyzed utilizing sequencing and alignment with reference sequences, or by enzymatic digestion analysis, for the presence of DNA modifications. In the selected studies, certain features of the gRNA and complementary DNA site were specified, such as the maximum number of mismatches allowed between the off-target site and complementary sequence of the gRNA with the presence and maximum size of DNA along with RNA bulges. Li et al. (2019b) used WGS to study the Cas9-induced off-target edits in cotton plants compared to somaclonal variation and inherent genetic variation. The WGS data revealed only four Cas9-induced off-target indels, with mutations present in two out of 182 predicted off-target sites for MYB44 sgRNA2 and two out of 341 for ARC sgRNA1, while higher numbers of spontaneous mutations were observed in an average of 466 SNPs and 77 indels per plant. The CRISPR-Cas system was also often stably integrated into the genome, while there are indications that delivery of the CRISPR gRNA and Cas nuclease molecules directly as a ribonucleoprotein (RNP) complex and not on a DNA vector can reduce the frequency of off-target modifications, as described by Svitashev et al. (2016).

5.6 BIOETHICAL ISSUES

Scientists and experts consider CRISPR-Cas9, the utmost groundbreaking findings of the 21st century. Its rapid development has raised concerns about ecological balance, social, bioethical, and legal issues in medication, farming, farm fauna production, and the environment. Off-target effects must be considered when using CRISPR-Cas9 for RNA-targeted gene expurgation. Gene drift makes off-target mutations likely and might upsurge over generations (Rodriguez, 2016; Hundleby and Harwood, 2019). Inadvertent gene transfer between species may introduce harmful traits to related organisms. Gene property variability across populations makes managing these issues difficult.

CRISPR-Cas9 makes precise genetic modifications, making genetically modified organisms (GMOs) hard to identify and manage once they are released to the market.

Food and Drug Administration (FDA) and European Medicine Agency (EMA) must address GMO safety for consumers. The most effective method for assessing new CRISPR-Cas9 crops remains unclear (Hundleby and Harwood, 2019). Patenting is another CRISPR-Cas9 issue. For therapeutic and industrial usage, many human gene sequences and transgenic organisms have been patented (Rodriguez, 2016).

5.7 THE GENE-EDITED PLANT SAFETY EVALUATION GUIDELINES

The guiding principle for the protection valuation of Gene-edited Plants for Agricultural utilization (Trial) was unconstrained by the Ministry of Agriculture and Rural Affairs of China. This guideline evaluates the safety of gene-edited plants deprived of foreign genes based on peril levels and is required for manufacture and appliance protection certificates. It prepares gene-edited crops for systematic development and deregulation. CRISPR and other gene-editing methods have advanced crop breeding, and future advances are expected to be even superior.

These advancements highlight that genome editing has the probability to significantly change macronutrient profiles, leading to improved nutritional value. The "Gene-edited Plant Safety Evaluation Guidelines" provide a framework for assessing the safety of plants modified using gene-editing technologies, like CRISPR-Cas9. These guidelines aim to address both the potential risks and benefits associated with gene-edited crops.

The "Gene-edited Plant Safety Evaluation Guidelines" are considered to guarantee that gene-edited crops are assessed comprehensively and systematically (Douglas, 2018). By obeying these guidelines, researchers and regulators can help mitigate risks and promote the safe adoption of gene-editing technologies in agriculture.

Important features of the guidelines comprise the following:

1. To evaluate both intended and unintended genetic modifications to estimate their effect on plant safety.
2. To understand how genetic modifications affect plant traits like development, growth, and reproduction.
3. To evaluate the ecological effects of gene-edited plants, as well as their interactions with non-target organisms and potential invasiveness.
4. To assess food Safety for consumption, and allergenicity, toxicity, and nutritional changes, with comparative analyses to conventional varieties to ensure no adverse effects.
5. The guidelines emphasize compliance with regulatory frameworks and a transparent process for documenting and reporting safety evaluations.

5.8 CONCLUSION AND FORTHCOMING SCENARIOS

Scientists are now more precisely and successfully modifying crop-specific features with the development of the CRISPR/Cas9-based gene-editing tool. It aided in the development of several crop types with the ideal agronomic qualities. The development of CRISPR systems has transformed the breeding of crops. Compared to traditional breeding, genome editing techniques offer numerous advantages because of their effectiveness, simplicity, and high specificity. The quality of derived foods and plant performance is impacted by the precise mutants with intriguing properties produced by genome editing techniques. These mutants can also be used to increase resilience to biotic and abiotic stress. Tools for modifying genes are a natural fit for creating more sustainable crops. They provide new avenues for achieving increased food production while conserving a significant number of non-renewable resources (such as water, energy, and soil) through the development of modified crops with higher yields or more effective water usage. Plant genome editing will be able to make a greater contribution to agricultural sustainability and global food security. Biosafety concerns, causing the need for a comprehensive risk assessment framework for CRISPR-edited plants. International collaboration is crucial for establishing unified biosafety guidelines for gene-edited crops, achieving a balance between innovative technological advancements and rigorous safety considerations. One major challenge is the potential for off-target effects, where unintended mutations occur in non-targeted regions of the genome. Off-target modifications by CRISPR-Cas genome editing in plants, provided novel insights into the occurrence, frequency, and nature of off-target edits and need to be evaluated.

It is predicted that gene-editing technology will become more frequently employed in the future and will unavoidably play a significant role in improving crop quality, notwithstanding the remaining obstacles that need to be overcome.

5.9 AI DECLARATION

NO AI tools have been used.

REFERENCES

Abdelrahman, M., Al-Sadi, A. M., Pour-Aboughadareh, A., Burritt, D. J., and Phan Tran, L.-S. (2018). Genome editing using CRISPR/Cas9–targeted mutagenesis: an opportunity for yield improvements of crop plants grown under environmental stresses. *Plant Physiol. Biochem. 131*, 31–36.

Al Murad, M., Khan, A. L., and Muneer, S. (2020). Silicon in horticultural crops: cross-talk, signaling, and tolerance mechanism under salinity stress. *Plants (Basel) 9* (4), E460. doi: 10.3390/plants9040460

Alam, M. S., Kong, J., Tao, R., Ahmed, T., Alamin, M., Alotaibi, S. S., et al. (2022). CRISPR/Cas9 mediated knockout of the OsbHLH024 transcription factor improves salt stress resistance in rice (Oryza sativa L.). *Plants (Basel) 11* (9), 1184. doi: 10.3390/plants11091184

Ali, Z., Abulfaraj, A., Idris, A., Ali, S., Tashkandi, M., and Mahfouz, M. M. (2015). CRISPR/Cas9-mediated viral interference in plants. *Genome Biol. 16*, 238. doi: 10.1186/s13059-015-0799-6

Andersson, M., Turesson, H., Nicolia, A., Fält, A.-S., Samuelsson, M., and Hofvander, P. (2017). Efficient targeted multiallelic mutagenesis in tetraploid potato (*Solanum tuberosum*) by transient CRISPR-Cas9 expression in protoplasts. *Plant Cell Rep. 36*, 117–128. doi: 10.1007/s00299-016-2062-3

Angon, P. B., and Habiba, U. (2023). Application of the CRISPR/Cas9 gene-editing system and its participation in plant and medical science. *Curr. Appl. Sci. Technol. 10*, 55003.

Baltes, N. J., Gil-Humanes, J., and Voytas, D. F. (2017). Genome engineering and agriculture: opportunities and challenges. In: *Progress in Molecular Biology and Translational Science. Vol. 149*. Ed. Weeks, D. P., and Yang, B., San Diego: Academic Press: 1–26.

Baltes, N. J., Hummel, A. W., Konecna, E., Cegan, R., Bruns, A. N., Bisaro, D. M., and Voytas, D. F. (2015). DNA replicons for plant genome engineering. *Plant Cell 26* (1), 15144. doi: 10.1038/nplants.2015.144

Bhat, J. A., Ali, S., Salgotra, R. K., Mir, Z. A., Dutta, S., Jadon, V., et al. (2016). Genomic selection in the era of next generation sequencing for complex traits in plant breeding. *Front. Genet. 7*, 221.

Budiani, A., Putranto, R. A., Riyadi, I., Sumaryono, M. H., and Faizah, R. (2018). Transformation of oil palm calli using CRISPR/Cas9 System: toward genome editing of oil palm. *IOP Conf. Ser. Earth Environ. Sci. 183*, 12003.

Bull, S. E., Seung, D., Chanez, C., Mehta, D., Kuon, J.-E., Truernit, E., et al. (2018). Accelerated ex situ breeding of GBSS- and PTST1-edited cassava for modified starch. *Sci. Adv. 4*, eaat6086. doi: 10.1126/sciadv.aat6086

Carroll, D. (2011). Genome editing: ZFNs, TALENs, and CRISPR-Cas9.*Nat. Methods 8* (12), 957–961. doi: 10.1038/nmeth.2107

Carter, M., and Shieh, J. (2015). Gene delivery strategies. In: *Guide to Research Techniques in Neuroscience* (2nd Edition). Ed. Carter, M., and Shieh, J., London: Academic Press: 239–252.

Cermák, T., Baltes, N. J., Cegan, R., Zhang, Y., and Voytas, D. F. (2015). High-frequency, precise modification of the tomato genome. *Genome Biol. 16*, 232. doi: 10.1186/s13059-015-0796-9

Chandrasekaran, A., Song, Y., and Davis, M. E. (2016). Involvement of non-homologous end joining (NHEJ) in double-strand break repair in cucumber (*Cucumis sativus* L.) and the role of elF4E (eukaryotic translation initiation factor 4E). *Plant Sci. 247*, 112–121. doi: 10.1016/j.plantsci.2016.03.007

Chang, H., Yi, B., Ma, R., Zhang, X., Zhao, H., Xi, Y., et al. (2016). CRISPR/cas9, a novel genomic tool to knock down microRNA *in vitro* and *in vivo*. *Sci. Rep. 6*, 22312. doi: 10.1038/srep22312

Chen, X., Zheng, S., and Li, H. (2023). CRISPR/Cas9-mediated knockout of GhHsfA2 enhances heat stress tolerance in cotton. *Front. Plant Sci. 14*, 103237.

Cheng, Q., Dong, L., Su, T., Li, T., Gan, Z., Nan, H., et al. (2019). CRISPR/Cas9-mediated targeted mutagenesis of GmLHY genes alters plant height and internode length in soybean. *BMC Plant Biol. 19*, 562.

Chu, C., Huang, R., Liu, L., Tang, G., Xiao, J., Yoo, H., et al. (2022). The rice heavy-metal transporter *OsNRAMP1* regulates disease resistance by modulating ROS homoeostasis. *Plant Cell Environ. 45* (4), 1109–1126. doi: 10.1111/pce.14263

Cong, L, Ran, F. A., Cox, D., Lin, S., Barretto, R., Habib, N., et al. (2013). Multiplex genome engineering using CRISPR/Cas systems. *Science 339*. doi: 10.1126/science.1231143

Deng, L., Wang, H., Sun, C., Li, Q., Jiang, H., Du, M., et al. (2018). Efficient generation of pink-fruited tomatoes using CRISPR/Cas9 system. *J. Genet. Genom. 45*, 51–54. doi: 10.1016/j.jgg.2017.10.002

Deutsch, C. A., Tewksbury, J. J., Tigchelaar, M., Battisti, D. S., Merrill, S. C., Huey, R. B., et al. (2018). Increase in crop losses to insect pests in a warming climate. *Science 361* (6405), 916–919. doi: 10.1126/science.aat3466

Dey, A. (2021). CRISPR/Cas genome editing to optimize pharmacologically active plant natural products. *Pharmacol. Res. Perspect., 9* (2), e00785. doi: 10.1002/prp2.785

Doudna, J. A., and Charpentier, E. (2014). The new frontier of genome engineering with CRISPR-Cas9. *Science 346*, 1077–1087.

Douglas, A. E. (2018). Strategies for enhanced crop resistance to insect pests. *Annu. Rev. Plant Biol. 69*, 637–660. doi: 10.1146/annurev-arplant-042817-040248

Doyon, Y., and Livak, K. J. (2009). Targeted genome editing. *Nat. Rev. Genet., 10* (8), 569–580. doi: 10.1038/nrg2638

Filler Hayut, S., Melamed Bessudo, C., and Levy, A. A. (2017). Targeted recombination between homologous chromosomes for precise breeding in tomato. *Nat. Commun. 8*, 15605. doi: 10.1038/ncomms15605

Fu, Y., Foden, J. A., Khayter, C., Maeder, M. L., Reyon, D., Joung, J. K., et al. (2013). High-frequency off-target mutagenesis induced by CRISPR-Cas nucleases in human cells. *Nat. Biotechnol. 31* (9), 822–826. doi: 10.1038/nbt.2623

Gao, C. (2018). The future of CRISPR technologies in agriculture. *Nat. Rev. Mol. Cell Biol. 19* (5), 275–276.

Gilbertson, R. L., Batuman, O., Webster, C. G., and Adkins, S. (2015). Role of the insect supervectors Bemisiatabaci and Frankliniella occidentalis in the emergence and global spread of plant viruses. *Ann. Rev. Virol. 2*, 67–93.

Gong, L., Han, S., Yuan, M., Ma, X., Hagan, A., and He, G. (2020). Transcriptomic analyses reveal the expression and regulation of genes associated with resistance to early leaf spot in peanut. *BMC Res. Notes 13*, 381. doi: 10.1186/s13104-020-05225-9

Hamada, H., Liu, Y., Nagira, Y., Miki, R., Taoka, N., and Imai, R. (2018). Biolistic-delivery-based transient CRISPR/Cas9 expression enables in planta genome editing in wheat. *Sci. Rep. 8* (1), 14422. doi: 10.1038/s41598-018-32714-6

Han, X., Chen, Z., Li, P., Xu, H., Liu, K., Zha, W., et al. (2022). Development of novel rice germplasm for salt-tolerance at seedling stage using CRISPR-Cas9. *Sustainability 14*, 2621. doi: 10.3390/su14052621

Hanley-Bowdoin, L., Bejarano, E. R., Robertson, D., and Mansoor, S. (2013). Geminiviruses: masters at redirecting and reprogramming plant processes. *Nat. Rev. Microbiol. 11* (11): 777–788. doi: 10.1038/nrmicro3117

Hooghvorst, I., López-Cristoffanini, C. and Nogués, S. (2019). Efficient knockout of phytoene desaturase gene using CRISPR/Cas9 in melon. *Sci. Rep. 9* (1), 1–7. doi: 10.1038/s41598-019-53710-4

Hua, K., Tao, X., Liang, W., Zhang, Z., Gou, R., and Zhu, J. K. (2020). Simplified adenine base editors improve adenine base editing efficiency in rice. *Plant Biotechnol. J. 18*, 770. doi: 10.1111/pbi.13244

Hundleby, P. A. C., and Harwood, W. A. (2019). Impacts of the EU GMO regulatory framework for plant genome editing. *Food Energy Secur. 8*, e00161. doi: 10.1002/fes3.161

Jagadish, S. V. K., Way, D. A., and Sharkey, T. D. (2021). Plant heat stress: concepts directing future research. *Plant Cell Environ. 44* (7), 1992–2005. doi: 10.1111/pce.14050

Jeon, J. E., Kim, J. G., Fischer, C. R., Mehta, N., Dufour-Schroif, C., Wemmer, K., et al. (2020). A pathogen-responsive gene cluster for highly modified fatty acids in tomato. *Cell 180* (1), 176–187. e119. doi: 10.1016/j.cell.2019.11.037

Ji, X., Zhang, H., Zhang, Y., Wang, Y., & Gao, C. (2015). Establishing a new plant breeding technology for resistance to beet severe curly top virus using CRISPR/Cas-mediated mutagenesis. *Nat. Biotechnol., 33* (3), 302–304. doi: 10.1038/nbt.3162

Jia, H., Zhang, Y., Orbovic, V., Xu, J., White, F. F., Jones, J. B., and Wang, N. (2017). Genome editing of the disease susceptibility gene CsLOB1 in citrus confers resistance to citrus canker. *Plant Biotechnol. J. 15*, 817–823.

Jiang, W. Z., Henry, I. M., Lynagh, P. G., Comai, L., Cahoon, E. B., and Weeks, D. P. (2017). Significant enhancement of fatty acid composition in seeds of the allohexaploid, *Camelina sativa*, using CRISPR/Cas9 gene editing. *Plant Biotechnol. J. 15*, 648–657. doi: 10.1111/pbi.12663

Jinek, M., Chylinski, K., Fonfara, I., Hauer, M., Doudna, J. A., and Charpentier, E. (2012). A programmable dual-RNA–guided DNA endonuclease in adaptive bacterial immunity. *Science 337* (6096), 816–821. doi: 10.1126/science.1225829

Joshi, R. K., Bharat, S. S., Mishra, R. (2020). Engineering drought tolerance in plants through CRISPR/Cas genome editing. *3 Biotech 10*, 400–414.

Jung, J. H., and Seo, Y. W. (2017). Challenges in wide implementation of genome editing for crop improvement. *J. Crop. Sci. Biotechnol. 20* (2), 129–135.

Kaur, N., Alok, A., Shivani Kumar, P., Kaur, N., Awasthi, P., Chaturvedi, S., et al. (2020). CRISPR/Cas9 directed editing of lycopene epsilon-cyclase modulates metabolic flux for β-carotene biosynthesis in banana fruit. *Metab. Eng. 59*, 76–86. doi: 10.1016/j.ymben.2020.01.008

Kim, D., Alptekin, B., and Budak, H. (2018). CRISPR/Cas9 genome editing in wheat. *Funct. Integr. Genom. 18* (1), 31–41.

Klap, C., Yeshayahou, E., Bolger, A. M., Arazi, T., Gupta, S. K., Shabtai, S., Usadel, B., Salts, Y., and Barg, R. (2017). Tomato facultative parthenocarpy results from SlAGAMOUS-LIKE 6 loss of function. *Plant Biotechnol. J. 15* (5), 634–647.

Kotagama, O. W., Jayasinghe, C. D., and Abeysinghe, T. (2019). Era of genomic medicine: a narrative review on CRISPR technology as a potential therapeutic tool for human diseases. *Biomed. Res. Int. 201*, 1–15. doi: 10.1155/2019/1369682

Kumar, V. S., Verma, R. K., Yadav, S. K., Yadav, P., Watts, A., Rao, M. V., et al. (2020). CRISPR-Cas9 mediated genome editing of drought and salt tolerance (OsDST) gene in indica mega rice cultivar MTU1010. *Physiol. Mol. Biol. Plants*. doi: 10.1007/s12298-020-00819-w. Epub 2020 May 10.

Lawrenson, T., Shorinola, O., Stacey, N., Li, C., Ostergaard, L., Patron, N., Uauy, C., and Harwood, W. (2015). Induction of targeted, heritable mutations in barley and Brassica oleracea using RNA-guided Cas9 nuclease. *Genome Biol. 16*, 258. doi: 10.1186/s13059-015-0826-7

Lemmon, Z. H., Reem, N. T., Dalrymple, J., Soyk, S., Swartwood, K. E., Rodriguez-Leal, D., et al. (2018). Rapid improvement of domestication traits in an orphan crop by genome editing. *Nat Plants 4*, 766–770. doi: 10.1038/s41477-018-0259-x

Li, A., Jia, S., Yobi, A., Ge, Z., Sato, S. J., Zhang, C., et al. (2018). Editing of an alpha-kafirin gene family increases, digestibility and protein quality in sorghum. *Plant Physiol. 177*, 1425–1438. doi: 10.1104/pp.18.00200

Li, C., Nguyen, V., Liu, J., Fu, W., Chen, C., Yu, K., and Cui, Y. (2019b). Mutagenesis of seed storage protein genes in soybean using CRISPR/Cas9. *BMC Res. Notes 12* (1), 176. doi: 10.1186/s13104-019-4207-2

Li, J., Manghwar H., Sun L., Wang P., Wang G., Sheng H., Zhang J., Liu H., Qin L., Rui H., Li B., Lindsey K., Daniell H., Jin S., and Zhang X. (2019a). Whole genome sequencing reveals rare off-target mutations and considerable inherent genetic or/and somaclonal variations in CRISPR/Cas9-edited cotton plants. *Plant Biotechnol. J. 17* (5), 858–868. doi: 10.1111/pbi.13020

Li, Y., Liu, M., and Zhang, X. (2023). Knockout of SlACS2 improves salt tolerance in tomato. *J. Plant Physiol. 274*, 153758.

Liang, G., Zhang, H., Li, Y., and Liu, X. (2014). CRISPR/Cas9-mediated mutagenesis of the maize LPK1, LPK2, and MRP4 genes reveals their roles in phytic acid synthesis. *Biochem. Biophys. Res. Commun. 450* (2), 551–555. doi: 10.1016/j.bbrc.2014.06.138

Liu, H., Ding, Y., Zhou, Y., Jin, W., Xie, K., and Chen, L.-L. (2017). CRISPR-P 2.0: an improved CRISPR-Cas9 tool for genome editing in plants. *Mol. Plant 10* (3), 530–532. doi: 10.1016/j.molp.2017.01.003

Liu, J., Zhang, Y., and Zhao, C. (2020). Overexpression of AtDREB1A enhances drought tolerance in Arabidopsis. *Plant Cell Rep. 39* (2), 215–228.

Liu, L., Zhang, J., Xu, J., Li, Y., Guo, L., Wang, Z., Zhang, X., Zhao, B., Guo, Y.-D., and Zhang, N. (2020a). CRISPR/Cas9 targeted mutagenesis of SlLBD40, a lateral organ boundaries domain transcription factor, enhances drought tolerance in tomato. *Plant Sci. 301*, 110683.

Liu, M., Rehman, S., Tang, X., Gu, K., Fan, Q., Chen, D., and Ma, W. (2019). Methodologies for improving HDR efficiency. *Front. Genet. 9*, 691. doi: 10.3389/fgene.2018.00691

Liu, X., Wu, D., Shan, T., Xu, S., Qin, R., Li, H., et al. (2020b). The trihelix transcription factor OsGTγ-2 is involved adaption to salt stress in rice. *Plant Mol. Biol. 103* (4–5), 545–560. doi: 10.1007/s11103-020-01010-1

Liu, Z., Ma, C., Hou, L., Wu, X., Wang, D., Zhang, L., et al. (2022). Exogenous SA affects rice seed germination under salt stress by regulating Na (+)/K (+) balance and endogenous GAs and ABA homeostasis. *Int. J. Mol. Sci. 23* (6), 3293. doi: 10.3390/ijms23063293

Macovei, A., Sevilla, N. R., Cantos, C., Jonson, G. B., Slamet-Loedin, I., Čermák, T., Voytas, D. F., Choi, I. R., and Chadha-Mohanty, P. (2018). Novel alleles of rice eIF4G generated by CRISPR/Cas9-targeted mutagenesis confer resistance to Rice tungro spherical virus. *Plant Biotechnol. J. 16* (11), 1918–1927. doi: 10.1111/pbi.12927

Mali, P., Yang, L., Esvelt, K. M., Aach, J., Guell, M., DiCarlo, J. E., et al. (2013). RNA-guided human genome engineering via Cas9. *Science 339* (6121), 823–826. doi: 10.1126/science.1232033

Mao, Y., Botella, J. R., Liu, Y., and Zhu, J. K. (2019). Gene editing in plants: progress and challenges. *Natl. Sci. Rev. 6* (3), 421–437. doi: 10.1093/nsr/nwz005

Martignago, D., Rico-Medina, A., Blasco-Escámez, D., Fontanet-Manzaneque, J. B., and Caño-Delgado, A. (2020). Drought resistance by engineering plant tissue-specific responses. *Front. Plant Sci. 10*, 1676.

Miao, C., Xiao, L., Hua, K., Zou, C., Zhao, Y., Bressan, R. A., et al. (2018). Mutations in a subfamily of abscisic acid receptor genes promote rice growth and productivity. *Proc. Natl. Acad. Sci. USA 115*, 6058–6063. doi: 10.1073/pnas.1804774115

Mikami, M., Toki, S., and Endo, M. (2015). Comparison of CRISPR/Cas9 expression constructs for efficient targeted mutagenesis in rice. *Plant Mol. Biol. 88* (6), 561–572. doi: 10.1007/s11103-015-0342-x

Miller, J. C., Tan, S., Qiao, G., Barlow, K. A., Wang, F., Xia, D. F., et al. (2011). A TALE nuclease architecture for highly specific genome editing. *Nature 500* (7461), 423–426. doi: 10.1038/nature12399

Morineau, C., Bellec, Y., Tellier, F., Gissot, L., Kelemen, Z., Nogué, F., et al. (2017). Selective gene dosage by CRISPR-Cas9 genome editing in hexaploid *Camelina sativa*. *Plant Biotechnol. J. 15*, 729–739. doi: 10.1111/pbi.12671

Naeem, M., Majeed, S., Hoque, M. Z., and Ahmad, I. (2020). Latest developed strategies to minimize the off-target effects in CRISPR-Cas-mediated genome editing. *Cells 9* (7), 1608.

Nawaz, G., Usman, B., Peng, H., Zhao, N., Yuan, R., Liu, Y., and Li, R. (2020). Knockout of *Pi21* by CRISPR/Cas9 and iTRAQ-based proteomic analysis of mutants revealed new insights into *M. oryzae* Resistance in Elite Rice Line. *Genes 11*, 735. doi: 10.3390/genes11070735

Ogata, T., Ishizaki, T., Fujita, M., and Fujita, Y. (2020). CRISPR/Cas9-targeted mutagenesis of OsERA1 confers enhanced responses to abscisic acid and drought stress and increased primary root growth under nonstressed conditions in rice. *PLoS One 15* (12), e0243376.

Okuzaki, A., Ogawa, T., Koizuka, C., Kaneko, K., Inaba, M., Imamura, J., et al. (2018). CRISPR/Cas9-Mediated genome editing of the fatty acid desaturase 2 gene in *Brassica napus*. *Plant Physiol. Biochem. 131*, 63–69. doi: 10.1016/j.plaphy.2018.04.025

Ortigosa, A., Gimenez-Ibanez, S., Leonhardt, N., Solano, R. (2019). Design of a bacterial speck resistant tomato by CRISPR/Cas9-mediated editing of *SlJAZ2*. *Plant Biotechnol. J. 17* (3), 665–673. doi: 10.1111/pbi.13006

Pacesa, M., Lin, C. H., Cléry, A., Saha, A., Arantes, P. R., Bargsten, K., et al. (2022). Structural basis for Cas9 off-target activity. *Cell 185* (22), 4067–4081. doi: 10.1016/j.cell.2022.09.026

Peng, A., Chen, S., Lei, T., Xu, L., He, Y., Wu, L., et al. (2017). Engineering canker-resistant plants through CRISPR/Cas9-targeted editing of the susceptibility gene *CsLOB1* promoter in citrus. *Plant Biotechnol. J. 15* (12), 1509–1519. doi: 10.1111/pbi.12733

Pompili, V., Dalla, C. L., Piazza, S., Pindo, M., and Malnoy, M. (2020). Reduced fire blight susceptibility in apple cultivars using a high-efficiency CRISPR/Cas9-FLP/FRT-based gene editing system. *Plant Biotechnol. J. 18* (3), 845–858. doi: 10.1111/pbi.13253

Ran, F. A., Hsu, P. D., Wright, J., Agarwala, V., Scott, D. A., & Zhang, F. (2013). Genome engineering using the CRISPR-Cas9 system. *Nature protocols, 8* (11), 2281–2308. https://doi.org/10.1038/nprot.2013.143

Ren, B., Liu, L., Li, S., Kuang, Y., Wang, J., Zhang, D., Zhou, X., Lin, H., and Zhou, H. (2019). Cas9-NG greatly expands the targeting scope of the genome-editing toolkit by Recognizing NG and other atypical PAMs in rice. *Mol. Plant 12* (7), 1015–1026. doi: 10.1016/j.molp.2019.03.010

Rodriguez, E. (2016). Ethical issues in genome editing using Crispr/Cas9 system. *J. Clin. Res. Bioeth. 7*, 266–266.

Roossinck, M. J., Martin, D. P., and Roumagnac, P. (2015). Plant virus metagenomics: advances in virus discovery. *Phytopathology 105* (6), 716–727. doi: 10.1094/phyto-12-14-0356-rvw

Sanfaçon, H. (2015). Plant translation factors and virus resistance. *Viruses 7* (7): 3392–3419. doi: 10.3390/v7072778

Santosh Kumar, R. K. Verma, S. K. Yadav, P. Yadav, A. Watts, M. V., and Rao, V. C. (2020). CRISPR-Cas9 mediated genome editing of drought and salt tolerance (OsDST) gene in indica mega rice cultivar MTU1010. *Physiol. Mol. Biol. Plants 26*, 1099–1110.

Schloss, P. D., and Handelsman, J. (2004). Status of the microbial census. *Microbiol. Mol. Biol. Rev. 68* (4), 686–691. doi: 10.1128/mmbr.68.4.686-691.2004

Shen, L., Wang, C., Fu, Y., Wang, J., Liu, Q., Zhang, X., et al. (2018). QTL editing confers opposing yield performance in different rice varieties. *J. Integr. Plant Biol. 60*, 89–93. doi: 10.1111/jipb.12501

Shenstone, E., Lippman, Z., and Van Eck, J. (2020). A review of nutritional properties and health benefits of physalis species. *Plant Foods Hum. Nutr. Dordr. Neth. 75*, 316–325. doi: 10.1007/s11130-020-00821-3

Shi, J., Gao, H., Wang, H., Lafitte, H. R., Archibald, R. L., Yang, M., et al. (2017). ARGOS8 variants generated by CRISPR-Cas9 improve maize grain yield under field drought stress conditions. *Plant Biotechnol. J. 15*, 207–216. doi: 10.1111/pbi.12603

Siddiqui, M. N., Mostofa, M. G., Akter, M. M., Srivastava, A. K., Sayed, M. A., Hasan, M. S., et al. (2017). Impact of salt-induced toxicity on growth and yield-potential of local wheat cultivars: oxidative stress and ion toxicity are among the major determinants of salt-tolerant capacity. *Chemosphere 187*, 385–394. doi: 10.1016/j.chemosphere.2017.08.078

Silva, C. J., van den Abeele, C., Ortega-Salazar, I., Papin, V., Adaskaveg, J. A., Wang, D., et al. (2021). Host susceptibility factors render ripe tomato fruit vulnerable to fungal disease despite active immune responses. *J. Exp. Bot. 72* (7), 2696–2709. doi: 10.1093/jxb/eraa601

Smith, M. R., and Myers, S. S. (2018). Impact of anthropogenic CO2 emissions on global human nutrition. *Nat. Clim. Chang. 8*, 834–839. doi: 10.1038/s41558-018-0253-3

Soares, E., Xu, Q., Li, Q., Qu, J., Zheng, Y., Raeven, H. H. M., et al. (2019).Single-cell RNA seq identifies a reversible mesodermal activation in abnormally specified epithelia of p63 EEC syndrome. *Proc. Natl. Acad. Sci. U S A 116*, 17361–17370.

Song, G., Balakrishnan, R., Binkley, G., Costanzo, M. C., Dalusag, K., Demeter, J., et al. (2016). *Integration of New Alternative Reference Strain Genome Sequences into the Saccharomyces Genome Database*. Oxford University Press, article ID baw074.

Songmei, L., Jie, J., Yang, L., Jun, M., Shouling, X., Yuanyuan, T., et al. (2019). Characterization and evaluation of *OsLCT1* and *OsNramp5* mutants generated through CRISPR/Cas9-mediated mutagenesis for breeding low Cd rice. *Rice Sci. 26*, 88–97.

Springmann, M., Mason-D'Croz, D., Robinson, S., Garnett, T., Godfray, H. C. J., Gollin, D., et al. (2016). Global and regional health effects of future food production under climate change: a modelling study. *Lancet 387*, 1937–1946. doi: 10.1016/S0140-6736(15)01156-3

Su, W., Xu, M., Radani, Y., and Yang, L. (2023). Technological development and application of plant genetic transformation. *Int. J. Mol. Sci. 24* (13), 10646. doi: 10.3390/ijms241310646

Sun, Y., Jiao, G., Liu, Z., Zhang, X., Li, J., Guo, X., et al. (2017). Generation of high-amylose rice through CRISPR/Cas9-mediated targeted mutagenesis of starch branching enzymes. *Front. Plant Sci. 8*, 298. doi: 10.3389/fpls.2017.00298

Sun, Y., Zhang, X., Wu, C., He, Y., Ma, Y., Hou, H., Guo, X., Du, W., Zhao, Y., and Xia, L. (2016). Engineering herbicide-resistant rice plants through CRISPR/Cas9-mediated homologous recombination of acetolactate synthase. *Mol. Plant 9*, 628–631. doi: 10.1016/j.molp.2016.01.001

Svitashev, S., Schwartz, C., Lenderts, B., Young, J. K., and Mark Cigan, A. (2016). Genome editing in maize directed by CRISPR-Cas9 ribonucleoprotein complexes. *Nat. Commun. 7*, 13274.

Tanaka, Y., and Ohmiya, A. (2008). Seeing is believing: engineering anthocyanin and carotenoid biosynthetic pathways. *Curr. Opin. Biotechnol. 19*, 190–197. doi: 10.1016/j.copbio.2008.02.015

Tang, L., Mao, B., Li, Y., Lv, Q., Zhang, L., Chen, C., et al. (2017). Knockout of OsNramp5 using the CRISPR/Cas9 system produces low Cd-accumulating indica rice without compromising yield. *Sci. Rep. 7*, 14438. doi: 10.1038/s41598-017-14832-9

Thomazella, D. P. T., Seong, K., Mackelprang, R., Dahlbeck, D., Geng, Y., Gill, U. S., et al. (2021). Loss of function of a *DMR6* ortholog in tomato confers broad-spectrum disease resistance. *Proc. Natl. Acad. Sci. USA 118* (27), e2026152118. doi: 10.1073/pnas.2026152118

Tran, M. T., Doan, D. T. H., Kim, J., Song, Y. J., Sung, Y. W., Das, S., et al. (2021). CRISPR/Cas9-based precise excision of SlHyPRP1 domain(s) to obtain salt stress-tolerant tomato. *Plant Cell Rep. 40* (6), 999–1011. doi: 10.1007/s00299-020-02622-z

Usman, B., Zhao, N., Nawaz, G., Qin, B., Liu, F., Liu, Y., and Li, R. (2021). CRISPR/Cas9 guided mutagenesis of grain size 3 confers increased rice (*Oryza sativa L.*) grain length by regulating cysteine proteinase inhibitor and ubiquitin-related proteins. *Int. J. Mol. Sci. 22* (6), 1–19. doi: 10.3390/IJMS22063225

Vlčko, T., and Ohnoutková, L. (2020). Allelic Variants of CRISPR/Cas9 induced mutation in an inositol trisphosphate 5/6 kinase gene manifest different phenotypes in Barley. *Plants (Basel) 9* (2), E195. doi: 10.3390/plants9020195

Walker, P. T. (1984). Quantification and economic assessment of crop losses due to pests, diseases and weeds. In: *Advancing Agricultural Production in Africa: Cabs First Scientific Conference*, Arusha, Tanzania, 12–18 February.

Wang, F., Wang, C., Liu, P., Lei, C., Hao, W., Gao, Y., Liu, Y. G., and Zhao, K. (2016a). Enhanced rice blast resistance by CRISPR/Cas9-targeted mutagenesis of the ERF transcription factor gene OsERF922. *PLoS One 11* (4), e0154027. doi: 10.1371/journal.pone.0154027

Wang, H., La Russa, M., and Qi, L. S. (2016b). CRISPR/Cas9 in genome editing and beyond. *J. Annu. Rev. Biochem. 85*, 227–264.

Wang, L., Chen, L., Li, R., Zhao, R., Yang, M., Sheng, J., and Shen, L. (2017). Reduced drought tolerance by CRISPR/Cas9-mediated SlMAPK3 mutagenesis in tomato plants. *J. Agric. Food Chem. 65* (39), 8674–8682. doi: 10.1021/acs.jafc.7b02745

Wang, M., Wang, S., Liang, Z., Shi, W., Gao, C., and Xia, G. (2018a). From genetic stock to genome editing: gene exploitation in wheat. *Trends Biotechnol. 36*, 160–172. doi: 10.1016/j.tibtech.2017.10.002

Wang, Q., Huang, Z., and Zhao, Y. (2021). ZmNAC111 overexpression improves cold tolerance in maize. *Plant Sci. 311*, 110876.

Wang, X., Tu, M., Wang, D., Liu, J., Li, Y., Li, Z., … & Wang, X. (2018b). CRISPR/Cas9-mediated efficient targeted mutagenesis in grape in the first generation. *Plant Biotechnology Journal, 16* (4), 844–855.

Wang, Y., Cheng, X., Shan, Q., Zhang, Y., Liu, J., Gao, C., and Qiu, J.-L. (2014). Simultaneous editing of three homoeoalleles in hexaploid bread wheat confers heritable resistance to powdery mildew. *Nat. Biotechnol. 32* (9), 947–951. doi: 10.1038/nbt.2969

Wolter, F., Schindele, P., and Puchta, H. (2019). Plant breeding at the speed of light: the power of CRISPR/Cas to generate directed genetic diversity at multiple sites. *BMC Plant Biol. 19* (1), 176.

Xie, Q., Lu, X., and Yu, T. (2022). Gene editing of HvP5CS enhances drought tolerance in barley. *Mol. Breed. 42*, 47.

Young, J., Zastrow-Hayes, G., Deschamps, S., Svitashev, S., Zaremba, M., Acharya, A., Paulraj, S., Peterson-Burch, B., Schwartz, C., Djukanovic, V., Lenderts, B., Feigenbutz, L., Wang, L., Alarcon, C., Siksnys, V., May, G., Chilcoat, N. D., and Kumar S. (2019). CRISPR-Cas9 editing in maize: systematic evaluation of off-target activity and its relevance in crop improvement. *Sci. Rep. 9* (1), 6729. doi: 10.1038/s41598-019-43141-6

Yue, E., Cao, H., and Liu, B. (2020). *OsmiR535*, a potential genetic editing target for drought and salinity stress tolerance in *Oryza sativa*. *Plants (Basel) 9* (10), E1337. doi: 10.3390/plants9101337

Zafar, S. A., Zaidi, S. S. E. A., Gaba, Y., Singla-Pareek, S., Dhankher, O. M., Li, X. et al. (2020). Engineering abiotic stress tolerance via CRISPR/Cas-mediated genome editing. *J. Exp. Bot. 71* (2), 470–479.

Zaidi, S. S., Tashkandi, M., Mansoor, S., and Mahfouz, M. M. (2016). Engineering plant immunity: using CRISPR/Cas9 to generate virus resistance. *Front. Plant Sci. 7*, 1673. doi: 10.3389/fpls.2016.01673

Zeng, Y., Wen, J., Zhao, W., Wang, Q., and Huang, W. (2019). Rational improvement of rice yield and cold tolerance by editing the three genes *OsPIN5b, GS3*, and *OsMYB30* with the CRISPR-Cas9 system. *Front. Plant Sci. 10*, 1663. doi: 10.3389/fpls.2019.01663

Zhang, P., Du, H., Wang, J., Pu, Y., Yang, C., Yan, R., Yang, H., Cheng, H., and Yu, D. (2020). Multiplex CRISPR/Cas9-mediated metabolic engineering increases soya bean isoflavone content and resistance to soya bean mosaic virus. *Plant Biotechnol. J. 18* (6), 1384–1395. doi: 10.1111/pbi.13302

Zhang, X. H., Tee, L. Y., Wang, X. G., Huang, Q. S., and Yang, S. H. (2015). Off-target effects in CRISPR/Cas9-mediated genome engineering. *Mol. Ther. Nucleic Acids 4* (11), e264.

Zhang, Y., Song, H., and Liu, J. (2021). CRISPR/Cas9-mediated knockout of OsWRKY45 enhances resistance to bacterial blight in rice. *Rice 14*, 34.

Zhang, Z., Xu, J., and Zhang, F. (2022). Mutation of TaPIP2;5 improves drought resistance in wheat. *Front. Plant Sci. 13*, 865476.

Zhao, C., Takeshima, R., Zhu, J., Xu, M., Sato, M., Watanabe, S., et al. (2016). A recessive allele for delayed flowering at the soybean maturity locus *E9* is a leaky allele of *FT2a*, a *Flowering Locus T* ortholog. *BMC Plant Biol. 16*, 20.

Zhou, H., He, Y., Xiong, W., Jing, S., Duan, X., Huang, Z., et al. (2022). MSC-based gene delivery methods and strategies improve the therapeutic efficacy of neurological diseases. *Bioact. Mater. 23*, 409–437. doi: 10.1016/j.bioactmat.2022.11.007

Zhou, J., Peng, Z., Long, J., Sosso, D., Liu, B., Eom, J. S., et al. (2015). Gene targeting by the TAL effector PthXo2 reveals cryptic resistance gene for bacterial blight of rice. *Plant J. 82* (4), 632–643. doi: 10.1111/tpj.12838

Zsögön, A., Čermák, T., Naves, E. R., Notini, M. M., Edel, K. H., Weinl, S., et al. (2018). *De novo* domestication of wild tomato using genome editing. *Nat. Biotechnol. 36*, 1211–1216. doi: 10.1038/nbt.4272

6 Mitigation of Plant Abiotic Stress Via Microbial Applications

Ghada Abd-Elmonsef Mahmoud, Mohamed A. Abdel-Sater, Nemmat A. Hussein, and Eshraq A. AL-Amery

6.1 INTRODUCTION

The habitat of plants is exposed to many potential biological and abiotic environmental challenges. Several microorganisms, like bacteria, viruses, fungi, and other herbivores, are present in wide numbers of various ecosystems, which also greatly reduces the production of crops (Mahmoud et al. 2019, 2021). More than 19,000 fungal varieties worldwide cause crop diseases. They may stay dormant on both live and dead tissues of the plant until conditions are favorable for their reproduction. Spores of fungi can easily spread through soil, wind, insects, water, and other invertebrates. In such a way, they can infect whole crops (Mahmoud 2021). Fungi are responsible for several plant diseases. These diseases involve leaf spot, rust, damping off, root rot, and wilting. Plant pathogens are the main cause of commercial losses and crop quality degradation. Such pathogens can also penetrate plant tissues by secreting toxic microbial poisons, thereby damaging plants (Iqbal et al. 2018; Bizuneh 2021).

In plant pathology, the disease triangle demonstrates the relationship of pathogens and plants with ecosystems. A sensitive plant host, a virulent pathogen, and appropriate environmental factors should be available so that the infection. The influence on the pathogens and plants of environmental variables may possess positive, neutral, or negative effects on plant disease growth. Pathogens and plants have optimal environments that are resistant to pathogens and have the best conditions for growth and reproduction. Resistant plants cannot stand powerless against infection. Therefore, they accumulate anti-microbial secondary metabolites, which are used to combat multiple microbial infections and control the resulting diseases and infections. This wealth of antimicrobial secondary metabolites is commonly referred to as phytoalexins (Bizuneh 2021).

The name phytotoxin was first used by Müller and Börger (1940), who clarified the aforementioned phenomenon in *Phytophthora-contaminated* potato tubers. Phytoalexins have been identified as compounds formed due to parasite invasion by living cells. After infection with *Sclerotinia fructicola* and seven other optional and obligatory phytopathogenic fungi, in 1960, it was isolated from pea (*Pisum sativum*)

 DOI: 10.1201/9781003714484-6

and reported the first characteristic phytoalexin pisatin (Cruickshank and Perrin 1960). The inhibitor growth of momilactones such as A (3-oxo-9β-pimara-7, 15-dien-19, 6β-olide) and B (3,20-epoxy-3α-hydroxy-9β-pimara-7, 15-dien-19, 6β-olide) separated from rice husk for the first time. Later, before rice blast infection, they were found to be phytoalexins in the rice leaves and straws treated with 2,2-dichloro-3,3-dimethylcyclopropanecarboxylic acid. Paxton (1980) identified that phytoalexins are a small molecular weight antibacterial active compound that will be produced and cumulated after the microorganism has infected plants or non-biological initiators, but it is most suitable in the scope of phytoalexins study. Through the years, with the acquisition of new knowledge about phytoalexins. This definition was changed by VanEtten et al. (1994), who insisted that the definition should reflect the need to re-synthesize phytoalexins to distinguish them from phytoanticipins. At present, the definition of phytoalexins is mainly dependent on its production and function, rather than based on chemical structure or biosynthetic source. Recently, research on phytoalexins not solely covers the protection of plant pathogens but also promotes human health (Egbuonu and Eneogwe 2018).

Many different kinds of chemicals reflect phytoalexins from very different plant families. Most phytoalexins developed by legumes belong to six isoflavones. Some of the pterocarpane phytoalexins are also extremely well known. Pisatin compound was the first phytoalexins isolated and defined from pea plants. In addition to such compounds, a small quantity of legumes includes the non-isoflavonoid phytoalexins, like furan-acetylene and stilbenes. Polturak et al. (2023) described a group of isoflavonoid compounds formed by wheat with special reference to isoflavone. Regarding the Solanaceous plant, five main phytoalexins were reported. They were phenylpropane-linked to compounds such as coumarin, steroid glycoalkaloids, and sesquiterpenes (Jeandet et al. 2014; Mahmoud et al. 2021, 2022). According to Sunilkumar et al. (2006), gossypol and its derivatives are considered to be the main Malvaceous phytoalexin. Besides, polyacetylene phytoalexins were isolated from Compositae and Umbelliferae. At the same time, diterpenes and furanosequiterpenes were noted in Convolvulaceae and Euphorbiaceae (Jeandet et al. 2014). Moreover amaryllidaceae produced (flavans and allixin), chenopodiaceae (flavanones, isoflavones, and betavulgarin), asteraceae (polyacetylenes), cruciferae (indole, camalexin, and brassinin), convolvulaceae (furanosesquiterpenes), euphorbiaceae (diterpenes and casbene), poaceae (diterpenoids, phytocassanes, zealexins, kauralexins, flavanones, phenylamides), Leguminosae (isoflavones, pterocarpenes, glyceollin, and stilbenes), Malvaceae (terpenoids), and Solanaceae (phenylpropanoid, steroid, terpenes, sesquiterpenoids, flavonoids, phenols, and coumarins) (Pedras et al. 2000; Sunilkumar et al. 2006; Jeandet et al. 2014; Kilgore and Kutchan 2016; Mahmoud et al. 2021, 2022; Hussein et al. 2022; Polturak et al. 2023) (Table 6.1).

6.2 PHYTOALEXINS ELICITORS

Elicitors are signal molecules that motivate the plant to synthesize phytoalexin compounds. They are chemical compounds that can be obtained from biotic or abiotic sources, which can stimulate plant stress responses, leading to increased or induced secondary metabolites (Naik and Al-Khayri 2016). Elicitor molecules are composed

TABLE 6.1
Common Phytoalexins in Different Plant Families

Plant Families	Phytoalexins	Group
Amaryllidaceae	Flavans	Flavonoids
	Allixin	Pyranones
Linaceae	Phenylpropanoids	Organic compounds
Brassicaceae	Brassinin	Indole
	Camalexin	Indole alkaloid
Chenopodiaceae	Flavanones	Aromatic flavonoids
	Isoflavones	Isomer of flavone
	Betavulgarin	Hydroxyisoflavone
Rutaceae	Methylated phenolic compounds	Phenolics
Convolvulaceae	Furanosesquiterpenes	Terpenoids
Vitaceae	Stilbenes	Phenolics
	Resveratrol	Polyphenoles
Euphorbiaceae	Casbene	Diterpene
Poaceae	Flavanones – sakuranetin	Flavonoids
	Oryzalexins – zealexins – kauralexins – phytocassanes	Terpenoids
Malvaceae	Naphthaldehydes gossypol	Terpenoids
Leguminosae	Isoflavones – pisatin-pterocarpans.	Flavonoids
	Maackiain – glyceollin	Prenylated pterocarpans
	Resveratrol – stilbenes	Phenolics
Orchidaceae	Loroglossol	Dihydrophenanthrenes
Moraceae	Moracins	Furanopterocarpans
Solanaceae	Coumarins	Phenolics
	Norsequi	Sesquiterpenoids
Umbelliferae	Falcarinol	Polyacetylenes
	Xanthotoxin – 6-Methoxymellein	Phenolics
Rosaceae	Auarperin	Biphenyls
	Cotonefurans	Dibenzofurans

of molecules like oligosaccharides, lipids, peptides, and proteins. These molecules can be officially released or created from pathogens that give plants signals (Hussein et al. 2022). Plants detect and recognize these pathogen signals on cell surface receptors, thus activating plant immunity.

Pathogen elicitors are classified into two categories: common and race-specific elicitors. In both host and non-host plants, well-known elicitors may activate defenses, while race-specific elicitors induce defense responses and only cause disease resistance in specific host species. The specific (gene-to-gene) resistance of the variety is determined by the specific pathogen race and another pair of genes in the host variety (Abdul Malik et al. 2020). Therefore, only in host plant varieties carrying the corresponding resistance gene, a variety-specific elicitor encoded or created by the action of virulence genes present in a particular breed of a pathogen will induce resistance. The removal of two gene products normally leads to illness.

Elicitors have existed in the pathogen's cell walls such as chitin, lipopolysaccharide, and glucan. The deletion of two gene products usually leads to disease. Some elicitors can recognize only a small number of plants and are only active in particular hosts. For example, the oligosaccharide elicitor of *Magnaporthe grisea* triggers the synthesis of phytoalexins in rice cells serving as specific hosts instead of soybean cotyledon cells. For example, the oligosaccharide elicitor of *Magnaporthe grisea* triggers the synthesis of phytoalexins in rice cells serving as specific hosts instead of soybean cotyledon cells (Hu et al. 2005). It also was found that using *Trichoderma harizainum* as a biological control agent enhanced phytoalexins production in tomatoes under biotic and abiotic stress (Hussein et al. 2021; Mahmoud et al. 2021).

The primary elicitor and the determinant of virulence in several plant pathogens may be cell wall-degrading enzymes that encourage penetration. These enzymes supply pathogens with nutrients and liberate pectic fragments, which can be used as endogenous elicitors (Abdul Malik et al. 2020). Elicitors of biological origin can be involved in the activity of plants with prospective pathogens. Biological elicitors may originate from invasive organisms and are referred to as exogenous elicitors, or they may be released from plants through interactions between pathogens and plants called endogenous elicitors. Molecule elicitors, such as polysaccharides, glycoproteins, lipids, lipopolysaccharides, oligosaccharides, and activity components enzymes, may have been reported via different structural forms, despite their activity can be attributed to their liberation from pathogens or host cell wall. Several abiotic elicitors are composed of complex molecules generated by unnatural sources like tissues of host or pathogens. Non-biological elicitors consist of abiotic substances that are categorized into chemical, physical, and hormonal parameters, such as fungicides, heavy metals, salts, and inserted DNA reagents (Naik and Al-Khayri 2016). Here, it can be noted that chitin is one of the molecules of a fungal elicitor according to the concept Pusztahelyi (2018). Chitin is an essential part of fungal pathogenicity, and pathogens with malfunctioning chitin production possess greatly decreased toxicity to susceptible hosts. Such molecules can be recognized by plants and stimulate a series of activities in them, which, in turn, lead to various defense responses to fungal infections. Besides, chitin induces ion outflow, generation of reactive oxygen species (ROS), increased levels of chitinase, and increased phytoalexins and hypersensitivity of infected cells (Pérez-Pérez et al. 2012). Chitinase normally exists in plants and deteriorates chitin into soluble compounds and activates immunity. Besides, toxins produced by some pathogens can also act as triggers at low concentrations and cause defensive reactions. For example, the pathogen of rice blast produces α-picolinic acid, which increases resistance to rice blast by inducing hydrogen peroxide production and death of cells (Zhang et al. 2004).

Fungi have unique molecular patterns that lead to plant defenses. The bulk of the molecular structures are either part of the pathogen structure or exudate. Through the elicitor-receptor relationship, it is very significant to understand the behavior between these species and the host to provide us with the knowledge we need to establish sustainable plant resistance. Likewise, it is important to study the role of beneficial microorganisms and synthetic elicitors in activating plant defenses and disease resistance (Abdul Malik et al. 2020).

6.3 PHYTOALEXINS STIMULATION MECHANISMS

The outer surface that prevents pathogens from entering the plant is the initial line of protection for plants. Cutin, cork, and wax are used to protect the stems and leaves that are exposed to the outer environment. These compounds not only avoid water loss due to evaporation but also block the entry of fungal spores into the implied tissues. Plant cell walls are also an important physical barrier to pathogens, which can be further enhanced by inducing defense reactions (Mahmoud et al. 2021).

Plant pathogens stimulate host plants to trigger a variety of chemical defense reactions. A variety of molecules that trigger molecular defenses, called elicitors, have been discovered. These molecules range from peptides produced by bacteria to fragments of fungi on the cell wall. The initiator may be produced from fragments of components of plant cell walls that are digested via pathogens. Currently studying the plant immune system includes two branches (Jones and Dangl 2006). The plant transmembrane pattern recognition (PRR) receptor recognizes and induces a reaction to a strongly retained, slowly evolving pathogen-related molecular pattern receptor in the first station. Transmembrane receptors named the signal confession receptors, which activate signaling routes leading to general immunity, are identified by PAPM. This stimulates the immunity caused by PAMP (PTI) and can significantly stimulate the defensive response of plants without the clear priority of the pathogen. PRR contains receptor-like protein (RLP) and receptor-like kinase (RLK). While PAMPs are highly preserved a recent study showed that different plant species, such as long and short b-1,3-linked glucans, may recognize various PAMPs (Wanke et al. 2020). This, in turn, increases the likelihood among plant species of different PRR variants. In general, the responses of multiple plant species to PAMP/effector combinations must be extensively studied to how these hormone signaling approaches work to influence the induction of phytoalexins. Since PRR and NB-LRR have a direct effect on protein phosphorylation and kinase signal transduction, hormones are likely to influence kinase signal transduction as they are widely involved in phytoalexin signal transduction (Ahmed and Kovinich 2020).

The second station, known as effector-mediated immunity, is caused by effector proteins unique to certain organisms or races of pathogens. To prioritize (speed up/increase in frequency) unique defense responses over others, effector-induced immunity acts on PTI. There are numerous effector proteins that are Vector-encoded. Proteins from the effector have typically evolved to inhibit PTI, creating a triggered susceptibility effector. In order to identify unique effector proteins and prevent effector-induced immunity, plant resistance genes have evolved. The R gene usually encodes a nucleotide-binding leucine-rich repeat sequence (NBLRR) protein. The function of the NB-LRR protein is to block effector activity, such as binding to PRR (Dangl and Jones 2001; Gao et al. 2005).

In plant genomes, there are several R genes, and the proteins they encode can identify effector proteins with specificity from gene to gene. However, typically, pathogen detection involves stimulating a complex network of partly overlapping PAMP and ETI-stimulated signal networks (Wu et al. 2018). Plants generate many PR proteins or proteins associated with the disease process. The cell walls of pathogenic fungi are dissolved by certain enzymes. In this case, chitinase enzyme demolishes chitin in fungal cell walls. In addition, the degradation products of the cell wall of the pathogen may activate a defensive response. PR proteins may send indications to plant cells that are not invaded so that they can initiate their defense responses. PR protein is not a weapon for rapid response. On the contrary, its action may be slower, maybe when the pathogen attack has been weakened by other mechanisms (Abdel-Ghany et al. 2015). Polymers such as lignin are sedimented on the cell wall to strengthen the cell wall. These polymers can also block cytoplasmic tumors and reduce the ability of pathogens to disperse between cells via such channels. Changes in gene cascade lead to a turn in the expression of the gene. The regulated genes involve disease-connected genes (PR) and genes encoding phytoalexin that produce antimicrobial compounds (Abdul Malik et al. 2020).

6.4 THREE MAIN BIOSYNTHETIC PATHWAYS OF PHYTOALEXINS

6.4.1 Phenylpropanoic-Polymalonic Acid Pathway

This pathway is used to biosynthesize flavonoids, stilbene phytoalexins, and their derivatives. It starts with lower phenylalanine and/or tyrosine. The p-coumarin acid is activated by 4-coumarin-CoA ligase to p-coumarin-CoA. Naringin chalcone has led to the production of all stilbene precursors. Biosynthetic pathways for phytoalexin from main flavonoids and stilbene in legumes are shown in Figure 6.1 (Jeandet et al. 2014; Tiku 2020).

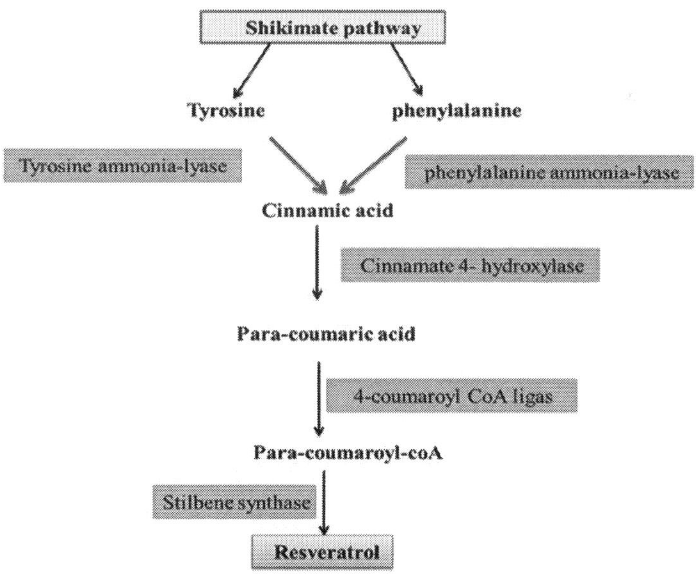

FIGURE 6.1 Resveratrol phytoalexin biosynthesis via the phenylpropanoid-polymalonate pathway.

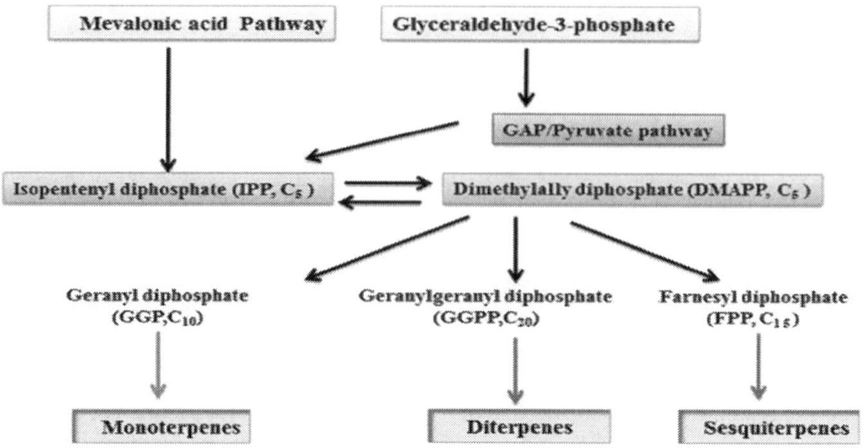

FIGURE 6.2 Biosynthetic pathway of terpenoid phytoalexins.

6.4.2 Mevalonoid Pathway

This path is necessary for the production of monoterpenes, diterpenes, and sesquiterpenes (Figure 6.2). Plant cells produce isopentenyl diphosphate (IPP) and allyl isomer dimethylallyl diphosphate (DMAPP) in one of two ways: (1) the mevalonate or (2) the currently identified methylerythritol phosphate (MEP) pathway. Then, the C5 basic units are compacted for producing three greater isoprenyl diphosphates, namely geranyl diphosphate, farnesyl diphosphate, and geranylgeranyl diphosphate. The geranyl and geranylgeranyl diphosphate require various rearrangements to produce the carbon skeleton of every terpenes. Geranylgeranyl diphosphate is transformed into monoterpenes, farnesyl diphosphate is turned to sesquiterpenes, and geranylgeranyl diphosphate is converted to diterpenes (Jeandet et al. 2014; Tiku 2020).

6.4.3 Pathway for Biosynthesis of Indole Phytoalexins

The camalexin indole ring is obtained from tryptophan. Camalexin is regulated by the P450 homologs CYP79B2 and CYP79B3 cytochromes to produce indole-3-acetaldoxime. Then, in the presence of CYP71A13 and P450, indole 3-acetaldoxime becomes indole 3-acetonitrile (IAN). The consequent binding of IAN and glutathione is achieved through the combined impact of glutathione transferase and cytochrome P450. Then, under the action of two γ-glutamyl transpeptidases 1 and 3 Acyl-cysteine IAN, the IAN glutathione transforms into IAN cysteine-glycine via plant chelatase synthase, or into γ-glutamine. The two complexes contribute to the binding of IAN cysteine. The last step is regulated by the CYP71B15 gene, which contains a multifunction of enzymes that produce kamatoxin through the dihydrocarbamate. This pathway primarily synthesizes camalexin in the Brassicaceae family (Figure 6.3) (Jeandet et al. 2014; Tiku 2020).

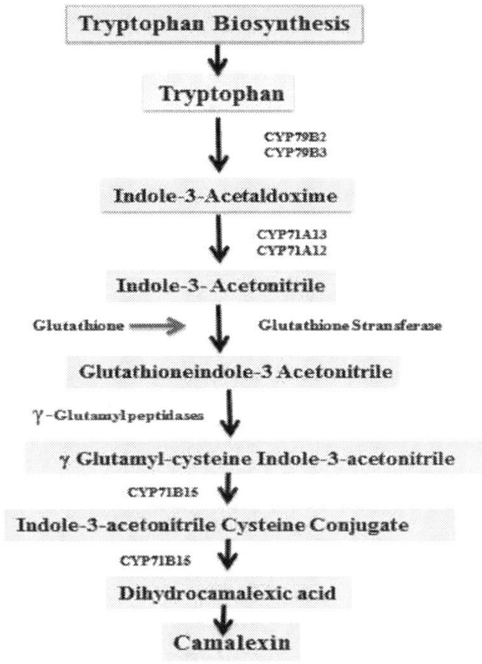

FIGURE 6.3 Biosynthetic pathway from tryptophan to camalexin.

6.5 ANTIMICROBIAL ROLES OF PHYTOALEXINS

Through the granulation of cytoplasm, disorder of cell contents, rupture of plasma membrane, inhibition of spore germination, and fungal cell enzymes, the fungal toxicity of phytoalexins is proved. It may also modify the morphology and physiological features of the fungal cells (Arruda et al. 2016). The *in vivo* activity of some coumarins and scopoletin in reducing of the symptoms of green mold caused by *Penicillium digitatum* in orange was studied (Sanzani et al. 2014). Similarly, resveratrol, scopoletin, and scoparone have been seen to considerably block the *Penicillium* growth and patulin cumulation in apples (Sanzani et al. 2014). In the spores of *Botrytis cinerea* treated with sublethal doses of resveratrol, asymmetric growth of germ tubes was also shown, resulting in the formation of "curved embryo tubes." The conidia of *Botrytis cinerea* was observed after treatment with stilbene and resveratrol, the mitochondria were completely disordered and the plasma membrane was destroyed (Adrian and Jeandet 2012). Furthermore, phytoalexins have antibacterial activity. For example, Rishinin reduces the viability of *Erwinia* cells. Also, resveratrol has activity against so many bacteria that infect humans, such as *Helicobacter pylori, Staphylococcus*, *Streptococcus*, *Pseudomonas*, and *Neisseria* (Jeandet et al. 2012) (Table 6.2). Therefore, it is clear that phytoalexins are toxic throughout most of the prokaryotic and eukaryotic (Zhang et al. 2015; Arruda et al. 2016).

TABLE 6.2
Antimicrobial Activities of Phytoalexins

Phytoalexins	Pathogen	Reference
Resveratrol	*Botrytis cinerea*	Adrian and Jeandet (2012)
	Helicobacter pylori, *Staphylococcus* sp., *Streptococcus* sp., *Pseudomonas* sp., and *Neisseria* sp.	Jeandet et al. (2012)
Coumarins Scopoletin	*Penicillium digitatum*	Sanzani et al. (2014)
Phenols Flavonoids	*Salmonella typhimurium*, *S. aureus*, *E. coli*, and *Listeria ivanovii*	Silva-Beltrán et al. (2015)
Flavonoids	*Colletotrichum* species	Schnippenkoetter et al. (2017)
Stilbenoids	*Erysiphe necator*, *B. cinerea*, *Aspergillus* sp., *Rhizopus*, and *Plasmopara viticola*	Jeandet et al. (2019)
Phenols Flavonoids	*Fusarium oxysporum*, *Glomerella cingulata*, and *Rhizoctonia solani*.	Kim et al. (2019)
Momilactones	*Bacillus cereus*, *E. coli*, *Pseudomonas* sp., *B. cinerea*, and *F. oxysporum*	Bizuneh (2020)
Indole phytoalexins	*Alternaria brassicicola*	Pastorczyk et al. (2020)
Phenols Flavonoids	*E. coli*, *Pseudomonas aeruginosa*, *Staphylococcus aureus*, *S. epidermidis*, *C. albicans*, *C. tropicalis*, *F. oxysporum*, *F. solani*, *and P. digitatum*	Mahmoud et al. (2021)

6.6 PHYTOALEXINS IN HUMAN DISEASES CONTROL

Based on the above-mentioned large number of observations, phytoalexins are manifested in a variety of plants as a variety of antifungal compounds or antioxidants, anticancer and cardioprotective agents, and are used in neurodegenerative diseases. In particular, the future application of phytoalexins in the field of health-promoting humans has been emphasized.

6.6.1 Phytoalexins as Antimicrobial Agents

Phytoalexins have antimicrobial potential. They accumulate in plants to inhibit microorganisms. Therefore, they can be used to resist pathogenic microorganisms and control infectious diseases that result (Egbuonu and Eneogwe 2018). Nowadays, many plant compounds include reports of antibiotic activity against a wide range of pathogenic fungi. Many types of antimicrobial drugs have been recorded, which include flavonoids, polyphenols and phenolic acids, phenanthrene, and terpenoids (Sharifi-Rad et al. 2018). The plant processing process of phytoalexins contributes to the production of active antibiotics in the infected site after the detection of the pathogen, which can be used to treat human illnesses caused by microbes (Bamji and Corbitt 2017). For example, the main phytoalexin in tobacco plants hydroxycoumarin scopoletin, epigallocatechin-gallate is related to the catechin in the bacterial wall. In addition, the phytoalexin in the extract of cranberry and garlic has antibacterial effects against a variety of pathogens (Puupponen-Pimiä et al. 2005).

Carvacrol is a monoterpene present in essential oil plants such as thyme, oregano, and pepperwort. It has a variety of bioactivities, especially for antibacterial agents and antioxidants, which are beneficial for medicinal applications (Sharifi-Rad et al. 2018). Many investigations have been performed to study the effect of carvacrol on different bacteria and fungi, especially plant pathogens, and their mechanisms of activity (Liu et al. 2019). The momilactones in rice have antibacterial activity against many bacterial species, such as *Bacillus cereus*, *E. coli*, and *Pseudomonas*, and they have antifungal action against *Botrytis cinerea*, *Fusarium oxysporum*, and *Fusarium* wilt (Bizuneh 2020).

6.6.2 Phytoalexins as Anticancer Agents and Anti-Inflammatory

Phytoalexins can have an impact on humans. Some of them have antioxidant properties and anti-tumor and anti-inflammatory behaviors, thereby enhancing the immune system. The Pezzuto research team was the first to demonstrate the chemical safety of phytoalexin. In several other human cancer models, this groundbreaking work was subsequently verified. Phytoalexin can inhibit cancer tumor growth due to processes induced by inflammatory reactions involving cytokine and chemokine activity, reactive oxygen species, and nitrogen compounds, which may also contribute to the emergence of abnormal cells (Zigová et al. 2024). The 3-deoxy anthocyanidin flavonoid phytoalexin compound produced by the Gramineae family helps minimize the occurrence of gastrointestinal tumors (Yang et al. 2009). Besides, indole phytoalexins, such as brassinin display considerable *in vitro* anti-proliferative action against colon cancer (Kello et al. 2014). Resveratrol has anti-tumor activity in the body, especially in local skin cancer. Although it has anti-leukemia effects in vitro, it does not seem to be sufficient to inhibit leukemia. It only has anti-cancer effects in laboratory experiments that cause breast cancer in large doses. Resveratrol has certain anti-cancer activity in the liver, lung, and gastrointestinal tract (Jeandet et al. 2012). Carvacrol is utilized in conventional drugs to treat many illnesses, such as anti-cancer and anti-inflammatory medications (Sharifi-Rad et al. 2018). With poor mechanized mechanisms, Phytoalexins are often used as anti-inflammatory drugs, including inhibition of inducible nitric oxide synthesis, which involves the process of inducible NOS for producing large amounts of nitric oxide during inflammation (Kowluru et al. 2004). Resveratrol can prevent inflammation. By inhibiting LPS-related signaling pathways, inhibiting phosphorylation, or reversing the degree and production and necrosis agent alpha of liver tumors to express the response induced by LPS lipopolysaccharide in colon cancer cells (Panaro et al. 2012). As anti-inflammatory, anti-cancer and an antioxidant, resveratrol (trans 3,5,4′-trihydroxystylbenzene), cis and trans piperidine, stilbenoid, and resveratrol dimers, trimers and tetramers have demonstrated significant potential in modulating cellular signaling pathways, reducing oxidative stress, and inhibiting tumor progression. They can also protect the liver, avoid elevated plasma glucose and triglycerides, and effectively fight Alzheimer's disease (Flamini and De Rosso 2019).

6.6.3 Phytoalexins as Anti-Apoptotic and Cardioprotective Agents

Phytoalexins can promote health in humans. It is proven that many of them have antioxidant, anticarcinogenic, and cardiovascular effects (Jeandet 2018; Sharma et al. 2022). Many current studies have provided proof of phytoalexins in cardioprotective agents such as stilbenes. In the study of isolated rat hearts, it has been observed that resveratrol can inhibit low-density lipoprotein peroxidation, has an effective role in preventing atherosclerosis, and can prevent platelet clumping in rabbits on a high-cholesterol diet (Jeandet et al. 2012). The antitumor, antidiabetic, antioxidant, heart and nerve protective, stimulator of defense, antiparasitic, and animal models of maslinic acid that exist in several productive resources (spices, edible vegetables, and fruits) have been studied by Lozano-Mena et al. (2014). It demonstrates important antiproliferative activity *in vitro*, such as brassinosteroids and their derivatives. Resveratrol can act as a cardioprotective agent against apoptosis by reducing the action of 3-caspase. Interestingly, earlier research has found that the cardiotoxicity of adriamycin is improved after resveratrol injection (Sin et al. 2015).

6.6.4 Phytoalexins in Oxidative Stress

Oxidative damage and increased free radical production play a crucial role in the pathogenesis of cardiotoxicity. By increasing the amount of superoxide dismutase and reducing the ratio of malondialdehyde, Phytoalexins minimize oxidative stress, thereby avoiding the cardiotoxicity caused by Adriamycin (Tatlidede et al. 2009; Sharma et al. 2022). It was found that the resveratrol may protect the lipids from the degradation of peroxide and inhibit the concentration-dependent absorption of LDL cholesterol oxidized to the blood vessel wall. In order to increase apoptosis, it interacts with molecular mechanisms that control oxidative stress, particularly, kinases and transcription factors. Probable resveratrol functions, such as the avoidance of genetic mutations as well as the inhibition of DNA synthesis, have been documented (Egbuonu et al. 2018). Scientific research has shown that resveratrol in the diet can diminish mitochondrial dysfunction and oxidation caused by aging and exercise (Sun et al. 2012). The potential application of phytoalexins would have a huge effect on human ailments through the regulation of chronic inflammation (Egbuonu and Oriji 2017).

6.7 MICROBIAL DETOXIFICATION OF PHYTOALEXINS BY PHYTOPATHOGENIC FUNGI

Antimicrobial mixtures called phytoalexins, which are plant-specific ingredients produced in response to microbial infections, despite this chemical defense, Pathogenic microbes can still destroy plants through enzymes that detoxify these plants quickly (Pedras et al. 2017). There was no due attention given to the ability of pathogenic microorganisms to detoxify phytoalexins into fewer dangerous chemicals that may inhibit the development of defense reactions (Jeandet 2018). Some fungal pathogens have been used in various reaction forms to detoxify phytoalexins, such as *Botrytis cinerea*, *Fusarium solani*, and *Sclerotinia sclerotiorum*. The sort of reaction conducted may be hydrolysis, oxidation, or reduction, determined by the type of pathogen and the chemical composition of the phytoalexin (Pedras et al. 2011). For instance, the main role of brassinosteroids is essential for plant adaptability, whether as active cruciferous phytoalexin, or as a biosynthetic intermediate of some other plant defenses. The detoxification of the brassinin into beta-indolylaldehyde will catalyze by brassinin oxidase, a detoxification enzyme generated by *Phoma lingam* (Jeandet 2018). The main inducers and plant resistance associated with disease progression are phenylphenalenones. However, it is possible to treat harmful strains of *Mycosphaerella fijiensis* (black leaf) by converting phenylphenalenones into a sulfate conjugate to overcome the problem of plant resistance (Hidalgo et al. 2016).

Camalexin phytoalexin was isolated from *Camelina sativa* crucifer leaves contaminated with *Alternaria brassicae*. It has been found in other members of the crucifers, also present in Arabidopsis and several different *Brassicaceae* species since it offers a special model for the defense response analysis of the phytoalexins (Bednarek et al. 2011). Camalexin was detoxified by some fungal species, *A. brasicicola*, *Botrytis cinerea*, *Rhizoctonia solani*, and *Sclerotinia sclerotiorum*. The hydroxylation of the indole ring or oxazoline derivative formation, deteriorates camalexin into glycosylated derivatives of indole-3-carbaldehyde (Pedras and Ahiahonu 2002). In each fungal species, the conversion of camalexin is different, and the end product becomes lower toxicity to the developing species than camalexin (Pedras and Khan 2000). Another problem with the use of phytoalexins in plant defense is that some pathogens tend to detoxify phytoalexins into other compounds. To resolve this problem, the discovery and understanding of phytoalexins detoxification enzyme inhibitors are necessary. Paldoxins, a new generation of chemicals originating from detoxification inhibitors of phytoalexins, offer up a whole world of biological processes aimed at providing safe handling of crops (Pedras et al. 2008). The use of paldoxin (Figure 6.4) will make natural plant defenses safer in the environment because the destructive power of priority inhibitors is getting weaker (Pedras and

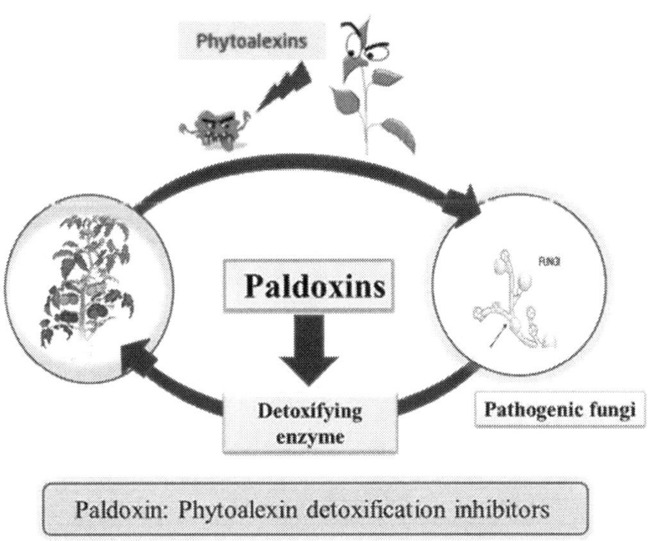

FIGURE 6.4 Detoxification of phytoalexins using phytopathogenic fungi.

Abdoli 2017). To humans, plants, or microorganisms, paldoxin is harmless. In addition, paldoxin mixtures may be engineered to function synergistically with plants' normal disease resistance factors to effectively preserve crops with minimal ecological effects (Pedras and Abdoli 2017).

6.8 CONCLUSION AND FUTURE PROSPECTS

In general, phytoalexins have recently received special attention as a health-promoting effect on humans. Although they have important antimicrobial effects. However, according to the knowledge collected in this article, there is an imperfection in some domains relevant to phytoalexin. There is a gap in information on the potential applications of phytoalexins in biotechnology. As it has been stated, phytoalexins have enormous impacts due to their chemical diversity in various plant sources. So the importance of phytoalexins includes both existing applications and potential compounds, providing an overview of phytoalexins and an in-depth insight into the future. Comprehensive scientific experimental studies must be carried out to profit from them on a broad scale and easy means of manufacturing phytoalexins on an industrial scale. It is proposed that future research should explore valuable future applications of phytoalexins. Besides, the protection of such phytoalexins must be promoted at sufficient doses to determine their possible use in human treatment. One of the most important topics that needs to be paid intensive attention to in future studies is the early forecasting of the likelihood of infection. That's when it's achieved; the plant will be alerted promptly to produce phytoalexins.

ARTIFICIAL INTELLIGENCE STATEMENT

The authors have not used any artificial intelligence tool during the writing of the chapter.

REFERENCES

Abdel-Ghany S.E., Golovkin M., Reddy A.S.N. 2015. Engineering of plants for the production of commercially important products: approaches and accomplishments. *Plant Biotechnol.* 11, 551–577.

Abdul Malik N.A., Kumar I.S., Nadarajah K. 2020. Elicitor and receptor molecules: orchestrators of plant defense and immunity. *Int. J. Mol. Sci.* 21(3), 1–34.

Adrian M., Jeandet P. 2012. Effects of resveratrol on the ultrastructure of *Botrytis cinerea* conidia and biological significance in plant/pathogen interactions. *Fitoterapia* 83(8), 1345–1350.

Ahmed S., Kovinich N. 2020. Regulation of phytoalexin biosynthesis for agriculture and human health. *Phytochem. Rev.* 20, 1–23.

Arruda R.L., Paz A.T.S., Bara M.T.F., Côrtes M.V.D.C.B., Filippi M.C.C.D., Conceição E.C.D. 2016. An approach on phytoalexins: function, characterization and biosynthesis in plants of the family Poaceae. *Ciênc. Rural* 46(7), 1206–1216.

Bamji S.F., Corbitt C. 2017. Glyceollins: soybean phytoalexins that exhibit a wide range of health-promoting effects. *J. Funct. Foods* 34, 98–105.

Bednarek P., Piślewska-Bednarek M., Ver Loren van Themaat E., Maddula R.K., Svatoš A., Schulze-Lefert P. 2011. Conservation and clade-specific diversification of pathogen-inducible tryptophan and indole glucosinolate metabolism in Arabidopsis thaliana relatives. *New Phytol.* 192(3), 713–726.

Bizuneh G.K. 2020. The chemical diversity and biological activities of phytoalexins. *Adv. Tradit. Med.* 18, 1–13.

Bizuneh G.K. 2021. The chemical diversity and biological activities of phytoalexins. *Adv. Tradit. Med. (ADTM)* 21, 31–43.

Cruickshank I.A.M., Perrin D.R. 1960. Isolation of a phytoalexin from Pisum sativum L. *Nature* 187(4739), 799–800.

Dangl J.L., Jones J.D. 2001. Plant pathogens and integrated defence responses to infection. *Nature* 411, 826–833.

Egbuonu A.C., Oriji S.O. 2017. Pulverized Mangifera indica (mango) seed kernel mitigated monosodium glutamate-intoxicated rats' kidney histology and bio-functions. *J. Nutrit. Health Food Sci.* 5, 1–7.

Egbuonu A.C.C., Eneogwe J.C. 2018. Phytoalexins: current and possible future applications in human health and diseases control. *IJMBOA* 3, 105–110.

Egbuonu, A. C. C., Opara, I. C., Onyeabo, C., & Uchenna, N. O. 2018. Proximate, functional, antinutrient and antimicrobial properties of avocado pear (Persea americana) Seeds. *J. Nutr. Health Food Eng.* 8(2), 00260.

Flamini R., De Rosso M. 2019. High-resolution mass spectrometry and biological properties of grapevine and wine stilbenoids. *Stud. Nat. Prod. Chem.* 61, 175–210.

Gao H., Narayanan N.N., Ellison L., Bhattacharyya M.K. 2005. Two classes of highly similar coiled coil-nucleotide binding-leucine rich repeat genes isolated from the Rps1-k locus encode Phytophthora resistance in soybean. *Mol. Plant-Microbe Interact.* 18, 1035–1045.

Hidalgo W., Chandran J.N., Menezes R.C., Otálvaro F., Schneider B. 2016. Phenylphenalenones protect banana plants from infection by Mycosphaerella fijiensis and are deactivated by metabolic conversion. *Plant Cell Environ.* 39, 492–513.

Hu G., DeHart A.K., Li Y., Ustach C., Handley V., Navarre R., Baker B. 2005. EDS1 in tomato is required for resistance mediated by TIR-class R genes and the receptor-like R gene Ve. *Plant J.* 42, 376–391.

Hussein N.A., Abdel-Sater M.A., AL-Amery E., Mahmoud G.A.-E. 2021. Effect of Trichoderma harzianum on enhancing the defense ability of tomato plants against Alternaria cerealis MT808477. *Assiut Univ. J. Bot. Microbiol.*, 50, 67–92.

Hussein N.A., Abdel-Sater M.A., AL-Amery E., Mahmoud G.A.-E. 2022. New bioactive compounds and phytoalexins from tomato plant (Solanum lycopersicum L.) as red card against phytopathogenic Alternaria cerealis MT808477. *J. Basic Appl. Mycol.* 13, 1–11.

Iqbal Z., Khan M.A., Sharif M., Shah J.H., Rehman M.H., Javed K. 2018. An automated detection and classification of citrus plant diseases using image processing techniques: a review. *Comput. Electron. Agric.* 153, 12–32.

Jeandet P. 2018. Structure, chemical analysis, biosynthesis, metabolism, molecular engineering, and biological functions of phytoalexins. *Molecules*, 23(61), 1–4.

Jeandet P., Clément C., Cordelier S. 2019. Regulation of resveratrol biosynthesis in grapevine: new approaches for disease resistance? *J. Exp. Bot.* 70(2), 375–378.

Jeandet P., Delaunois B., Aziz A., Donnez D., Vasserot Y., Cordelier S., Courot E. 2012. Metabolic engineering of yeast and plants for the production of the biologically active hydroxystilbene, resveratrol. *J. Biomed. Biotechnol.* 2(3), 1–14.

Jeandet P., Hébrard C., Deville M.A., Cordelier S., Dorey S., Aziz A., Crouzet J. 2014. Deciphering the role of phytoalexins in plant-microorganism interactions and human health. *Molecules* 19(11), 18033–18056.

Jones J.D., Dangl J.L. 2006. The plant immune system. *Nature* 444(7117), 323–329.

Kello M., Drutovic D., Chripkova M., Pilatova M., Budovska M., Kulikova L., Mojzis J. 2014. ROS-dependent antiproliferative effect of brassinin derivative homobrassinin in human colorectal cancer Caco2 cells. *Molecules* 19(8), 10877–10897.

Kilgore M.B., Kutchan T.M. 2016. The Amaryllidaceae alkaloids: biosynthesis and methods for enzyme discovery. *Phytochem. Rev.* 15(3), 317–337.

Kim D.S., Kwack Y., Lee J.H., Chun C. 2019. Antimicrobial activity of various parts of tomato plants varied with different solvent extracts. *Plant Pathol. J.* 35, 149–155.

Kowluru R.A., Abbas S.N., Odenbach S. 2004. Reversal of hyperglycemia and diabetic nephropathy: effect of reinstitution of good metabolic control on oxidative stress in the kidney of diabetic rats. *J. Diabetes Complicat.* 18(5), 282–288.

Liu Q., Qiao K., Zhang S. 2019. Potential of a small molecule carvacrol in management of vegetable diseases. *Molecules* 24(10), 1–8.

Lozano-Mena G., Sánchez-González M., Juan M.E., Planas J.M. 2014. Maslinic acid, a natural phytoalexin-type triterpene from olives a promising nutraceutical. *Molecules* 19(8), 11538–11559.

Mahmoud G.A.-E. 2021. Biotic stress to legumes: fungal diseases as major biotic stress factor. In: Guleria P., Kumar V., Lichtfouse E. (eds) *Sustainable Agriculture Reviews*, vol 51. Springer, Cham.

Mahmoud G.A.-E., Abdel-Sater M.A., AL-Amery E., Hussein N.A. 2021. Controlling Alternaria cerealis MT808477 tomato phytopathogen by trichoderma harzianum and tracking the plant physiological changes. *Plants* 10, 1846.

Mahmoud G.A.-E., Abdel-Sater M.A., AL-Amery E., Hussein N.A. 2022. Production of terpenoid phytoalexins as a plant response against *Alternaria cerealis* AUMC 14484 controlled by *Trichoderma harzianum*. *J. Basic Appl. Mycol.* 13, 12–33.

Mahmoud G.A.-E., Zidan A.S., Aly A.M., Mosbah H.K., Ibrahim A.B.M. 2019. Calcium and strontium anthranilato complexes as effective Fusarium moniliforme controlling agents. *Appl. Organomet. Chem.* 33(2), e4740.

Müller K.O., Börger H. 1940. Experimentelle Untersuchungen über die Phytophtora-Resistenz der Kartoffel. *Arbeiten aus der Biologischen Reichsanstalt für Land- und Forstwirtschaft, Berlin-Dahlem* 23, 189–231. Berlin-Dahlem: Biologische Reichsanstalt.

Naik P.M., Al-Khayri J.M. 2016. Abiotic and biotic elicitors–role in secondary metabolites production through in vitro culture of medicinal plants. In: Shanker A.K., Shanker C. (eds) *Abiotic and Biotic Stress in Plants-Recent Advances and Future Perspectives*, 247–277. InTech, Rijeka, Croatia.

Panaro M.A., Carofiglio V., Acquafredda A., Cavallo P., Cianciulli A. 2012. Anti-inflammatory effects of resveratrol occur via inhibition of lipopolysaccharide-induced NF-KB activation in Caco-2 and SW480 human colon cancer cells. *Br. J. Nutr.* 108(9), 1623–1632.

Pastorczyk M., Kosaka A., Pis'lewska-Bednarek M., López G., Frerigmann H., Kułak K., Glawischnig E., Molina A., Takano Y., Bednarek P. 2020. The role of CYP 71A12 monooxygenase in pathogen-triggered tryptophan metabolism and Arabidopsis immunity. *New Phytol.* 225, 400–412.

Paxton J.D. 1980. A new working definition of the term "phytoalexin". *Plant Dis.* 64, 734.

Pedras M.S.C., Abdoli A. 2017. Pathogen inactivation of cruciferous phytoalexins: detoxification reactions, enzymes and inhibitors. *R. Soc. Chem. Adv* 7(38), 23633–23646.

Pedras M.S.C., Abdoli A., Sarma-Mamillapalle V.K. 2017. Inhibitors of the detoxifying enzyme of the phytoalexin brassinin based on quinoline and isoquinoline scaffolds. *Molecules* 22(8), 2–15.

Pedras M.S.C., Ahiahonu P.W. 2002. Probing the phytopathogenic stem rot fungus with phytoalexins and analogues: unprecedented glucosylation of camalexin and 6-methoxycamalexin. *Bioorg. Med. Chem.* 10(10), 3307–3312.

Pedras M.S.C., Khan A.Q. 2000. Biotransformation of the phytoalexin camalexin by the phytopathogen Rhizoctonia solani. *Phytochemistry* 53(1), 59–69.

Pedras M.S.C., Minic Z., Jha M. 2008. Brassinin oxidase, a fungal detoxifying enzyme to overcome a plant defense–purification, characterization and inhibition. *FEBS J.* 275(14), 3691–3705.

Pedras M.S.C., Okanga F.I., Zaharia I.L., Khan A.Q. 2000. Phytoalexins from crucifers: synthesis, biosynthesis, and biotransformation. *Phytochemistry* 53(2), 161–176.

Pedras M.S.C., Yaya E.E., Glawischnig E. 2011. The phytoalexins from cultivated and wild crucifers: chemistry and biology. *Nat. Prod. Rep.* 28(8), 1381–1405.

Pérez-Pérez M.E., Lemaire S.D., Crespo J.L. 2012. Reactive oxygen species and autophagy in plants and algae. *Plant Physiol.* 160, 156–164.

Polturak G., Misra R.C., El-Demerdash A. 2023. Discovery of isoflavone phytoalexins in wheat reveals an alternative route to isoflavonoid biosynthesis. *Nat. Commun.* 14, 6977.

Pusztahelyi T. 2018. Chitin and chitin-related compounds in plant–fungal interactions. *J. Mycol.*, 9(3), 189–201.

Puupponen-Pimiä R., Nohynek L., Alakomi H.L., Oksman-Caldentey K.M. 2005. Bioactive berry compounds -novel tools against human pathogens. *Appl. Microbiol. Biotechnol.* 67(1), 8–18.

Sanzani S.M., Schena L., Ippolito A. 2014. Effectiveness of phenolic compounds against citrus green mould. *Molecules* 19(8), 12500–12508.

Schnippenkoetter W., Lo C., Liu G., Dibley K., Chan W.L., White J., Milne R., Zwart A., Kwong E., Keller B., Godwin I., Krattinger S.G., Lagudah E. 2017. The wheat Lr34 multipathogen resistance gene confers resistance to anthracnose and rust in sorghum. *Plant Biotechnol. J.* 15(11), 1387–1396.

Sharifi-Rad M., Varoni E.M., Iriti M., Martorell M., Setzer W.N., del Mar Contreras J. 2018. Carvacrol and human health: a comprehensive review. *PTR* 32(9), 1675–1687.

Sharma I., Thakur A., Sharma A., Singh N., Kumar R., Sharma A. 2022. Phytoalexins: implications in plant defense and human health. In: Sharma A.K., Sharma A. (eds) *Plant Secondary Metabolites*. Springer, Singapore.

Silva-Beltrán N.P., Ruiz-Cruz S., Cira-Chávez L.A., Estrada-Alvarado M.I., Ornelas-Paz J.D.J., López-Mata M.A., Márquez-Ríos E. 2015. Total phenolic, flavonoid, tomatine, and tomatidine contents and antioxidant and antimicrobial activities of extracts of tomato plant. *Int. J. Anal. Chem.* 284071.

Sin T.K., Tam B.T., Yung B.Y., Yip S.P., Chan L.W., Wong C.S., Siu P.M. 2015. Resveratrol protects against doxorubicin-induced cardiotoxicity in aged hearts through the SIRT1-USP7 axis. *J. Physiol.* 593(8), 1887–1899.

Sun M., Qian F., Shen W., Tian C., Hao J., Sun L., Liu J. 2012. Mitochondrial nutrients stimulate performance and mitochondrial biogenesis in exhaustively exercised rats. *Scand. J. Med. Sci. Sports* 22(6), 764–775.

Sunilkumar G., Campbell L.M., Puckhaber L., Stipanovic R.D., Rathore K.S. 2006. Engineering cottonseed for use in human nutrition by tissue-specific reduction of toxic gossypol. *Proc. Natl. Acad. Sci.* 103(48), 18054–18059.

Tatlidede E., Şehirli Ö., Velioğlu-Öğünç A., Çetinel Ş., Yeğen B.Ç., Yarat A., Şener G. 2009. Resveratrol treatment protects against doxorubicin-induced cardiotoxicity by alleviating oxidative damage. *Free Radic. Res.* 43(3), 195–205.

Tiku A.R. 2020. Antimicrobial compounds (phytoanticipins and phytoalexins) and their role in plant defense. In: Mérillon J.-M., Ramawat K.G. (eds) *Co-Evolution of Secondary Metabolites*, 845–868. Springer, Cham.

VanEtten H.D., Mansfield J.W., Bailey J.A., Farmer E.E. 1994. Two classes of plant antibiotics: phytoalexins versus "phytoanticipins". *Plant Cell* 6(9), 1191–1192.

Wanke A., Rovenich H., Schwanke F., Velte S., Becker S., Hehemann J.H., ... Zuccaro A. 2020. Plant species-specific recognition of long and short β-1, 3-linked glucans is mediated by different receptor systems. *Plant J.* 102, 1142–1156

Wu C.H., Derevnina L., Kamoun S. 2018. Receptor networks underpin plant immunity. *Science* 360(6395), 1300–1301.

Yang L., Browning J.D., Awika J.M. 2009. Sorghum 3-deoxyanthocyanins possess strong phase II enzyme inducer activity and cancer cell growth inhibition properties. *J. Agric. Food Chem.* 57(5), 1797–1804.

Zhang H., Huang L., Dai Y., Liu S., Hong Y., Tian L., Song F. 2015. Arabidopsis AtERF15 positively regulates immunity against Pseudomonas syringae pv. tomato DC3000 and Botrytis cinerea. *Front. Plant Sci.* 6, 686.

Zhang H.K., Zhang X., Mao B.Z., Qun L.I., Zu Hua H.E. 2004. Alpha-picolinic acid, a fungal toxin and mammal apoptosis-inducing agent, elicits hypersensitive-like response and enhances disease resistance in rice. *Cell Res.* 14(1), 27–33.

Zigová M., Michalková R., Mojžiš J. 2024. Anticancer potential of indole phytoalexins and their analogues. *Molecules* 29(10), 2388.

7 Plant–Microbe Abetted Mitigation of Hazardous Pollutants

Toward a Green and Clean Agroecosystem

Charu Sharma, Vishal Rajput, Nisha Dhillon, Vijay Kumar, Usha Rani, Ashutosh Kumar Choudhary, Bindu Naik, and Vivek Kumar

7.1 INTRODUCTION

The exponential growth of the world's population has created an insatiable demand for resources such as food, water, energy, timber, and other ecosystem services. The environment and natural resources are under unprecedented stress as a result of this (United Nations Environment Programme, 2011). Through a variety of processes, interactions between plants and microbes are essential for reducing environmental contaminants. According to Weyens et al. (2015), these interactions may improve the elimination, breakdown, or immobilization of pollutants in soil and water. Utilizing plants and the microorganisms they are connected with, phytoremediation aims to eliminate or neutralize pollutants from soil and water (Yan et al., 2020). The removal of pollutants from the soil through phytoremediation involves a number of processes, including rhizofiltration, phytoextraction, phytostabilization (also known as phytosequestration), and phytovolatilization. Plants have the ability to remove contaminants through a single mechanism or by combining one or more mechanisms. Through their roots, plants take up pollutants, especially heavy metals, which they then store in their shoots (Kafle et al., 2022).

Both plants and bacteria create enzymes that aid in this process by breaking down complicated contaminants and aiding in the overall cleanup of damaged areas (Agarwal et al., 2020). The rhizosphere is a nutrient-rich environment that contains higher numbers of microbes (and these are metabolically more active) than bulk soils (Ling et al., 2022). Additionally, some plants secrete compounds that can stimulate the degradation activities required for the elimination of contaminants. These plants immobilize contaminants in the soil, preventing their dispersion by reducing erosion and leaching. Microbes can sequester contaminants, restricting their mobility and

DOI: 10.1201/9781003714484-7

thus reducing the bioavailability of toxic contaminants (Alves et al., 2022). Some plants can absorb pollutants and release them into the atmosphere in a less harmful form. For instance, certain plants can take up mercury and release it as elemental mercury gas (Priya et al., 2023). Utilization of efficient microorganisms in the elimination of soil contaminants using bioaugmentation has been employed with various degrees of success. Several factors influence the efficiency or effectiveness of bioremediation, such as availability of nutrients in the soil, soil surface properties, toxicity, and reduced bioavailability of the contaminants. Many times, the presence of competing Indigenous microbial communities and failure to express the necessary metabolism required for the degradation of contaminants may influence the bioremediation efficiency of the microbes (Alori et al., 2022).

The rhizoremediation process specifically involves the plant root system and the surrounding soil, known as the rhizosphere (Correa García et al., 2018). The roots exude a wide variety of compounds, which help the rhizospheric microbiome take up nutrients more effectively. Root exudates are one of the crucial factors influencing the growth of microbial communities, which in turn enhance the degradation of pollutants through various mechanisms. Enhanced degradation-root exudates stimulate microbial activity and diversity, leading to more effective degradation of organic pollutants such as hydrocarbons and pesticides (Rohrbacher and St-Arnaud, 2016). Microbes also carry out the process of biotransformation; here, the microbes in the rhizosphere can transform pollutants into less toxic or more easily degradable forms. Another important process is bioaugmentation, in this process, the introduction of specific pollutant-degrading microbes into the rhizosphere can positively impact the degradation process (Bala et al., 2022).

The application of fungi (mycorrhizae) to reduce pollution is known as mycoremediation (Deshmukh et al., 2016). The enzymes produced by fungi can break down various contaminants, including Persistent Organic Pollutants (POPs), and can also convert agro-industrial waste into products. The synergy between plants and microbes can improve nutrient uptake and stress resistance, supporting plant growth in contaminated environments. By utilizing these natural processes, we can clean up contaminated sites, improve soil health, and safeguard water quality (Iqbal et al., 2023). In this article, we will go over the most recent research on the rhizoremediation of hazardous pollutants, with a particular emphasis on the interactions between microbes and plants.

7.2 MICROORGANISMS AND BIOREMEDIATION

Diverse microorganisms are used in bioremediation, a natural, environmentally benign process, to break down environmental contaminants. Plants and microorganisms, such as bacteria and fungi, are the active ingredients in this practical and effective method of cleaning up environmental pollutants (Hlihor et al., 2022). Numerous *Aeromonas* species, *Alcaligens, Aspergillus, Bacillus species, Escherichia, Flavobacterium, Ganoderma, Methosinus, Nocardia, Mycobacterium, Phormidium, Pseudomonas species, Rhodococcus, Spingomonas*, and *Stereum* are among the various microbial species that exhibit significant bioremediation potential and are typically used for site-specific pollution remediation (Ojuederie and Babalola, 2017; Saha et al., 2021). However, various fungal species such as *Phanaerochaete chrysosporium, Pleurotus species, Agaricus bisporus, Trametes versicolor*, and *Irpex lacteus* (Byss et al., 2008; Fulekar et al., 2013; García-Delgado et al., 2015; Pezzella et al., 2017; Chun et al., 2019). These fungi are considered to be very efficient in bioremediation as their digestive enzymes degrade or transform a variety of xenobiotics such as hydrocarbons, dyes, heavy metals, industrial solvents, and pesticides. Microbes utilize different ways to interact with inorganic metals. It includes biotransformation, enzyme utilization, extrusion, production of exopolysaccharides (EPS), and the synthesis of metallothionein (Sharma et al., 2022; Ayilara and Babalola, 2023). Table 7.1 shows the microbial-plant cooperative system for improving pollutant removal from soil environments.

7.2.1 Endophytic Microbes

Endophytes are microorganisms, such as fungi and bacteria, that reside inside living plants without producing any instantaneous harm. The expression "endophytes" was first mentioned by De Bary in 1886 to describe fungi, yeast, and bacteria that inhabit the plant's internal tissue (Vega et al., 2008) "Endophyte" is a term that describes an organism that inhabits the internal tissues of a plant host for a certain period of time. Endophytic microorganisms make up one of the largest and most diverse groups of living organisms, with an estimated one million species (Burragoni and Jeon, 2021). After conducting research for over a decade, it has been discovered that a huge number of plant species in natural ecosystems harbor various types of endophytic microorganisms. This finding has led to the realization that endophytic microbes represent one of the most important and yet untapped natural resources for bioprospecting of biosynthetic enzymes and secondary metabolites (Manganyi and Ateba, 2020).

The extensive documentation in the literature reveals the widespread presence of endophyte organisms in nearly all plants. Different plant components, including weeds and vegetables, can be used to isolate these endophytes (Catambacan and Cumagun, 2021; Khalil et al., 2021).

Plant growth-promoting endophytes (PGPE), like rhizobacteria, assist their host plants' development and multiplication by means of a variety of strategies. These processes include the manufacture of unique enzymes, the solubilization of minerals, the fixation of N2, the synthesis of phytohormones, and the siderophores that chelate iron. By using one or more of these techniques, certain PGPEs can lessen metal toxicity and promote plant development (Beneduzi et al., 2012; Backer et al., 2018; Khanna et al., 2019). Endophytes are useful for bioremediation processes such as methane absorption, nitrogen fixation, and the breakdown of different man-made substances since they use a variety of metabolic pathways (Stępniewska and Kuźniar, 2013; Qin et al., 2022).

Furthermore, endophytes produce antimicrobial secondary metabolic products and precursors of plant hormones, vitamins B12 and B1, and bioprotectants (Singh et al., 2017; Chaudhary et al., 2023; Sharma et al., 2023). By generating phytohormones, solubilizing nutrients, and fostering systemic resistance, endophytes promote plant development and lessen the toxicity of heavy metals (Kumar and Nautiyal, 2022). They appear to have the potential for bioremediation in polluted areas, but

TABLE 7.1
Microorganism-Plant Collaborative System for Improving Pollutant Removal from Soil

Plant Species	Microbes	Pollutants	Reference
Coffea arabica	*Bacillus lentimorbus*	Caffeine	Molina Delgado et al. (2021)
Echinochloa crus-galli	*Pseudomonas* sp.	Nicotine	Asemoloye et al. (2019)
Liriope muscari	*Bacillus aryabhattai*	Monocrotophos	Dash and Osborne (2020)
Pinus thunbergii	*Hebeloma vinosophyllum* *Pisolithus* sp.	Cr(VI)	Shi et al. (2019)
Maize root exudates	*Streptomyces* sp. *Streptomyces* sp.	Lindane and Cr(VI)	Sola et al. (2019)
Lolium perenne	*Hyphomicrobium* sp.	17α-ethynylestradiol	He et al. (2019)
Dracaena sanderiana	*Bacillus thuringiensis* *Pantoea, Dispersa*	Bisphenol A	Suyamud et al. (2018)
Transgenic Arabidopsis plant	*Sphingobium* sp.	Phenylurea herbicide-isoproturon	Yan et al. (2018)
Phragmites australis	*Proteobacteria, Firmicutes*	Nitrobenzene	Chi et al. (2018)
Acacia saligna, Eucalyptus camaldulensis	Rhizosphere microorganisms	As, Cd, Pb, and Zn	Guarino and Sciarrillo (2017)
Brassica juncea	*Pseudomonas aeruginosa*	Ni, Cr, and Cd	Ndeddy Aka and Babalola (2016)
Ryegrass	*Microbacterium* sp. *Candida tropicalis*	Polycyclic aromatic hydrocarbons (PAHs) and heavy metals (Cd and Zn)	Chen et al. (2016)
Sedum plumbizincicola	*Bacillus* sp.	Lead (Pb)/zinc (Zn)/ cadmium (Cd)	Ma et al. (2015)
Sedum plumbizincicola-	*Achromobacter piechaudii*	Cadmium, Zinc, and Lead	Lei et al. (2015)
Poplar (clone "I-214")	*Pseudomonas fluorescens*	Cadmium (Cd)	Cocozza et al. (2014)
Cytisus striatus	*Rhodococcus erythropolis*	Lindane	Becerra-Castro et al. (2013)
Lupinus luteus	*Serratia* sp.	As, Cd, Cu, Pb, and Zn	El Aafi et al. (2012)
Scorzonera mongolica, Atriplex centralasiatica, Limonium bicolor	Growth-promoting rhizobacteria	Crude oil	Ying et al. (2011)
Trifolium repens	*Comamonas* sp.	Dioxin	Wang and Oyaizu (2011)

more studies are required to verify their practicality. Due to their mutualistic evolution, endophytic bacteria and plants are able to withstand both biotic and abiotic stressors (Gowtham et al., 2024). By removing or transforming metal ions and storing them inside or on the surface of their cells, endophytic bacteria reduce the toxicity of metals (Tiwari and Bae, 2023).

By increasing nutrient absorption and lessening the impacts of metal stress, adding metal-resistant endophytic bacteria to plants can improve growth and phytoremediation in soil polluted with metal. *Paenibacillus, Bacillus, Exiguobacterium, Alcaligenes, Pantoea, Brevibacillus*, and *Pseudomonas* are among the effective bacteria (Jan et al., 2019; Wang et al., 2022). By boosting physiological and biochemical processes and raising stress-inhibiting hormones, inoculating plants with *Enterobacter ludwigii* and *Exiguobacterium indicum* can increase growth under metal stress (Jan et al., 2019; Wang et al., 2020).

Through a variety of mechanisms, including changing the valence of metals, precipitating them both inside and outside of cells, and actively absorbing them, fungi are among the best candidates for tolerating and detoxifying metals (Amobonye et al., 2023). They can also break down and render toxic substances harmless, such as heavy metals, which makes them a promising option for environmental remediation (Vaksmaa et al., 2023). Fungi also possess the biochemical and ecological ability to lessen the risks associated with metals, metalloids, and radionuclides by altering their chemical properties or by influencing their bioavailability (Chan et al., 2016; Lopez-Fernandez et al., 2021).

Plant physiology and biodiversity, microbial predominance, and soil aggregate absorption may all play a role in the early colonization of plants by endophytes. Key parameters governing microbial colonization are regulated by dynamics such as the plant's genotype, developmental stage, physiological characteristics, plant tissue types, soil state, and agricultural practice (Chauhan et al., 2023; Nadarajah and Abdul Rahman, 2023).

7.2.2 Rhizospheric Microbes

The rhizosphere, the soil zone surrounding plant roots, is a hotspot for plant-bacteria interactions and is influenced by soil type, texture, environmental factors, and plant processes. Plant root exudates attract beneficial and sometimes harmful bacteria to this area (McNear, 2013). Plant growth-promoting rhizobacteria (PGPB) inhabit the rhizosphere, where they boost plant growth, health, and pathogen resistance across various crops. The bacterial species diversity in the rhizosphere is a key indicator of soil quality and fertility, aiding in N_2 fixation, phytohormones creation, and nutrient acquisition. These bacteria also impact the pH of soil, the composition of water, organic carbon contents, and soil porosity (Kumar et al., 2001a; Rani et al., 2019).

Rhizosphere soils, richer in bacterial biomass than bulk soils, support high bacterial densities (10^8–10^{12} cells per gram) due to root exudates providing abundant substrates (Wang et al., 2022). These exudates, which include enzymes, ions, mucilage, and various carbon-containing metabolites, attract bacteria and stimulate their motility, aiding colonization. Common rhizosphere bacteria include *Microbacterium, Burkholderia, Bacillus, Azospirillum, Erwinia, Aeromonas, Serratia, Pseudomonas,*

Mesorhizobium, Rahnella, Enterobacter, and *Acinetobacter* (Nwachukwu et al., 2021; Gómez-Godínez et al., 2023). Harnessing these bacteria can improve soil vitality, productivity, and crop productivity through biocontrol, bioremediation, and biofertilization, promoting Ecological sustainability (Iqbal et al., 2023).

Plant growth-promoting rhizobacteria (PGPR) are defined as bacteria beneficial to plants due to their ability to produce secondary metabolites for instance, siderophores, 3-indoleacetic acid (IAA), and 1-aminocyclopropane-1-carboxylic acid (ACC), along with deaminase activity and capabilities of solubilizing phosphate (Kumar and Nautiyal, 2022). As they reduce stress and metal toxicity in plants and change metal mobilization to promote plant development, PGPR is essential to phytoremediation activities and directly contributes to the detoxification of metals (Qin et al., 2024).

Numerous studies have shown how effective rhizobacteria are in removing pollutants from the environment, particularly Bacillus and Pseudomonas species. It has been demonstrated that in a variety of environmental conditions, these bacteria may efficiently aid in the cleanup of pollutants (Viesser et al., 2020). About 80%–90% of vascular plants have symbiotic relationships with arbuscular mycorrhizal fungi (AMF), which may be responsible for up to half of the soil's total microbiological biomass. By spreading their mycelial networks across the soil and establishing connections between plant roots and rhizospheric bacteria, these fungi play a significant role in enhancing phytoremediation efforts (Solanki et al., 2011). AMF promotes the absorption of nutrients and water by plants, especially micronutrients, phosphates, and nitrogen. Their pervasiveness and mutually advantageous relationships highlight their importance in environmentally conscious farming and environmental management (Tyagi et al., 2014).

7.3 FACTORS AFFECTING MICROBIAL BIOREMEDIATION

Microbial metabolic characteristics and the physico-chemical characteristics of pollutants control bioremediation processes. The efficiency of bioremediation is greatly influenced by environmental conditions such as the kind of contaminant, its concentration, pH, temperature, moisture content, soil structure, and oxygen availability (Alori et al., 2022). Biodegradation rates are influenced by variations in microbial diversity and metabolic activities, such as the generation of enzymes and oxygen needs. The type, texture, pH, and nutrient content of the soil all have an impact on the dispersion and breakdown of hydrocarbon pollutants (Kebede et al., 2021).

The diversity of microbes determines whether an aerobic or anaerobic system is in place, and as a result, their efficiency in promoting oxygen-based biodegradation rates depends on the availability of oxygen (Smagin et al., 2023). The environmental factors that affect microbial growth and activity at a site include temperature, moisture, soil structure, pH, water solubility, nutrient availability, site characteristics, oxygen levels, and redox potential. The ideal pH range for biodegradation is between 6.5 and 8.5 in the majority of aquatic and terrestrial environments, though degradation can occur over a wider range of pH. Moisture levels have an impact on contaminant metabolism, which is determined by the availability of soluble material, pH, and osmotic pressure both in aquatic and terrestrial systems (Adedeji et al., 2022).

The metabolic diversity of bacteria plays a crucial role in degrading, transforming, and mineralizing various hydrocarbon pollutants in soil (Das and Chandran, 2011). Certain microbial species can dominate contaminated sites by utilizing toxic pollutants (Kumar et al., 2016; Upadhyay et al., 2017).

Numerous microorganisms inhabit soil, which is a complex habitat that is impacted by several physicochemical parameters. These factors include soil type, texture, moisture, temperature, nutrients, availability of oxygen, and pH (Abdul Rahman et al., 2021). These factors crucially determine the composition of microbial communities and functional genes, impacting the stability, bioavailability, and movement of hydrocarbon pollutants (Kebede et al., 2021). Nutrient availability, especially nitrogen, phosphorus, and potassium, enhances microbial metabolism and the degradation of hydrocarbons in contaminated soils (Das and Chandran, 2011). Oxygen availability governs aerobic bioremediation, acting as the final electron acceptor for enzymatic reactions that break down aromatic hydrocarbons. Temperature influences bacterial growth rates, enzyme activities, and the overall efficiency of hydrocarbon pollutant removal (Ghosal et al., 2016). Optimal pH conditions (neutral to slightly alkaline) promote microbial biomass and enzyme activity, crucial for enhancing hydrocarbon biodegradation (Kumar et al., 2017). Adjusting soil pH within the range of 5–8 is recommended for maximizing bioremediation effectiveness in hydrocarbon-contaminated sites (Rawat and Kumar, 2021).

The effectiveness of bioremediation projects depends on the chemical composition of organic molecules found at contaminated sites, which can be affected by xenobiotic groups such as Cl, NO_2, CN, and SO_3 that inhibit microbial metabolism (Rawat et al., 2022). Enzymes crucial to metabolic reactions include lyases, transferases, isomerases, oxidoreductases, ligases, and hydrolases. Oxygenase enzymes, such as toluene dioxygenase, exhibit broad substrate specificity, capable of degrading numerous aromatic compounds like TCE and chlorobenzene (Mousavi et al., 2021). Other enzymes like esterases, depolymerases, dehalogenases, decarboxylases, hydratases, glutathione S-transferase, racemases, and CoA-ligase play key roles in specific biochemical transformations essential for bioremediation (Karigar and Rao, 2011).

Rhizospheric microbes create biofilms around plant roots, offering protection and optimizing pollutant access. These biofilms aid in genetic material exchange and cellular communication among diverse microorganisms like bacteria, fungi, algae, and archaea. The formation of stable biofilms depends on environmental factors, electron acceptor/donor presence, and nutrient levels (Mishra et al., 2022).

7.4 PHYTOREMEDIATION

Phytoremediation essentially describes the employment of efficient plants and their accompanying microbiome to lessen the concentration or lethal outcomes of pollutants in the ecosystem. Phytoremediation is generally established as an economical and viable environment reinstatement technology. Plants can remove, build up, and decontaminate the pollutants from the substratum (soil, air, and water) using physicochemical, or biological approaches. Plants utilizes diverse approaches (Figure 7.1) to remove the toxicants from soils, these diverse approaches are discussed in the following section.

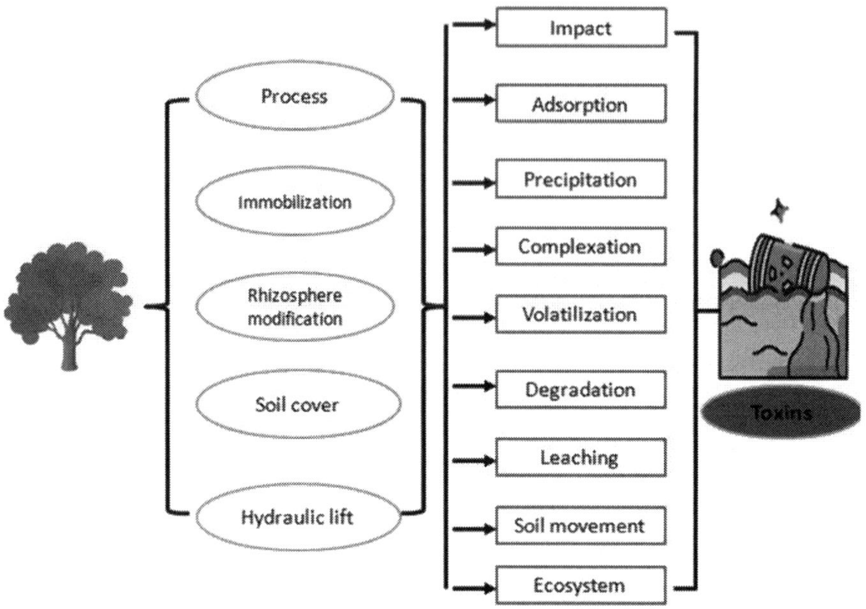

FIGURE 7.1 Diverse approaches utilized by plants to mitigate the contaminants.

7.4.1 Phytoextraction/Phytoaccumulation

Phytovolatilization is the term used about the mechanism of converting a pollutant from a fixed form into a volatile form that may be released from lithosphere and enter the atmosphere. This does not entail the buildup of heavy metal contaminants within plant components. Phytoremediation, in which plants absorb and store heavy metals from the soil, can be divided into two main methods: phytoextraction and phytostabilization (Yao et al., 2020).

In heavily contaminated environments, phytostabilization is utilized to prevent corrosion from air or water and to generate a dense and full plant cover. This mostly includes immobilizing and stabilizing soil contaminants. Because the foundation of phytostabilization is the ability to produce dense vegetation, plants that can cover compacted soils with large, dense root formations and systems are considered a better alternative for this type of phytoremediation. Phytoextraction is a more complicated procedure, but it is more effective and practical. Due to the application of this method, tolerant plants are grown in contaminated soils. Plants absorb different toxicants from the lithosphere, which causes the target pollutants to disappear and reach the plant's aerial parts (Rafati et al., 2020). Following a significant build-up of toxicants in plant parts, the plants are harvested, dried, and then burned, producing concentrated material that is disposed of separately and contains pollutants. This method works because highly concentrated bulk material has even higher concentrations of pollutants than dirt. The positive aspect of this methodology is energy production during incineration, which can be helpful for experimenting with other energy-related activities.

The leaves of hyperaccumulators typically contain concentrations of trace metals named Cd, Ni, and are reported many times higher than those in contaminated soil, which amply demonstrates how well-polluted soils are cleaned. Heavy metal deposition is greater in shoot regions than in root systems (Becher et al., 2019). Plants that provide little biomass as they mature are good candidates for ideal phytoextraction. Moreover, commercially significant nutrients found at trace levels in the lithosphere may be absorbed and concentrated by this method, a phenomenon known as phytomining (Tan et al., 2023).

7.4.2 Phytofiltration

Due to the inappropriate natural and manmade processes, inorganic and organic toxicants are contaminating the agroecosystem, including groundwater and surface water. Contaminants effects on humans and wildlife are cytotoxic, mutagenic, and carcinogenic; therefore, their release into the environment poses a major risk (Olguín and Sánchez-Galván, 2020). Different methodologies such as ion exchange, chemical precipitation, flotation, coagulation-flocculation, and the application of membrane filtration, have been used to remove the toxicants from contaminated wastewater. Nevertheless, their intrinsic limits stem mostly from the fact that treating vast volumes of contaminated water through lower metal concentrations is not economically feasible. Biological metal treatment, particularly phytoremediation, is starting to be a more appealing option. One important type of phytoremediation technique is phytofiltration, which involves the application of plants to filter pollutants from water bodies (Srivastava et al., 2019). There have been descriptions of three distinct phytofiltration systems: (1) built wetlands/constructed wetlands, and lagoons; (2) rhizofiltration (using the roots of hydroponically grown plants); and (3) bioadsorbent-based systems (Rai, 2020). Various factors, such as wastewater type, pH, and the makeup of the supporting media and sediment, affect how well these systems work. Nevertheless, the design of a phytofiltration system involves other important considerations. The application of techniques for the ultimate use and disposal of contaminated plant biomass is often a topic of interest to potential users of phytofiltration technology. However, sufficient information is not available about this specific problem; therefore, more investigations are needed. The selection of plant species, particularly those from tropical and subtropical regions, the choice of a suitable phytofiltration system, the environmental factors influencing plant uptake of metals, the handling and disposal of biomass containing metals, and, whenever feasible, the recovery of metals at the end of the process are the main factors influencing the design of sustainable processes for phytofiltration of heavy metals (Mishra et al., 2019).

7.4.3 Phytostimulation

Enhanced microbial involvement in the plant root zone or rhizosphere facilitates the breakdown of organic toxicants in the lithosphere. Such a method is named phytostimulation, but it is also called rhizodegradation, rhizosphere biodegradation, or plant-assisted bioremediation/degradation. There are multiple ways in which the rhizosphere stimulates microbial activity: (1) compounds emanate by the roots, named enzymes, amino acids, and acetates and enrich native microbe populations; (2) oxygen brought to the rhizosphere by root systems ensures aerobic transformations; (3) the fine-rooted biomass elevates the organic carbon; (4) because of special enzymatic pathways, mycorrhizae fungi, which grow in the rhizosphere, can breakdown organic pollutants that bacteria alone are unable to change; and (5) plants improve the environment to increase microbial populations and activity. This methodology is effective for clearing soil and sediments from organic pollutants such as pesticides, aromatic compounds, and polynuclear aromatic hydrocarbons (PAHs).

7.4.4 Phytostabilization

Employing root accumulation or rhizosphere-based precipitation, phytostabilization seeks to keep pollutants out of the vadose zone. By doing so, the movement of pollutants through soil dispersion, leaching, and erosion from wind and water is prevented. Adding plant cover to the surface of polluted soils to lessen their exposure to elements, wind, and direct human or animal contact is another definition of phytostabilization. Using any plant type with the capacity to withstand the higher concentration of pollution and soil additives that are efficient at immobilizing metal(loid)s can improve phytostabilization. Without eliminating inorganic pollutants from the area, phytostabilization lowers their risk by reducing their mobility. There is no polluted secondary trash produced by this method, which further needs advanced processing. Additionally, it improves soil fertility, which leads to the regeneration of ecosystems (Berti and Cunningham, 2020). However, the site needs to be regularly monitored to ensure that the ideal stabilizing conditions are maintained because the toxins are left *in situ*. If soil amendments are applied to improve immobilization, their efficacy might need to be improved regularly.

7.4.4.1 Contaminant Removal

Plant removal of contaminants may still be important in this methodology, even though plant uptake of toxins may not be the primary activity engaged in phytostabilization. According to Barton et al. (2018), growing crops can produce biomass rich in zinc that may be applied as nutrient-rich stock feed while also assisting in the phytostabilization of soil contaminated with metals. High amounts of boron (B) were removed by poplars utilized in the phytostabilization of a sawdust pile polluted with B (Ruttens et al., 2019). Here, biomass might be applied to neighboring orchards as a B-rich mulch. The plants' main function in both situations was to stabilize the area. The biomass was greater when the metal(loid)s was extracted (Figure 7.2). To improve this technology, hyperaccumulating plants that are efficient at removing metal(loid)s may be employed (Mendez and Maier, 2018).

Plants use different processes to absorb metal(loid)s from the soil. For instance, in the rhizosphere of plants lacking iron, phytosiderophores have been found in quantities up to molar levels under nonsterile conditions. Ruttens et al. (2019) observed increased iron immobilization at a 4 mm distance from wheat root and concluded that phytosiderophores are very successful at acquiring Fe from the soil to the plants.

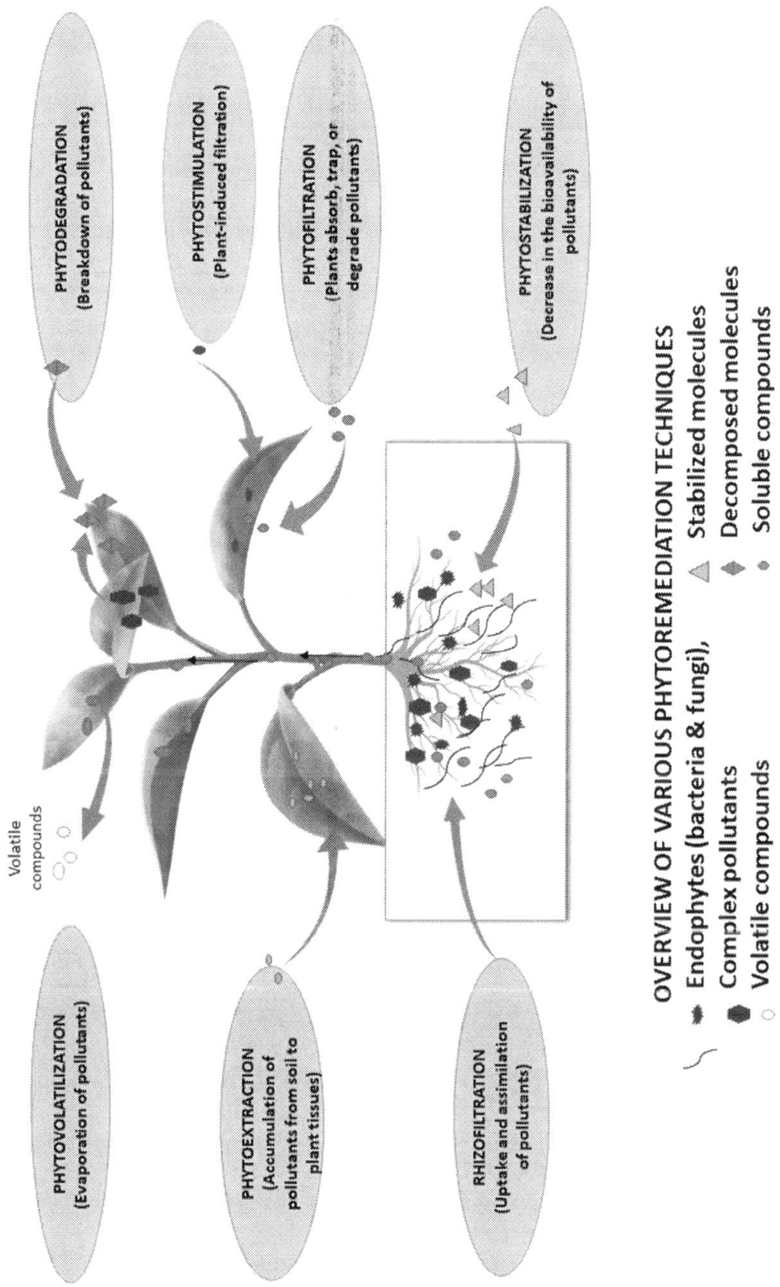

FIGURE 7.2 Phytostabilizat on of toxicants in agro-ecosystem.

7.4.5 Phytovolatilization

Since phytovolatilization usually results in significant dilution and photochemical degradation of pollutants in the atmosphere, phytovolatilization is frequently considered an advantageous approach. Numerous pollutants, both inorganic and organic, have been shown to undergo phytovolatilization. Several research attempts mentioned the two modes of phytovolatilization: as indirect and direct phytovolatilization (Cantamessa et al., 2020). The more logical and well-researched form, known as direct phytovolatilization, arises from pollutants being absorbed and translocated by plants, ultimately causing the compound to volatilize through the leaves and other parts of the tree. Although some of the substances that are phytovolatilized are mildly hydrophobic and may diffuse through hydrophobic natured barriers such as the epidermis (cutin), the direct phytovolatilization pathway frequently involves transpiration. Numerous pollutants volatilize from plants through the methodologies that may have been clarified by research on the phytovolatilization of contaminants. Nonetheless, there is still uncertainty regarding the relative significance of several of these elimination mechanisms, particularly for matters with lower research data backup. Research involving these highly volatile elements exhibits that when log KOA decreases, diffusive transport becomes increasingly important for transportation (Arnold et al., 2019). However, additional research using a larger range of substances is needed. Furthermore, substances that are hydrophobic and may not translocate within plants may nevertheless be reported in the tissues of aboveground plants by particle deposition or airborne partitioning. Thus, appropriate care is required when designing experiments to ensure that aboveground biomass contamination is caused by translocation. Further comprehension of the mechanical dynamics governing gas transport in tree trunks is required (Burken and Schnoor, 2018). Variations in lenticel permeability, aerenchyma formation, and trunk gas concentration are seen over many temporal periods. Since measuring these fundamental parameters makes it possible to anticipate diffusivities for a wide range of toxicants, it is more effective than merely tracking the fluxes of a specific pollutant over time (Marr et al., 2019).

7.4.6 Phytodegradation

Organic substances can undergo phytodegradation either in the plant or in its rhizosphere. This technique may eliminate a wide variety of materials and chemical types from the environment, including solvents in groundwater, petroleum in soils, and volatile chemicals in the air. The subject of pesticide phytodegradation has recently yielded some surprising discoveries. Kafle et al. (2022) investigated the rate at which ryegrass absorbed lindane and trifluralin. They discovered that whereas trifluralin was metabolized by the ryegrass plant, lindane was poorly metabolized by the plant, resulting in binding pesticide residues on roots. Various variables have been shown to modify the process of pesticide phytodegradation.

Simazine (2-chloro-4,6-bis (ethylamino)-1,3,5-triazine) was more readily absorbed by younger plants (two weeks old) than by older plants (two weeks), according to Knuteson et al. (2020). The uptake of DDE was enhanced by organic acids. Although the absorption capacity and relocation of toxicants have long been understood to be species- and ecotype-specific, White et al. (2020) revealed that DDE uptake was also species-specific, with some cultivars demonstrating noticeably greater uptake.

7.4.6.1 Phytodegradation of Petroleum Compounds

Under fresh plants, there is little rhizosphere breakdown of chlorinated solvents; however, research by Godsy et al. (2021) suggests that as trees become old, microbic populations may shift and promote anaerobic biodegradation. Rhizosphere involvement is highly anticipated during the dealing with substances such as polycyclic aromatic hydrocarbons (PAHs), petrochemicals, and polychlorinated biphenyls (PCBs).

Effects of plants on PAH breakdown in soil are still difficult to detect, and multiple interactions between the soils, rhizospheric zone, microbial community, and host plant system make it difficult to analyze the effects of phytoremediation treatments. The various aspects of the system are regularly interacting with one another. These interactions change depending on the temperature, moisture content, and growth stage of the microbes and plants. It is difficult to comprehend and keep track of phytoremediation because of these interactions. Joner et al. (2019), reported a noticeable priming effect linked to the experimental procedures, which might have tended to obscure the positive outcomes of the combined impacts of planting and fertilization. Reports of accelerated plant degradation are still being supported by substantial evidence. Chen et al. (2020) reported that tall fescue (*Festuca arundinacea*) and switchgrass (*Panicum virgatum L.*) exhibited 38% and 30% pyrene mineralization, respectively, in response to 14C-labelled pyrene compared to 4.3% in the control set.

7.4.7 Rhizofiltration

Comparing this developing technology to existing chemical and physical-chemical cleanup solutions, it is more economical, energy-efficient, and environmentally benign. Rhizofiltration is a process that is widely utilized in phytoremediation methods to remove contaminated groundwater. Utilizing plant roots, rhizofiltration concentrates and precipitates pollutants on or within the roots by adsorbing them from the surrounding root zone (rhizosphere) as mentioned in Table 7.2. When plant roots secrete exudates of roots (secondary metabolites), they undergo biogeochemical processes that cause pollutants to accumulate on the root surface or in the water body. Then, depending on the type of contamination and concentration, the contaminants will either adsorb on the roots, may move inside the roots, or translocate to the phyllosphere (above-ground plant parts) (Sharma and Juwarkar, 2020).

For contaminated surface water, *in situ* rhizofiltration could be employed; for polluted groundwater resources, ex situ rhizofiltration, which uses an engineered tank system to supply polluted water to plants, can be used. Rhizofiltration is thus a viable method for extracting and filtering contaminated plumes underground (Tsao, 2021).

Singh and Santal (2019) reported that compared with regular plant-based systems, hydroponically grown terrestrial plant roots have a greater capacity for remediation for toxicants and heavy metals (HMs). The initial step in the rhizofiltration process is

TABLE 7.2
Synergistic Plant–Microbe Interactions in Heavy Metal Phytoremediation

Heavy Metal	Plant	Microbes	Removal Percentage	References
Cd	*Medicago sativa*	*Bacillus subtilis*	139%; phytoextraction	Li et al. (2021)
Pb	*Pelargonium hortorum*	*Aspergillus flavus and Microbacterium paraoxydans*	Two-to-five-folds more uptake than control; Phytoextraction	Manzoor et al. (2021)
Hg	*Zea mays*	*Serratia marcescens Bac156 and Pseudomonas* sp. *Bac138*	47.16% and 62.42%; endophytic; volatilization	Mello et al. (2020)
Cr	*Sesbania sesban*	*Bacillus xiamenensis PM14*	56% Phytoextraction	Din et al. (2020)
Cd and trichlorfon (pesticide)	*Brassica juncea*	*Aspergillus sydowii*	10% and 4% phytoextraction and phytodegradation	Zhang et al. (2020)
Cu	*Medicago sativa*	*Paenibacillus mucilaginosus*	55% and 76%; phytoextraction	Ju et al. (2019)
Pb and Cd	*Phragmites communis*	*Simplicillium chinense*	29%–48.0%; mycorrhizal; biosorption	Jin et al. (2019)
Cr(VI), Ni, Zn, Cu, Cd	*Vetiveria zizanioides* L.	*Bacillus cereus*	130%–211% Cr(VI); 31%–40% (Ni); 30%–61% (Zn); 65%–178% (Cu); 84%–107% (Cd); phytostabilization	Nayak et al. (2018)
As	*Betula celtiberica*	*Rhodococcus erythropolis, Ensifer adhaerens, Variovorax paradoxus*, and *Phyllobacterium myrsinacearum*	16%–35%; endophytic; phytoextraction	Mesa et al. (2017)
Cd	*Solanum nigrum*	*Glomus versiforme*	90.0%; mycorrhizal; phytoextraction	Liu et al. (2015)

to use the P and T system to locate and extract groundwater. Plants are grown hydroponically in greenhouses, once appropriate plant species with high HMs adsorption potential from contaminated zones have been chosen. At the contaminated locations, plants showing high HM removal effectiveness and acclimatized to the contaminated groundwater were planted.

7.5 BIOREMEDIATION EXPLOITING GENETICALLY ENGINEERED ORGANISMS

To have a healthy ecosystem, it is crucial to Break down and restrict the dispersion of hazardous substances. Designing modified microbes has emerged as a pocket-friendly and eco-friendly solution to HM pollution. Using genetically modified microbes to break down pollutants and waste, bio-augmentation offers an answer to the issue of slow and inefficient degradation of hazardous contaminants and waste (Pande et al., 2022). Bacteria exhibit the ability to break down environmental contaminants. Several studies have identified bacterial strains that can digest various hazardous compounds (e.g., nitro-aromatics, chlor-aromatics, and polycyclic aromatic compounds). These microbes can be used to facilitate the bioremediation of contaminated areas by utilizing their metabolic capabilities (Mohapatra and Phale, 2021). Many microbes, however, are unable to metabolize specific persistent and hazardous xenobiotic compounds, namely strongly nitrated or halogenated arene compounds, insect repellents, and explosive substances, owing to their chemical inertness under natural conditions. This leads to an ineffective decomposition process (Rafeeq et al., 2023). Complexities arising from pollutant combinations, coupled with their toxicity to existing microbial communities, make efficient biodegradation challenging. Genetically modified microorganisms with efficient catabolic pathways can overcome these constraints, thus helping to improve the bioremediation potential (Rebello et al., 2021). These specifically designed microorganisms can effectively eliminate most of the pollutants that wild indigenous microbes cannot break down. Several molecular techniques can be used for the development of engineered organisms such as biological ballistics, electropermeabilization, conjugation, horizontal transfer of bacterial DNA, recombinant DNA technology, and transformation of protoplasts (Ahmad et al., 2021). The transfer and activation of new genes with superior degradation abilities also reduce the time needed for cleanup. Modified microorganisms can be used for remediating a wide range of toxic compounds, such as toluene, octane, naphthalene, salicylate, and xylene, facilitated by the expression of plasmid genes found in bacterial species (Varner et al., 2022).

Four different methods have been proposed by the researchers: (1) changing the enzyme's attraction and selectivity to the substrate; (2) modifying genes and regulatory pathways; (3) bioremediation process development, regulation, and monitoring; and (4) application of bio-affinity reporters for the detection of contaminants, reduction of toxicity levels and envisage the outcomes (Rebello et al., 2021). The incorporation of multiple genes can enhance a microorganism's potential to break down a wide range of xenobiotics, which can increase its degradation potential (Miglani et al., 2022).

7.5.1 Bioremediation of Pesticides

The discovery of numerous genes with a superior ability to break down organic contaminants provides the potential for generating genetically modified organisms (GMOs) capable of degrading pesticides (Ruomeng et al., 2023). As we transition to ecological farming methods and the utilization of genetically modified plants to increase production, biological pesticides have become integral to viable agricultural practices. The genetically altered microbe plays a vital role in revitalizing soil health by breaking down pesticide traces that would otherwise persist in the soil for ages (Sudheer et al., 2020). Atrazine chlorohydrolase expressed by gene atzA can readily break down atrazine, which is one of the widely used pesticides (Fernandes et al., 2020). Another field-scale study highlighted the ability of genetically transformed *Escherichia coli* with atrazine chlorohydrolase to remediate atrazine-contaminated soil (Strong et al., 2000). In a parallel study, researchers unequivocally expressed the gene tpd responsible for triazophos hydrolase from *Ochrobactrum* sp. mp-4 in *Pseudomonas putida* KT2440, demonstrating its efficacy in breaking down pesticides of the organophosphorus category and other arene hydrocarbons (Zhang et al., 2006).

7.5.2 Removal of Heavy Metal

The removal of heavy metals using microorganisms relies on the mechanisms of biosorption and bioaccumulation to decontaminate polluted environments (Igiri et al., 2018). Genetically transformed microbes for heavy metal remediation employ diverse strategies, such as modifying the transport proteins associated with metal transport through microbial membranes and expressing various proteins with binding affinity for metals such as ferritin, metallothionenin, polyphosphates, and phytoalexins. These metal-binding proteins function as storage for metals in microbial cytoplasm (Iravani and Varma, 2022).

7.5.3 Bioremediation of Hydrocarbons

Remediating oil-contaminated locations has shown to be more successful when done with genetically modified bacteria. Due to their complexity, native microbial strains are unable to break down pollutants derived from oil (Das and Chandran, 2011). Plasmids containing numerous genes expressing degrading enzymes can be introduced into an organism to provide better features. The *Acinetobacter baumannii* S30 pJES strain has been created by researchers, and it is more successful in breaking down total petroleum hydrocarbons (TPH). A reporter lux gene for bioremediation site monitoring is also present in this strain (Mishra et al., 2004). Similar to this, scientists altered *Streptomyces coelicolor* M145 to enhance n-hexadecane breakdown by upregulating the expression of the alkB gene, which is in charge of producing the enzyme alkane monooxygenase (Gallo et al., 2012). In another work, the xylE gene, which codes for the catechol 2,3-dioxygenase enzyme from *Pseudomonas putida* strain BNF1, was effectively inserted into *Acinetobacter* sp. BS3. The breakdown of aromatic hydrocarbons is catalyzed by this enzyme (Xie et al., 2014). The altered strain outperformed the wild type in terms of remediation efficiency for n-alkanes and other arene chemicals, as well as substrate specificity (Mesarch et al., 2000).

7.6 ADVANTAGES OF EMPLOYING GEMS

GEMs are very effective in degrading certain pollutants because they may be designed to specifically target and target those pollutants. Compared to naturally existing microorganisms, GEMs' enhanced metabolic pathways enable the quicker and more thorough breakdown of contaminants (Wu et al., 2021). It is possible to modify GEMs so that they function well under a range of environmental circumstances, including as high or low pH, the presence of competing species, or both. Greenhouse gas extraction methods (GEMs) provide a more environmentally friendly and sustainable alternative to chemical approaches (Urgun-Demirtas et al., 2006).

7.6.1 Challenges and Considerations

GEMs run the danger of escaping their intended confines and maybe upsetting nearby ecosystems or giving native bacteria access to their DNA. Deployment of genetically modified organisms (GEMs) may be impeded by strict restrictions and public concerns about the discharge of GMOs into the environment (Bawa and Anilakumar, 2013). The difficulty in using bioengineered microorganisms to bioremediation of polluted areas stems from their frequent incapacity to endure extremes in temperature, salinity, pH, and other parameters, as well as to perform efficiently in the geological environment for prolonged periods of time (Kebede et al., 2021). Unlike plasmids, biodegradation genes must be integrated into the chromosomal genome in order for target proteins to be expressed effectively and steadily.

Validating how well this integration facilitates the bioremediation process is crucial, though (Tsai et al., 2022). Furthermore, public involvement and open risk assessment are necessary to resolve ethical concerns about the use of GMOs (Yaashikaa et al., 2022). To sum up, genetically modified microorganisms have a lot of promise to tackle some of the most difficult environmental toxins. They have the potential to be extremely important in environmentally sustainable management with more study and prudent implementation.

7.7 CONCLUSION AND FUTURE PERSPECTIVES

The contamination of the environment is a critical issue and is regarded as one of the most significant concerns of the current century. Designing efficient ways to protect the environment for future generations is a time-consuming undertaking. Environmental contamination, particularly contamination with heavy metals, has a significant impact on soil quality, fertility, microbiological biodiversity, and the degradation of vegetative cover. Researchers have utilized numerous physicochemical approaches to bioremediate soils polluted with toxicants. Although these treatments are quite ubiquitous and suitable for removing toxicants, they are difficult to implement, very expensive, non-specific, inefficient in a given scenario, and can dramatically alter soil quality. As an alternative, new biological-based approaches for remediating polluted soil have been devised. Biological approaches, exploiting plants and microbes' potential to depollute and stabilize the toxicants and heavy metals, are emerging as inventive know-how, demonstrating growing prospects to reinstate heavy metals and toxicants polluted soils. Microbial and phytoremediation are dependable, economical, effective, and eco-friendly substitutions. Microbes and plants acquire intrinsic means of tolerating contaminated components and eradicating toxicants from the ecosystem. Microorganisms employ diverse procedures, comprising extra- and intracellular sequestration, production of metal chelators, volatilization, precipitation, and detoxification by enzymes. The remediation strategy of microbes can be combined with phytoremediation processes (phyto-stabilization, phyto-extraction, and phyto-volatilization) for a successful replenishment. Engineered plants and microorganisms can augment the effectiveness of bioremediation that could exceed the constraints that each strategy experiences.

A single technique is neither successful nor sufficient for the practical remediation of polluted soils. The amalgamation of dissimilar tactics, physico-chemical, and biological is indispensable for extremely successful and complete phytoremediation in the future and must be encouraged. The research must concentrate on examining the effect of integrating dissimilar microorganisms on phytoremediation efficacy, such as combined microbial bioremediation with organic and inorganic chelating additives. Moreover, metagenomics methodologies and microbic metabolic exploration need to be investigated to identify ensuring metals and toxicants resistance and decontamination genes that can be introduced in other strains to enhance their particular functioning. Metagenomics methods might also emphasize the microbial community's evolution during remediation processes and decide the reliable approaches to increase their existence owing to the probable competition with aboriginal microbial communities. Furthermore, issues such as host plants and potential microbial compatibility, as well as the ability of microorganisms to function and proliferate in polluted soils, need to be examined under natural field circumstances for longer periods of time. In this background, it is proposed that microbial inoculations might begin to substitute other exercises employed in phytoremediation, for instance, chemical fertilizers, phytohormones, and chelating materials, subsequently lessening the consequent prices. Furthermore, additional genomic study is necessitated to completely comprehend the metabolic pathways and the processes intricate in microbial and plants' decontamination and tolerance of toxicants.

Realistically, a specific method is neither promising nor appropriate for successfully cleaning contaminated soils. Combining diverse tactics, comprising recombinant DNA technology, microbial-assisted and chelate-assisted methodologies, is indispensable for very effectual and comprehensive phytoremediation in the time ahead.

ARTIFICIAL INTELLIGENCE USAGE DECLARATION

The authors have not used regenerative artificial intelligence tools during writing.

BIBLIOGRAPHY

Aafi, N.E., Brhada, F., Dary, M., Maltouf, A.F., Pajuelo, E., 2012. Rhizostabilization of metals in soils using *Lupinus luteus* inoculated with the metal resistant rhizobacterium *Serratia* sp. MSMC541. *Int. J Phytoremediation* 14(3), 261–274. https://doi.org/10.1080/15226514.2011.604693

Abdul Rahman, N.S.N., Abdul Hamid, N.W., Nadarajah, K., 2021. Effects of abiotic stress on soil microbiome. *Int. J. Mol. Sci.* 22(16), 9036. https://doi.org/10.3390/ijms22169036

Adedeji, J.A., Tetteh, E.K., Opoku Amankwa, M., Asante-Sackey, D., Ofori-Frimpong, S., Armah, E.K., Rathilal, S., Mohammadi, A.H., Chetty, M., 2022. Microbial bioremediation and biodegradation of petroleum products-A mini review. *Appl. Sci.* 12(23), 12212. https://doi.org/10.3390/app122312212

Agarwal, P., Giri, B.S., Rani, R., 2020. Unravelling the role of rhizospheric plant-microbe synergy in phytoremediation: A genomic perspective. *Curr. Genomics* 21(5), 334–342. https://doi.org/10.2174/1389202921999200623133240

Ahmad, S., Cui, D., Zhong, G., Liu, J., 2021. Microbial technologies employed for biodegradation of neonicotinoids in the agroecosystem. *Front. Microbiol.* 12. https://doi.org/10.3389/fmicb.2021.759439

Alori, E.T., Gabasawa, A.I., Elenwo, C.E., Agbeyegbe, O.O., 2022. Bioremediation techniques as affected by limiting factors in soil environment. *Front. Soil Sci.* 2, 937186. https://doi.org/10.3389/fsoil.2022.937186

Alves, A.R.A., Yin, Q., Oliveira, R., Silva, E.F., Novo, L.A.B., 2022. Plant growth-promoting bacteria in phytoremediation of metal-polluted soils: Current knowledge and future directions. *Sci. Total Environ.* 838, 156435. https://doi.org/10.1016/j.scitotenv.2022.156435

Amobonye, A., Aruwa, C.E., Aransiola, S., Omame, J., Alabi, T.D., Lalung, J., 2023. The potential of fungi in the bioremediation of pharmaceutically active compounds: A comprehensive review. *Front. Microbiol.* 14, 1207792. https://doi.org/10.3389/fmicb.2023.1207792

Arnold, C.W., Parfitt, D.G., Kaltreider, M., 2019. Phytovolatilization of oxygenated gasoline-impacted groundwater at an underground storage tank site via conifers. *Int. J. Phytoremediation* 9(1), 53–69.

Asemoloye, M.D., Jonathan, S.G., Ahmad, R., 2019. Synergistic plant-microbes' interactions in the rhizosphere: A potential headway for the remediation of hydrocarbon polluted soils. *Int. J. Phytoremediation* 21(2), 71–83. https://doi.org/10.1080/15226514.2018.1474437

Ayilara, M.S., Babalola, O.O., 2023. Bioremediation of environmental wastes: The role of microorganisms. *Front. Agron.* 5, 1183691. https://doi.org/10.3389/fagro.2023.1183691

Backer, R., Rokem, J.S., Ilangumaran, G., Lamont, J., Praslickova, D., Ricci, E., Subramanian, S., Smith, D.L., 2018. Plant growth-promoting rhizobacteria: Context, mechanisms of action, and roadmap to commercialization of biostimulants for sustainable agriculture. *Front. Plant Sci.* 9, 1473. https://doi.org/10.3389/fpls.2018.01473

Bala, S., Garg, D., Thirumalesh, B.V., Sharma, M., Sridhar, K., Inbaraj, B.S., Tripathi, M., 2022. Toxics recent strategies for bioremediation of emerging pollutants: A review for a green and sustainable environment. *Toxics* 10(8), 484. https://doi.org/10.3390/toxics10080484

Banks, M.K., Kulakow, P., Schwab, A.P., Chen, Z., Rathbone, K., 2021. Degradation of crude oil in the rhizosphere of *Sorghum bicolor*. *Int. J. Phytoremediation* 5, 225–234.

Barton, C., Marx, D., Adriano, D.C., Koo, B.J., Newman, L., Czapka, S., Blake, J., 2018. Phytostabilization of a landfill containing coal combustion waste. *Environ. Geosci.* 12, 251–265.

Bawa, A.S., Anilakumar, K.R., 2013. Genetically modified foods: Safety, risks and public concerns-A review. *J. Food Sci. Technol.* 50(6), 1035–1046. https://doi.org/10.1007/s13197-012-0899-1

Becerra-Castro, C., Kidd, P.S., Rodríguez-Garrido, B., Monterroso, C., Santos-Ucha, P., Prieto-Fernández, Á., 2013. Phytoremediation of hexachlorocyclohexane (HCH)-contaminated soils using *Cytisus striatus* and bacterial inoculants in soils with distinct organic matter content. *Environ. Pollut.* 178, 202–210. https://doi.org/10.1016/j.envpol.2013.03.027

Becher, M., Talke, I.N., Krall, L., Kramer, U., 2019. Cross-species microarray transcript profiling reveals high constitutive expression of metal homeostasis genes in shoots of the zinc hyperaccumulator *Arabidopsis halleri. Plant J.* 37(2), 251–268.

Beneduzi, A., Ambrosini, A., Passaglia, L.M., 2012. Plant growth-promoting rhizobacteria (PGPR): Their potential as antagonists and biocontrol agents. *Genet. Mol. Biol.* 35(4), 1044–1051. https://doi.org/10.1590/s1415-47572012000600020

Berti, W.R., Cunningham, S.D., 2020. Phytostabilization of metals. In *Phytoremediation of Toxic Metals: Using Plants to Clean Up the Environment*, Raskin, I.B., Ensley, D., Eds., Wiley: New York, 122–136.

Burken, J.G., Schnoor, J.L., 2018. Distribution and volatilization of organic compounds following uptake by hybrid poplar trees. *Int. J. Phytoremediation* 1(2), 139–151.

Burragoni, S.G., Jeon, J., 2021. Applications of endophytic microbes in agriculture, biotechnology, medicine, and beyond. *Microbiol. Res.* 245, 126691. https://doi.org/10.1016/j.micres.2020.126691

Byss, M., Elhottová, D., Tříska, J., Baldrian, P., 2008. Fungal bioremediation of the creosote-contaminated soil: Influence of *Pleurotus ostreatus* and *Irpex lacteus* on polycyclic aromatic hydrocarbons removal and soil microbial community composition in the laboratory-scale study. *Chemosphere* 73(9),1518–1523. https://doi.org/10.1016/j.chemosphere.2008.07.030

Cantamessa, S., Massa, N., Gamalero, E., Berta, G., 2020. Phytoremediation of a highly arsenic polluted site, using *Pteris vittata* L. and Arbuscular Mycorrhizal Fungi. *Plants (Basel, Switz)* 9(9), 1211. https://doi.org/10.3390/plants9091211

Catambacan, D.G., Cumagun, C.J.R., 2021. Weed associated fungal endophytes as biocontrol agents of *Fusarium oxysporum* f. sp. cubense TR4 in Cavendish banana. *J. Fungi* 7(3), 224. https://doi.org/10.3390/jof7030224

Chan, W.K., Wildeboer, D., Garelick, H., Purchase, D., 2016. Mycoremediation of heavy metal/metalloid-contaminated soil: Current understanding and future prospects. In *Fungal Applications in Sustainable Environmental Biotechnology. Fungal Biology*, Purchase, D., Ed., Springer: Cham, 189–205. https://doi.org/10.1007/978-3-319-42852-9_10

Chaudhary, R., Kumar, V., Gupta, S., Naik, B., Prasad, R., Mishra, S., Saris, P.E.J., Kumar, V., 2023. Finger millet (*Eleusine coracana*) plant-endophyte dynamics: Plant growth, nutrient uptake, and zinc biofortification. *Microorganisms* 11, 973. https://doi.org/10.3390/microorganisms11040973

Chauhan, P., Sharma, N., Tapwal, A., Kumar, A., Verma, G.S., Meena, M., Seth, C.S., Swapnil, P., 2023. Soil microbiome: Diversity, benefits and interactions with plants. *Sustainability* 15(19), 14643. https://doi.org/10.3390/su151914643

Chen, F., Tan, M., Ma, J., Zhang, S., Li, G., Qu, J., 2016. Efficient remediation of PAH-metal co-contaminated soil using microbial-plant combination: A greenhouse study. *J. Hazard. Mater.* 302, 250–261. https://doi.org/10.1016/j.jhazmat.2015.09.068

Chen, Y.C., Banks, M.K., Schwab, A.P., 2020. Pyrene degradation in the rhizosphere of tall fescue (*Festuca arundinacea*) and switchgrass (*Panicum virgatum* L.). *Environ. Sci. Technol.* 37, 5778–5782.

Chi, X., Zhang, Y., Wang, D., Wang, F., Liang, W., 2018. The greater roles of indigenous microorganisms in removing nitrobenzene from sediment compared with the exogenous *Phragmites australis* and strain JS45. *Front. Environ. Sci. Eng.* 12, 1–11. https://doi.org/10.1007/s11783-018-1016-0

Chun, S.C., Muthu, M., Hasan, N., Tasneem, S., Gopal, J., 2019. Mycoremediation of PCBs by *Pleurotus ostreatus*: Possibilities and prospects. *Appl. Sci.* 9(19), 4185. https://doi.org/10.3390/app9194185

Cocozza, C., Vitullo, D., Lima, G., Maiuro, L., Marchetti, M., Tognetti, R., 2014. Enhancing phytoextraction of Cd by combining poplar (clone "I-214") with *Pseudomonas fluorescens* and microbial consortia. *Environ. Sci. Pollut. Res.* 21, 1796–1808. https://doi.org/10.1007/s11356-013-2073-3

Correa García, S., Pande, P., Séguin, A., St Arnaud, M., Yergeau, E., 2018. Rhizoremediation of petroleum hydrocarbons: A model system for plant microbiome manipulation. *Microb. Biotechnol.* 11(5), 819–832. https://doi.org/10.1111/1751-7915.13303

Das, N., Chandran, P., 2011. Microbial degradation of petroleum hydrocarbon contaminants: An overview. *Biotechnol. Res. Int.* https://doi.org/10.4061/2011/941810

Dash, D.M., Osborne, J.W., 2020. Biodegradation of monocrotophos by a plant growth promoting Bacillus aryabhattai (VITNNDJ5) strain in artificially contaminated soil. *Int. J. Environ. Sci. Technol.* 17(3), 1475–1490. https://doi.org/10.1007/s13762-019-02432-1

de la Rosa, G., Peralta-Videa, J.R., Montes, M., Parsons, J.G., Cano-Aguilera, I., Gardea-Torresdey, J.L., 2004. Cadmium uptake and translocation in tumbleweed (*Salsola kali*), a potential Cd-hyperaccumulator desert plant species: ICP/OES and XAS studies. *Chemosphere* 55, 1159–1168. https://doi.org/10.1016/j.chemosphere.2004.01.028

Deshmukh, R., Khardenavis, A.A., Purohit, H.J., 2016. Diverse metabolic capacities of fungi for bioremediation. *Indian J. Microbiol.* 56(3), 247–264. https://doi.org/10.1007/s12088-016-0584-6

Din B.U., Rafique M., Javed M.T., Kamran M.A., Mehmood S., Khan M., et al., 2020. Assisted phytoremediation of chromium spiked soils by *sesbania sesban* in association with *bacillus xiamenensis* PM14: A biochemical analysis. *Plant Physiol. Biochem.* 146, 249–258. https://doi.org/10.1016/j.plaphy.2019.11.010

Fernandes, A.F.T., Wang, P., Staley, C., Aparecida Silva Moretto, J., Miguel Altarugio, L., Chagas Campanharo, S., Guedes Stehling, E., Jay Sadowsky, M., 2020. Impact of atrazine exposure on the microbial community structure in a Brazilian tropical latosol soil. *Microbes Environ.* 35(2), ME19143. https://doi.org/10.1264/jsme2.ME19143

Fulekar, M.H., Pathak, B., Fulekar, J., Godambe, T., 2013. Bioremediation of organic pollutants using *Phanerochaete chrysosporium*. In *Fungi as Bioremediators. Soil Biology*, vol 32, Goltapeh, E., Danesh, Y., Varma, A., Eds., Springer, Berlin: Heidelberg, 265–279. https://doi.org/10.1007/978-3-642-33811-3_6

Gallo, G., Lo Piccolo, L., Renzone, G., La Rosa, R., Scaloni, A., Quatrini, P., Puglia, A.M., 2012. Differential proteomic analysis of an engineered *Streptomyces coelicolor* strain reveals metabolic pathways supporting growth on n-hexadecane. *Appl. Microbiol. Biotechnol.* 94(5), 1289–1301. https://doi.org/10.1007/s00253-012-4046-8

García-Delgado, C., Yunta, F., Eymar, E., 2015. Bioremediation of multi polluted soil by spent mushroom (*Agaricus bisporus*) substrate: Polycyclic aromatic hydrocarbons degradation and Pb availability. *J. Hazard. Mater.* 300, 281–288. https://doi.org/10.1016/j.jhazmat.2015.07.008

García-Salgado, S., García-Casillas, D., Quijano-Nieto, M.A., Bonilla-Simón, M.M., 2012. Arsenic and heavy metal uptake and accumulation in native plant species from soils polluted by mining activities. *Water Air Soil Pollut.* 223, 559–572. https://doi.org/10.1007/s11270-011-0882-x

Ghosal, D., Ghosh, S., Dutta, T.K., Ahn, Y., 2016. Current state of knowledge in microbial degradation of polycyclic aromatic hydrocarbons (PAHs): A review. *Front. Microbiol.* 7, 1369. https://doi.org/10.3389/fmicb.2016.01369

Godsy, E.M., Warren, E., Paganelli, V.V., 2021. The role of microbial reductive dechlorination of TCE at a phytoremediation site. *Int. J. Phytoremediation* 5, 73–88.

Gómez-Godínez, L.J., Aguirre-Noyola, J.L., Martínez-Romero, E., Arteaga-Garibay, R.I., Ireta-Moreno, J., Ruvalcaba-Gómez, J.M., 2023. A look at plant-growth-promoting bacteria. *Plants (Basel, Switz)*, 12(8), 1668. https://doi.org/10.3390/plants12081668

Gowtham, H.G., Hema, P., Murali, M., Shilpa, N., Nataraj, K., Basavaraj, G.L., Singh, S.B., Aiyaz, M., Udayashankar, A.C., Amruthesh, K.N., 2024. Fungal endophytes as mitigators against biotic and abiotic stresses in crop plants. *J. Fungi* 10(2),116. https://doi.org/10.3390/jof10020116

Guarino, C., Sciarrillo, R., 2017. Effectiveness of in situ application of an integrated phytoremediation system (IPS) by adding a selected blend of rhizosphere microbes to heavily multi-contaminated soils. *Ecol. Eng.* 99, 70–82. https://doi.org/10.1016/j.ecoleng.2016.11.051

He, S., Guo, H., He, Z., Yang, C., Yu, T., Chai, Q., Lu, L., 2019. Interaction of *Lolium perenne* and *Hyphomicrobium* sp. GHH enhances the removal of 17α-ethinyestradiol (EE2) from soil. *J. Soils Sediments* 19, 1297–1305. https://doi.org/10.1007/s11368-018-2116-y

Hlihor, R., Cozma, P., Gavrilescu, M., 2022. Removal of heavy metals from the environment by phytoremediation and microbial remediation. In *Sustainable Solutions for Environmental Pollution*, Wiley: Hoboken, NJ, USA; Scrivener Publishing LLC: Beverly, MA, USA, 231–245.

Igiri, B.E., Okoduwa, S.I.R., Idoko, G.O., Akabuogu, E.P., Adeyi, A.O., Ejiogu, I.K., 2018. Toxicity and bioremediation of heavy metals contaminated ecosystem from tannery wastewater: A review. *J. Toxicol.*, 2568038. https://doi.org/10.1155/2018/2568038

Iqbal, B., Li, G., Alabbosh, K.F., Hussain, H., Khan, I., Tariq, M., Javed, Q., Naeem, M., Ahmad, N., 2023. Advancing environmental sustainability through microbial reprogramming in growth improvement, stress alleviation, and phytoremediation. *Plant Stress* 10, 100283. https://doi.org/10.1016/j.stress.2023.100283

Iravani, S., Varma, R.S., 2022. Genetically engineered organisms: Possibilities and challenges of heavy metal removal and nanoparticle synthesis. *Clean Technol.* 4(2), 478. https://doi.org/10.3390/cleantechnol4020030

Jan, R., Khan, M.A., Asaf, S., Lubna, Lee, I.J., Kim, K.M., 2019. Metal resistant endophytic bacteria reduces cadmium, nickel toxicity, and enhances expression of metal stress related genes with improved growth of *Oryza Sativa*, via regulating its antioxidant machinery and endogenous hormones. *Plants (Basel, Switz)* 8(10), 363. https://doi.org/10.3390/plants8100363

Jia, Y., Huang, H., Sun, G.X., Zhao, F.J., Zhu, Y.G., 2018. Pathways and relative contributions to arsenic volatilization from rice plants and paddy soil. *Environ. Sci. Technol.* 46(15), 8090–8096.

Jin, Z., Deng, S., Wen, Y., Jin, Y., Pan, L., Zhang, Y., et al., 2019. Application of *simplicillium chinense* for cd and Pb biosorption and enhancing heavy metal phytoremediation of soils. *Sci. Total Environ.* 697, 134148. https://doi.org/10.1016/j.scitotenv.2019.134148

Joner, E.J., Hirmann, D., Szolar, O.H., Todorovic, D., Leyval, C., Loibner, A.P., 2019. Priming effects on PAH degradation and ecotoxicity during a phytoremediation experiment. *Environ. Pollut.* 128, 429–435.

Ju W., Liu L., Fang L., Cui Y., Duan C., Wu H., 2019. Impact of co-inoculation with plant-growth-promoting *rhizobacteria* and *rhizobium* on the biochemical responses of alfalfa-soil system in copper contaminated soil. *Ecotoxicol. Environ. Saf.* 167, 218–226. https://doi.org/10.1016/j.ecoenv.2018.10.016

Kafle, A., Timilsina, A., Gautam, A., Adhikari, K., Bhattarai, A., Aryal, N., 2022. Phytoremediation: Mechanisms, plant selection and enhancement by natural and synthetic agents. *Environ. Adv.* 8, 100203. https://doi.org/10.1016/j.envadv.2022.100203

Karigar, C.S., Rao, S.S., 2011. Role of microbial enzymes in the bioremediation of pollutants: A review. *Enzyme Res.*, 805187. https://doi.org/10.4061/2011/805187

Kebede, G., Tafese, T., Abda, E.M., Kamaraj, M., Assefa, F., 2021. Factors influencing the bacterial bioremediation of hydrocarbon contaminants in the soil: Mechanisms and impacts. *J. Chem.*, 1–17. https://doi.org/10.1155/2021/9823362

Khalil, A.M.A., Hassan, S.E., Alsharif, S.M., Eid, A.M., Ewais, E.E., Azab, E., Gobouri, A.A., Elkelish, A., Fouda, A., 2021. Isolation and characterization of fungal endophytes isolated from medicinal plant *Ephedra pachyclada* as plant growth-promoting. *Biomolecules* 11(2), 140. https://doi.org/10.3390/biom11020140

Khanna, K., Jamwal, V.L., Gandhi, S.G., Ohri, P., Bhardwaj, R., 2019. Metal resistant PGPR lowered Cd uptake and expression of metal transporter genes with improved growth and photosynthetic pigments in *Lycopersicon esculentum* under metal toxicity. *Sci. Rep.* 9, 5855. https://doi.org/10.1038/s41598-019-41899-3

Knuteson, S.L., Whitwell, T., Klaine, S.J., 2020. Influence of plant age and size on simazine toxicity and uptake. *J. Environ. Qual.* 31, 2096–2103.

Koptsik, G., 2014. Problems and prospects concerning the phytoremediation of heavy metal polluted soils: A review. *Eurasian Soil Sci.* 47, 923–939. https://doi.org/10.1134/S1064229314090075

Kubota, H., Takenaka, C., 2003. Field note: *Arabis gemmifera* is a hyperaccumulator of Cd and Zn. *Int. J. Phytoremediation* 5, 197–201. https://doi.org/10.1080/713779219

Kumar, M., Kumar, V., Varma, A., Prasad, R., Sharma, A., Pal, A., Arshi, A., Singh, J., 2016. An efficient approach towards the bioremediation of Copper, Cobalt and Nickel contaminated field samples. *J. Soils Sediments* 17(4), 2118. https://doi.org/10.1007/s11368-016-1398-1

Kumar, M., Prasad, R., Goyal, P., Teotia, P., Tuteja, N., Varma, A., Kumar, V., 2017. Environmental biodegradation of xenobiotics: Role of potential microflora. In: *Xenobiotics in the Soil Environment. Monitoring, Toxicity and Management*, Hashmi, M.Z., Kumar, V., Verma, A., Eds., Springer International Publishing, 319–334.

Kumar, V., Behl, R.K., Narula, N., 2001a. Establishment of P-solubilizing analogue resistant strains of *Azotobacter chroococcum* in rhizosphere and their effect on wheat under greenhouse conditions. *Microbial. Res.* 156(1), 87–94. https://doi.org/10.1078/0944-5013-00081

Kumar, V., Behl, R.K., Narula, N., 2001b. Effect of P-solubilizing *Azotobacter chroococcum* on yield traits and their survival in the rhizosphere of wheat genotypes under field conditions. *Acta. Agron. Hung.* 49(2), 141–149.

Kumar, V., Nautiyal, C.S., 2022. Plant abiotic and biotic stress alleviation: From an endophytic microbial perspective. *Curr. Microbiol.* 79, 311–325. https://doi.org/10.1007/s00284-022-03012-2

Lei, M., Zhang, L., Lei, J., Zong, L., Li, J., Wu, Z., Wang, Z., 2015. Overview of emerging contaminants and associated human health effects. *Biomed. Res. Int.* 2015(1), 404796. https://doi.org/10.1155/2015/404796

Leong, Y.K., Chang, J.S., 2020. Bioremediation of heavy metals using microalgae: Recent advances and mechanisms. *Bioresour. Technol.* 303, 122886.

Li, H., Sheng, G., Sheng, W., Xu, O., 2019. Uptake of trifluralin and lindane from water by ryegrass. *Chemosphere* 48, 335–341.

Li Q., Xing Y., Fu X., Ji L., Li T., Wang J., et al., 2021. Biochemical mechanisms of rhizospheric bacillus subtilis-facilitated phytoextraction by alfalfa under cadmium stress–microbial diversity and metabolomics analyses. *Ecotoxicol. Environ. Saf.* 212, 112016. https://doi.org/10.1016/j.ecoenv.2021.112016

Li, X., Zhang, X., Yang, Y., Li, B., Wu, Y., Sun, H., Yang, Y., 2016. Cadmium accumulation characteristics in turnip landraces from China and assessment of their phytoremediation potential for contaminated soils. *Front. Plant Sci.* 7, 1862. https://doi.org/10.3389/fpls.2016.01862

Ling, N., Wang, T., Kuzyakov, Y., 2022. Rhizosphere bacteriome structure and functions. *Nat. Commun.* 13, 836–845. https://doi.org/10.1038/s41467-022-28448-9

Liu, H., Yuan, M., Tan, S., Yang, X., Lan, Z., Jiang, Q., et al., 2015. Enhancement of arbuscular mycorrhizal fungus (*Glomus versiforme*) on the growth and cd uptake by cd-hyper-accumulator *solanum nigrum*. *Appl. Soil Ecol.* 89, 44–49. https://doi.org/10.1016/j.apsoil.2015.01.006

Lopez-Fernandez, M., Jroundi, F., Ruiz-Fresneda, M.A., Merroun, M.L., 2021. Microbial interaction with and tolerance of radionuclides: Underlying mechanisms and biotechnological applications. *Microb. Biotechnol.* 14(3), 810–828. https://doi.org/10.1111/1751-7915.13718

Ma, Y., Oliveira, R.S., Wu, L., Luo, Y., Rajkumar, M., Rocha, I., Freitas, H., 2015. Inoculation with metal-mobilizing plant-growth-promoting rhizobacterium *Bacillus sp.* SC2b and its role in rhizoremediation. *J. Toxicol. Environ. Health A* 78(13–14), 931–944. https://doi.org/10.1080/15287394.2015.1051205

Manganyi, M.C., Ateba, C.N., 2020. Untapped potentials of endophytic fungi: A review of novel bioactive compounds with biological applications. *Microorganisms* 8(12), 1934. https://doi.org/10.3390/microorganisms8121934

Manzoor, M., Gul, I., Manzoor, A., Kallerhoff, J., Arshad, M., 2021. Optimization of integrated phytoremediation system (IPS) for enhanced lead removal and restoration of soil microbial activities. *Chemosphere* 277, 130243. https://doi.org/10.1016/j.chemosphere.2021.130243

Marr, L.C., Booth, E.C., Andersen, R.G., Widdowson, M.A., Novak, J.T., 2019. Direct volatilization of naphthalene to the atmosphere at a phytoremediation site. *Environ. Sci. Technol.* 40(17), 5560–5566.

McNear Jr, D.H., 2013. The rhizosphere – Roots, soil and everything in between. *Nat. Edu. Knowl.* 4(3), 1–10.

Mello, I.S., Targanski, S., Pietro-Souza, W., Stachack, F.F.F., Terezo, A.J., Soares, M.A., 2020. Endophytic bacteria stimulate mercury phytoremediation by modulating its bioaccumulation and volatilization. *Ecotoxicol. Environ. Saf.* 202, 110818. https://doi.org/10.1016/j.ecoenv.2020.110818

Mendez, M.O., Maier, R.M., 2018. Phytostabilization of mine tailings in arid and semiarid environments-an emerging remediation technology. *Environ. Health Perspect.* 116, 278–283.

Mesa V., Navazas A., González-Gil R., González A., Weyens N., Lauga B., et al., 2017. Use of endophytic and rhizosphere bacteria to improve phytoremediation of arsenic-contaminated industrial soils by autochthonous *betula celtiberica*. *Appl. Environ. Microbiol.* 83, e03411–e03416. https://doi.org/10.1128/AEM.03411-16

Mesarch, M.B., Nakatsu, C.H., Nies, L., 2000. Development of catechol 2,3-dioxygenase-specific primers for monitoring bioremediation by competitive quantitative PCR. *Appl. Environ. Microbiol.* 66(2), 678–683.

Miglani, R., Parveen, N., Kumar, A., Ansari, M.A., Khanna, S., Rawat, G., Panda, A.K., Bisht, S.S., Upadhyay, J., Ansari, M.N., 2022. Degradation of xenobiotic pollutants: An environmentally sustainable approach. *Metabolites* 12(9), Article 9. https://doi.org/10.3390/metabo12090818

Mishra, S., Huang, Y, Li, J, Wu, X, Zhou, Z, Lei, Q, Bhatt, P, Chen, S., 2022. Biofilm-mediated bioremediation is a powerful tool for the removal of environmental pollutants. *Chemosphere* 294, 133609. https://doi.org/10.1016/j.chemosphere.2022.133609

Mishra, S., Sarma, P.M., Lal, B., 2004. Crude oil degradation efficiency of a recombinant *Acinetobacter baumannii* strain and its survival in crude oil contaminated soil microcosm. *FEMS Microbiol. Lett.* 235(2), 323–331. https://doi.org/10.1016/j.femsle.2004.05.002

Mishra, S., Srivastava, S., Tripathi, R.D., 2019. Lead detoxification by coontail (*Ceratophyllum demersum* L.) involves induction of phytochelatins and antioxidant system in response to its accumulation. *Chemosphere* 65(6), 1027–1039.

Mohapatra, B., Phale, P.S., 2021. Microbial degradation of naphthalene and substituted naphthalenes: Metabolic diversity and genomic insight for bioremediation. *Front. Bioeng. Biotechnol.* 9. https://doi.org/10.3389/fbioe.2021.602445

Molina Delgado, L., Wittich, R.M., Van Dillewijn, P., Segura, A., 2021. Plant-bacteria interactions for the elimination of atmospheric contaminants in cities. *Agronomy* 11(3), 493 https://doi.org/10.3390/agronomy11030493

Mousavi, S.M., Hashemi, S.A., Iman Moezzi, S.M., Ravan, N., Gholami, A., Lai, C.W., Chiang, W.H., Omidifar, N., Yousefi, K., Behbudi, G., 2021. Recent advances in enzymes for the bioremediation of pollutants. *Biochem. Res. Int.*, 5599204. https://doi.org/10.1155/2021/5599204

Nadarajah, K., Abdul Rahman, N.S.N., 2023. The microbial connection to sustainable agriculture. *Plants (Basel, Switz)* 12(12), 2307. https://doi.org/10.3390/plants12122307

Nayak A.K., Panda S.S., Basu A., Dhal N.K., 2018. Enhancement of toxic cr (VI), fe, and other heavy metals phytoremediation by the synergistic combination of native *bacillus cereus strain* and *vetiveria zizanioides* l. *Int. J. Phytoremediation* 20, 682–691. https://doi.org/10.1080/15226514.2017.1413332

Ndeddy Aka, R.J., Babalola, O.O., 2016. Effect of bacterial inoculation of strains of *Pseudomonas aeruginosa, Alcaligenes feacalis* and *Bacillus subtilis* on germination, growth and heavy metal (Cd, Cr, and Ni) uptake of *Brassica juncea. Int. J. Phytoremediation* 18(2), 200–209. https://doi.org/10.1080/15226514.2015.1073671

Nwachukwu, B.C., Ayangbenro, A.S., Babalola, O.O., 2021. Elucidating the rhizosphere associated bacteria for environmental sustainability. *Agriculture* 11(1), 75. https://doi.org/10.3390/agriculture11010075

Ojuederie, O.B., Babalola, O.O., 2017. Microbial and plant-assisted bioremediation of heavy metal polluted environments: A review. *Int. J. Environ. Res. Public Health* 14(12), 1504. https://doi.org/10.3390/ijerph14121504

Olguín, E.J., Sánchez-Galván, G., 2020. Aquatic phytoremediation: Novel insights in tropical and subtropical regions. *Pure Appl. Chem.* 82(1), 27–38.

Pande, V., Pandey, S.C., Sati, D., Bhatt, P., Samant, M., 2022. Microbial interventions in bioremediation of heavy metal contaminants in agroecosystem. *Front. Microbiol.* 13, 824084. https://doi.org/10.3389/fmicb.2022.824084

Pérez-Sanz, A., Millán, R., Sierra, M.J., Alarcón, R., García, P., Gil-Díaz, M., Vazquez, S., Lobo, M.C., 2012. Mercury uptake by Silene vulgaris grown on contaminated spiked soils. *J. Environ. Manag.* 95, S233–S237. https://doi.org/10.1016/j.jenvman.2010.07.018

Pezzella, C., Macellaro, G., Sannia, G., Raganati, F., Olivieri, G., Marzocchella, A., Schlosser, D., Piscitelli, A., 2017. Exploitation of *Trametes versicolor* for bioremediation of endocrine disrupting chemicals in bioreactors. *PLoS One* 12(6), e0178758. https://doi.org/10.1371/journal.pone.0178758

Priya, A.K., Muruganandam, M., Ali, S.S., Kornaros, M., 2023. Clean-up of heavy metals from contaminated soil by phytoremediation: A multidisciplinary and eco-friendly approach. *Toxics* 11(5), 422. https://doi.org/10.3390/toxics11050422

Qin, H., Wang, Z., Sha, W., Song, S., Qin, F., Zhang, W., 2024. Role of plant-growth-promoting rhizobacteria in plant machinery for soil heavy metal detoxification. *Microorganisms* 12(4), 700. https://doi.org/10.3390/microorganisms12040700

Qin, Y., Xie, X.Q., Khan, Q., Wei, J.L., Sun, A.N., Su, Y.M., Guo, D.J., Li, Y.R., Xing, Y.X., 2022. Endophytic nitrogen-fixing bacteria DX120E inoculation altered the carbon and nitrogen metabolism in sugarcane. *Front. Microbiol.* 13, 1000033. https://doi.org/10.3389/fmicb.2022.1000033

Rafati, M., Khorasani, N., Moattar, F., Shirvany, A., Moraghebi, F., Hosseinzadeh, S., 2020. Phytoremediation potential of *Populus alba* and *Morus alba* for cadmium, chromuim and nickel absorption from polluted soil. *Int. J. Environ. Res.* 5, 961–970.

Rafeeq, H., Afsheen, N., Rafique, S., Arshad, A., Intisar, M., Hussain, A., Bilal, M., Iqbal, H.M.N., 2023. Genetically engineered microorganisms for environmental remediation. *Chemosphere* 310, 136751. https://doi.org/10.1016/j.chemosphere.2022.136751

Rai, P.K., 2008. Phytoremediation of Hg and Cd from industrial effluents using an aquatic free floating macrophyte *Azolla pinnata. Int. J. Phytoremediation* 10, 430–439. https://doi.org/10.1080/15226510802100606

Rai, P.K., 2020. Phytoremediation of Hg and Cd from industrial effluents using an aquatic free floating macrophyte *Azolla pinnata. Int. J. Phytoremediation* 10(5), 430–439.

Rani, U., Sharma, S., Kumar, V., 2019. *Bacillus* species: A potential plant growth regulator. In *Bacilli and Agrobiotechnology: Phytostimulation and Biocontrol. Bacilli in Climate Resilient Agriculture and Bioprospecting*, Islam, M., Rahman, M., Pandey, P., Boehme, M., Haesaert, G., Eds., Springer: Cham. https://doi.org/10.1007/978-3-030-15175-1_2

Rawat, G., Choudhary, R., Kumar, V., Kumar, V., 2022. Biosurfactants: Promising biomolecules for environmental cleanup. In: *Sustainable Solutions for Environmental Pollution Volume 2: Air, Water and Soil Reclamation*, El-Gendy, N.S., Ed., Scrivener Publishing LLC, pp 293–320. https://doi.org/10.1002/9781119827665.ch6

Rawat, G., Kumar, V., 2021. Contributions of biosurfactants in the environment: A green and clean approach. In: *Bioprocessing of Agri-Food Residues for Production of Bioproducts*, Flores-Gallegos, A.C., Rodriguez-Jasso, R.M., Aguilar, C.N., Eds., Apple Academic Press, 201–224.

Rebello, S., Nathan, V.K., Sindhu, R., Binod, P., Awasthi, M.K., Pandey, A., 2021. Bioengineered microbes for soil health restoration: Present status and future. *Bioengineered* 12(2), 12839–12853. https://doi.org/10.1080/21655979.2021.2004645

Rochín-Hernández, L.S., Rochín-Hernández, L.J., Flores-Cotera, L.B., 2022. Endophytes, a potential source of bioactive compounds to curtail the formation-accumulation of advanced glycation end products: A review. *Molecules* 27(14), 4469. https://doi.org/10.3390/molecules27144469

Rodriguez, L., Lopez-Bellido, F., Carnicer, A., Alcalde-Morano, V., 2003. Phytoremediation of mercury-polluted soils using crop plants. *Fresenius Environ. Bull.* 9, 328–332. https://doi.org/10.1007/s11356-019-06563-3

Rohrbacher, F., St-Arnaud, M., 2016. Root exudation: The ecological driver of hydrocarbon rhizoremediation. *Agronomy* 6(1), 1. https://doi.org/10.3390/agronomy6010019

Ruomeng, B., Meihao, O., Siru, Z., Shichen, G., Yixian, Z., Junhong, C., Ruijie, M., Yuan, L., Gezhi, X., Xingyu, C., Shiyi, Z., Aihui, Z., Baishan, F., 2023. Degradation strategies of pesticide residue: From chemicals to synthetic biology. *Synth. Syst. Biotechnol.* 8(2), 302–313. https://doi.org/10.1016/j.synbio.2023.03.005

Ruttens, A., Mench, M., Colpaert, J.V., Boisson, J., Carleer, R., Vangronsveld, J., 2019. Phytostabilization of a metal contaminated sandy soil. I: Influence of compost and/or inorganic metal immobilizing soil amendments on phytotoxicity and plant availability of metals. *Environ. Pollut.* 144, 524–532.

Saha, L., Tiwari, J., Bauddh, K., Ma, Y., 2021. Recent developments in microbe–plant-based bioremediation for tackling heavy metal-polluted soils. *Front. Microbiol.* 12, 731723. https://doi.org/10.3389/fmicb.2021.731723

Sas-Nowosielska, A., Galimska-Stypa, R., Kucharski, R., Zielonka, U., Małkowski, E., Gray, L., 2008. Remediation aspect of microbial changes of plant rhizosphere in mercury contaminated soil. *Environ. Monit. Assess.* 137, 101–109. https://doi.org/10.1007/s10661-007-9732-0

Sharma, A., Kumar, P., Pahal, V., Kumar, J., Pandey, S.S., 2023. Endophytic phytohormone production and utilization of functional traits in plant growth promotion. In: *Plant Microbiome for Plant Productivity and Sustainable Agriculture. Microorganisms for Sustainability*, vol 37, Chhabra, S., Prasad, R., Maddela, N.R., Tuteja, N., Eds., Springer: Singapore. https://doi.org/10.1007/978-981-19-5029-2_15

Sharma, J.K., Juwarkar, A.A., 2020. Phytoremediation: General account and its application. In *Plant Biology and Biotechnology: Volume II: Plant Genomics and Biotechnology*, Bahadur, B., Venkat Rajam, M., Sahijram, L., Krishnamurthy, K.V., Eds., Springer India: New Delhi, 673–684.

Sharma, P., Bano, A., Singh, S.P., Dubey, N.K., Chandra, R., Iqbal, H.M.N., 2022. Recent advancements in microbial-assisted remediation strategies for toxic contaminants, *Cleaner Chem. Eng.* 2, 100020, https://doi.org/10.1016/j.clce.2022.100020

Shi, L., Deng, X., Yang, Y., Jia, Q., Wang, C., Shen, Z., Chen, Y., 2019. A Cr (VI)-tolerant strain, *Pisolithus* sp1, with a high accumulation capacity of Cr in mycelium and highly efficient assisting *Pinus thunbergii* for phytoremediation. *Chemosphere* 224, 862–872. https://doi.org/10.1016/j.chemosphere.2019.03.015

Singh, M., Kumar, A., Singh, R., Pandey, K.D., 2017. Endophytic bacteria: A new source of bioactive compounds. *3 Biotech* 7(5), 315. https://doi.org/10.1007/s13205-017-0942-z

Singh, N.P., Santal, A.R., 2019. Phytoremediation of heavy metals: The use of green approaches to clean the environment. In *Phytoremediation: Management of Environmental Contaminants*, vol. 2, Ansari, A.A., Gill, S.S., Gill, R., Lanza, G.R., Newman, L., Eds., Springer International Publishing: Cham, 115–129. https://doi.org/10.1007/978-3-319-10969-5

Smagin, A.V., Sadovnikova, N.B., Budnikov, V.I., 2023. Biodegradation of aqueous superabsorbents: Kinetic assessment using biological oxygen demand analysis. *J. Compos. Sci* 7(4), 164. https://doi.org/10.3390/jcs7040164

Solá, M.Z.S., Lovaisa, N., Costa, J.S.D., Benimeli, C.S., Polti, M.A., Alvarez, A., 2019. Multi-resistant plant growth-promoting actinobacteria and plant root exudates influence Cr (VI) and lindane dissipation. *Chemosphere* 222, 679–687. https://doi.org/10.1016/j.chemosphere.2019.01.197

Solanki, A.S., Kumar, V., Sharma, S., 2011. AM fungi, *A. chroococcum*, yield, nutrient uptake and economics of *Chlorophytum borivillianum* in Indian arid region. *Int. J. Agric. Technol.* 7(4), 983–991.

Srivastava, M., Ma, L.Q., Santos, J.A.G., 2006. Three new arsenic hyperaccumulating ferns. *Sci. Total Environ.* 364, 24–31. https://doi.org/10.1016/j.scitotenv.2005.11.002

Srivastava, S., Mishra, S., Dwivedi, S., 2019. Evaluation of zinc accumulation potential of *Hydrilla verticillata. Biol. Plant.* 53(4), 789–792.

Stępniewska, Z., Kuźniar, A., 2013. Endophytic microorganisms-promising applications in bioremediation of greenhouse gases. *Appl. Microbiol. Biotechnol.* 97(22), 9589–9596. https://doi.org/10.1007/s00253-013-5235-9

Strong, L.C., McTavish, H., Sadowsky, M.J., Wackett, L.P., 2000. Field-scale remediation of atrazine-contaminated soil using recombinant *Escherichia coli* expressing atrazine chlorohydrolase. *Environ. Microbiol.* 2(1), 91–98. https://doi.org/10.1046/j.1462-2920.2000.00079.x

Sudheer, S., Bai, R.G., Usmani, Z., Sharma, M., 2020. Insights on engineered microbes in sustainable agriculture: Biotechnological developments and future prospects. *Curr. Genomics* 21(5), 321–333. https://doi.org/10.2174/1389202921999200603165934

Suyamud, B., Thiravetyan, P., Panyapinyopol, B., Inthorn, D., 2018. Dracaena sanderiana endophytic bacteria interactions: Effect of endophyte inoculation on bisphenol A removal. *Ecotoxicol. Environ. Saf.* 157, 318–326. https://doi.org/10.1016/j.ecoenv.2018.03.066

Tan, H.W., Pang, Y.L., Lim, S., Chong, W.C. 2023. A state-of-the-art of phytoremediation approach for sustainable management of heavy metals recovery, *Environ. Technol. Innov.* 30, 103043, https://doi.org/10.1016/j.eti.2023.103043

Tiwari, P., Bae, H., 2023. Trends in harnessing plant endophytic microbiome for heavy metal mitigation in plants: A perspective. *Plants* 12(7), 1515. https://doi.org/10.3390/plants12071515

Tsai, H.C., Pietrobon, V., Peng, M., Wang, S., Zhao, L., Marincola, F.M., Cai, Q., 2022. Current strategies employed in the manipulation of gene expression for clinical purposes. *J. Transl. Med.* 20(1), 535. https://doi.org/10.1186/s12967-022-03747-3

Tsao, D.T., 2021. Overview of phytotechnologies. In *Phytoremediation*, Tsao, D.T., Ed., Springer: Berlin, Heidelberg, 1–50. https://doi.org/10.1007/3-540-45991-X_2

Tyagi, S., Kumar, V., Singh, J., Teotia, P., Bisht, S., Sharma, S., 2014. Bioremediation of pulp and paper mill effluent by dominant aboriginal microbes and their consortium. *Int. J. Environ. Res.* 8(3), 561–568.

United Nations Environment Programme. 2011. *Decoupling Natural Resource Use and Environmental Impacts from Economic Growth.* International Resource Panel, UNEP. https://wedocs.unep.org/xmlui/handle/20.500.11822/9816

Upadhyay, N., Vishwakarma, K., Singh, J., Mishra, M., Kumar, V., Rani, R., Mishra, R.K., Chauhan, D.K., Tripathi, D.K., Sharma, S., 2017. Tolerance and reduction of Chromium (VI) by Bacillus sp. MNU16 isolated from contaminated coal mining soil. *Front. Plant Sci.* 8, 778–791. https://doi.org/10.3389/fpls.2017.00778

Urgun-Demirtas, M., Stark, B., Pagilla, K., 2006. Use of genetically engineered microorganisms (GEMs) for the bioremediation of contaminants. *Crit. Rev. Biotechnol.* 26(3), 145–164. https://doi.org/10.1080/07388550600842794

Vaksmaa, A., Guerrero-Cruz, S., Ghosh, P., Zeghal, E., Hernando-Morales, V., Niemann, H., 2023. Role of fungi in bioremediation of emerging pollutants. *Front. Mar. Sci.* 10, 1070905. https://doi.org/10.3389/fmars.2023.1070905

Varner, P.M., Allemann, M.N., Michener, J.K., Gunsch, C.K., 2022. The effect of bacterial growth strategies on plasmid transfer and naphthalene degradation for bioremediation. *Environ. Technol. Innov.* 28, 102910. https://pubmed.ncbi.nlm.nih.gov/37091576/

Vega, F.E., Posada, F., Aime, M.C., Pava-Ripoll, M., Infante, F., Rehner, S.A., 2008. Entomopathogenic fungal endophytes, *Biol. Control* 46(1), 72–82, https://doi.org/10.1016/j.biocontrol.2008.01.008

Viesser, J.A., Sugai-Guerios, M.H., Malucelli, L.C. Pincerati, M.R., Karp, S.G., Maranho, L.T., 2020. Petroleum tolerant rhizospheric bacteria: Isolation, characterization and bioremediation potential. *Sci. Rep.* 10, 2060. https://doi.org/10.1038/s41598-020-59029-9

Wang, G., Ren, Y., Bai, X., Su, Y., Han, J., 2022a. Contributions of beneficial microorganisms in soil remediation and quality improvement of medicinal plants. *Plants (Basel, Switz)* 11(23), 3200. https://doi.org/10.3390/plants11233200

Wang, Q.F., Li, Q., Lin, Y., Hou, Y., Deng, Z., Liu, W., Wang, H., Xia, Z.M., 2020. Biochemical and genetic basis of cadmium biosorption by *Enterobacter ludwigii* LY6, isolated from industrial contaminated soil, *Environ. Pollut.* 264, 114637, https://doi.org/10.1016/j.envpol.2020.114637

Wang, Y., Jiao, P., Guo, W., Du, D., Hu, Y., Tan, X., Liu, X., 2022. Changes in bulk and rhizosphere soil microbial diversity and composition along an age gradient of Chinese Fir (*Cunninghamia lanceolate*) plantations in subtropical China. *Front. Microbiol.* 12, 777862. https://doi.org/10.3389/fmicb.2021.777862

Wang, Y., Oyaizu, H., 2011. Enhanced remediation of dioxins-spiked soil by a plant–microbe system using a dibenzofuran-degrading Comamonas sp. and Trifolium repens L. *Chemosphere* 85(7), 1109–1114. https://doi.org/10.1016/j.chemosphere.2011.07.028

Weyens, N., Thijs, S., Popek, R., Witters, N., Przybysz, A., Espenshade, J., Gawronska, H., Vangronsveld, J., Gawronski, S.W., 2015. The role of plant–microbe interactions and their exploitation for phytoremediation of air pollutants. *Int. J. Mol. Sci.* 16(10), 25576–25604. https://doi.org/10.3390/ijms161025576

White, J.C., Wang, X., Gent, M.P., Iannucci-Berger, W., Eitzer, B.D., Schultes, N.P., Arienzo, M., Mattina, M.I., 2020. Subspecies level variation in the phytoextraction of weathered DDE by *Cucurbita pepo*. *Environ. Sci. Technol.* 37, 4368–4373.

Wu, C., Li, F., Yi, S., Ge, F., 2021. Genetically engineered microbial remediation of soils co-contaminated by heavy metals and polycyclic aromatic hydrocarbons: Advances and ecological risk assessment. *J. Environ. Manag.* 296, 113185. https://doi.org/10.1016/j.jenvman.2021.113185

Xie, Y., Yu, F., Wang, Q., Gu, X., Chen, W., 2014. Cloning of catechol 2,3-dioxygenase gene and construction of a stable genetically engineered strain for degrading crude oil. *Indian J. Microbiol.* 54(1), 59–64. https://doi.org/10.1007/s12088-013-0411-2

Xiong, Y.H., Yang, X.E., Ye, Z.Q., He, Z.L., 2004. Characteristics of cadmium uptake and accumulation by two contrasting ecotypes of *Sedum alfredii* Hance. *J. Environ. Sci. Health A* 39, 2925–2940. https://doi.org/10.1081/LESA-200034269

Yaashikaa, P.R., Devi, M.K., Kumar, P.S., 2022. Engineering microbes for enhancing the degradation of environmental pollutants: A detailed review on synthetic biology. *Environ. Res.* 214, 113868. https://doi.org/10.1016/j.envres.2022.113868

Yan, A., Wang, Y., Tan, S.N., Yusof, M.L.M., Ghosh, S., Chen, Z., 2020. Phytoremediation: A promising approach for revegetation of heavy metal-polluted land. *Front. Plant Sci.* 11. https://doi.org/10.3389/fpls.2020.00359

Yan, X., Huang, J., Xu, X., Chen, D., Xie, X., Tao, Q., Jiang, J., 2018. Enhanced and complete removal of phenylurea herbicides by combinational transgenic plant-microbe remediation. *Appl. Environ. Microbiol.* 84(14), e00273-18. https://doi.org/10.1128/AEM.00273-18

Yao, Z., Li, J., Xie, H., Yu, C., 2020. Review on remediation technologies of soil contaminated by heavy metals. *Procedia Environ. Sci.* 16, 722–729.

Ying, X., Dongmei, G., Judong, L., Zhenyu, W., 2011. Plant-microbe interactions to improve crude oil degradation. *Energy Procedia* 5, 844–848. https://doi.org/10.1016/j.egypro.2011.03.149

Zhang, J., Zhang, M., Song, H., Zhao, J., Shabala, S., Tian, S., et al., 2020. A novel plasma membrane-based NRAMP transporter contributes to cd and zn hyper accumulation in *sedum alfrediihance*. *Environ. Exp. Bot.* 176, 104121. https://doi.org/10.1016/j.envexpbot.2020.104121

Zhang, R., Cui, Z., Zhang, X., Jiang, J., Gu, J.D., Li, S., 2006. Cloning of the organophosphorus pesticide hydrolase gene clusters of seven degradative bacteria isolated from a methyl parathion contaminated site and evidence of their horizontal gene transfer. *Biodegradation* 17(5), 465–472. https://doi.org/10.1007/s10532-005-9018-6

8 Phytoremediation of Polluted Soil and Waters

Arezoo Dadrasnia and Rosazlin Abdullah

8.1 INTRODUCTION

The environment is a defined and distinct area consisting of interconnected living and non-living elements. The environment is maintained and operational through intricate interactions that are oriented toward physical, biological, and chemical processes. Nevertheless, the natural condition of the environment is significantly impacted by a multitude of anthropogenic activities. Unfortunately, human needs and associated economic interests are insatiable, rendering some degree of direct environmental damage inevitable. However, the rehabilitation of the compromised or contaminated environment is a significant topic for many experts. Removing contaminants from the environment is an essential approach for returning any environmental medium to its natural or original state, often termed "environmental restoration" (Kuppan et al., 2024). Although the introduction of toxins or pollution into the air, water, and soil can occur easily, quickly, and persistently, their removal is frequently a challenging endeavor (Dadrasnia et al., 2015). Various removal strategies involve physical, chemical, and biological methods. These processes have been extensively analyzed in terms of their advantages and disadvantages. However, phytoremediation is increasingly becoming recognized as one of the most widely used systems for environmental restoration. Therefore, the objective of this chapter is to clarify the potential of phytoremediation as a viable alternative for environmental restoration.

8.2 PLANT-BASED REMEDIAL TECHNIQUES

The 20th century is recognized as a period characterized by the progressive acquisition of environmental knowledge aimed at safeguarding the planet on a global scale. Inadequate legislation and government intervention have resulted in extensive soil and groundwater pollution at various locations worldwide. Pollution has been prevalent in the environment since 1950, necessitating significant public attention. Various private and public entities prioritized the proper disposal of hazardous garbage. The restoration of contaminated lands in the United States costs billions of dollars. Nevertheless, the absence of an affordable and appropriate technique has restricted the clean-up of trash in contaminated areas, hence exacerbating the potential threat to human health. Remediation techniques vary depending on the source of the contaminants. While, among diverse sorts of approaches, phytotechnology is applicable to all types. Phytoremediation is a biological approach that relies on plants

 DOI: 10.1201/9781003714484-8

to eliminate contaminants from soil, water, and air. The term "phyto" is derived from the Greek word "phyton," which means "that which has grown" and is associated with the color green (Kafle et al., 2022). Therefore, the phenomenon is the utilization of plants to eliminate, modify, or relocate pollutants that are present in the environment, particularly in terrestrial and aquatic systems. Phytoremediation is perceived as an eco-friendly and cost-effective alternative to other conventional soil remediation technologies, such as chemical processing, which is frequently preceded by excavation (Yadav et al., 2022). The purity of effluents discharged from various mines and industries has been significantly enhanced by phytoremediation, as evidenced by investigations (Atakhanova, 2023). Occasionally, this remediation method has been employed to purify wastewater that is contaminated with heavy metals. Furthermore, phytoremediation offers other advantages. The biomass derived from plants cultivated in a polluted region can be utilized as biofuel once it is harvested. In certain circumstances, plants may be allowed to grow on polluted sites, serving as prospective species for the regeneration of the ecosystem. This process contributes to the increase of local biodiversity, enhances the absorption of atmospheric CO_2, and helps to contain the contamination within the soil. Various air pollutants, such as rainfall, the oxides of nitrogen, ozone, and sulfur dioxide, can be effectively reduced by plants, making them an excellent instrument for mitigating the levels of these toxins in contaminated air. The advanced techniques employed to eliminate toxins, including DNT, TNT, heavy metals, and organic substances from groundwater include complex oxidation and granular carbon processes (Schneider & Bláha, 2020). The methods used to clean up groundwater are costly. Phytotechnology has the potential to be used for treatment in this context.

Nevertheless, aquatic plants are utilized in wastewater treatment to remove heavy metals, nutrients, leachate, and organic compounds (Bayuo et al., 2024). *Ceratophyllum demersum, Eichhornia crassipes, Elodea canadensis*, etc., have been extensively utilized in the field of phytotoxicology. Aquatic plants have a crucial role in regulating water quality, the availability of metals in the environment, and the cycling of nutrients. Various plant enzymes, such as glucosyltransferases, aromatic dehalogenases, peroxygenases, and phosphatases, are involved in phytodegradation (Wolfe & Hoehamer, 2003). These enzymes facilitate the transformation of polluted compounds into less harmful ones.

Furthermore, phytodegradation technology has been employed to treat radioactive contamination resulting from nuclear fuel testing, unintentional spills, and emissions (Nedjimi, 2021). Phytoremediation can augment environmental restoration by increasing efficiency, as plants and microorganisms frequently have a synergistic relationship with their environment. However, the efficacy of phytoremediation does not indicate that it is without drawbacks. A notable challenge linked to phytoremediation is the existence of "uncertainty." This uncertainty stems from the differing tolerances of diverse plant species to contaminated environments, the effectiveness of the process is contingent upon the kind and concentration of contaminants, and the efficiency is affected by specific microbial interactions (Chandra et al., 2017). In summary, phytotechnology can be employed for the remediation of heavy metals, radioactive substances, and petroleum hydrocarbons. The challenges in assessing the early stage can be attributed to the method's sluggishness and protracted duration, often exceeding ten years. The practice of employing plants for the remediation and purification of water and soil is an age-old technique that has been utilized for over three centuries. Multiple studies have been conducted to determine the ability of plants to remove pollutants and toxic chemicals.

Researchers first became interested in phytoremediation research, and Rufus L. Chaney, a USDA agronomic, was one of the initial researchers to disseminate information regarding the potential of hyperaccumulator plants. In 1978, he co-authored a paper titled "The Physiology of Metal Toxicity in Plants," which examined the mechanisms by which certain plants endure and sequester metals. Subsequently, in 1995, Chaney et al. (2007) and associates examined the prospective application of metal hyperaccumulators in environmental management. These works substantially enhanced the comprehension and utilization of hyperaccumulator plants in phytoremediation and phytomining. Aquatic plants have been utilized more recently to detoxify radioactively contaminated waterways. Currently, numerous laboratories in both industry and academia are engaged in extensive phytoremediation initiatives. Evaluations of the efficacy of soil and groundwater plants for remedial purposes have been conducted through field studies. The viability of using plants as a remediation technology at a contaminated site will be covered in the sections that follow.

8.3 PHYTOTECHNOLOGY AS A SUSTAINABLE ENVIRONMENTAL BIOTECHNOLOGY

It is inevitable to clean up contaminated soils using an inexpensive technique, especially in poor nations. When physical or chemical procedures fail to remove oil contamination from soil, biological solutions can be the most successful (Park and Oh, 2023). This makes biological processes seem low-risk to the environment and inexpensive, yet sometimes they take longer. Utilizing plants, phytoremediation is a relatively new green method that cleans up contaminated areas in soil, sediment, surface water, and groundwater (Paz-Alberto 2013).

A remediation strategy can be applied wherever that can sustain plant growth, as phytoremediation is both quite simple to apply and economical with low maintenance requirements. Compared to other cleanup techniques, phytoremediation seems to be efficient, affordable, and visually appealing. The perspective strategy has a positive reputation and is frequently more affordable than other methods.

Utilizing phytotechnology can help clean up contamination from radioactive elements, heavy metals, and petroleum hydrocarbons. It is challenging to assess this technique's potential early on because it is slow and frequently takes longer than ten years. Phytoremediation studies requires understanding of identity (common and scientific names), experimental conditions (field or laboratory plant morphology, growth form and duration. The plant survial under high level of contaminations requires, specific smechanism of phytoremediation such as uptake, transformation, and stabilization which can further be enhanced by the interactions with rhizospheric microbes. The efficiency can also be linked with the age of plants when exposed to the contamination, nutrient availability for plant growth, site of contamination in the plant tissues (i.e., root, stem, leaf, or no storage). Further, understanding of the cultural practices related to plant establishment and management is necessary to provide critical insights into the feasibility of large-scale phytoremediation's applications. (Kafle et al., 2022).

Not all phytotoxic substances can be treated using phytoremediation; some compounds or large concentrations of pollutants may not be suitable for this method. Moreover, depending on the key removal processes in a given phytoremediation application, phytoremediation is only applicable to contamination inside the rhizosphere or the depth of evapotranspiration's effect.

One of the main disadvantages of phytoremediation is that it might take a long time for organic pollutants to be detoxified. If the decomposition process is not finished, hazardous substances can build up in plant tissue and leak through the environment or food chains.

8.4 METHODS FOR CLEANING UP OIL SPILLS AND THE FUNCTION OF PHYTOREMEDIATION

Petroleum hydrocarbons are the foundation of and govern the modern industrial society. Without petroleum oil, our modern global networked economy would not be able to function, which would cause calamity for a large number of people. Nonetheless, oil spills caused by industry and human activity happen all the time around the world. The significance of these discharges stems from the possibility that a leaky underground storage tank could contaminate groundwater supplies that are used as sources of drinking water with hazardous waste or petroleum. Oil pollution is increasingly mostly caused by vehicles leaking petroleum hydrocarbons across the road and washing oil into coastal areas. Cars, ships, lorries, and tractors run on gasoline, diesel, and kerosene. To understand the extent of unintentional hydrocarbon leakage, let's examine some global statistics. Over 63.5 million barrels of petroleum were consumed daily worldwide in 2003. According to US Energy Information Administration (EIA) forecasts, 98 million barrels of oil were used globally each day in 2006. According to EIA estimates, oil consumption will reach 118 million barrels per day by 2030.

According to Sonawdekar (2012), there is an estimated 600,000 metric tons of natural crude oil spill annually, with a 200,000 metric ton uncertainty range (Sonawdekar, 2012).

Globally, there has been a lot of news of oil leaks. For example, in 1989, there was a 0.04 megatons crude oil spill into Prince William Sound, Alaska (Zapata, 2021). In addition, the Prestige oil disaster in 2002 spilled 63,000 tons of fuel oil over 1,900 km of coastlines in northern and northern regions of Spain as well as western France, 209 kilometers offshore (Penela-Arenaz et al., 2009). Certain developing nations, such as Iran, the first oil-producing nation in the Middle East, have begun producing more than four million barrels of petroleum products and 80,000 m^3 of diesel fuel each day. The release of crude oil into the surrounding environment has contaminated approximately 1.5×10^6 m^3 of soil surrounding the Tehran refinery (Adeniran et al., 2023).

Growing global awareness of this problem is increasing the number of polluted sites that will need to be remedied shortly (Weyens et al., 2015). Additionally, the location where the spill has occurred will determine the capability of the remediation approaches. Many biological methods, such as phytoremediation, have recently been developed to speed up the rate of breakdown. This technique is under investigation for the remediation of oil-contaminated sites. Phytoremediation may be a preferable option for remediation due to its applicability in situ and lower cost compared to other current approaches (Kuppan et al., 2024). This strategy minimizes environmental disruption that may arise during the cleanup procedure. Various plants, including legumes and grasses, have been acknowledged for their potential in this regard. In contaminated environments, they exhibit a remarkable ability to endure and accumulate petroleum hydrocarbons.

8.5 TECHNIQUES FOR USING PHYTOTREATMENT

8.5.1 In Situ Phytotreatment

Vegetation is situated in polluted areas that engage with contaminated groundwater throughout this hydrotreatment process to restore the ecosystem. This method inhibits the decomposition of contaminated material before phytoremediation occurs. If the Phyto treatment method involves the accumulation and absorption of pollutants, plants may be collected from contaminated areas and harvested for recovery after remediation. For the in situ method to be effective, the pollution must be physically accessible to the roots (Susarla et al., 2002).

8.5.2 In Vivo Phytotreatment

An alternative approach to phytoremediation is viable in areas where contaminants are beyond the reach of crops, such as contamination in deep aquifers (Chandra et al., 2017). This process involves the mechanical removal of toxins and their subsequent relocation to a temporary treatment area for exposure to specific plants specialized for phytoremediation. Following treatment, the soil or water may be restored to its original site, and if required, the vegetation can be removed for disposal. This method is typically more costly than in situ phytoremediation.

8.5.3 In Vitro Phytotreatment

This process often involves the extraction of enzymes from living plant tissues. The utilization of plant extracts in a polluted pond or wetland, or the implementation of enzyme-impregnated porous barriers within a contaminated groundwater plume, exemplifies potential in situ applications of this technology (Susarla et al., 2002). Because of the expenses associated with preparing and extracting the plant enzymes, this approach is theoretically the most expensive form of phytoremediation; nevertheless, certain plants, like tarragon (*Artemisia dracunculas* var satiya), release exudates when stressed, which may lead to lower costs of production.

8.6 HOW PHYTOREMEDIATION WORKS

Plants can remediate polluted regions through various methods (Khan et al., 2022). Plants operate as solar-powered pumping and filtration systems by absorbing pollutants, primarily water-soluble ones, through their roots and transporting them across various plant tissues for digestion, volatilization, or sequestration (Figure 8.1.)

Different strategies are employed by plants to deal with environmental contaminants found in soil. Phytoremediation methods include the biophysical and

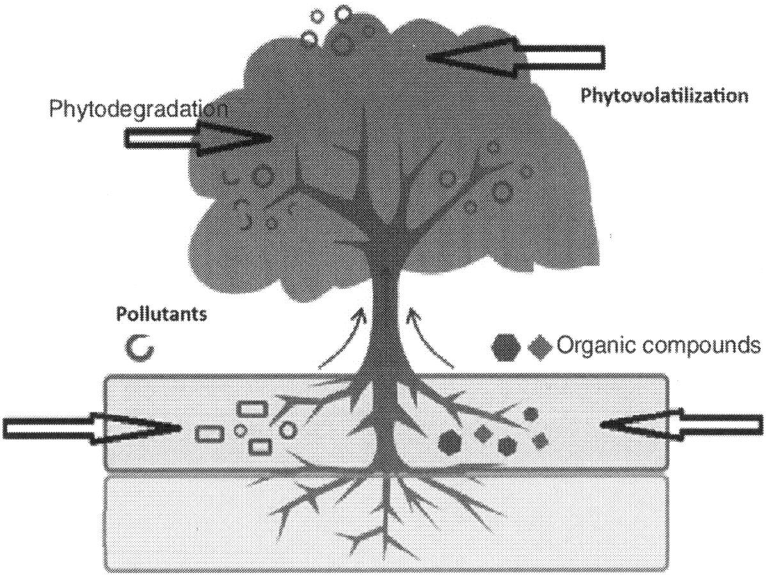

FIGURE 8.1 Schematic representation of phytoremediation strategies (Abdel-Shafy & Mansour, 2018; Gul & Marakli, 2023).

biochemical processes of adsorption, transport, translocation, transformation, and mineralization of chemicals by plant enzymes (Table 8.1) (Kafle et al., 2022). It has been demonstrated that plants may break down heterologous chemicals, such as trichloroethylene (TCE), through oxidative decomposition procedures that involve plant-specific dehalogenases.

Even after the plants died, dehalogenase activity was seen to persist. When plants die, they decompose and are buried in the sediment, which causes enzymes to become bound to the organic matrix of the sediment and increases the dehalogenase action seen in chemical-rich sediments (Nzengung et al., 1999). Plants include a wide range of enzymes that break down contaminants. These comprise laccases, phosphatases, dehalogenases, nitrilases, nitroreductases, peroxidases, dioxygenases, and P450 monooxygenases. The fundamental biological procedures that higher plants and related microbes go through, like transpiration, photosynthesis, digestion, and substance nutrition, are the foundation of phytoremediation. When roots interact with microorganisms in the rhizosphere, they can stabilize and bind chemicals on their external surfaces (Farraji et al., 2017). Plants grow their roots in soils, sediments, and water. Roots can also take up organic and inorganic substances (Centenaro et al., 2018). The many cells and tissues of the plant can transport, store, transform, or accumulate uptake chemicals("Plant–Water Relations (1): Uptake and Transport," 2014). Finally, the plant's aerial parts may engage in gas exchange with the atmosphere to enable absorption or molecular release (Centenaro et al., 2018).

TABLE 8.1
Different Types of Phytoremediation with Examples and Mechanisms

Phytoremediation Type	Description	Examples of Plants	Pollutants Treated	Mechanisms	Reference
Phytoextraction	Plants absorb pollutants from the soil and translocate them to the shoots	Indian mustard (*Brassica juncea*)	Heavy metals (e.g., Pb, Cd)	Uptake and accumulation	Lim et al. (2004)
Phytostabilization	Plants immobilize contaminants in the root zone, reducing mobility	Poplar trees (*Populus* spp.)	Heavy metals, radionuclides	Adsorption and stabilization	Guerra et al. (2011)
Phytodegradation	Plants and their associated microorganisms degrade organic pollutants	Sunflower (Helianthus annuus)	Pesticides, hydrocarbons	Biodegradation	Al-Jobori and Kadhim (2019)
Phytovolatilization	Plants take up pollutants and release them into the atmosphere as vapors	Certain species of eucalyptus	Mercury, some organic compounds	Volatilization	Tonelli et al. (2020)

8.6.1 Phytoextraction

Phytoextraction is the method by which plants extract and sequester contaminants. It is frequently employed to extract and preserve metals that can undergo speciation in plants but are indigestible. Phytoremediation refers to the ability of plants to absorb toxins via their roots and translocate them to aerial parts, such as stems or leaves (Yan et al., 2020). Upon absorption, a chemical can be sequestered in the plant by binding it or its by-products to the lignin inside the plant biomass via a process known as lignification (Alaoui et al., 2022). Alternatively, the chemical may be sequestered in the cell vacuoles of aerial tissues instead of root cells, as part of a process termed phytosequestration. More precisely, studies have demonstrated that organic substances with log K_{ow} ratios ranging from 1 to 3.5 are capable of permeating plants (Gimeno et al., 2024). Consequently, hydrophobic compounds with a log K_{ow} value over 3.5 exhibit insufficient solubility in the transpiration stream or are firmly adhered to the root surface, impeding their transit into the plant (Gimeno et al., 2024). In contrast, roots do not absorb chemicals with low log K_{ow} due to their elevated polarity.

8.6.2 Phytotransformation

Phytodegradation denotes a series of mechanisms via which plants metabolize contaminants, effectively breaking down and assimilating these compounds inside their metabolic systems. Plants subjected to herbicides frequently utilize the elements of this widely researched system. Phytodegradation, or phytotransformation, is the process whereby plants absorb contaminants and then decompose, convert them into minerals, or metabolize them via internal enzymatic reactions and metabolic pathways. Plant enzymes are crucial in the phytodegradation process, breaking down pollutants such as metals, herbicides, and chlorinated solvents found in soil, sediment, and groundwater (Kafle et al., 2022). A study discovered 70 chemical molecules that can be assimilated and sequestered by plants and trees. Nevertheless, the profundity of the root system can restrict phytodegradation. Pollutant degradation can occur in environments without microorganisms, such as those where microorganisms have been eradicated by elevated pollutant levels (Singh & Kumar Mishra, 2021). This is due to the enzymes that are produced by the plant.

8.6.3 Phytostabilization

Phytostabilization is the mechanism by which plants immobilize or render pollutants insoluble in contaminated settings via their roots or the rhizosphere (Bolan et al., 2011). This process reduces the movement of pollutants and prevents their migration into groundwater (Pandey et al., 2021). The high amount of water transpired by plants, however, allows for hydraulic control to limit leachate migration (Dadrasnia et al., 2015). Plant growth may be inhibited by toxic effects at high concentrations. As a result, plants should be able to withstand high contamination concentrations, produce large amounts of root biomass that can immobilize pollutants, and retain pollutants in their roots.

8.6.4 Rhizodegradation

Rhizodegradation is the process via which contaminants in the plant rhizosphere, the microbe-dense area next to the root vascular system, are converted by Indigenous bacteria (Yang et al., 2023). Vegetation on polluted terrain can substantially influence soil moisture, organic matter, and redox conditions, mostly by the excretion of acidic chemicals from their roots (Hernández-Cáceres et al., 2022). Rhizodegradation is also known as rhizoremediation or microbe-assisted phytoremediation. It's becoming clear that this is one of the best ways that plants can improve the removal of organic pollutants, especially big, resistant substances.

8.6.5 Phytovolatilization

One of the primary mechanisms that can occur throughout the phytoremediation process is phytovolatilization. The absorption and evaporation of a pollutant by a plant, together with the release of that pollutant or a modified variant into the atmosphere due to the plant's transpiration, metabolism, and pollutant uptake. Plant transpiration emits metabolic chemical constituents into the environment during the phytovolatilization process.

8.7 INTERACTION OF PLANTS AND MICROORGANISMS

The interplay between bacteria and plants is particularly advantageous for the extraction of pollutants from soil, especially via phytoremediation mechanisms. Microorganisms, comprising bacteria, fungi, and other soil-dwelling entities, are crucial for the degradation or modification of harmful substances in the soil, including organic pollutants and heavy metals. A notable group of microorganisms includes hydrocarbon-degrading bacteria, which thrive in the rhizosphere of plants, where they metabolize detrimental chemicals like petroleum hydrocarbons. These microorganisms transform pollutants into less deleterious compounds via enzymatic processes, thereby remediating the soil. The existence of these bacteria in the rhizosphere enhances the efficacy of phytoremediation by degrading pollutants and facilitating plant growth. Microbial metabolic activities improve soil structure and nutrient availability, creating a more favorable environment for plant growth. In exchange, plants release organic molecules that support and perpetuate the microbial community, fostering a mutually beneficial symbiotic connection (Prakash et al., 2019).

The appropriate formation of contacts between plants and bacteria is essential for effective phytodegradation, as these bacterial-plant interactions can directly or indirectly influence plant growth. Plants can support hydrocarbon-degrading bacteria that aid in phytoremediation within their "rhizosphere" in the root zone (Siddique et al., 2024). Nonetheless, there is limited comprehension regarding the influence of organic compounds, such as root exudates, on these microbial processes. Further research is required to enhance the conditions facilitating the collaboration between bacteria and plants for the removal of contaminants. Petroleum hydrocarbons can significantly disturb microbial populations and impede plant growth. Hydrocarbon-resistant microbes are essential for alleviating this issue; nevertheless, the difficulty is in sustaining their activity and viability over time. Microorganisms not only facilitate breakdown but also augment the bioavailability of pollutants, enabling plants to absorb and metabolize dangerous compounds more efficiently. By breaking down complex contaminants into simpler molecules, microbes make it easier for plants to detoxify their environment. By doing so, they immediately facilitate the elimination of contaminants and aid in the restoration of soil health. To enhance phytoremediation strategies, continuous research should concentrate on augmenting microbial resilience and investigating methods to boost plant–microbe interactions, so ensuring the sustained efficacy of these natural remediation systems (Montreemuk et al., 2024).

The accessibility of soil nutrients to the plants can then be enhanced by microorganisms. Petroleum hydrocarbons are detrimental to the community of microbes as well as the growth of plants. To enhance the interactions between petroleum-based hydrocarbons and microbes, it is necessary to enhance the viability and durability of phytoremediation.

8.8 GROUNDWATER POLLUTION AND PHYTOREMEDIATION

Subsurface or groundwater courses are easily penetrated by pollutants or toxins. Groundwater cleanup is still feasible, particularly when using a phytoremediation strategy. Pollution of groundwater can typically be addressed by phytovolatization and phytodegradation. The level of the groundwater and the polluted zone are the two main factors to take into account while analyzing the contamination of groundwater. According to the in situ situation, phytoremediation of groundwater is restricted to the area of pollution in the outside portion of the water table that is also accessible to plant roots and to uncontaminated aquifers where deep levels of the water tables are reachable by plant roots. In essence, plant roots cannot spread via pure groundwater to a more deeply contaminated area. Whenever on-site phytoremediation of greater contaminated water is required, it is important to ascertain whether groundwater movement toward the root can be induced through modelling, or if the water table can be reduced by the vegetation or through pumping (Islam and Quaff, 2024).

Nevertheless, modelling may be hampered by the seasonality and unpredictability of plant water consumption rates. Therefore, to validate the modelling results, careful field measurements and cautious estimations of water uptake must be carried out in conjunction with appropriate monitoring of alterations in the water table. When a plant's roots cannot access the depth of groundwater, extraction wells are employed to extract water from the subsurface before the implementation of phytoremediation (Prieto et al., 2012). To avoid groundwater migration beyond the vegetation, it is often necessary for the pace at which plants absorb water to correspond with the rate at which groundwater flows into a phytoremediation area (Fernandez et al., 2023).

8.9 SOIL POLLUTION AND PHYTOREMEDIATION

Many materials, including organics like pesticides and heavy metals, can pollute soil. Phytoremediation is one of the most widely used remediation techniques for cleaning up contaminated soil. The remediation of heavy metal-contaminated soils is the foremost application of phytoremediation efficacy in soil restoration (Bhat et al., 2022).

Employing plants that sequester metals or possess the ability to extract harmful metals from the soil, such as Cd, Cr, Ni, Pb, and Zn, is essential for the efficient phytoremediation of contaminated soil (heavy metals) (Lavanya et al., 2024). Some plants translocate the metals absorbed by their roots to other parts, including their shoots and leaves, while also collecting them in their roots (Wang et al., 2003). Rascio and Navari-Izzo (2011) assert that numerous plants possess the capacity to accumulate exceptionally large levels of metals in their shoots, a phenomenon known as "hyperaccumulators" (Rascio & Navari-Izzo, 2011).

Consequently, rather than engaging in the labor-intensive and challenging excavation, expensive disposal, and loss of topsoil typically associated with traditional remediation methods, the metal-rich plant can be harvested and extracted from the site upon completion of treatment. Climate and pollutant bioavailability significantly influence the phytoremediation of contaminated soil, especially for certain heavy metals (Nie et al., 2024). The mobilization of metallic pollutants in soil and vegetation improves the efficacy of phytoremediation (Montreemuk et al., 2024). Consequently, lowering the soil pH and including synthetic chelates are two significant amendment strategies that enhance the availability of metals, particularly lead (Pb). A case study on lead (Pb) removal was undertaken in the Philippines with Vetiveria zizanioides L. (vetiver grass), *Imperata cylindrica* L. (cogon grass), and *Paspalum conjugatum* L. (carabao grass) cultivated in soils with differing Pb concentrations (Paz-Alberto et al., 2007). The V. zizanioides species exhibited the highest dry matter and survival rates. The maximum Pb absorption occurred in the sequence of V. zizanioides, I. cylindrica, and *P. conjugatum*, while the Pb levels in the shoots and roots of the grasses remained relatively unchanged with varying Pb concentrations in the soil (Egendorf et al., 2020). The other two, *I. cylindrica* and *P. conjugatum* showed similar trends in that order after *V. zizanioides* (Prima et al., 2014). Consequently, it suggests that plant's capacity to absorb lead is increased when its biomass grows. Additionally, plant's large root system increases its ability to absorb Pb because it allows for greater contact with soil nutrients (Tamma et al. 2025).

8.10 PHYTOREMEDIATION AND POLLUTION FROM OIL SPILLS

Some settings are prone to oil spills because of the usual exploration operations carried out there. The petroleum product is essentially a blend of several chemicals with varying volatility, the ability to dissolve, and susceptibility to microbiological degradation (Kuok Ho, 2024). While crude oil contains aromatic hydrocarbons, the rate of deterioration diminishes as the concentration of ring structures inside the oil increases (Sharma et al., 2024a, 2024b). The asphaltic percentage of hydrocarbons has more double covalent bonds and branching, which facilitates the decomposition of the aliphatic fraction while safeguarding the aesthetic fraction from microbial degradation (Montreemuk et al., 2024). Therefore, it is concerning that while the anaerobic breakdown of crude oil does proceed, it does so slowly and incompletely. The majority of oil spillage happens in marshy wetland habitats. These types of habitats contain low oxygen content soils, called hypoxia or anoxic soils. Plants in wetlands can expedite the restoration process by facilitating the diffusion of oxygen from their shoots to their roots and the surrounding soil, where it is utilized by anaerobic soil bacteria for respiration (Cahoon et al., 2021).

Certain fresh-marsh plant species, including *S. lancfolia*, *P. hemitomon*, *A. philoxeroides*, and *P. australis*, have been studied for their potential to restore environments contaminated by oil spills. Because of oiling, *P. australis* and *A. philoxeroides* had lower photosynthetic rates. Likewise, every plant species' above- and belowground biomass was impacted, except *S. lanternifolia*. However, it was discovered that *S. lancfolia* and P. hemitomon are more suited for revegetating oiled fresh-marsh wetlands in the event that the native vegetation is unable to recover.

8.11 IMPACT OF CIRCUMSTANCE FACTORS

Numerous environmental variables influence the process of the phytodegradation. The amount of water in the soil has an impact on the microbial growth of plants as well as the accessibility of oxygen needed for aerobic conditions (Sharma et al., 2024a,b). Other factors to take into account include the type of soil, the age and kind of plants, substances, the toxicology of the contaminants, the accessibility of water and oxygen, the chemical properties of the soil (pH and CEC), and the depth of pollution, which affects where the contaminants can be treated in the rhizosphere or by plant uptake (Delgado-Baquerizo et al., 2020). Numerous data indicate that phosphorus and nitrogen are the inorganic elements that most commonly inhibit the degradation of petroleum hydrocarbons in soil (Zeng et al., 2024). Petroleum-derived hydrocarbons may not readily desorb, rendering them inappropriate for phytoremediation (Zeng et al., 2024).

8.12 GUIDELINES FOR SELECTION OF PLANTS

A critical factor influencing the effectiveness of the phytoremediation project is the selection of plants. Selecting plants that can thrive at the site comes next after the growth parameters at each location have been assessed. An elementary comprehension of plant biology can be advantageous while conducting phytodegradation studies. The particular and popular names of plants, their development habits, and their adaptability to different environmental factors like humidity, temperature, and pathogens are some examples of the normal knowledge that is required about plants. Due to their adaptation to the local climate, native crops and herbs may be regarded as alternatives for the phytodegradation process. A multitude of plant species have been identified for their potential use in phytoremediation processes. Leguminous and grassy plants are the most prevalent and have demonstrated potential for phytoremediation. Grasses have a large root surface area (per m^3 of soil) and can reach depths of up to 3 meters in the soil, making them a good choice for phytoremediation (Huot et al., 2020). Due to the superior nitrogen-fixing capability of leguminous plants compared to other species, there exists a limitation on the nitrogen levels in oil-contaminated areas.

Certain characteristics of plants render them suitable for the remediation of hydrocarbon compounds, including those with elevated phytotoxicity, as well as species capable of acclimatizing to diverse climatic conditions and being removed post-remediation; Plants capable of adapting to many environmental conditions and can be eliminated after cleanup; Plants capable of effectively transporting oxygen to their stems, roots, and leaves; Plants capable of collecting and absorbing detrimental chemicals.

8.13 PROPOSED SOLUTIONS AND RECOMMENDATIONS

8.13.1 Climate Influence

Considering that climate dependency significantly limits the efficacy of phytoremediation, it is essential to examine plant biodiversity throughout diverse global regions to comprehend its function as a remediation agent. Access to a complete data inventory of plant species' properties would enable the identification of potential phytoremediators that are well adapted to their environments based on their climatic tolerance.

8.13.2 The Nature of Chemicals

The efficacy of phytoremediation is significantly affected by the properties of the chemicals or pollutants being addressed. Consequently, a proficient approach to tackle this issue is to do comprehensive pre-characterization of polluted areas, considering the distinct properties of the pollutant or chemical in its original state. This will enhance the process of structural development and the identification of suitable phytoremediation crops.

8.13.3 Fluctuations in Outcomes

The variability of results noted in previous environmental restoration initiatives employing phytoremediation has elicited considerable apprehension among specialists. This has impeded the capacity to achieve the intended result. Therefore, it is essential to meticulously evaluate prior environmental issues before executing phytoremediation at a designated contaminated site.

8.13.4 Restriction to Areas with a Lower Concentration of Contaminants

Phytoremediation is more efficacious in regions with reduced pollution levels, owing to the heightened sensitivity of plants to the detrimental impacts of pollutants. Therefore, to effectively employ plants for environmental restoration, it is essential to evaluate the pollution levels in the surrounding environment. Moreover, augmenting the density of flora in a designated location may accelerate biological transformation, as plants can engage with and degrade contaminants.

8.14 ECONOMIC OUTLOOKS AND FUTURE PROSPECTS

The economic prospects for phytoremediation are favorable, propelled by increasing environmental consciousness, governmental endorsement, and technological developments. With the growing emphasis on sustainable practices, phytoremediation is expected to assume a more prominent role in mitigating contamination issues while offering economic advantages.

Phytoremediation has established itself as a specialized method in the effort to restore damaged and polluted ecosystems using a biological approach. Nevertheless, the presence of specific gaps and ambiguities in this remediation process has highlighted the necessity for additional investigations and comprehensive research that is both effective and enduring. Hence, additional exploration might be pursued in the following areas of research and study on phytoremediation:

- Seeking non-edible plants that are also unappealing to humans and most other large species, such as animals and birds, for the purpose of using them in phytoremediation of contaminated areas. This will promote sustainability as certain restoration plants serve as a food source for people or other macro-organisms. Therefore, their use can potentially lead to food scarcity or disruption of the food supply chain and web.
- Alleviating time limitations. The factor of time has frequently impeded the acceptance of the biological approach for environmental restoration, as well as pollution solutions like treatment of wastewater and detoxifying of sewage solids. Phytoremediation, in contrast to other physico-chemical methods, requires a significant amount of time. Hence, future studies should incorporate potential genetic alterations to expedite plant development without compromising the plant's bio-transformation, translocation, and conversion capabilities.
- Further research is required to understand the mechanisms of phyto-removal. This understanding will help enhance prediction models. Depending on the variety of plants, the ideal concentration of contaminants varies. However, it is important not to overlook this factor when promoting technology.

8.15 CONCLUSION

As previously stated in this chapter, human actions are responsible for the decline of the environment. Therefore, rehabilitation is necessary in order to prevent the catastrophic collapse of the ecosystem. The purpose of discussing phytoremediation as the best tool for environmental restoration was not only to showcase the technique as the best choice but also to support its sustainability and viability when weighed against other biological methods.

When promoting phytotechnology, it is important to highlight its notable advantages, such as being a green technology that can be easily implemented in any location where plants can thrive. When adopting a phytoremediation technique to restore or reclaim a damaged environment, it is important to examine certain fundamental needs. These include factors such as the physical characteristics of the plants and the compatibility of the soil, as well as understanding the process of phytoremediation and the age of the plants. Environmentalists involved in the restoration exercise are interested in specific methodologies, including in situ, in vivo, and in vitro applications. The interaction between plants and their immediate environmental components, particularly microorganisms, is of great importance. Research has shown that a strong and effective interaction between plants and microbes is crucial for the success of phytoremediation. The capacity of microbes to improve soil nutrient availability to the remediation programs is largely responsible for this. Ultimately, a large portion of the population supports phytoremediation since it is environmentally benign. Its comparative affordability gives it an advantage over other remediation methods, particularly when one takes into account that the majority of anthropogenic actions are likewise motivated by cost. Although there have been many concerns about its reduced speed, it still has multiple techniques for removing pollutants. Further study into methods for modifying time and producing sustainable conventional plants, nevertheless, will be a positive step toward using phytoremediation to restore the environment.

ARTIFICIAL INTELLIGENCE DECLARATION

The author has used AI tools to improve the language during writing the chapter. However, the final interpretation, synthesis, and critical analysis of the literature remain the responsibility of the authors. Authors are accountable for the originality, validity, and integrity of the content of their submissions.

REFERENCES

Abdel-Shafy, H., & Mansour, M. (2018). Phytoremediation for the elimination of metals, pesticides, PAHs, and other pollutants from wastewater and soil. In V. Kumar, M. Kumar, & R. Prasad (Eds.), *Phytobiont and Ecosystem Restitution* (pp. 101–136). Springer, Singapore. https://doi.org/10.1007/978-981-13-1187-1_5

Adeniran, M. A., Oladunjoye, M. A., & Doro, K. O. (2023). Soil and groundwater contamination by crude oil spillage: A review and implications for remediation projects in Nigeria [Review]. *Frontiers in Environmental Science*, 11. https://doi.org/10.3389/fenvs.2023.1137496

Alaoui, I., Ghadraoui, O., Serbouti, S., Ahmed, H., Mansouri, I., el Kamari, F., Taroq, A., Ousaaid, D., Squalli, W., & Farah, A. (2022). The mechanisms of absorption and nutrients transport in plants: A review. *Tropical Journal of Natural Product Research*, 6, 8–14. https://doi.org/10.26538/tjnpr/v6i1.2

Al-Jobori, K., & Kadhim, A. (2019). Evaluation of sunflower (Helianthus annuus L.) for phytoremediation of lead contaminated soil. *Journal of Pharmaceutical Sciences and Research*, 11, 847–854. https://www.semanticscholar.org/paper/Evaluation-of-sunflower-(Helianthus-annuus-L.)-for-Al-Jobori-Kadhim/e662b4115980629423bc6fd87e4b1bc669c34122

Atakhanova, Z. (2023). Water impacts and effluent quality regulations of Canadian mining. *Sustainability*, 15(22), 16086. https://www.mdpi.com/2071-1050/15/22/16086

Bayuo, J., Rwiza, M. J., Choi, J. W., Njau, K. N., & Mtei, K. M. (2024). Recent and sustainable advances in phytoremediation of heavy metals from wastewater using aquatic plant species: Green approach. *Journal of Environmental Management*, 370, 122523. https://doi.org/10.1016/j.jenvman.2024.122523

Bhat, S. A., Bashir, O., Ul Haq, S. A., Amin, T., Rafiq, A., Ali, M., Américo-Pinheiro, J. H. P., & Sher, F. (2022). Phytoremediation of heavy metals in soil and water: An eco-friendly, sustainable and multidisciplinary approach. *Chemosphere*, 303, 134788. https://doi.org/10.1016/j.chemosphere.2022.134788

Bolan, N. S., Park, J. H., Robinson, B., Naidu, R., & Huh, K. Y. (2011). Phytostabilization: A green approach to contaminant containment. In D. L. Sparks (Ed.), *Advances in Agronomy* (Vol. 112, pp. 145–204). Academic Press. https://doi.org/10.1016/B978-0-12-385538-1.00004-4

Cahoon, D. R., McKee, K. L., & Morris, J. T. (2021). How plants influence resilience of salt marsh and mangrove wetlands to sea-level rise. *Estuaries and Coasts*, 44(4), 883–898. https://doi.org/10.1007/s12237-020-00834-w

Centenaro, G., Hudek, C., Zanella, A., & Crivellaro, A. (2018). Root-soil physical and biotic interactions with a focus on tree root systems: A review. *Applied Soil Ecology*, 123, 318–327. https://doi.org/10.1016/j.apsoil.2017.09.017

Chandra, R., Dubey, N. K., & Kumar, V. (2017). *Phytoremediation of Environmental Pollutants*. 1-554. https://doi.org/10.4324/9781315161549

Chaney, R. L., Angle, J. S., Broadhurst, C. L., Peters, C. A., Tappero, R. V., & Sparks, D. L. (2007). Improved understanding of hyperaccumulation yields commercial phytoextraction and phytomining technologies. *Journal of Environmental Quality*, 36(5), 1429–1443. https://doi.org/10.2134/jeq2006.0514

Dadrasnia, A., Emenike, C., & Ismail, S. (2015). Restoration of environment through phytoremediation. In Shiv Om Singh and Kajal Srivastava (Eds.), *Handbook of Research on Uncovering New Methods for Ecosystem Management through Bioremediation* (pp. 172–195). https://doi.org/10.4018/978-1-4666-8682-3.ch008

Delgado-Baquerizo, M., Reich, P. B., Bardgett, R. D., Eldridge, D. J., Lambers, H., Wardle, D. A., Reed, S. C., Plaza, C., Png, G. K., Neuhauser, S., Berhe, A. A., Hart, S. C., Hu, H.-W., He, J.-Z., Bastida, F., Abades, S., Alfaro, F. D., Cutler, N. A., Gallardo, A., … Fierer, N. (2020). The influence of soil age on ecosystem structure and function across biomes. *Nature Communications*, 11(1), 4721. https://doi.org/10.1038/s41467-020-18451-3

Egendorf, S. P., Groffman, P., Moore, G., & Cheng, Z. (2020). The limits of lead (Pb) phytoextraction and possibilities of phytostabilization in contaminated soil: A critical review. *International Journal of Phytoremediation*, 22(9), 916–930. https://doi.org/10.1080/15226514.2020.1774501

Farraji, H., Zaman, N., Zahed, M., & Faraji, H. (2017). *Role of Rhizoremediation in Decontaminating Some Hazardous Pollutants*. IGI Global. http://doi.org/10.4018/978-1-5225-2325-3.ch009

Fernandez, L., Castro, R., & Nunez, J. (2023). Influence of environmental factors on groundwater phytoremediation efficiency. *Journal of Hydrology*, 620, 129874. https://doi.org/10.1016/j.jhydrol.2023.129874

Gimeno, S., Allan, D., Paul, K., Remuzat, P., & Collard, M. (2024). Are current regulatory log Kow cut-off values fit-for-purpose as a screening tool for bioaccumulation potential in aquatic organisms? *Regulatory Toxicology and Pharmacology*, 147, 105556. https://doi.org/10.1016/j.yrtph.2023.105556

Guerra, F., Gainza-Cortés, F., Pérez-Castro, R., & Zamudio, F. (2011). Phytoremediation of heavy metals using poplars (Populus Spp.). A glimpse of the plant responses to Copper, Cadmium and Zinc stress. In Ivan A. Golubev (Ed.), *Handbook of Phytoremediation* (pp. 387–414). Nova Science Publishers, Hauppauge, New York.

Gul, B., & Marakli, S. (2023). A review on the phytoremediation techniques of potentially toxic elements: Understanding of principles and practices. In Koray Özrenk, Selçuk (Eds.), *International Studies in Agriculture, Forestry and Aquaculture Sciences* (pp. 1–23). Serüven Yayınevi, Ankara, Turkey.

Hernández-Cáceres, D., Stokes, A., Angeles-Alvarez, G., Abadie, J., Anthelme, F., Bounous, M., Freschet, G. T., Roumet, C., Weemstra, M., Merino-Martín, L., & Reverchon, F. (2022). Vegetation creates microenvironments that influence soil microbial activity and functional diversity along an elevation gradient. *Soil Biology and Biochemistry*, 165, 108485. https://doi.org/10.1016/j.soilbio.2021.108485

Huot, C., Zhou, Y., Philp, J. N. M., & Denton, M. D. (2020). Root depth development in tropical perennial forage grasses is related to root angle, root diameter and leaf area. *Plant and Soil*, 456(1), 145–158. https://doi.org/10.1007/s11104-020-04701-2

Islam, I., & Quaff, A. R. (2024). *Phytoremediation of groundwater contaminants: A systematic review*. Groundwater Contamination and Remediation: A Review. *African Journal of Biomedical Research*, 27(3), 3855–3867. https://doi.org/10.53555/AJBR.v27i3S.3062

Kafle, A., Timilsina, A., Gautam, A., Adhikari, K., Bhattarai, A., & Aryal, N. (2022). Phytoremediation: Mechanisms, plant selection and enhancement by natural and synthetic agents. *Environmental Advances*, 8, 100203. https://doi.org/10.1016/j.envadv.2022.100203

Khan, A. A., Qamar, Z., Ali, S., & Zubair, M. (2022). Nanotechnology-enabled phytoremediation: A comprehensive review of prospects and limitations. *Plant Science Today*, 9(2), 43–61.

Kuok Ho, D. T. (2024). Phytoremediation of petroleum hydrocarbons: An update of its recent progress. *Tropical Environment Biology and Technology* 2, 106–123. http://doi.org/10.53623/tebt.v2i2.532

Kuppan, N., Padman, M., Mahadeva, M., Srinivasan, S., & Devarajan, R. (2024). A comprehensive review of sustainable bioremediation techniques: Eco friendly solutions for waste and pollution management. *Waste Management Bulletin*, 2(3), 154–171. https://doi.org/10.1016/j.wmb.2024.07.005

Lavanya, M. B., Viswanath, D. S., & Sivapullaiah, P. V. (2024). Phytoremediation: An eco-friendly approach for remediation of heavy metal-contaminated soils-A comprehensive review. *Environmental Nanotechnology, Monitoring & Management*, 22, 100975. https://doi.org/10.1016/j.enmm.2024.100975

Lim, J.-M., Salido, A. L., & Butcher, D. J. (2004). Phytoremediation of lead using Indian mustard (Brassica juncea) with EDTA and electrodics. *Microchemical Journal*, 76(1), 3–9. https://doi.org/10.1016/j.microc.2003.10.002

Montreemuk, J., Stewart, T. N., & Prapagdee, B. (2024). Bacterial-assisted phytoremediation of heavy metals: Concepts, current knowledge, and future directions. *Environmental Technology & Innovation*, 33, 103488. https://doi.org/10.1016/j.eti.2023.103488

Nedjimi, B. (2021). Phytoremediation: A sustainable environmental technology for heavy metals decontamination. *SN Applied Sciences*, 3(3), 286. https://doi.org/10.1007/s42452-021-04301-4

Nie, X., Huang, X., Li, M., Lu, Z., & Ling, X. (2024). Advances in soil amendments for remediation of heavy metal-contaminated soils: Mechanisms, impact, and future prospects. *Toxics*, 12(12), 872. https://doi.org/10.3390/toxics12120872

Nzengung, V. A., Wolfe, L. N., Rennels, D. E., McCutcheon, S. C., & Wang, C. (1999). Use of aquatic plants and algae for decontamination of waters polluted with chlorinated alkanes. *International Journal of Phytoremediation*, 1(3), 203–226. https://doi.org/10.1080/15226519908500016

Pandey, N., Chandra, J., Xalxo, R., & Sahu, K. (2021). Concept and types of phytoremediation. In M. Hasanuzzaman (Ed.), *Approaches to the Remediation of Inorganic Pollutants* (pp. 281–302). Springer Singapore. https://doi.org/10.1007/978-981-15-6221-1_14

Park, J. K., & Oh, K. (2023). Advancements in phytoremediation research for soil and water resources: Harnessing plant power for environmental cleanup. *Sustainability*, 15(18), 13901. https://doi.org/10.3390/su151813901

Paz-Alberto, A. M. (2013). Phytoremediation: A green technology to remove environmental pollutants. *American Journal of Climate Change* 2, 71–86. http://doi.org/10.4236/ajcc.2013.21008

Paz-Alberto, A. M., Sigua, G. C., Baui, B. G., & Prudente, J. A. (2007). Phytoextraction of lead-contaminated soil using vetivergrass (Vetiveria zizanioides L.), cogongrass (Imperata cylindrica L.) and carabaograss (Paspalum conjugatum L.). *Environmental Science and Pollution Research International*, 14(7), 498–504. https://doi.org/10.1065/espr2007.05.415

Penela-Arenaz, M., Bellas, J., & Vázquez, E. (2009). Effects of the prestige oil spill on the biota of NW Spain: 5 years of learning. In David W. Sims (Ed.), *Advances in Marine Biology* (Vol. 56, pp. 365–396). Academic Press. https://doi.org/10.1016/S0065-2881(09)56005-1

Plant–Water Relations (1): Uptake and Transport (2014). *The Plant Cell*, 26(1). https://doi.org/10.1105/tpc.114.tt0114

Prakash, G., Soni, R., Mishra, R., & Sharma, S. (2019). *Role of Plant-Microbe Interaction in Phytoremediation*. Springer. http://doi.org/10.1007/978-981-32-9824-8_6

Prieto, I., Armas, C., & Pugnaire, F. I. (2012). Water release through plant roots: New insights into its consequences at the plant and ecosystem level. *New Phytologist*, 193(4), 830–841. https://doi.org/10.1111/j.1469-8137.2011.04039.x

Prima, E. C., Yuliarto, B., Suendo, V., & Suyatman. (2014). Improving photochemical properties of *Ipomea pescaprae*, *Imperata cylindrica* (L.) Beauv, and Paspalum conjugatum Berg as photosensitizers for dye sensitized solar cells. *Journal of Materials Science: Materials in Electronics*, 25(10), 4603–4611. https://doi.org/10.1007/s10854-014-2210-x

Rascio, N., & Navari-Izzo, F. (2011). Heavy metal hyperaccumulating plants: How and why do they do it? And what makes them so interesting? *Plant Science*, 180(2), 169–181. https://doi.org/10.1016/j.plantsci.2010.08.016

Schneider, M., & Bláha, L. (2020). Advanced oxidation processes for the removal of cyanobacterial toxins from drinking water. *Environmental Sciences Europe*, 32(1), 94. https://doi.org/10.1186/s12302-020-00371-0

Sharma, K., Shah, G., Singh, H., Bhatt, U., Singhal, K., & Soni, V. (2024a). Advancements in natural remediation management techniques for oil spills: Challenges, innovations, and future directions. *Environmental Pollution and Management*, 1, 128–146. https://doi.org/10.1016/j.epm.2024.08.003

Sharma, P., Jha, A. B., & Dubey, R. S. (2024b). Enhancing phytoremediation efficacy in plants cultivated in heavy metal-contaminated soil under drought stress: Understanding plant responses and genetic engineering strategies. *Water, Air, & Soil Pollution* 235, 451 https://doi.org/10.1007/s11270-024-07239-6

Siddique, A., Al Disi, Z., AlGhouti, M., & Zouari, N. (2024). Diversity of hydrocarbon-degrading bacteria in mangroves rhizosphere as an indicator of oil-pollution bioremediation in mangrove forests. *Marine Pollution Bulletin*, 205, 116620. https://doi.org/10.1016/j.marpolbul.2024.116620

Singh, A., & Kumar Mishra, V. (2021). Biodegradation of organic pollutants for its effective remediation from the environment and the role of various factors affecting the biodegradation process. In V. Kumar Mishra & A. Kumar (Eds.), *Sustainable Environmental Clean-up* (pp. 1–27). Elsevier. https://doi.org/10.1016/B978-0-12-823828-8.00001-3

Sonawdekar, S. (2012). Bioremediation: A boon to hydrocarbon degradation. *International Journal of Environmental Sciences*, 2.

Susarla, S., Medina, V. F., & McCutcheon, S. C. (2002). Phytoremediation: An ecological solution to organic chemical contamination. *Ecological Engineering*, 18(5), 647–658. https://doi.org/10.1016/S0925-8574(02)00026-5

Tamma, A. A., Lejcuś, K., Fiałkiewicz, W., & Marczak, D. (2025). Advancing phytoremediation: A review of soil amendments for heavy metal contamination management. *Sustainability*, 17(13), 5688. https://doi.org/10.3390/su17135688

Tonelli, F. M. P., Tonelli, F. C. P., de Melo Nunes, N. A., & Lemos, M. S. (2020). Mechanisms and importance of phytoremediation. In R. A. Bhat & K. R. Hakeem (Eds.), *Bioremediation and Biotechnology, Vol 4: Techniques for Noxious Substances Remediation* (pp. 125–141). Springer International Publishing. https://doi.org/10.1007/978-3-030-48690-7_6

Wang, Q. R., Cui, Y. S., Liu, X. M., Dong, Y. T., & Christie, P. (2003). Soil contamination and plant uptake of heavy metals at polluted sites in China. *Journal of Environmental Science and Health. Part A, Toxic/Hazardous Substances & Environmental Engineering*, 38(5), 823–838. https://doi.org/10.1081/ESE-120018594

Weyens, N., Thijs, S., Popek, R., Witters, N., Przybysz, A., Espenshade, J., Gawronska, H., Vangronsveld, J., & Gawronski, S. W. (2015). The role of plant-microbe interactions and their exploitation for phytoremediation of air pollutants. *International Journal of Molecular Sciences*, 16(10), 25576–25604. https://doi.org/10.3390/ijms161025576

Wolfe, N. L., & Hoehamer, C. F. (2003). Enzymes used by plants and microorganisms to detoxify organic compounds. In Steven C. McCutcheon and Jerald L. Schnoor (Eds.), *Phytoremediation: Transformation and Control of Contaminants* (pp. 159–187). https://doi.org/10.1002/047127304X.ch5

Yadav, R., Singh, S., Kumar, A., & Singh, A. N. (2022). Phytoremediation: A wonderful cost-effective tool. In S. Kathi, S. Devipriya, & K. Thamaraiselvi (Eds.), *Cost Effective Technologies for Solid Waste and Wastewater Treatment* (pp. 179–208). Elsevier. https://doi.org/10.1016/B978-0-12-822933-0.00008-5

Yan, A., Wang, Y., Tan, S. N., Mohd Yusof, M. L., Ghosh, S., & Chen, Z. (2020). Phytoremediation: A promising approach for revegetation of heavy metal-polluted land. *Frontiers in Plant Science*, 11, 359. https://doi.org/10.3389/fpls.2020.00359

Yang, K. M., Poolpak, T., & Pokethitiyook, P. (2023). Rhizodegradation: The plant root exudate and microbial community relationship. In L. Newman, A. A. Ansari, S. S. Gill, M. Naeem, & R. Gill (Eds.), *Phytoremediation: Management of Environmental Contaminants*, (Vol. 7, pp. 209–229). Springer International Publishing. https://doi.org/10.1007/978-3-031-17988-4_11

Zapata, K. (2021). Arctic oil spill cleanup costs could reach $9.4B over 5 years, says risk analyst. CBC News Jul 15. https://www.cbc.ca/news/canada/north/arctic-oil-spill-study-1.6103155

Zeng, P., Zhao, Q., Hu, J.-Y., Zhang, X., Mao, B., Sun, Q.-Y., & Wu, W.-G. (2024). Nitrogen addition has divergent effects on phosphorus fractions in four types of soils. *Ecological Processes*, 13(1), 43. https://doi.org/10.1186/s13717-024-00523-7

9 Agricultural Bioinputs in Soil Bioremediation and Enhancing Phytoremediation

Yanca Araujo Frias, Victor Hugo Cruz, Bruno Rafael de Almeida Moreira, Thalia Silva Valério, Paulo Henrique Frata Ferreira, Matheus Mantuanelli Roberto, and Paulo Renato Matos Lopes

9.1 INTRODUCTION

To fully comprehend the agricultural bioinputs role associated with the phytoremediation of pesticides in soil, it is crucial to define the fundamental concepts that guide this innovative approach. In this context, this topic explores the definitions of agricultural biological products and plant treatment processes, highlighting their characteristics, benefits, and applications in sustainable agriculture.

Agricultural bioinputs are derived from microorganisms, plants, natural extracts, and other living organisms or their parts, used to enhance acting on different fronts in agricultural fields with great sustainable appeal. They are vital in maintaining soil health, controlling pests and diseases, and promoting plant growth (Rocha et al. 2024). There are several compounds applied as bioinputs in modern agriculture divided into five main groups. They are biological control agents (biopesticides), biostimulants, biofertilizers, bioinoculants, and soil conditioners (Brazil 2021).

According to their composition, agricultural bioinputs also become a bioremediation agent as they are mostly composed of microorganisms and enzymes that can degrade soil contaminants, including pesticides (Tornisielo et al. 2019), heavy metals, and hydrocarbons (Seo et al. 2009), thereby helping to restore environmental quality.

These soil phytoremediation strategies are extremely important for sustainable agriculture, such as converting cultivation areas from the conventional system more quickly to the organic system, making the situation very attractive for farmers in this sector, as this slow conversion process becomes a major obstacle.

DOI: 10.1201/9781003714484-9

Additionally, phytoremediation is defined as a bioremediation technology that uses plants to remove, degrade, and/or stabilize contaminants from soil, water, and air. It is an ecological and sustainable approach that leverages the natural ability of plants to absorb, transform, and accumulate toxic substances (Greipsson 2011; Kafle et al. 2022).

Phytoremediation is usually more cost effective than traditional remediation methods (physical and chemical) based mainly on a lower environmental impact, the maintenance of soil structure and vegetation cover, and its potential on large contaminated areas (Daurova et al. 2023; Islam et al. 2024).

The integration of agricultural bioinputs into phytoremediation enhances the efficiency of this technique, whose combination creates a more favorable environment for the action of plants and microorganisms. Beneficial microbiotas present in bioinputs can accelerate the degradation of pesticides and improve nutrient uptake by plants, which increases the remediation success and promotes soil recovery in a faster, cheaper, less toxic, and more eco-friendly manner. Therefore, this synergy between agricultural biological products and plant treatment processes offers a promising approach to addressing the challenges of pesticide residues in soil, contributing to sustainable agriculture and environmental health (Figure 9.1).

9.2 AGRICULTURAL BIOINPUTS

Agricultural bioinputs have emerged as essential components for sustainable agriculture and offer natural and less impactful alternatives to conventional chemical products. These biocompounds are derived from living organisms or their metabolic processes, which are applied in agriculture to: promote plant growth, control pests and diseases, improve soil fertility, or remediate contaminated environments. They are classified as follows.

9.2.1 Biopesticides

These substances are used to control pests, pathogens, and weeds as an alternative to synthetic pesticides. The main concept of biopesticides revolves around using natural organisms or substances derived from living organisms as bacteria, fungi, viruses, protozoa, and plant extracts, to manage agricultural pests (Kumar et al. 2021; Ayilara et al. 2023; Hezakiel et al. 2024).

9.2.2 Biostimulants

Compounds that are applied to stimulate natural plant processes, improving nutrient uptake, metabolic efficiency, stress resistance, and crop quality (Kumar and Aloke 2020). Based on their origin, biostimulants are categorized into two broad groups: non–microbial (chitosan, humic and fulvic acids, protein hydrolysates, phosphites, seaweed extracts, and silicon) and microbial (arbuscular mycorrhizal fungi, plant growth–promoting rhizobacteria (Bhupenchandra et al. 2022).

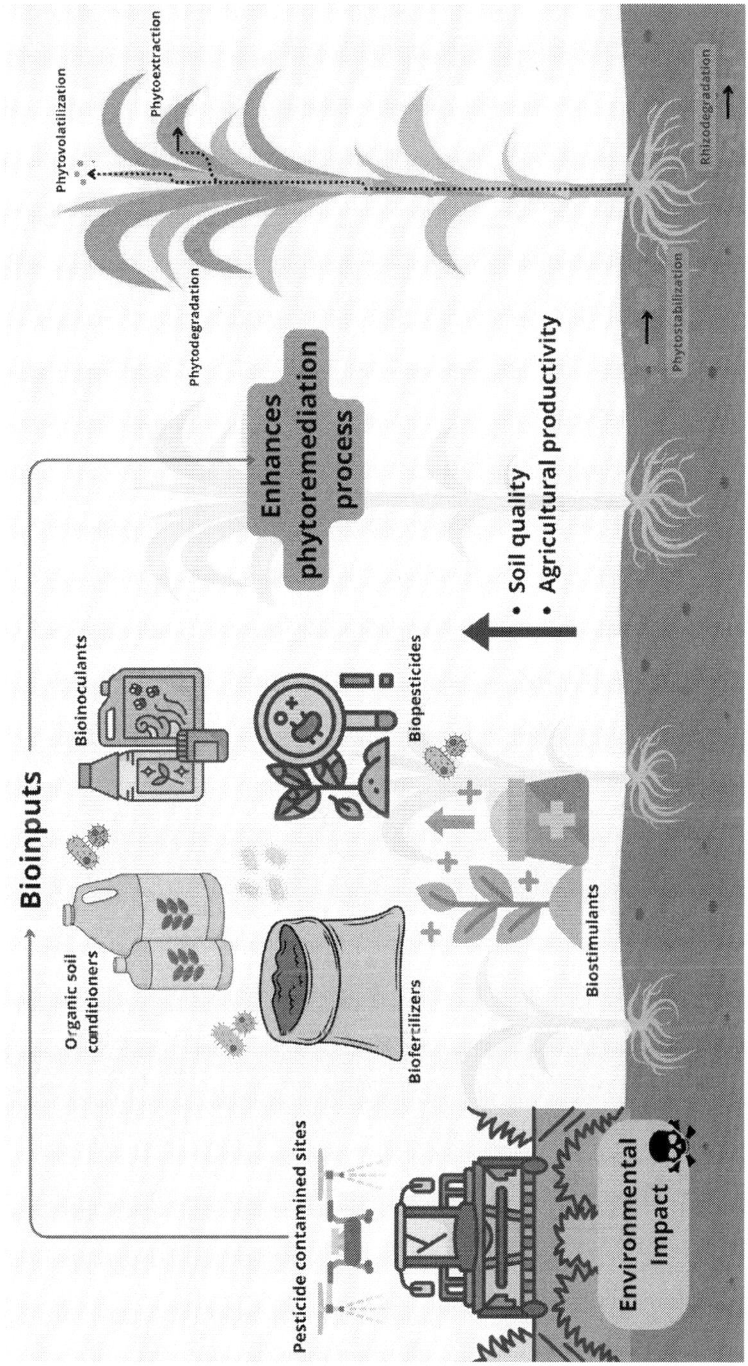

FIGURE 9.1 Agricultural bioinputs as biotechnological tools associated with phytoremediation and soils with pesticides.

9.2.3 Biofertilizers

Products containing beneficial microorganisms, such as nitrogen-fixing bacteria, phosphorus and potassium solubilizers, mycorrhizal fungi, and cyanobacteria, enhance plant-growth promoting (PGP) (Sun et al. 2020). They can be prepared by using sustainable and reusable materials as agro-industrial wastes, such as vinasse from sugarcane ethanol production (Senatore et al. 2023), castor bean cake (Candian et al. 2024), neem vegetable cake (Silva et al. 2024), and bokashi (Narayanan and Ma 2024). They serve as a sustainable alternative to synthetic fertilizers, helping to improve soil structure, nutrient retention, and overall plant health while potentially reducing the need for chemical inputs (Santos et al. 2024).

9.2.4 Bioinoculants

Preparations containing single or consortia of microorganisms, when introduced to the crop, are dedicated to carrying out particular functions like growth promotion, improved soil health, and reducing the use of chemical fertilizers and pesticides (Santos et al. 2024). They include bacteria, fungi, protozoa, and endophytes that interact with plants in various ways to promote growth, nutrient uptake, and stress tolerance (Kamath et al. 2023).

9.2.5 Organic Soil Conditioners

Materials are derived from living organisms (e.g., plants and animals) and also from agro-industrial residues, manures, peat, biochar, bone meal, blood meal, coffee grounds, compost tea, coir, sewage sludges, farmyard manure, and sawdust (Babla et al. 2022). Regulation of these compounds varies globally, and each country determines standards to ensure their safety, efficacy, and quality. In Brazil, the legislation (Decree n. 10.375/2020) establishes criteria for their registration, production, and commercialization. The regulatory framework for agricultural bioinputs is evolving to keep pace with market growth and technological innovations.

The production and commercialization of agricultural bioinputs involve specific technological processes that ensure their viability and effectiveness, as well as compliance with local and international regulations. Its global market has experienced significant growth, driven by the increased demand for more sustainable agricultural practices. Thus, global market projections indicate a continued expansion based on the goals of sustainable agricultural practices and regulatory pressure for agrochemicals (Mafra 2023). Some recent studies prove the importance of bioinputs in modern agriculture, as can be seen in Table 9.1.

The market for biological inputs is growing worldwide, with Brazil playing a major role due to the size of the country's agricultural production. Among the various types, the global market for biopesticides was estimated at US$ 6.51 billion in 2022, with an estimate of US$ 18.15 billion in 2029 (Mafra 2023), a growth of 64.13% over the period. For bio fertilizers, the estimate for 2022 was US$ 2.02 billion, rising to US$ 4.47 billion in 2029. For biostimulants, the estimate was US$ 3.14 billion in 2022, rising to US$ 6.69 billion in 2029 (Mafra 2023).

TABLE 9.1
Summary of Recent Work on the Use and Importance of Bioinputs in Agriculture

Type	Overview of Discussion	References
Biopesticides	The types of biopesticides, their mode of action, formulation strategies, recent advancements of biopesticides focusing mainly on improvement of its action spectra to thereby replace chemical pesticides, and finally, the future aspects of biopesticides have been discussed in these reviews.	Ayilara et al. (2023), Hezakiel et al. (2024), and Assadpour et al. (2023)
Biostimulants	These reviews present the state of the art about the agricultural and environmental applications of biostimulants to soils as a sustainable agronomic tool to improve soil health and help plant growth under stressed conditions.	Grammenou et al. (2023), Gupta et al. (2023), and Rakkammal et al. (2023)
Biofertilizers	The articles address the production of biofertilizer from the hydrothermal carbonization of microalgae biomass grown in wastewater, as well as the perspective in relation to the concept of circular bioeconomy and sustainable development goals. And also, the composition of the microbiome of a commercial biofertilizer known for its plant growth-promoting activity.	Castro et al. (2023), Cao et al. (2023), and Cristóbal Hernández-Álvarez et al. (2023)
Bioinoculants	This review provides information on how bioinoculants based on microorganisms have been shown to increase crop yield and quality through strategies, such as increasing nutrient uptake or levels of certain phytohormones.	Díaz-Urbano et al. (2023)
Organic soil conditioners	These articles shown how biochar/humic substance/fertilizer compound soil conditioner can efficient improve soil fertility and heavy metals remediation in acidic soils; and how soil conditioners improve Cd-contaminated farmland soil microbial communities to inhibit Cd accumulation in rice.	Liu et al. (2023) and Zhao et al. (2023)

Advancements in the production and regulation of bioinputs, combined with global market growth, denote a continued shift toward more sustainable practices. Bioinputs are thus consolidating as a crucial tool for future agriculture due to their valuable contributions to soil health, food security, and the mitigation of environmental impacts in agroecosystems.

9.3 INTEGRATION OF BIOINPUTS AND PLANTS

The combination of agricultural biological products and plant treatment processes performs a strategic approach to maximize productivity and minimize the environmental impacts of pesticides. Some examples are included in Table 9.2.

The selection of plant species regards various criteria to ensure the treatment effectiveness and the intervention sustainability. Hence, the plant's ability to tolerate and accumulate contaminants, its compatibility with bioinputs, and its adaptability to the local environment and climate conditions should be considered. Moreover, the consistency between organisms is crucial, as some plants form symbiotic associations with specific microorganisms to enhance nutrient uptake and contaminant degradation, such as legumes with *Rhizobium* spp. with mycorrhizal fungi (Calderon and Dangi 2024; Hnini et al. 2024).

Hejazirad et al. (2024) evaluated the impact of phytoremediation on reducing metolachlor residues in soils treated at doses of 530.7 and 1,061.4 g ha^{-1} and analyzed its effects on soil microbial biodiversity in contaminated areas. *Avena sativa* and *Medicago sativa* demonstrated significant herbicide dissipation rates, achieving 54.5% and 36.4% efficacy, respectively, with higher effectiveness at elevated doses. Despite an initial reduction in soil microbial diversity due to metolachlor, phytoremediating plants (especially *M. sativa*) enhanced microbial richness and

TABLE 9.2
Combinations of Agricultural Biological Products and Plant Treatment Processes

Combination	Concept	References
Biofertilizers and Rhizodegradation	Biofertilizers with microorganisms can promote rhizodegradation by degrading contaminants, a process that is facilitated by the exudation of organic compounds from the roots that stimulate microbial activity.	Song et al. (2020) and Suryatmana et al. (2022)
Biopesticides and Antagonistic Action	Biopesticides can act as bioremediation agents by suppressing soil pathogens and decomposing contaminant residues. For instance, *Trichoderma* spp. have antagonistic properties against plant-parasitic nematodes, including parasitism and secreting the nematicidal compounds that affect nematode survival, thereby improving conditions for plant growth and enhancing the effectiveness of phytoremediation.	Javeed et al. (2016) and Boakye et al. (2022)
Biostimulants and Stress Resistance	Biostimulants can increase resistance to stress caused by contaminants, which allows plants to maintain their metabolic activity and contribute to ongoing remediation.	Van Oosten et al. (2017) and Jia et al. (2023)

distribution, which counteracted the herbicide's adverse effects. Furthermore, key microbial groups responded differently to herbicide levels in the rhizosphere. For both plant species, *Proteobacteria* thrived at higher doses, while *Planctomycetota*, *Mortierellomycota*, and *Basidiomycota* tolerated elevated metolachlor in *A. sativa* rhizosphere, unlike in *M. sativa*. Conversely, *Rozellomycota* and *Actinobacteriota* were prevalent in *M. sativa* at the highest herbicide dose, but only in *A. sativa* under the highest dose or no herbicide exposure. The authors offer valuable insights for selecting microbial groups for bioinput production in soil decontamination efforts, with specific microbial recommendations based on the chosen phytoremediating plant.

Hence, endophytic microorganisms play a crucial role in plant's biochemical and physiological functions, contributing to improved plant production, nutrient cycling, bioremediation, and energy storage (Lopez et al. 2022). In general, endophytes are microorganisms transmitted by seeds and begin their activities shortly after germination, establishing themselves in the plant's tissues and offering benefits such as growth support and protection against pathogens (Bhatt et al. 2022). In this case, endophytes can help with phytoextraction and phytodegradation mechanisms in plants with the potential for phytoremediation.

Fast-growing plants with short life cycles are preferred for phytoremediation, as they allow for multiple harvests and remediation cycles within a short period, as seen with legumes (Ferreira et al. 2021). When selecting species, it is important to consider characteristics such as ease of management, resistance to pests, and minimal resource requirements (Kafle et al. 2022).

Research on bioinputs that promote phytoremediation remains limited in the literature. However, it is established that the microorganisms found in these bioinputs are effective plant growth-promoting rhizobacteria (PGPR), which aids in the phytoremediation process (Sharma et al. 2024). The term PGPR refers to bacteria that inhabit the rhizospheric region of plants, directly enhancing nutrient uptake through various mechanisms while also indirectly supporting plants by suppressing phytopathogen growth and activity (Gurikar et al. 2016).

Direct promotion of plant growth by PGPR includes the production and release of secondary metabolites, such as plant growth regulators, as well as facilitating the uptake of specific nutrients from the root environment. The application of PGPR has been shown to improve seed germination rates, increase root and shoot length and weight, enhance chlorophyll content, and boost crop yield. Additionally, these beneficial bacteria contribute to plants' tolerance to drought and salt stress and help delay leaf senescence (Sharma et al. 2024).

Zhou et al. (2013) demonstrated that *Ensifer adhaerens* TMX-23 has significant bioaugmentation potential, promoting soybean growth in soil contaminated with thiamethoxam. Bhatt et al. (2022) mention that PGPRs improve the availability of essential nutrients, such as zinc and iron, by solubilizing them in the soil. For example, they produce organic acids that help convert insoluble forms of these nutrients into soluble forms that plants can easily absorb. This greater availability of nutrients contributes to better plant growth and development, leading to an increase in crop yields, and consequently improves the plant's ability to extract contaminants from the soil.

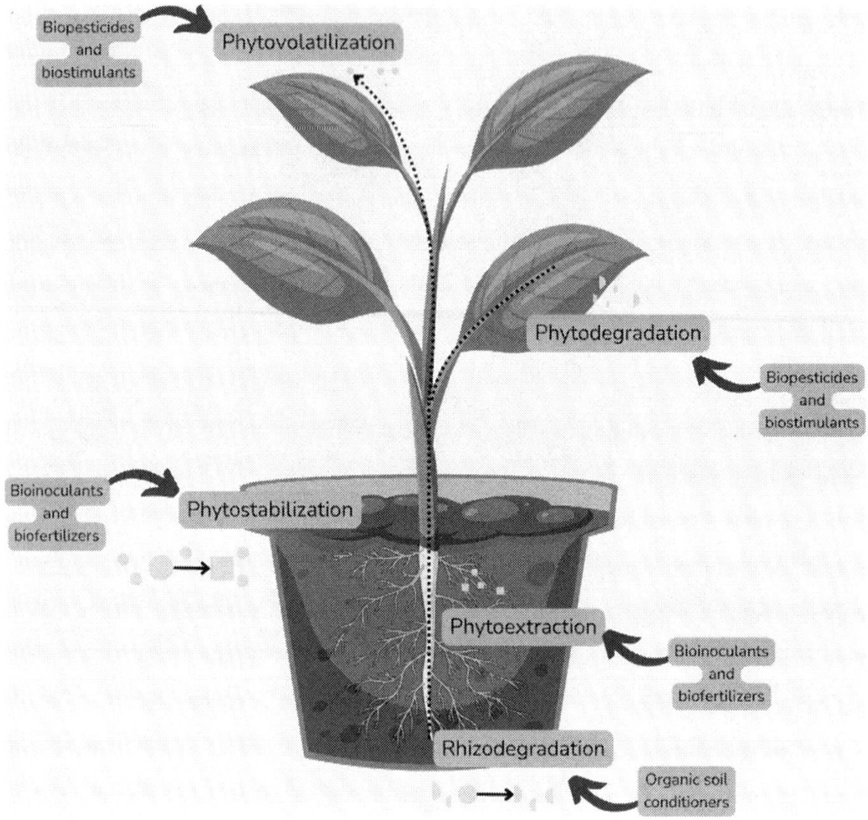

FIGURE 9.2 Schematic representation of phytoremediation mechanisms in plants and target sites for bioinput action.

9.4 PHYTOREMEDIATION

Phytoremediation is a biological technique that employs plants, often in conjunction with microorganisms, to remove, degrade, and/or stabilize contaminant compounds in the environment (Tonelli et al. 2022). Figure 9.2 shows a schematic of the phytoremediation mechanisms that occur in the plant and some examples in which bioinputs can act.

The effectiveness of this biotechnology depends on a range of natural mechanisms involving both plants and microorganisms. These pathways work synergistically to transform pollutants into less toxic or harmless forms.

9.4.1 Phytoextraction: Absorption and Accumulation

Plant roots absorb contaminants from the environmental matrix and transport them through the vascular system to various parts such as leaves, stems, and roots (Yadav et al. 2018; Kafle et al. 2022).

9.4.2 Phytodegradation: Enzymatic Degradation

Plant enzymes – peroxidases, laccases, and nitroreductases – degrade contaminants directly in the environmental matrix or within their tissues, whose compounds originated can be incorporated into the natural nutrient cycle (Yadav et al. 2018; Kafle et al. 2022).

9.4.3 Rhizodegradation: Stimulation of Microbial Activity

Plant roots exude organic compounds that serve as substrates for soil microbiota and stimulate their activity, whose biochemical processes – oxidation, reduction, and hydrolysis – degrade the contaminants (Rungwa et al. 2013; Kafle et al. 2022).

9.4.4 Phytostabilization: Immobilization and Stabilization

Plants capable of immobilizing contaminants in the environmental matrix can reduce their mobility and bioavailability. The process is based on root adsorption or soil chemical properties changes, which prevents pollutants from leaching and dispersion, by confining them into a less harmful scenario (Kumpiene et al. 2008; Kafle et al. 2022).

9.4.5 Phytovolatilization: Absorption and Evapotranspiration

Plants can absorb contaminants and convert them into volatile forms released into the atmosphere through transpiration (Yadav et al. 2018; Kafle et al. 2022).

9.5 CASE STUDIES

Ferreira et al. (2021) evaluated the tolerance of five green manure species in the phytoremediation of tebuthiuron under vinasse application in soil. Their findings revealed that *Mucuna pruriens* and *Pennisetum glaucum* were the species with higher development with herbicide, but vinasse presence enhanced the ecotoxic effects at the initial time. However, the fertigation strategy commonly used in sugarcane cultivation proved effective for biomass production. Hence, the reduction of phytotoxicity through time in soil with vinasse was attributed to its composition rich in soluble carbon compounds, which likely stimulated microbial metabolism and facilitated the degradation of tebuthiuron. This suggests that fertigation enhances microbial activity in the rhizosphere, promoting rhizodegradation, where plant roots stimulate microbes to break down contaminants. Therefore, vinasse acts as a biofertilizer that can biostimulate soil microorganisms, enhancing the degradation process of pollutants like tebuthiuron.

Frias et al. (2024) investigated the ecotoxicity of soil samples contaminated with tebuthiuron and thiamethoxam, evaluating the effects of sugarcane vinasse application following phytoremediation with *Pennisetum glaucum*. Their findings revealed that the addition of vinasse supplied essential nutrients to *Pennisetum glaucum*, enhancing the plants' ability to cope with the stress induced by tebuthiuron and thiamethoxam. This benefit was evident in the improved hypocotyl growth of *Lactuca sativa* (ecotoxicity test organism) in the presence of sugarcane vinasse, indicating that it effectively provided necessary nutrition to the plants.

Nurzhanova et al. (2020) isolated 580 microbial strains from the rhizosphere of *Cucurbita pepo* and *Xanthium strumarium* in soil contaminated with the insecticide dichlorodiphenyltrichloroethane (DDT). Two bacterial isolates (*Bacillus vallismortis* and *Bacillus aryabhattai*) stood out for their ability to grow in media containing 0.5–5.0 mg L^{-1} of the DDT main metabolite – dichlorodiphenyldichloroethylene (DDE) – as the sole carbon source. Thus, both species demonstrated a capacity for consuming over 90% of this pollutant after 21 days of cultivation. Hence, their inoculation to *C. pepo* and *X. strumarium* resulted in a significant reduction of plant stress, as evidenced by increased aerial and root biomass. These microorganisms improved the absorption and phytostabilization potential and opened new perspectives for the treatment of soil contaminated with DDT and/or DDE.

Although two recent research are mentioned above, there are few studies in the literature involving the use of bioinputs in the phytoremediation of organic compounds. On the other hand, it is possible to find interesting results in scientific databases for biotechnological processes in the remediation of environments contaminated with inorganic compounds.

Zaidi et al. (2006) observed nickel accumulation in *Brassica juncea* when using the strain *Bacillus subtilis* SJ-101. The authors observed that plants with bioinoculant accumulated significantly more nickel. The bioconcentration factor was determined to be 2.2 after 140 days, indicating a similar Ni translocation into their aerial parts, regardless of the bioinoculant presence. Hence, *B. subtilis* SJ-101 not only facilitated higher accumulation but also improved the growth parameters of *B. juncea*, even under Ni stress. The bioinoculated plants exhibited a significant increase in biomass and growth attributes compared to uninoculated controls, suggesting that the bacteria strain mitigated the adverse effects of Ni toxicity. Thus, *B. subtilis* was found to produce indole acetic acid (IAA) and solubilize inorganic phosphate, which may enhance nutrient availability and promote plant growth, thereby facilitating greater Ni uptake. The study suggests that bioinoculant could be effectively utilized for bacteria-assisted phytoaccumulation of nickel from contaminated soils, highlighting its potential role in phytoremediation strategies.

Ma et al. (2009) researched the copper-resistant plant growth-promoting bacterial (PGPB) *Achromobacter xylosoxidans* strain Ax10 in *Brassica juncea*. This microbial strain isolated from a copper mine soil tolerated concentrations up to 600 mg Cu L^{-1} on agar medium. The bacterium has also exhibited the capability of producing IAA and solubilizing inorganic phosphate. Thus, bioinoculantion significantly increased the root lengths and fresh and dry weights of *B. juncea* compared to the control. Moreover, the input of *A. xylosoxidans* strain improved Cu uptake by plants. Therefore, this bacterium can be an effective bioinoculant in promoting growth and sequestering metals in copper-contaminated soils. The study suggests that the interaction between growth-promoting bacteria and root exudates enhances metal solubilization and absorption, highlighting phytoremediation as a cost-effective and environmentally friendly method for soil decontamination. Specifically, PGPB can significantly boost phytoextraction by improving both plant growth and metal uptake.

The presented case studies highlight the substantial potential of phytoremediation strategies enhanced by bioinputs and microbial inoculants for effective soil decontamination. These findings emphasize the adaptability of bioinoculants in not only promoting plant growth but also enhancing the efficiency of phytoextraction processes. This approach offers a cost-effective, sustainable, and eco-friendly solution for remediating soils contaminated with both organic and inorganic pollutants. The success of these methods points to the importance of further research into microbial-plant interactions to refine and optimize phytoremediation techniques across diverse contaminated environments.

Finally, evidence of bioinputs and microbial abilities associated with agricultural plant cultivation in decontamination are easily found for inorganic substances. Therefore, there are gaps in scientific knowledge to study and discover new sustainable strategies in the remediation of agroecosystems with organic contaminants, such as pesticides.

9.6 CHALLENGES AND FUTURE PERSPECTIVES

The success of agricultural bioinputs in soil phytoremediation with pesticides can be limited by environmental conditions, compatibility between plants and microorganisms, and the complexity of contaminants (Tarla et al. 2020; Arantza et al. 2022; Karimi et al. 2022). For instance, the degradation of certain pesticides may require specific combinations that are not yet widely developed or commercialized.

However, the cost of biocompounds can be a significant barrier, especially on a large-scale treatment. Inadequate infrastructure for bioinputs production and distribution also limits their availability and accessibility. However, these products are considered agro-industrial residues, whose low-cost and high-generation volume allow overcoming these difficulties. A simple and widespread example in agricultural practices is the fertigation of sugarcane crops with vinasse (Soto et al. 2017). As it is available in large quantities, this residue presents a low-cost alternative for waste reuse, soil fertility, and plant production (Pinto and Araujo 2019), and to increase the biological treatment of agricultural soils with pesticides (Aparicio et al. 2017; Nantes et al. 2023).

Therefore, the development of customized bioinputs tailored to specific conditions provides an efficient and cost-effective solution, particularly when they are produced directly on the agricultural chain. This approach has the potential to significantly reduce production costs based on the main guidelines of sustainable development. Additionally, the bioprospection of microorganisms able to degrade a wide range of pesticides represents one of the pillars for the development of biotechnological inputs to revolutionize soil treatment and agricultural practices with pesticides (Lima et al. 2022).

Despite the significant challenges, phytoremediation with agricultural bioinputs holds substantial potential for growth and innovation, contributing to more sustainable and resilient agriculture.

9.7 CONCLUSION

Integrating agricultural bioinputs in phytoremediation presents a promising solution for mitigating the environmental impacts of pesticides. Combining microorganisms, plants, and/or natural compounds can enhance the efficiency of soil remediation while promoting more sustainable agriculture. However, this research area is still little explored in scientific studies and some challenges must be considered to maximize its potential. Therefore, the continuous development of bio-tools and strategies tailored to local conditions is closely associated with in-depth knowledge of plant treatment processes and biological products to achieve modern and sustainable agricultural practices.

ACKNOWLEDGMENTS

São Paulo Research Foundation (FAPESP, Process 2022/14132-5), Coordination for Improvement of Higher Education Personnel (CAPES-Brazil, financing code n.001), National Council for Scientific and Technological Development (CNPq-Brazil, 313530/2021-1), and the Fundação Agrisus (Agrisus Foundation – Brazil, PA 3740/24).

ARTIFICIAL INTELLIGENCE DECLARATIONS

The authors declare that AI tools for language improvement, as all authors are non-native English speakers. The final interpretation, synthesis, and critical analysis of the literature remain the responsibility of the authors. Authors are accountable for the originality, validity, and integrity of the content of their submissions.

REFERENCES

Aparicio, J.D., Benimeli, C.S., Almeida, C.A., Polti, M.A., Colin, V.L., 2017. Integral use of sugarcane vinasse for biomass production of actinobacteria: potential application in soil remediation. *Chemosphere* 181, 478–484.

Arantza, S.J., Hiram, M.R., Erika, K., Chávez-Avilés, M.N., Valiente-Banuet, J.I., Fierros-Romer, G., 2022. Bio- and phytoremediation: plants and microbes to the rescue of heavy metal polluted soils. *SN Appl. Sci.* 4(59). https://doi.org/10.1007/s42452-021-04911-y

Assadpour, E., Can Karaça, A., Fasamanesh, M., Mahdavi, S.A., Shariat-Alavi, M., Feng, J., Jafari, S.M., 2023. Application of essential oils as natural biopesticides; recent advances. *Crit. Rev. Food Sci. Nutr.* 64(19), 6477–6497.

Ayilara, M.S., Adeleke, B.S., Akinola, S.A., Fayose, C.A., Adeyemi, U.T., Gbadegesin, L.A., Omole, R.K., Johnson, R.M., Uthman, Q.O., Babalola, O.O., 2023. Biopesticides as a promising alternative to synthetic pesticides: a case for microbial pesticides, phytopesticides, and nanobiopesticides. *Front. Microbiol.* 14, 1040901.

Babla, M., Katwal, U., Yong, M., Jahandari, S., Rahme, M., Chen, Z., Tao, Z., 2022. Value-added products as soil conditioners for sustainable agriculture. *Resour. Conserv. Recycl.* 178, 106079.

Bhatt, K., Suyal, D.C., Kumar, S., Singh, K., Goswami, P., 2022. New insights into engineered plant-microbe interactions for pesticide removal. *Chemosphere* 309, 136635.

Bhupenchandra, I., Chongtham, S.K., Devi, E.L.R.R., Choudhary, A.K., Salam, M.D., Sahoo, M.R., Bhutia, T.L., Devi, S.H., Thounaojam, A.S., Behera, C.M.N.H., Kumar, A., Dasgupta, M., Devi, Y.P., Singh, D., Bhagowati, S., Devi, C.P., Singh, H.R., Khaba, C.I., 2022. Role of biostimulants in mitigating the effects of climate change on crop performance. *Front. Plant Sci.* 13, 967665.

Boakye, T.A., Anning, D.K., Li, H., Osei, R., Boamah, S., Anim, S., 2022. Mechanism of antagonistic bioagents in controlling root-knot nematodes (*Meloidogyne* sp.): a review. *Asian Res. J. Agric.* 15(2), 27–44.

Brazil, 2021. *Bioinputs: learn about the conceptual basis of the National Bioinputs Program* (Bioinsumos: Conheça a base conceitual do Programa Nacional de Bioinsumos). Ministry of Agriculture, Livestock and Food Supply (Ministério da Agricultura, Pecuária e Abastecimento – MAPA). Brasília-DF, Brazil: MAPA. https://www.planalto.gov.br/ccivil_03/_ato2019-2022/2020/decreto/d10375.htm

Brazil, 2020. *Decree No. 10,375 of May 26, 2020. Institutes the National Bioinputs Program and the Strategic Council of the National Bioinputs Program (Institui o Programa Nacional de Bioinsumos e o Conselho Estratégico do Programa Nacional de Bioinsumos).* Brasília-DF, Brazil: Presidency of the Republic. https://www.gov.br/agricultura/pt-br/assuntos/inovacao/bioinsumos/o-programa/conceitos

Calderon, R.B., Dangi, S.R., 2024. Arbuscular mycorrhizal fungi and rhizobium improve nutrient uptake and microbial diversity relative to dryland site-specific soil conditions. *Microorganisms* 12(4), 667.

Candian, J.S., Nakada-Freitas, P.G., Silva, A.B.R., Cardoso, A.I.I., 2024. Impact of castor bean cake fertilizer as side-dressing on cauliflower yield and quality in organic farming systems. *Aust. J. Crop Sci.* 18, 240–247.

Cao, T.N.D., Mukhtar, H., Le, L.T., Tran, D.T.H., Ngo, M.T.T., Pham, M.D.T., Nguyen, T.B., Vo, T.K.Q., Bui, X.T., 2023. Roles of microalgae-based biofertilizer in sustainability of green agriculture and food-water-energy security nexus. *Sci. Total Environ.* 870, 161927.

Castro, J.S., Ferreira, J., Magalhães, I.B., Jesus Junior, M.M., Marangon, B.B., Pereira, A.S.A.P., Lorentz, J.F., Gama, R.C.N., Rodrigues, F.A., Calijuri, M.L., 2023. Life cycle assessment and techno-economic analysis for biofuel and biofertilizer recovery as by-products from microalgae. *Renew. Sust. Energ. Rev.* 187, 113781.

Daurova, D., Zhambakina, K., Shamekovaa, M., 2023. Phytoremediation as a way to clean technogenically polluted areas of Kazakhstan. *Braz. J. Biol.* 83, e271684.

Díaz-Urbano, M., Goicoechea, N., Velasco, P., Poveda, J., 2023. Development of agricultural bio-inoculants based on mycorrhizal fungi and endophytic filamentous fungi: co-inoculants for improve plant-physiological responses in sustainable agriculture. *Biol. Control* 182, 105223.

Ferreira, L.C., Moreira, B.R.A., Montagnolli, R.N., Prado, E.P., Viana, R.S., Tomaz, R.S., Cruz, J.M., Bidoia, E.D., Frias, Y.A., Lopes, P.R.M., 2021. Green manure species for phytoremediation of soil with tebuthiuron and vinasse. *Front. Bioeng. Biotechnol.* 8, 613642.

Frias, Y.A., Valério, T.S., Moreira, B.R.A., Cruz, V.H., Santos, A.L., Andriolli, V., Pincerato, G.M.T., Lopes, P.R.M., 2024. *Pennisetum glaucum* in reducing ecotoxicity in soil with tebuthiuron, thiamethoxam and vinasse, *J. Agric. Food Res.* 18, 101470.

Grammenou, A., Petropoulos, S.A., Thalassinos, G., Rinklebe, J., Shaheen, S.M., Antoniadis, V., 2023. Biostimulants in the soil–plant interface: agro-environmental implications – a review. *Earth Syst. Environ.* 7, 583–600.

Greipsson, S., 2011. Phytoremediation. *Nat. Educ. Knowledge* 3(10), 7.

Gupta, S., Bhattacharyya, P., Kulkarni, M.G., Doležal, K., 2023. Editorial: growth regulators and biostimulants: upcoming opportunities. *Front. Plant Sci.* 14, 1209499.

Gurikar, C., Naik, M.K., Sreenivasa, M.Y., 2016. *Azotobacter*: PGPR activities with special reference to Effect of pesticides and biodegradation. In: Singh, D., Singh, H., Prabha, R. (eds) *Microbial Inoculants in Sustainable Agricultural Productivity*. Springer, New Delhi.

Hejazirad, S.P., de Abreu, C.M., Carneiro, G.H.F., Gomes, C.R., Spinola Filho, P.R.D.C., da Costa, M.R., Santos, J.B.D., 2024. The impact of metolachlor applications and phytoremediation processes on soil microorganisms: insights from functional metagenomics analysis. *J. Xenobiot.* 14, 970–988.

Hernández-Álvarez, C., Peimbert, M., Rodríguez-Martin, P., Trejo-Aguilar, D., Alcaraz, L.D., 2023. A study of microbial diversity in a biofertilizer consortium. *PLoS One* 18(8), e0286285.

Hezakiel, H.E., Thampi, M., Rebello, S., Sheikhmoideen, J.M., 2024. Biopesticides: a green approach towards agricultural pests. *Appl. Biochem. Biotechnol.* 196, 5533–5562.

Hnini, M., Rabeh, K., Oubohssaine, M., 2024. Interactions between beneficial soil microorganisms (PGPR and AMF) and host plants for environmental restoration: a systematic review. *Plant Stress* 11, 100391.

Islam, M.M., Saxena, N., Sharma, D., 2024. Phytoremediation as a green and sustainable prospective method for heavy metal contamination: a review. *RSC Sustain.* 2, 1269–1288.

Javeed, M.T., Al-Hazmi, A.S., Molan, Y.Y., 2016. Antagonistic effects of some indigenous isolates of *Trichoderma* spp. against *Meloidogyne javanica*. *Pak. J. Nematol.* 34(2), 183–191.

Jia, Y., Kang, L., Wu, Y., Zhou, C., Li, D., Li, J., Pan, C., 2023. Review on pesticide abiotic stress over crop health and intervention by various biostimulants. *J. Agric. Food Chem.* 71(37), 13595–13611.

Kafle, A., Timilsina, A., Gautam, A., Adhikari, K., Bhattarai, A., Aryal, N., 2022. Phytoremediation: mechanisms, plant selection and enhancement by natural and synthetic agents. *Environ. Adv.* 8, 100203.

Kamath, A., Shukla, A., Saiyed, T., Bhatt, S., Rathod, H., Makwana, V., Soni, D., Banerjee, S., Patel, D., 2023. Bioinoculants: the agrarian avengers. *Symbiosis* 91, 151–166.

Karimi, H., Mahdavi, S., Lajayer, B.A., Moghiseh, E., Rajput, V.D., Minkina, T., Astatkie, T., 2022. Insights on the bioremediation technologies for pesticide-contaminated soils. *Environ. Geochem. Health* 44, 1329–1354.

Kumar, H.D., Aloke, P., 2020. Role of biostimulant formulations in crop production: an overview. *Int. J. Appl. Res. Vet. Med.* 8, 38–46.

Kumar, J., Ramlal, A., Mallick, D., Mishra, V., 2021. An overview of some biopesticides and their importance in plant protection for commercial acceptance. *Plants* 10, 1185.

Kumpiene, J., Lagerkvist, A., Maurice, C., 2008. Stabilization of As, Cr, Cu, Pb and Zn in soil using amendments – a review. *Waste Manag.* 28(1), 215–225.

Lima, E.W., Brunaldi, B.P., Frias, Y.A., Moreira, B.R.A., Alves, L.S., Lopes, P.R.M., 2022. A synergistic bacterial pool decomposes tebuthiuron in soil. *Sci. Rep.* 12, 9225.

Liu, M., Tan, X., Zheng, M., Yu, D., Lin, A., Liu, J., Wang, C., Gao, Z., Cui, J., 2023. Modified biochar/humic substance/fertiliser compound soil conditioner for highly efficient improvement of soil fertility and heavy metals remediation in acidic soils. *J. Environ. Manag.* 325, 116614.

Lopez, A.M.Q., Silva, A.L.D.S., Maranhão, F.C.D.A., Ferreira, L.F.R., 2022. Plant growth promoting bacteria: aspects in metal bioremediation and phytopathogen management. In: Kumar, A. (ed) *Microbial Biocontrol: Sustainable Agriculture and Phytopathogen Management*. Springer, Cham, pp. 51–78.

Ma, Y., Rajkumar, M., Freitas, H., 2009. Inoculation of plant growth promoting bacterium *Achromobacter xylosoxidans* strain Ax10 for the improvement of copper phytoextraction by *Brassica juncea*. *J. Environ. Manage.* 90(2), 831–837.

Mafra, E., 2023. Where is the bioinputs market headed in Brazil? Understand the challenges, what they are and why their adoption responds to a global demand for sustainability in the field (O mercado de bioinsumos vai para onde no Brasil? Entenda os desafios, o que são e por que a sua adoção responde a uma demanda global por sustentabilidade no campo). *Forbes Agro*. Available in: https://forbes.com.br/forbesagro/2023/01/o-mercado-de-bioinsumos-vai-para-onde-no-brasil/

Nantes, L.S., Aragão, M.B., Moreira, B.R.A., Frias, Y.A., Valério, T.S., Lima, E.W., Viana, R.S., Lopes, P.R.M., 2023. Synergism and antagonism in environmental behavior of tebuthiuron and thiamethoxam in soil with vinasse by natural attenuation. *Int. J. Environ. Sci. Technol.* 20, 4883–4892.

Narayanan, M., Ma, Y., 2024. Recent progress on conservation and restoration of soil fertility for horticulture. *Chemosphere* 362, 142599.

Nurzhanova, A., Mukasheva, T., Berzhanova, R., Kalugin, S., Omirbekova, A., Mikolasch, A., 2020. Optimization of microbial assisted phytoremediation of soils contaminated with pesticides. *Int. J. Phytoremediation* 23(5), 482–491.

Pinto, L.E.V., Araujo, F.F., 2019. Use of vinasse as biofertilizer: effect on nodulation, growth and nutrient accumulation in soybean cultivation (Uso de vinhaça como biofertilizante: efeito na nodulação, crescimento e acúmulo de nutrientes no cultivo da soja). *Colloq. Agrar.* 15(5), 97–109.

Rakkammal, K., Maharajan, T., Ceasar, S.A., Ramesh, M., 2023. Biostimulants and their role in improving plant growth under drought and salinity. *Cereal Res. Commun.* 51, 61–74.

Rocha, T.M., Marcelino, P.R.F., Da Costa, R.A.M., Rubio-Ribeaux, D., Barbosa, F.G., da Silva, S.S., 2024. Agricultural bioinputs obtained by solid-state fermentation: from production in biorefineries to sustainable agriculture. *Sustainability* 16(3), 1076.

Rungwa, S., Arpa, G., Sakulas, H., Harakuwe, A., Timi, D., 2013. Phytoremediation – an eco-friendly and sustainable method of heavy metal removal from closed mine environments in Papua New Guinea. *Procedia Earth Planet. Sci.* 6, 269–277.

Santos, F., Melkani, S., Oliveira-Paiva, C., Bini, D., Pavuluri, K., Gatiboni, L., Mahmud, A., Torres, M., McLamore, E., Bhadha, J.H., 2024. Biofertilizer use in the United States: definition, regulation, and prospects. *Appl. Microbiol. Biotechnol.* 108, 511.

Senatore, D., Queirolo, A., Monza, J., Bajsa, N., 2023. Using sugarcane vinasse as a biofertilizer: effects on soil microbial community due to periodical application. *Environ. Sustain.* 6, 173–182.

Seo, J.S., Keum, Y.S., Li, Q.X., 2009. Bacterial degradation of aromatic compounds. *Int. J. Environ. Res. Public Health* 6(1), 278–309.

Sharma, I., Sharma, S., Sharma, V., Singh, A.K., Sharma, A., Kumar, A., Singh, J., Sharma, A., 2024. PGPR-Enabled bioremediation of pesticide and heavy metal-contaminated soil: a review of recent advances and emerging challenges. *Chemosphere* 362, 142678.

Silva, J.S.L., Silva, J.H.B., Silva, A.V., Silva, L.K.P., Ferreira, J.S., Guedes, J.V.S., Santiago, M.S.T., Costa, E.N., Freitas, P.V.M., Mielezrski, F., 2024. Use of neem vegetable cake (*Azadirachta indica* A. Juss) increases corn productivity. *Braz. J. Biol.* 84, e281515.

Song, C., Sarpong, C.K., He, J., Shen, F., Zhang, J., Yang, G., Long, L., Tian, D., Zhu, Y., Deng, S., 2020. Accelerating phytoremediation of degraded agricultural soils utilizing rhizobacteria and endophytes: a review. *Environ. Rev.* 28(1), 115–127.

Soto, M.A., Basso, J.B., Kiang, C.H., 2017. Impact of sugarcane fertigation by vinasse on soil physical, chemical and hydraulic properties (Impacto da fertirrigação da cana-de-açúcar por vinhaça nas propriedades físicas, químicas e hidráulicas do solo). In: Fontanetti, C.S., Bueno, O.C. (eds) *Sugarcane and Its Impacts: An Academic View (Cana-de-açúcar e seus impactos: uma visão acadêmica)*. Canal 6, Bauru-SP, Brazil, pp. 103–124.

Sun, B., Gu, L., Bao, L., Zhang, S., Wei, Y., Bai, Z., Zhuang, G., Zhuang, X., 2020. Application of biofertilizer containing Bacillus subtilis reduced the nitrogen loss in agricultural soil. *Soil Biol. Biochem.* 148, 107911.

Suryatmana, P., Setiawati, M., Herdiyantoro, D., Fitriatin, B., Kamaluddin, N., 2022. Characterization and potential of plant growth-promoting rhizobacteria (PGPR) isolates capacity correlating with their hydrocarbon biodegradation capability. *J. Agric. Sci.* 44(3), 559–574.

Tarla, D.N., Erickson, L.E., Hettiarachchi, G.M., Amadi, S.I., Galkaduwa, M., Davis, L.C., Nurzhanova, A., Pidlisnyuk, V., 2020. Phytoremediation and bioremediation of pesticide-contaminated soil. *Appl. Sci.* 10(4), 1217.

Tonelli, F.M.P., Bhat, R.A., Dar, G.H., Hakeem, K.R., 2022. The history of phytoremediation. In: Bhat, R.A., Tonelli, F.M.P., Dar, G.H., Hakeem, K. (eds) *Phytoremediation*. Academic Press, Cambridge-MA, USA, pp. 1–18.

Tornisielo, V.L., Vilca, F.Z., Guimarães, A.C.D., Mendes, K.F., 2019. *Organic Contaminants: From Analysis to Bioremediation (Contaminantes orgânicos: da análise à biorremediação)*. FEALQ, Piracicaba-SP, Brazil, 268 p.

Van Oosten, M.J., Pepe, O., De Pascale, S., Silletti, S., Maggio, A., 2017. The role of biostimulants and bioeffectors as alleviators of abiotic stress in crop plants. *Chem. Biol. Technol. Agric.* 4(5). https://doi.org/10.1186/s40538-017-0089-5

Yadav, K., Gupta, N., Kumar, A., Reece, L.M., Singh, N., Rezania, S., Khan, S.A., 2018. Mechanistic understanding and holistic approach of phytoremediation: a review on application and future prospects. *Ecol. Eng.* 120, 274–298.

Zaidi, S., Usmani, S., Singh, B.R., Musarrat, J., 2006. Significance of *Bacillus subtilis* strain SJ-101 as a bioinoculant for concurrent plant growth promotion and nickel accumulation in Brassica juncea. *Chemosphere* 64(6), 991–997.

Zhao, J., Lu, H., Qin, S., Pan, P. Tang, S., Chen, L., Wang, X., Tang, F., Tan, Z., Wen, R., He, B., 2023. Soil conditioners improve Cd-contaminated farmland soil microbial communities to inhibit Cd accumulation in rice, *J. Integr. Agric.* 22, 8, 2521–2535.

Zhou, G., Wang, Y., Zhai, S., Ge, F., Liu, Z., Dai, Y., Yuan, S., Hou, J., 2013. Biodegradation of the neonicotinoid insecticide thiamethoxam by the nitrogen-fixing and plant-growth-promoting Rhizobacterium *Ensifer adhaerens* strain TMX-23. *Appl. Microbiol. Biotechnol.* 97, 4065–4074.

10 Agricultural Land Pollution in Indonesia

Land Restoration and Reclamation Strategies

Qomarudin Helmy and Pujawati Suryatmana

10.1 INTRODUCTION

The agriculture sector remains a major contributor to the economic growth of developing countries such as Indonesia. Being one of the largest agricultural countries in the world, Indonesia has a small percentage of agricultural land, which is less than 5% of its total land area (Figure 10.1). The rapid increase in population has increased the demand for land for housing, services, industry, and other public facilities. Land needs for non-agricultural activities include the development of real estate, industrial estate, commercial districts, and other services that require a large enough area of land, some of which come from originated agricultural land. This is understandable, considering that the location is chosen in such a way that it is close to the service users who are concentrated in urban and surrounding areas. The regions surrounding the city traditionally associated with agriculture are being made subject to non-agricultural uses on account of their lower price and adequate infrastructure amenities (Sitorus and Pravitasari, 2017).

Conversion of agricultural land into urban settlements and industrial areas is increasingly common in many regions around the world. This shift, often driven by the fast urbanization and economic growth of the countries, can lead to significant threats to soil health and agricultural productivity. As industrial areas expand, the quality of agricultural soils can deteriorate, resulting in long-term ecological, economic, and social consequences. When agricultural land is converted to industrial estate, the immediate impacts on soil quality are striking and diverse. One of the problems that arises from land-use change is soil compaction, where the use of heavy machinery in the construction process and in ongoing industrial activities compresses the soil structure, reducing its porosity and ability to absorb water. This soil compaction not only inhibits root development for remaining vegetation but also reduces the soil capacity to support life, including important microorganisms that contribute to soil fertility (Magdoff and van Es, 2021). Compacted soils can lead to increased surface runoff and erosion, further degrading the soil quality. The introduction and release of foreign compounds associated with industrial activities further degrade soil.

DOI: 10.1201/9781003714484-10

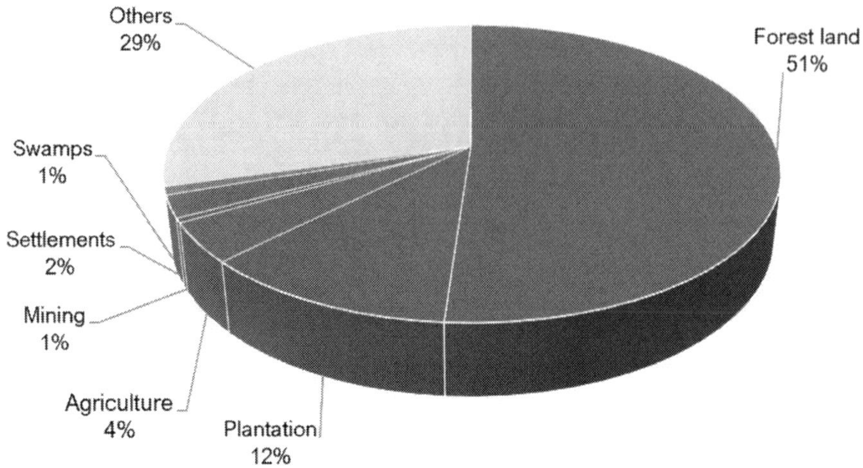

FIGURE 10.1 Land use data in Indonesia in 2022 (Indonesian Central Bureau of Statistics, 2024).

Persistent organics, heavy metals, and other hazardous substances can leach into soil, changing its characteristics and affecting its health. For example, industrial wastewater runoff may contain toxic metals such as chromium, lead, cadmium, and mercury, which can persist in the soil for years, posing a risk not only to the surrounding environment but also to the food chain system. Vegetations growing in contaminated soil can absorb and accumulate these hazardous substances, which can ultimately have serious impacts on human health (Priya et al., 2023).

Topsoil, which is rich in organic matter and nutrients essential for plant growth, is often removed during industrial development. Loss of topsoil can lead to decreased soil fertility, making it more difficult for agricultural activities to continue. The loss of organic matter reduces the soil's ability to hold water and retain nutrients, increasing the decline in agricultural production. The change from agricultural to industrial use also results in loss of biodiversity, as agricultural landscapes often support a wide variety of plant and animal species, which contribute to a healthy ecosystem. Loss of biodiversity disrupts natural processes such as pollination and natural control of pests, which are essential for maintaining healthy ecosystems and effective agricultural practices. Diverse plant and animal populations help create dynamic soil ecosystems where bacteria, fungi, and other organisms play important roles in recycling nutrients and breaking down organic waste. The loss of these populations can lead to imbalances that damage soil structure and fertility. The long-term impacts of deteriorating soil quality are very significant and have a wide impact because healthy soil is a basic requirement for food production very important for the sustainability of agricultural production. Farmers experience lower agricultural yields when soil quality deteriorates, resulting in food poverty and economic stress. To compensate for poor soil conditions, farmers may rely more on artificial fertilizer and pesticides; although these inputs may provide short-term benefits, they often lead to severe soil degradation over time. Excessive use of fertilizer can cause nutrient runoff, which contributes

FIGURE 10.2 Left: rice fields next to a textile factory in Majalaya Regency, West Java, Indonesia (Kusumadirezza, 2023); right; irrigation contamination due to wastewater discharge is frequently reported by landowners around the industrial area (Dachlan, 2013).

to water pollution and eutrophication in surrounding water bodies. Furthermore, reliance on chemical fertilizer inputs can reduce the natural resilience of the soil, making the soil more susceptible to erosion and degradation (Wolf et al., 2023).

Conversion of agricultural land into industrial areas is a multidimensional challenge that endangers soil quality and agricultural sustainability. To address this problem, both regulators and communities must work together to recognize the inherent value of fertile soil. Implementing sustainable land-use techniques is essential, including restoring degraded land, implementing regenerative farming practices, and incorporating green spaces into industrial estates. These technique scan help offset the harmful effects of industrialization while improving biodiversity and soil health. By prioritizing farmland health, we can ensure a more sustainable future for our food security system and environment. The difficulty of balancing industrial progress and agricultural integrity is considerable, but with integrated cooperation, it is possible to achieve harmonious coexistence that benefits both economic development and environmental sustainability (Arifin et al., 2024) (Figure 10.2).

10.2 CONTAMINATION OF AGRICULTURAL LAND

Urban development has various positive impacts on the wider community, for example, industrial development that can create new jobs for the surrounding population. However, this success often also has negative impacts that are detrimental to the community and surrounding environment. The development of industrial areas in productive agricultural areas causes a reduction in agricultural land area (land conversion), soil pollution that reduces agricultural yields, and disrupts the comfort and health of humans or other living things. It must be acknowledged that land as a final disposal site for waste is an easy alternative. The large numbers of factories, industries, and mines built around agricultural land have caused contamination of planting fields. When polluted air falls to the soil with rainwater, the water will pollute agricultural land. In agricultural practice, the use of agrochemical fertilizer and pesticides, especially on vegetable crops, is very intensive and given in high doses, this will result in high levels of pesticide residues in the soil and farm commodities. The causes of pollution in agricultural land can be classified into natural and anthropogenic activities, in the form of excessive use of agrochemical and non-agricultural activities, including industrial and mining activities (Notodarmojo, 2005; Rashid et al., 2023; Angon et al., 2024; Moldavan et al., 2024; Li et al., 2025).

10.2.1 Natural Source of Agricultural Land Contamination

Naturally, soil contains metals and heavy metals at levels that depend on the content of the parent rock, how the soil was formed, and the translocation of heavy metals in the soil (Table 10.1). From the perspective of plant nutrition, metals are micronutrients required by plants in small amounts. Essential metals for plant growth are Boron (B), Zinc (Zn), Copper (Cu), Chloride (Cl), Iron (Fe), Manganese (Mn), and Molybdenum (Mo), which are known as trace elements. In addition to metals, soil contains heavy metals that are not needed in plant metabolism. High concentrations and availability of heavy metals in the soil have the potential to poison plants. Heavy

TABLE 10.1
The Natural Presence of Heavy Metals in the Minerals that Make Up Rocks and Parent Rocks that Form Soil

Parent Rocks/Minerals	Heavy Metals	Concentration (mg/kg)	Reference
Carbonaceous rocks	Nickel	383–3,599	Koralay and Koralay (2022)
Carbonaceous rocks	Chromium	225–1,922	
Carbonaceous rocks	Cobalt	24–182	
Itabirite rocks	Manganese	<330–6,850	Zinn et al. (2020)
Itabirite rocks	Nickel	36–43	
Mafic and Ultramafic rocks	Chromium	210–258	
Sedimentary rocks (shales, mudstone, slate)	Arsenic	<1–900	O'Neill (1995)
Sedimentary rocks (sand, limestone)	Arsenic	<1–20	
Igneous rocks (granite, basalt)	Cadmium	0.01–1.6	Alloway (1995)
Sedimentary rocks (shales, clay)	Cadmium	0.017–11	
Ultramafic igneous rocks	Chromium and Nickel	1,000–3,400 and 270–3,600	McGrath (1995)
Granite igneous rocks	Chromium and Nickel	2–90 and 2–20	
Andesite igneous rock	Lead	1.9	Davies (1995)
Sedimentary rocks, limestone, and dolomite	Lead	71	
Metacinnabar sedimentary rock	Mercury	110,700	Steinnes (1995)
Basalt igneous rock	Mercury	0.2–17.7	

metals can poison plants from the germination phase to the growth and development phase (Alengebawy et al., 2021).

Heavy metals exist in soils in various forms, including as primary minerals in rock fragments, organic chelates, and complexes, substituting ions in crystal lattices of secondary minerals, immobilized in microorganisms and plants, sorbed onto mineral and organic surfaces, and leached into the soil matrix. Physical and chemical weathering cause the degradation of rock and minerals. Weathering mechanisms induce substantial modifications in mineral chemistry and rock composition. The soil material is the primary source of heavy metals in pristine environments, despite the fact that the forms and availabilities of heavy metals in rocks minerals are highly variable (Zinn et al., 2020). Rock minerals such as serpentine, which contains nickel (Ni), and gabbro which contain cobalt (Co) and chromium (Cr) is contributing to the amounts of these elements in the soil through weathering process (Kabata-Pendias and Mukherjee 2007). Volcanic activity has significant impact on the deposition of large amounts of heavy metals, such as cadmium, arsenic, and lead, through the fallout of volcanic ash (Varekamp and Thomas, 1998). The deposition of metal-rich sediments carried from mineral-rich mountainous areas is also influenced by the natural input from rivers. Eventually, the mobilization of materials across rivers might result in the accumulation of heavy metals in floodplains (Forstner and Wittmann, 2012). Another natural source of heavy metals in soils is the biogeochemical cycles that occur and contribute to the dispersion and concentration of these metals. Soil microorganisms and plant absorption can move elements such as zinc, Manganese, and mercury throughout the soil profile (Alloway, 2013; Moeller and Wall, 2017).

10.2.2 Anthropogenic Source of Agricultural Land Contamination

In recent decades, there has been increasing awareness of the dangers posed by toxic heavy metal pollution, which can be toxic due to the adverse effects they have on living organisms. Heavy metal contamination of soils and agricultural environments is accelerated by the industrialization of modern societies, the world's rapidly growing population, and intensive agriculture. Soil pollution is a significant problem because soil is a natural resource essential for human life (Al-Taai, 2021).

10.2.2.1 Agrochemical Pollution

The level of pollution and environmental damage in agricultural environments can be caused by the overuse of agrochemicals, for example, fertilizers and pesticides. The green revolution in agriculture in the last century has succeeded in changing agricultural productivity spectacularly with the introduction of the use of agrochemicals both in the form of chemical fertilizers and pesticides. Indeed, with the green revolution, world food production has increased sharply overcoming world concerns about food crisis issues. However, the impact of the use of agrochemicals has begun to be felt, including in the form of water, soil, and agricultural product pollution, farmer health problems, decreased in biodiversity, and disrupt the food chain in the ecosystem. Excessive pesticide use over time will have an influence on the life and survival of natural insect and disease enemies, as well as the life of soil biota. This leads to an increase of pests and diseases as well as the deterioration of soil biota. It is necessary to think that at this time pesticide residues will be a determining factor in the competitiveness of agricultural products that will enter the global market.

Heavy metal levels from phosphate fertilizers and nitrate fertilizers, including agricultural lime, cannot be ignored as one of the sources of contaminants in agricultural soil (Table 10.2). Phosphate rock is the basic material for making phosphate fertilizers which also contain heavy metals Pb, Cd, Cr, Co, Hg, Ni, and As. The natural metal content in phosphate fertilizers is quite high so that its content in fertilizers must be limited, as regulated by the Indonesian Minister of Agriculture number 261 of 2019 regarding technical requirements for organic fertilizer and soil amendment, where the limits of heavy metal contamination in natural phosphates as ameliorants and fertilizers are limited: <10 mg/kg As, <1 mg/kg Hg, <50 mg/kg Pb, <2 mg/kg Cd, <180 mg/kg Cr, and <50 mg/kg Ni

Other contaminant from agricultural activities that can cause soil pollution are the remains of pesticide used to eradicate plant pest, for example, DDT (Figure 10.3). Two properties that make DDT very dangerous to the environment are the apolar nature of DDT where it is insoluble in water but very soluble in fat. The more soluble a pesticide in the fat (the more lipophilic) the higher its apolar properties. This is one factor that causes DDT to penetrate animal and human skin easily. Then, the very stable and persistent nature of DDT makes it difficult to decompose so it tends to survive in the environment, capable of entering the food chain through the fatty material of living tissue. That is why DDT is bioaccumulative and biomagnificant.

TABLE 10.2
The Range of Heavy Metal Content Contained in Inorganic and Organic Fertilizers

	Heavy Metal Levels (mg/kg)			
Heavy Metals	**Phosphate Fertilizer**	**Nitrogen Fertilizer**	**Manure, Biosolids**	**Agricultural Lime**
Arsenic	2–1200	2.2–120	3–25	0.1–25
Cadmium	0.1–170	0.05–85	0.1–0.8	0.04–1
Chromium	66–245	3.2–19	1.1–55	10–15
Lead	7–225	2–27	1.1–27	20–1250
Mercury	0.01–0.05[a]	0.023–0.032[b]	1–56[c]	5.1 ± 0.3[d]
Nickel	0.06–29.7[a]	0.40–0.78[b]	2–5,300[c]	d.n.a
Vanadium	1.92–430[a]	1.53[a]	d.n.a	d.n.a

Source: Adapted from Alloway (1995).
Note: d.n.a: data not available.
[a] Murniasih et al. (2022).
[b] Gambus and Wieczorek (2012).
[c] Mortvedt (1996).
[d] Zhao and Wang (2010).

FIGURE 10.3 Rice paddy farmers spray insecticide in Okinawa, Japan (left), DDT spraying used by the US Department of Agriculture in 1955 for forest insect and disease detection surveys project (right).

Because of its stable and persistent nature, DDT lasts a very long time in the soil, DDT can even bind to organic matter in soil particles. DDT is included in the third order of Persistent Organic Pollutants/POPs, which have the following properties:

1. Difficult to degrade through photolysis, biologically or chemically
2. Halogenated (usually with chlorine compounds)
3. Very low solubility in water
4. Very soluble in fat
5. Semi-volatile, in the air can be moved by wind over long distances
6. Bioaccumulative.

Because of these properties, DDT is classified as a PBT (persistent, bioaccumulative, and toxic) material.

10.2.2.2 Industrial and Mining Pollution

Agricultural land contamination can result from the discharge of waste and contaminants into the surrounding environment as a result of industrial and mining activities in close range to agricultural land. These activities frequently introduce heavy metals and other toxic substances released into soil, which has a detrimental impact on agricultural output and soil quality. Industrial waste that can cause soil pollution can come from manufacturing factories and small-medium industries in the form of solid and liquid waste. Solid industrial waste in the form of solids or sludge can be generated during the product manufacturing process, for example, the waste from sugar cane factories, pulp and paper, rayon, plywood, and minerals processing industries. Wastewater containing harmful substances released during the manufacturing process without proper treatment before discharge also contributed to soil pollution. Contamination by hazardous and toxic industrial chemicals has been proven by a number of studies. The main attention is paid to toxic chemicals found both in irrigation waters and agricultural land in the form of persistent organic compounds and

TABLE 10.3
Major Pollutant Content in Wastewater from Various Industrial Activities

Industry	Solid	Organics	Nitrogen, Oil and Grease	Metals
Electroplating	TDS	—	—	Cu, Zn, Cr, Cd, Pb, Ni, CN, Ag
Leather Tanning	TSS	COD, BOD	N, O&G	Cr
Palm Oil	TSS	COD, BOD	N, O&G	—
Natural Rubber	TSS	COD, BOD	N	—
Plywood	TSS	COD, BOD	N	—
Dairy	TSS	COD, BOD	N	—
Soap, Detergent	TSS	COD, BOD	O&G	—
Battery	TSS	COD, BOD	O&G	Pb, Cu
Farm, Husbandry	TSS	COD, BOD	N	—
Electronics	TSS	COD, BOD	N, O&G	Cu, Zn, Cr, Cd, Hg, Pb, Ni
Refined Sugar	TSS, TDS	COD, BOD	—	—
Petrochemical	TSS	COD, BOD	O&G	Cr, Cu, Zn, Ni
Rayon	TSS	COD, BOD	—	Zn
Ceramics	TSS	—	—	Co, Pb, Cd, Cr
Polyethylene Terephthalate	TSS	COD, BOD	O&G	Cr, Cu, Zn
Oleochemical	TSS	COD, BOD	N, O&G	—
Pulp Paper	TSS	COD, BOD	—	—
Dry Battery	TSS	COD	N, O&G	Zn, Hg, Mn, Cr, Ni
Paints	TSS	BOD	O&G	Hg, Zn, Pb, Cu, Cr, Ti, Cd
Pharmaceutical	TSS	COD, BOD	N	—
Textile	TSS	COD, BOD	N, O&G	Cr
Gold Mining	TSS	—	—	Cu, Cd, Zn, Pb, As, Ni, Cr, Hg

Source: Adapted from the Indonesian Ministry of Environment (2014).
Note: TSS – total suspended solid; TDS – total dissolved solid; COD – chemical oxygen demand; BOD – biochemical oxygen demand; N – nitrogen compound; O&G – Oil and Grease.

heavy metals. Table 10.3 summarizes the pollutants contained in wastewater from various industrial activities (IME, 2014).

Gold mine tailings are typically disposed of by artisanal gold miners in storage ponds or agricultural land that is used for cultivation. Nevertheless, the gold mine sediments still contain heavy metals, including cadmium, copper, Lead, and zinc which can result in bioaccumulation in food chains. Cases of heavy metal contamination from illegal gold mining in Indonesia that still uses mercury as an extraction or amalgamation method. Mineral mining produces tailings in the form of crushed parent material or soil rocks containing associated heavy metals, eroded especially in the rainy season can contaminate water bodies and agricultural land around it. Illegal gold mining in Indonesia such as in the Pongkor area, West Java; Gunung Botak on

Buru Island; Poboya in Central Sulawesi; Bombana in Southeast Sulawesi; Cisitu in Lebak Banten has increased the levels of mercury and arsenic metals in agricultural land. Mercury in agricultural land on the Waeapo Buru plains adjacent to the gold mining location contains 1.64–4.22 mg/kg Hg while tailings deposited on topsoil rice fields reach levels of 9.92–166.1 mg/kg Hg (Hindersah et al., 2018). The Hg levels in the tailings of gold processing waste in the Poboya area, Central Sulawesi contain 621.37 mg/kg Hg (Zulfikah et al., 2014) and the former mining area in Kalimantan contains 700–2,645 mg/kg Cr (Saidy and Badruzsaufari, 2009).

Threats to the quality of agricultural land affected by mining are decreasing macronutrient content, increasing types and quantities of micronutrients, and decreasing physical fertility and extreme soil acidity. Piles of tailing on agricultural land will form new soil, Entisol, which is infertile. The biological availability of heavy metal is the ability of the metal to dissolve in the soil and then be absorbed by plants. The bioavailability of heavy metals is associated with plant absorption known as phytoavailability, which is closely related to adsorption affinity, complex formation, and organic ligands (Kim et al., 2015). Noviardi et al. (2023) reported that sweet potato roots and shoots can absorb the heavy metals once grown on land contaminated with gold mine tailings. Cadmium accumulation was identified with an increase of 29.41–47.38 mg/kg in the roots and 24.27–65.76 mg/kg in the shoots over the control. Copper content in sweet potato roots increased by 152.54–553.77 mg/kg over the control, while Lead content in roots increased by 8.04–144.71 mg/kg and in the shoots 36.86–87.07 mg/kg over the control. Zinc accumulation in sweet potato roots increased by 4,423.5 mg/kg and in shoot increased by 1,050–2,936.1 mg/kg over control. This study demonstrated that sweet potatoes could be employed for the phytoremediation of heavy metals in polluted soil, however, the plants may not be suitable for human consumption.

Other pollution from industrial activities is petroleum hydrocarbon compounds which mainly occur in the process of petroleum exploration and production through petroleum spills into the environment in production wells, piping networks, collection tanks, and the process of transporting petroleum to processing plants (Figure 10.4) (Helmy and Kardena, 2024). Leakage and pollution of petroleum in the

FIGURE 10.4 Petroleum oil contamination on palm oil plantations due to leaks from exploration well activities in Minas Oil Field working area, Indonesia.

environment are caused by the intensity of use exceeding the life span of the equipment and poor maintenance processes. Polycyclic Aromatic Hydrocarbons (PAHs) are unique contaminants in the environment due to incomplete combustion of organic materials, such as forest fires, home heating, transportation, or waste burning. In addition, PAHs can be produced from volcanic activity, industrial processes, gasification, and incineration of plastic waste. PAHs are also produced from power plants, coal, carbon black, asphalt, and combustion engines. PAHs through natural sources tend to have lower concentrations than anthropogenic sources. Most PAHs are carcinogenic, mutagenic, and teratogenic; and have high bioaccumulation to endanger human health so remediation is a top priority to reduce PAHs in petroleum-contaminated soil. In soil media, the accumulation of PAHs can cause biomagnification in the human food chain because they are highly soluble in fat through direct or indirect exposure to the digestive system. In aquatic environments, PAHs are absorbed into suspended solids and accumulate as sediment due to their hydrophobic and persistent properties. In addition to their effects on humans, PAHs can cause toxicity to animals, such as fish and birds, through disruption of cell membrane function and enzyme systems. PAHs metabolites can possibly bind to proteins and deoxyribonucleic acid/DNA, causing cell damage, biochemical disorders, immune system disorders, fertility disorders, and apoptosis (Kurniawan et al., 2018).

In the Minas oil exploration area, in addition to Dipterocarpaceae forest tree vegetation, there are also many oil palm (*Elaeis guineensis*) plantations. The results of field observations of oil palm plants were able to survive, even though some of their rhizosphere was contaminated by oil. The growth of the crown and roots of oil palm plants in oil-contaminated plant media was slower than the growth of plants that were not contaminated by oil (Evrayenri, 2007). The results of this study emphasize that oil inhibits plant growth. Cabello (2001) stated that soil contaminated with hydrocarbons has a very negative effect on plant communities. Inhibitions in the growth and development of test plants caused by oil pollution can occur through two mechanisms. The first mechanism is that plants experience direct poisoning by oil, which consists of various hydrocarbon compounds. Contact poisoning can occur because hydrocarbon compounds can dissolve the fat structure of cell membranes. The second mechanism is that the soil is not optimal as a planting medium due to oil pollution. Pollution by petroleum products has an indirect effect on plants through its interaction with biotic and abiotic components.

10.2.2.3 Domestic Activities Pollution

Agricultural land pollution can also contribute to daily human activities, such as the disposal of refuse and the discharge of wastewater from households. Wastewater or activities that include all water from toilets, kitchens, restaurants, hotels, hospitals, laundries, and others that are discharged into irrigation systems can damage soil fertility. This wastewater mainly consists of organic materials, including harmful bacteria, and detergents. Organic materials are generally waste that can rot or be degraded by microorganisms. Disposal of organic materials into water bodies can increase the population of microorganisms, so it is possible that anoxic conditions will occur due to oxygen depletion. The presence of excessive detergents and similar materials (soap, shampoo, and other cleaning materials) if disposed of carelessly can disrupt

FIGURE 10.5 High-strength organic waste comes from the Tofu home industry (left) which is dumped directly into the rice field irrigation system (right).

(Photo courtesy of Antoni and Fitriani, 2023; Wuri, 2024)

the life of water and soil microorganisms and damage the surrounding environment. Harmful substances are frequently present in household products, which can contaminate soil and water resources, thereby deteriorating the quality of agricultural land. High organic waste from household activities such as home-scale tofu making and home-scale cattle, chicken, and goat farming also contribute significantly to water and agricultural land pollution if not managed properly (Figure 10.5).

10.3 REMEDIATION OF CONTAMINATED AGRICULTURAL LAND

Agricultural land polluted with industrial organic substances and inorganic chemicals poses a significant environmental remediation challenge in developing nations, where the expense of decontaminating these sites to safeguard human and environmental health will be a considerable limitation in the process. Food crops cultivated in polluted soil containing hazardous organics and/or heavy metals may bioaccumulate these pollutants and biomagnify through higher trophic levels, potentially jeopardizing the well-being of both organisms and humans. U.S. Federal Remediation Technologies Roundtable (FRTR) has led to technology development and demonstration of a variety of clean-up technologies as summarized in Table 10.4 (FRTR, 2007).

This analysis of treatment technologies will be categorized into three classifications: physical, chemical, and biological treatment, respectively (Figure 10.6). A brief description of the physical and chemical treatment methods will be presented, as well as a more comprehensive discussion of biological treatment methods, especially phytoremediation of soil contaminated with organic compounds and heavy metals.

10.3.1 Physical Remediation

The physical cleanup of agriculturally contaminated land primarily encompasses soil replacement, soil washing, thermal desorption, and soil encapsulation. Soil replacement involves excavating contaminated soil, removing it from its original location, transporting it to a landfill or subjected to additional ex situ treatment, and

TABLE 10.4
Soil Contamination Clean-Up Technologies Screening Matrix

Technology	Type of Contaminants							
	Nonhalogenated VOCs	Halogenated VOCs	Nonhalogenated SVOCs	Halogenated SVOCs	Fuels	Inorganics, Metals	Radionuclides	Explosive
A.1. In Situ Physical/Chemical Treatment								
1. Chemical Oxidation	A	A	BA	A	BA	SC	BA	A
2. Electrokinetic	A	A	A	A	BA	AA	A	BA
3. Fracturing	A	A	A	A	A	BA	BA	BA
4. Soil Flushing	AA	AA	A	A	A	AA	BA	BA
5. Soil Vapor Extraction	AA	AA	BA	BA	AA	BA	BA	BA
6. Solidification/ Stabilization	BA	BA	A	A	BA	AA	AA	BA
A.2. In Situ Thermal Treatment								
1. Thermal Treatment	AA	AA	AA	AA	AA	BA	BA	BA
A.3. In Situ Biological Treatment								
1. Bioventing	AA	SC	AA	BA	AA	BA	SC	BA
2. Enhanced Bioremediation	AA	AA	AA	SC	AA	SC	SC	AA
3. Phytoremediation	A	A	A	SC	A	A	BA	BA
B.1. Ex Situ Physical/Chemical Treatment								
1. Chemical Extraction	A	A	AA	AA	A	AA	A	BA
2. Chemical Redox	A	A	A	A	A	AA	BA	A
3. Dehalogenation	BA	AA	BA	AA	BA	BA	BA	A
4. Separation	A	A	A	A	BA	A	BA	BA
5. Soil Washing	A	A	A	A	A	A	BA	BA
6. Solidification/ Stabilization	BA	BA	A	A	BA	AA	AA	BA

B.2. En Situ Thermal Treatment								
1. Hot Gas Decontamination	BA	BA	BA	BA	BA	BA	BA	BA
2. Incineration	AA	AA	AA	AA	AA	BA	BA	AA
3. Open Burn	BA	BA	BA	BA	BA	BA	BA	AA
4. Pyrolysis	A	A	AA	AA	A	BA	BA	BA
5. Thermal Desorption	AA	AA	AA	AA	AA	BA	BA	AA
B.3. Ex Situ Biological Treatment								
1. Biopiles	AA	AA	A	SC	AA	SC	BA	BA
2. Composting	A	A	A	SC	AA	BA	BA	AA
3. Landfarming	A	A	AA	A	AA	BA	BA	SC
4. Slurry Phase	A	AA	AA	SC	AA	SC	BA	AA
C Containment								
1. Landfill Cap	A	A	A	A	A	A	BA	A
D. Other Treatment								
1. Off-site Disposal	A	A	A	A	A	A	BA	A

Note: VOCs: Volatile organic compounds; SVOCs: semi-volatile organic compounds.
AA: Above Average, a well-documented final remedy that has been used at several locations.
A: Average, full-scale implementation has been made, but testing and enhancements are still required.
BA: Below Average, hasn't been used completely but has shown promise in pilot, bench, and lab scale testing.
SC: The particular contaminant and its application/design have a significant impact on the efficacy level.

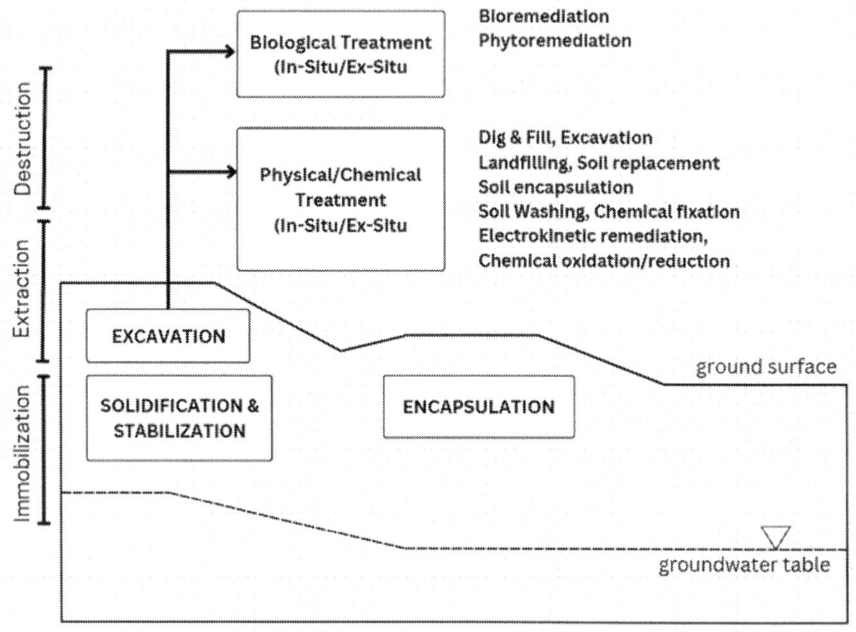

FIGURE 10.6 Classification of soil contamination remedial technologies.

substituting it with clean soil to lessen the concentration of contaminants (dig and fill method). Soil washing is the process of isolating contaminants adhered to the soil matrix and cleansing it with an aqueous solution, such as water or surfactants, to eliminate pollutants by either suspending or dissolving them in the wash solution, while separating the soil based on particle size and/or by gravity separation. Thermal desorption Soil remediation employs heat to enhance the volatility of contaminants, facilitating their separation from the soil. The volatilized pollutants are subsequently either thermally oxidized or captured in a vapor recovery unit. Soil encapsulation of contaminated soil will inhibit the dispersion and leaching of pollutants by enveloping the contamination source with layers of geomembrane, concrete, and compacted clay to restrict the migration of contaminants from the isolated area (Effendi et al., 2019).

10.3.2 Chemical Remediation

Chemical treatment, such as stabilization, employs oxidation and reduction to transform polluted soil into a non-hazardous condition. Stabilization entails the mixing of reagents to chemically reduce the mobility and toxicity of soil contaminants, frequently resulting in a dry solid upon completion of the procedure. Chemical oxidation utilizes chemical reagents on the soil through equipment such as pumps, mixers, and/or injection systems. Chemical oxidation with hydrogen peroxide and Fenton reagent is commonly used to breakdown and oxidize some recalcitrant organic compounds, especially for biological treatment. The use of chemical oxidants as a pre-oxidation step before biological treatment shows quite promising results in its application to degrade complex organic compounds PAHs and POPs. This method transforms contaminants into simpler compounds, more stable, less mobile, and less hazardous molecules through chemical processes. The chemical fixation process, also known as immobilization, is a practical approach to remediating soil contaminated with metals both ex situ and in situ. This process is carried out by adding chemical reagents to contaminated soil to make it into an insoluble or immobile form, low toxic, thus reducing the migration of heavy metals from the soil matrix to water, absorbed by plants and other environmental media, and achieving the remediation of soil. A chemical washing process can also be done with the aim of dissolving heavy metal ions from the soil matrix into the washing liquid so that the extracted metal can be recovered or further processed (Effendi et al., 2022). Chelating agents such as EDTA (Ethylenediaminetetraacetic acid) are widely used to extract Zn, Cd, Cu, and Pb metals from contaminated soil through chelation and covalent bonding mechanisms. The solubility and mobility of heavy metals in agricultural soil can also be reduced by materials such as flyash, silica, and lime which are easy and cheap. After 30 days of treatment, other materials like hazelnut hard shell can reduce the lead content in parsley leaves to 87.7% and the cadmium content to 100%. Comparatively, parsley root increased to 100% in every treatment, and lead and cadmium levels decreased as well. Hazelnut waste may work well as an adsorbent to extract heavy metals from soil amendments while simultaneously enhancing the safety of crops and soil (Fatahi et al., 2020). The addition of organic soil conditioners such as biochar, compost, and humic acid was reported by Mulyani et al., 2022 to be effective in immobilizing heavy metals Cr, Cd, and Pb. The results showed that the metal adsorption process by biochar reached an efficiency of 95.4–99.3% for Cd, 99% for Pb, and 99% for Cd. Metal adsorption by compost obtained results of 82–83.4% for Cr, 94.7–97.3% for Pb, and 95–97.4% for Cd. Meanwhile, metal adsorption by humic acid obtained results of 90% for Cr, 90.9–92.4% for Pb, and 90% for Cd, respectively.

10.3.3 Biological Remediation

The main principle of biological methods in remediation is to create and control an optimal environment for the proliferation, development, and activity of degrading agents (microbes, plants). In optimal conditions, degrading agents will produce auxiliary compounds (enzymes, secondary metabolites) that catalyze reactions in bioremediation. In general, the mechanism of biological pollutant degradation is by utilizing the pollutant as a carbon and energy source for the degrader. Therefore, pollutants that can be optimally utilized as a food source are pollutants that contain organic compounds. The degradation mechanisms of organic contaminants and inorganic contaminants such as heavy metals have different pathways, even though both types of contaminants are used in the degrader's metabolic process (Figure 10.7).

Several factors that must be considered in bioremediation are the type and nature of soil, the type and level of contaminants, the level of organic matter as an energy source, and the availability of nutrients for microbial metabolism (Kapahi and Sachdeva, 2019).

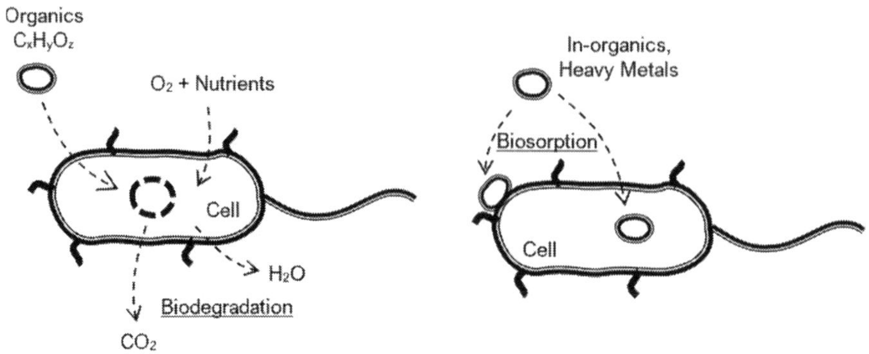

FIGURE 10.7 Illustration of the biodegradation process of organic pollutants (left), and biosorption process of Inorganic pollutants (right).

10.3.3.1 Soil Characteristics

Soil is an ecosystem with a diverse composition that supports numerous types and populations of microorganisms. The soil's physical properties will influence the chemical stability, bioavailability, and mobility of hydrocarbon pollutants. The surface soil (topsoil) typically exhibits a large bacterial population due to the presence of organic matter from flora and fauna, in addition to sufficient oxygen availability. The dimensions of soil particles influence soil permeability and the velocity of hydrocarbon biodegradation. Fine soil particles, such as clay, will hold hydrocarbons on the soil surface, thereby reducing the availability of nutrients and oxygen. Conversely, coarser soil particles facilitate the drainage of hydrocarbon pollutants more effectively. Enhancing soil porosity is essential to augment the rate of biodegradation, hence improving bioavailability and facilitating optimal metabolism and multiplication of microorganisms (Kebede et al., 2021).

10.3.3.2 Bioavailability of Contaminants in Soil

In general, the bioavailability of hydrocarbon contaminants indicates the amount of hydrocarbon substrate that can be accessed by degrading microorganisms. The bioavailability of contaminants can be limited due to the nature of the contaminant which is a hydrophobic compound, and has low solubility due to high molecular weight, low concentration, low diffusion rate, certain chemical structure, temperature, and soil characteristics as the contaminant medium. For example, at low temperatures, viscosity, solubility of contaminants, oxygen and nutrient transfer will decrease. This will cause the accessibility of microorganisms to decrease. The lack of bioavailability of hydrocarbon contaminants is the main factor inhibiting the rate of hydrocarbon degradation. To overcome these limitations, biosurfactants can be used to dissolve contaminants so that they can increase their bioavailability.

10.3.3.3 pH

Soil pH can affect the growth of microorganisms and the bioremediation process. Most hydrocarbon-degrading bacteria can work well in neutral to slightly alkaline pH conditions. Accumulation of biomass formed by bacterial growth can change the pH in the soil, thereby affecting nutrient availability, contaminant solubility, bioavailability, and bacterial activity. To obtain optimal hydrocarbon biodegradation results, soil pH needs to be maintained in the range of 5–8 to regulate bacterial biomass production and enzyme activity.

10.3.3.4 Nutrients

Bacterial metabolism and growth require nutrients and accessibility to these nutrients in petroleum-contaminated soil. This suggests that the availability or addition of macronutrients such as nitrogen, phosphate, and potassium (N, P, K) will increase biostimulation and determine the rate of degradation of hydrocarbon contaminants. The availability of carbon from hydrocarbon contaminants in the soil will cause rapid nutrient depletion due to rapid bacterial metabolic activity. This condition will limit the rate of biodegradation and biotransformation of hydrocarbons. To avoid this, it is necessary to adjust the carbon/nitrogen/phosphorus (C/N/P) ratio by 100:10:1 to 100:20:1 for high hydrocarbon levels. This can be done by adding urea, phosphate, ammonium, and phosphate salt fertilizers.

10.3.3.5 Oxygen Availability

Hydrocarbon degradation by microorganisms can occur with oxygen availability (aerobic) and without oxygen (anaerobic or facultative). In aerobic degradation, oxygen acts as the final electron acceptor and chemical reactant in the enzymatic oxygenation reaction for terminal, subterminal oxidation, and ring cleavage for aromatic hydrocarbon contaminants. Meanwhile, in anaerobic degradation, the final electron acceptor is played by nitrate, sulfate, or iron molecules which results in a slower biodegradation rate than aerobic. Therefore, hydrocarbon degradation by microorganisms can occur faster and more efficiently under aerobic conditions (Pandolfo et al., 2023).

10.3.3.6 Soil Moisture

Soil moisture acts as a medium for transporting soil nutrients and eliminating microbial metabolic products in soil particles. Soil moisture affects the bioavailability of pollutant compounds to microbes, oxygen levels, dissolved substances, diffusion processes, soil toxicity levels, and soil pH. Contaminated soil will have varying porosity and water-holding capacity compared to soil without contaminants. Soil moisture must be maintained at the appropriate amount because a small amount will inhibit microbial activity and too high an amount will inhibit the oxygen transfer process, thus triggering anaerobic conditions. The optimum water content or soil moisture value for the bioremediation process is in the range of 15%–25%.

10.3.3.7 Temperature

Temperature has a significant effect on the growth rate of microorganisms and their enzyme activities, the chemical structure of contaminants, physiology, and the variety of bacterial communities that play a role in the degradation process of pollutant compounds. However, contaminant-degrading bacteria have a wide range of work effectiveness. The rate of hydrocarbon degradation depends on the temperature and climate conditions of the contaminated area, such as tropical, temperate, and arctic.

Increasing environmental temperatures will increase the solubility, bioavailability, diffusion, and enzymatic activity of microorganisms. However, at higher temperatures (above 40°C), oxygen solubility will decrease and the metabolic activity of aerobic bacteria will be disrupted because the bacterial cell membrane will become more susceptible to contaminant toxicity, thereby reducing its biodegradation rate.

10.4 BIOLOGICAL OPTIONS FOR ORGANIC AND HEAVY METAL CONTAMINATED SOIL

Biological options in pollutant removal in the form of organic and inorganic compounds will be discussed in this sub-chapter. In general, biological decomposition using microbes is more widely used in the degradation of organic compounds because it utilizes the metabolic function of the microbes in decomposing pollutant compounds. Inorganic removal in this case heavy metals by microbes is limited in its application (i.e., Bioleaching and Biosorption) because the ability of microbes to uptake heavy metals in their metabolic pathways is limited, not as large as organic compounds used by microbes as carbon and energy source (Kardena et al., 2020). For this reason, the biological option for removing or reducing heavy metal levels is phytoremediation, where this technique uses hyperaccumulating plants to absorb heavy metals from contaminated soil. Although these heavy metal absorbing plants must be further processed by storing them in landfills, burning them, or utilizing them as other by-products.

10.4.1 Bioremediation

Microorganisms are used in biological remediation approaches to break down organic pollutants in solids, sludge, groundwater, or soil into nontoxic chemicals. By co-metabolizing pollutants with an energy source or using them as an energy source, microorganisms degrade pollutants. Energy is produced in microbial cells through redox reactions in bioremediation. Respiration and other biological processes required for cell maintenance and reproduction are examples of these responses. Generally speaking, a delivery system must supply one or more of the following: nutrients, an electron acceptor, and an energy source (electron donor). To stimulate and enhance microbial activity, it is essential to consider various factors such as the microorganisms themselves, the availability of air, organic substrates or other electron donors/acceptors, as well as nutrients and additional substances that influence their growth. Microorganisms and other materials, such as air, organic substances, or other electron donors/acceptors, nutrients, and other substances that impact their growth, must be taken into consideration in order to stimulate and improve microbial activity. Biostimulation can be used when the bacteria needed to degrade the contaminant are present, but environmental conditions do not support their growth (e.g., anaerobic bacteria in aerobic aquifers, aerobic bacteria in anaerobic aquifers, or lack of nutrients or electron donors/acceptors). Bioaugmentation can be used when the bacteria needed to degrade contaminants do not occur naturally at a site or are present in very small populations. Biostimulation and bioaugmentation can be used to treat soil and other solids, groundwater, or surface water (FRTR, 2020).

Biological organisms, including microscopic organisms such as bacteria, fungi, and algae both indigenous and non-indigenous, can play a role in the removal of environmental pollutants through their metabolic processes. There are three basic processes involved in this bioremediation technology; they are as follows:

1. Biotransformation, is the process of changing from one compound to another (this change can be simpler or more complex than the initial compound). In this biotransformation process, hazardous pollutant components can change into less hazardous products or even change into more hazardous compounds.
2. Biodegradation, is the process of breaking down a more complex molecule to produce simpler organic or inorganic molecules.
3. Mineralization, is a process that involves the decomposition of all organic materials into inorganic compounds such as carbon dioxide and H_2O.

Each of these processes can occur at the pollutant site (in situ) or outside the pollutant site (ex situ) and can take place aerobically or anaerobically. The nature of pollutants is very diverse, both based on their chemical structure and their properties, so there is no single bioremediation technique that functions as the ultimate technology in restoring a polluted environment.

10.4.1.1 In Situ Biological Remediation

In situ bioremediation is the remediation of soil and/or groundwater by utilizing naturally occurring microorganisms to biologically degrade contaminants. The primary benefit of in situ treatment is that it eliminates the need for soil excavation and transportation, which might save a substantial amount of expense. However, because soil and aquifer features vary and because it is more difficult to confirm the procedure efficacy, in situ treatment typically takes longer and offers less assurances on treatment uniformity. The following are types of in situ bioremediation (FRTR, 2020).

10.4.1.1.1 Bioventing

Bioventing aims to stimulate the natural biodegradation of any aerobically degradable compound in the soil by providing oxygen to the soil microorganisms present. Bioventing uses low airflow rates to provide sufficient oxygen to sustain microbial activity. Oxygen is most often supplied by direct air injection into the soil contamination. Volatile compounds will biodegrade as the vapor moves slowly through the biologically active soil (FRTR, 2020).

10.4.1.1.2 Biosparging

In the biosparging process, the air is supplied to the saturated zone below the soil surface. The air supply causes the movement of organic pollutants to the upper or surface zone. This process enhances the removal of pollutants from contaminated sites by stimulating microbial activity, where the biodegradation process takes place in the soil. This process is influenced by two main factors, namely soil permeability and pollutant biodegradability (Kumar et al., 2018). This biosparging process has been widely used for the treatment of hydrocarbon-contaminated soil such as benzene, toluene, ethylbenzene, and xylene as well as fuel oil products.

10.4.1.1.3 Biostimulation

The bioremediation method involves the addition of materials that function to stimulate or encourage the growth of indigenous microbial populations and activities in contaminated soil sites. Materials that function as biostimulants are mainly nutrient supply (carbon, nitrogen, phosphorus), and optimization of environmental conditions (temperature, pH, oxygen, redox content, and types of organic pollutants) to accelerate the growth and succession of pollutant-degrading microbes. Another strategy in biostimulation is to add biodegradable compounds as secondary substrates for degrading microbes (co-metabolism/co-substrate).

10.4.1.1.4 Bioaugmentation

In microbiology and bioremediation, the term bioaugmentation refers to the addition of bacterial or archaeal cultures (bacteria that live in extreme environments) to accelerate the degradation process of contaminants or change the availability of pollutant compounds at contaminated sites. The process of removing pollutants using the bioaugmentation method involves native microbial communities, microbes from other places, or genetically engineered microbes that are added to contaminated sites. Sometimes there is competition between indigenous microbes and added microbes, therefore proper land treatment unit management planning is needed. Three limiting factors in the use of additional microbial cultures in contaminated land treatment units are as follows:

1. Non-indigenous microbial cultures are rarely able to compete well enough with indigenous microbial populations to develop and maintain functional population levels, so the effectiveness of bioaugmentation is not optimal.
2. Indigenous microbes are often effective decomposer microbes and dominate the added microbes because the indigenous microbes have adapted well to the pollutant compounds.
3. The density of microbes needed is generally not sufficient to degrade the target compounds in contaminated land. Therefore, to achieve success, a mixed culture is needed, a consortium consisting of many strains and species of microbes to work together to overcome the limitations of the ability to degrade one or several species in the consortium.

10.4.1.2 Ex Situ Biological Remediation

Ex situ remediation techniques are the process of removing contaminated material for treatment and/or contaminant removal elsewhere, either on-site or at a different location.. The contaminated media is excavated or extracted (e.g., groundwater pumping) and transported to a process site that may be in or adjacent to the zone of contamination. These destruction or transformation techniques are directed at encouraging microorganisms to grow and use the contaminant as a food and energy source by creating an environment favorable to the microorganisms. The primary advantage of ex situ treatment is that it generally requires a shorter time frame than in situ treatment. However, it does require excavation which results in increased equipment and engineering costs, potential permitting, and material handling/worker exposure considerations (FRTR, 2020).

10.4.1.2.1 Landfarming

Landfarming is a bioremediation technique that generally integrates bed-laying and additional measures to manage the leaching of contaminants, necessitating the excavation and placing of contaminated soil. The contaminated media is configured into bed-laying structures that are routinely rotated or tilled to facilitate the aeration of the waste. Soil conditions are frequently regulated to enhance the rate of pollutant breakdown (Helmy and Kardena, 2024).

10.4.1.2.2 Composting

Treatment involves excavating polluted soil and mixing it with bulking agents and organic substances, such as wood chips, straw, manure, and plant waste biomass, with the expectation that microbes will transform organic pollutants into harmless and stable by-products. Thermophilic conditions must typically be sustained for optimal composting efficacy. The rise in temperature results from the heat generated by microbes during the decomposition of organic matter (FRTR, 2020).

10.4.1.2.3 Biopiles

Treatment is constructed by combining excavated soil with soil additives and subsequently situating it in a treatment area equipped with a leachate collecting system and a method of aeration. The objective is to diminish the concentration of petroleum compounds in the excavated soil via a biodegradation method. The treatment area is typically enclosed with a waterproof barrier to reduce the likelihood of contaminants infiltrating the uncontaminated soil. These soil piles typically have an underground air distribution system to facilitate air circulation through the soil via suction. The soil piles in this instance may reach heights of up to 6 meters. The soil piles may be enveloped in plastic to regulate runoff, minimize evaporation, and promote solar heating to the optimal temperature range for microorganisms. If volatile organic compounds (VOCs) are present in the soil and may volatilize into the air, the air exiting the soil can be processed to eliminate or neutralize these VOCs prior to atmospheric release (FRTR, 2020).

10.4.1.2.4 Slurry Phase Biological Remediation

The treatment involves the regulated treatment of excavated soil within a bioreactor. The excavated soil is initially processed to physically eliminate rocks and trash. The soil is subsequently combined with water to a specified concentration based on the level of pollutants, the biodegradation rate, and the soil's physical characteristics. The objective of the soil is to wash the soil to concentrate the impurities. The cleaned sand can thereafter be eliminated, retaining solely the particles and polluted wash water for biological remediation. The slurry generally has 10%–30% solids by weight. The solids are maintained in suspension within the reactor and combined with nutrients and oxygen. An acid or base is introduced to regulate the pH as required. Microorganisms may be introduced if suitable populations are not present. Upon the completion of biodegradation, the slurry undergoes dewatering (FRTR, 2020).

10.4.2 Phytoremediation

Phytoremediation is defined as an in situ remediation strategy using vegetation, which is often symbiotic with soil microbial activity, as well as the addition of organic/inorganic materials and other agronomic techniques in an effort to reduce or change the availability of contaminants in the soil. Plants provide an eco-friendly solution for pollutant remediation, requiring less energy and expense compared to traditional environmental cleanup procedures, as they harness solar energy and utilize physiological mechanisms. The capacity of plants to contain, sequester, or detoxify pollutants from contaminated soil and water is regarded as economically and environmentally advantageous. This is achieved through various mechanisms, including degradation (rhizodegradation and phytodegradation for organic contaminants), while accumulation (phytoextraction, rhizofiltration), dissipation (phytovolatilization), and immobilization (hydraulic control and phytostabilization) to reduce, remove, or immobilize inorganic contaminants (Yao et al., 2012; Praveen and Nagalakshmi, 2022). Furthermore, plants can modify the quantity and content of root exudates to enhance bioremediation by altering microbial community structure, promoting the proliferation of symbiotic microorganisms, or enhancing the metabolic activity of degrading microbes. The method of reducing the concentration and availability of heavy metals in the soil through phytoremediation includes several mechanisms considering that these heavy metals cannot be degraded (Wang et al., 2012; Adiloglu, 2018). The octanol-water partition coefficient, K_{ow}, of organic pollutants serves as a proxy for the bioavailability of the pollutant, which affects a plant's ability to metabolize it. Using the values of K_{ow} in log scale, Cunningham and Berti (1993) explained that this can be used as an indication of the depending on factors of plants to absorb, translocate, and metabolize the contaminant.

1. If log $K_{ow} \leq 1$, plants can absorb, translocate, and metabolize hydrophilic contaminants. Due to these contaminants' high bioavailability, water influx controls how well they can enter the plant, which could contaminate groundwater.
2. If the log K_{ow} value is between 1 and 4, the plant can still absorb, translocate, and metabolize contaminants, but the higher the range, the lower the ability.
3. If log K_{ow} levels exceed 4, plants typically cannot absorb, translocate, or metabolize pollutants due to elevated hydrophobicity, resulting in the contaminant adsorbing to lipids on the plant's root surface.

The general mechanisms of the phytoremediation process include the following:

***Phytoextraction*:** This method is used in the absorption of organic and inorganic contaminants by plant roots and crowns. Hyperaccumulator plants containing metals will be harvested and burned, then the ash containing the metal is extracted, this technique is called phytomining. The phytoextraction mechanism can occur first, based on the natural ability of hyperaccumulator plants to absorb heavy metals (its ability is more than 100 times greater than ordinary plants). Plant families known as metal accumulators include Brassicaceae, Fabaceae, Aserraceae, Lamiaceae, Euphorbiaceae, and Scrophulariaceae. The second mechanism is chelate-induced phytoextraction, where when chelates are applied to the soil, they will form a metal-chelate complex that can be absorbed by plants through the passive apoplastic pathway. Natural chelators produced by plants are involved in the absorption and detoxification of heavy metals by plants. The chemical chelate that is commonly applied and has been tested to mobilize Cd, Pb, and Hg metals in the soil is Ethylenediaminetetraacetic acid (EDTA) (Gul et al., 2019; Hasan et al., 2019).

Phytostabilization (Stabilization and immobilization at the root)**:** Phytostabilization, also known as phytoimmobilization, is a phytoremediation process that involves the inactivation or immobilization of contaminants within the roots of metal-tolerant plants or their rhizosphere. This process is intended to reduce the bioavailability of these contaminants and mitigate the risk of their mobilization within the ecosystem and food chain. The roots of plants or the rhizospheric environment inhibit the mobility of pollutants, preventing them from inducing phytotoxicity. This process occurs through three mechanisms: (1) absorption, (2) root accumulation, and (3) surface adsorption by roots, subsequently integrating into humic matter inside the plant rhizosphere. This method is ideally employed for metal toxicants in soils, sediments, and sludge.

Phytovolatilization**:** Contaminants can be removed from plants through transpiration or evaporation. Water flows from the roots to the shoots with the help of the vascular system, because water contains dissolved metals so that contaminants can be released into the air through volatilization. Rugh et al. (1996) reported that Hg-resistant transgenic plants can release Hg into the atmosphere. The contaminant that the plant has absorbed may move to different plant parts where it is partially or completely metabolized or incorporated into cellular components and volatilized. This whole process includes absorption, translocation, and volatilization, called phytovolatilization.

Rhizofiltration**:** Contaminants can be present in the roots or adsorbed by the roots through biotic and abiotic processes. This technique is similar to phytoextraction but is generally used to absorb or adsorb metals from groundwater or wastewater. Rhizofiltration first leads to the containment of pollutants, wherein these substances are immobilized or deposited on or within the plant rhizosphere. Contaminants are subsequently removed through the physical removal of the plant.

Phytoextraction, phytostabilization, phytovolatilization, and rhizofiltration will be effective if the metals in the soil are exchangeable and bioavailable for plant uptake (Figure 10.8). The obstacle to reducing heavy metal concentrations in the soil through phytoremediation is the low availability of metals in the soil. Plants that accumulate more than 0.1% of metals in roots, stems, leaves, or other tissues without showing visible symptoms of toxicity are categorized as hyperaccumulator plants. The effectiveness of phytoremediation has been proven by several plants (Table 10.5), such as *Cyrtomium macrophyllum* which is able to grow naturally in soil with a Mercury content of 225.73 mg/kg, has a Hg translocation factor of 2.62 and accumulates 36.44 mg/kg Hg in its crown (Xun et al., 2017). The extraction ratio of Pb from soil to the crown of *Brassica juncea* plants is 24.52 greater than the extraction ratio of Chromium of 0.37 (Gayatri et al., 2019). Wild plants *Tanacetum vulgare* are effective in absorbing Cd, Cu, Cr, Pb, Ni, and Zn among wild plants growing in landfill areas, able to accumulate metals in contaminated areas (Vaverkova and Adamcova, 2014).

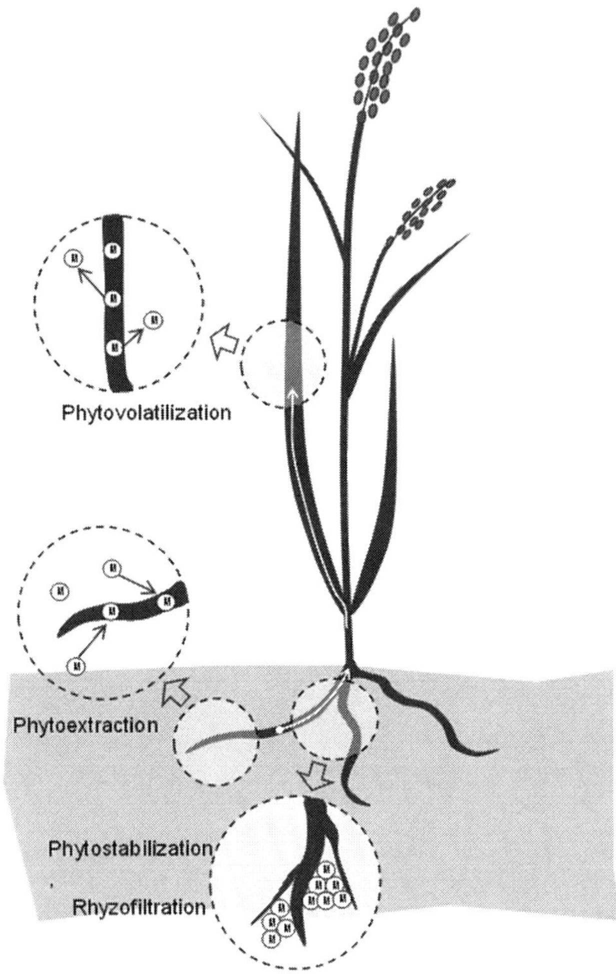

FIGURE 10.8 Mechanism of phytoremediation in heavy metals uptake.

The plant family with the greatest potential to absorb heavy metals is Brassicaceae. It was reported by Dar et al. (2015) that about 25% of members of the Brassicaceae family consisting of about 90 species are known to be metal hyperaccumulators. Hyperaccumulator plants develop mechanisms to avoid stress caused by heavy metals, one of which is through the energetic (active) efflux mechanism: a mechanism for removing metals from cells catalyzed by ATPase (Papoyan and Kochian, 2004; Chau et al., 2023). The weakness of phytoremediation is that the type of heavy metal accumulator plant is generally a shrub with limited biomass so the absorption of heavy metals adsorbed per plant is not large. Several *Brassicaceae* cultivars can sequester heavy metals, positioning them as potential agents for remediating heavy metal contamination in agricultural soils, despite their relatively modest biomass yield.

TABLE 10.5
Hyperaccumulator Plants for Phytoextraction of Heavy Metals

Family	Species	Heavy Metal
Azollaceae	*Azolla filiculoides*	Hg(II), Pb(II)
Asteraceae	*Berkheya coddii*	Ni
	Helianthus annuus	Pb, Cd, Zn
Brassicaceae	*Alyssum bertolonii*	Ni
	Alyssum murale	Ni
	Arabidopsis halleri	Zn, Cd
	Brassica juncea L.	Cd, Cu, Zn
	Brassica rapa L.	U
	Brassica nigra	Pb
Caryophyllaceae	*Minuarita verna*	Zn, Pb, Cd
Crassulaceae	*Sedum alfredii*	Pb
Cyperaceae	*Eliocharis acicularis*	Cu
Euphorbiaceae	*Euphorbia cheiradenia*	Cu, Fe, Pb, Zn
	Jatropha curcas	Al, Cd, Fe, Cr, Pb, Zn, Ni, Cu
Fabaceae	*Atralagu recemosus*	Se
	Medicago sativa	Pb
	Vigna mungo	Cd, Pb, Zn, Mn
Lemnaceae	*Lemna minor*	Pb, Cd, Ni, Cr
Poaceae	*Sparina argentinensis*	Cr
	Melica jacquemontii	Fe
	Oryza sativa	Cd, Zn, Fe, Cu, Pb, Cr, Mn
	Sorghum bicolor	As, Cd, Co, Cu, Zn
Pteridaceae	*Pteris vittata*	As, Hg
Salicaceae	*Populus* sp.	Cd
Violaceae	*Viola boashanensis*	Pb, Zn, Cd

Source: Adapted from Hindersah et al. (2023), Priya et al. (2023), Chau et al. (2023), Aksoy (2021), and Marchiol et al. (2007).

10.5 CONCLUSION

Soil pollution can result from diverse human activities, including agricultural practices, household actions, and industrial operations. The soil is often contaminated by industrial waste, which includes persistent organic pollutants, heavy metals, hazardous chemicals, and pesticides. This contamination arises when waste from various sources above is discarded without adequate treatment. Furthermore, domestic waste, including household refuse and medical waste from hospitals, constitutes an additional contributor to soil pollution. Agricultural practices utilizing agrochemicals, including various fertilizers and pesticides, may lead to the accumulation of detrimental substances in the soil. To address this contamination, various methods for the remediation of impacted soil have been developed. Bioremediation employs microorganisms, such as bacteria and fungi, to degrade potentially toxic substances present in the soil. Plants capable of absorbing and neutralizing soil toxins, known as hyperaccumulator plants, utilized in phytoremediation. The next remediation method is chemical stabilization which involves the addition of certain chemicals to reduce the solubility and mobility of the contaminants. This remediation technology not only restores soil quality but also enhances agricultural land productivity. The soil's structure and the activity of beneficial bacteria can be enhanced through the application of biofertilizers and liquid organic fertilizers. Rehabilitated land can be utilized for agricultural purposes if the restoration procedure is executed effectively. This improves the productivity of food crops while contributing in environmental preservation.

ARTIFICIAL INTELLIGENCE DECLARATION

In the preparation of this book chapter, we have utilized AI (i.e., Grammarly) to refine with the initial draft language. All content generated by the AI tool has been reviewed and edited before used to ensure accuracy and originality.

REFERENCES

Adiloglu, S. 2018. Heavy metal removal with phytoremediation. In: Shiomi, N., Ed., *Advances in Bioremediation and Phytoremediation*, IntechOpen. doi:10.5772/intechopen.70330

Aksoy, A. 2021. Accumulation of cadmium, copper and zinc in selected natural Viola taxa in Turkish Mediterranean serpentine soils. *Anatolian Journal of Botany* 5(2): 138–143. doi:10.30616/ajb.999023

Alengebawy, A., Abdelkhalek, S.T., Qureshi, S.R., and Wang, M.Q. 2021. Heavy metals and pesticides toxicity in agricultural soil and plants: Ecological risks and human health implications. *Toxics* 9(3): 42. doi:10.3390/toxics9030042

Alloway, B.J. 1995. Cadmium. In: Alloway, B.J., Ed., *Heavy Metals in Soils* (2nd Edition). Blackie, London, 368p.

Alloway, B.J. 2013. *Heavy Metals in Soils: Trace Metals and Metalloids in Soils and Their Bioavailability*. Springer, Dordrecht.

Al-Taai, S.H.H. 2021. Soil pollution – Causes and effects. *IOP Conference Series: Earth and Environmental Science* 790: 012009. doi:10.1088/1755-1315/790/1/012009

Angon, P.B., Islam, M.S., Shreejana, K.C., Das, A., Anjum, N., Poudel, A., and Suchi, S.A. 2024. Sources, effects and present perspectives of heavy metals contamination: Soil, plants and human food chain, *Heliyon* 10(7). doi:10.1016/j.heliyon.2024.e28357

Antoni, E., and Fitriani, R. 2023. 3 Pemilik Pabrik Tahu Tempe Terancam Pidana. https://radarkepahiang.disway.id/read/657680/3-pemilik-pabrik-tahu-tempe-terancam-pidana/15

Arifin, R., Masyhar, A., Sumardiana, B., Ramada, D.P., Kamal, U., and Fikri, S. 2024. Indonesian sustainable development policy: How the government ensures the environment for future generations. *IOP Conference Series: Earth and Environmental Science* 1355(2024): 012005. doi:10.1088/1755-1315/1355/1/012005

Cabello, M.N. 2001. Mycorrhizas and hydrocarbons. In: Gadd, G.M., Ed. *Fungi in Bioremediation*. British Mycological Society, Surrey, UK, 456–471.

Chau, T.P., Devanesan, S., Farhat, K., Liu, X., and Jhanani, G.K. 2023. Phytoremediation efficiency of Vigna mungo with the amalgamation of indigenous metal tolerant bacterial strains on metal polluted agricultural soil. *Environmental Research* 232: 116291. doi:10.1016/j.envres.2023.116291

Cunningham, S.D., and Berti, W.R. 1993. Remediation of contaminated soils with green plants: An overview. *In Vitro Cellular & Developmental Biology. Plant* 29P(4): 207–212. http://www.jstor.org/stable/4293003

Dachlan, D. 2013. Cita Citarum-Warna Warni Citarum di Majalaya. Photo Report, www.citarum.org

Dar, M.I., Khan, F.A., Rehman, F., Mashoodi, A., Ansari, A.A., Varshney, D., Naushin, F., and Naiko, M.I. 2015. Roles of Brassicaceae in phytoremediation of metals and metalloids. In: Ansari, A., Gill, S., Gill, R., Lanza, G., and Newman, L., Eds., *Phytoremediation*. Springer, Cham. doi:10.1007/978-3-319-10395-2_14

Davies, B.E. 1995. Leads. In: Alloway, B.J., Ed., *Heavy Metals in Soils* (2nd Edition). Blackie, London, 177–198.

Effendi, A.J., Kardena, E., and Helmy, Q. 2019. Biosurfactant-enhanced petroleum oil bioremediation. In: *Microbial Action on Hydrocarbons*. Springer. doi:10.1007/978-981-13-1840-5_7

Effendi, A.J., Ramadan, B.S., and Helmy, Q. 2022. Enhanced remediation of hydrocarbons contaminated soil using electrokinetic soil flushing – Landfarming processes. *Bioresource Technology Reports* 17: 100959. doi:10.1016/j.biteb.2022.100959

Evrayenri, E. 2007. The impact of crude oil pollution on oil palm plant (Elaeis guineensis). *Jurnal Ilmiah Pertanian* 4(1):19–25.

Fatahi, A., Ziarati, P., Jafarpour, A., and Rodriguez, L.C. 2020. Heavy metal removal from edible leafy vegetable by low cost novel adsorbents: Hazelnut shell. *Journal of Scientific Discovery* 4(2): jsd20039.

Forstner, U., and Wittmann, G.T.W. 2012. *Metal Pollution in the Aquatic Environment*. Springer-Verlag, Berlin.

FRTR. 2007. Treatment Technology Screening Matrix. Federal Remediation Technologies Roundtable. https://www.frtr.gov/matrix2/section3/table3_2.pdf

FRTR. 2020. Remediation Technology Screening Matrix and Reference Guide, Version 4.0. Federal Remediation Technologies Roundtable. https://www.frtr.gov/matrix2/section1/toc.html#Sec4

Gambus, F., and Wieczorek, J. 2012. Pollution of fertilizer with heavy metals. *Ecological Chemistry and Engineering A* 19(4–5): 353–360. doi:10.2428/ecea.2012.19(04)036

Gayatri, N., Sailesh, A.R., and Srinivas, N. 2019. Phytoremediation potential of Brassica juncea for soil removal of selected heavy metals in urban areas amended with cow dung. *Journal of Materials and Environmental Science* 10(5): 463–469.

Gul, I., Manzoor, M., Silvestre, J., Rizwan, R., Hina, K., Kallerhoff, J., and Arshad, M. 2019. EDTA-assisted phytoextraction of lead and cadmium by Pelargonium cultivars grown on spiked soil. *International Journal of Phytoremediation* 21(2): 101–110. doi:10.1080/15226514.2018.1474441

Hasan, M.M., Uddin, M.N., Ara-Sharmeen, I., Alharby, F.H., Alzahrani, Y., Hakeem, K.R., and Zhang, L. 2019. Assisting phytoremediation of heavy metals using chemical amendments. *Plants* 8(9): 295. doi:10.3390/plants8090295

Helmy, Q., and Kardena, E. 2024. Enhancing field-scale bioremediation of weathered petroleum oil-contaminated soil with biocompost as a bulking agent. *Case Studies in Chemical and Environmental Engineering* 9: 100735. doi:10.1016/j.cscee.2024.100735

Hindersah, R., Risamasu, R., Kalay, A.M., Dewi, T., and Makatita, I. 2018. Mercury contamination in soil, tailing, and plants on agricultural fields near a closed gold mine in Buru Island, Maluku. *Journal of Degraded and Mining Lands Management* 5(2): 1027–1034. doi:10.15243/jdmlm.2018.052.1027

Hindersah, R., Suryatmana, P., and Mulyani, O. 2023. *Bioremediation of Agricultural Land Contaminated with Heavy Metals* (1st Edition). Unpad Press, Jatinangor, 62–67.

Indonesian Central Bureau of Statistics. 2024. *Indonesia's Land Cover Area Inside and Outside Forest Areas 2014-2022*. Central Bureau of Statistics, Indonesia.

Indonesian Ministry of Environment (IME). 2014. *Wastewater Quality Standard. Regulation number 5 Year 2014*. IME, Indonesia.

Kabata-Pendias, A., and Mukherjee, A.B. 2007. *Trace Elements from Soil to Human*. Springer, Berlin Heidelberg.

Kapahi, M., and Sachdeva, S. 2019. Bioremediation options for heavy metal pollution. *Journal of Health and Pollution* 9(24): 191203. doi:10.5696/2156-9614-9.24.191203

Kardena, E., Panha, Y., Helmy, Q., and Hidayat, S. 2020. Application of mercury resistant bacteria isolated from artisanal small-scale gold tailings in biotransformation of mercury (II)-Contaminated soil. *International Journal of Geomate* 19(71): 106–114. doi:10.21660/2020.71.68786

Kebede, G., Tafese, T., Abda, E.M., Kamaraj, M., and Assefa, F. 2021. Factors influencing the bacterial bioremediation of hydrocarbon contaminants in the soil: Mechanisms and impacts. *Journal of Chemistry* 2021: 1–17. doi:10.1155/2021/9823362

Kim, R.Y., Yoon, J.K., Kim, T.S., Yang, J.E., Owens, G., and Kim, K.R. 2015. Bioavailability of heavy metals in soil: Definitions and practical implementation-a critical review. *Environmental Geochemistry and Health* 37(6): 1041–1061. doi:10.1007/s10653-015-9695-y

Koralay, T., and Koralay, D.B. 2022. Geochemical properties and heavy metal contents of carbonaceous rocks in the Pliocene siliciclastic rock sequence from southeastern Denizli-Turkey. *Open Geosciences* 14(1): 1324–1346. doi:10.1515/geo-2022-0431

Kumar, V., Shahi, S.K., and Singh, S. 2018. Bioremediation: An eco-sustainable approach for restoration of contaminated sites. In: Singh, J., Sharma, D., Kumar, G., and Sharma, N., Eds., *Microbial Bioprospecting for Sustainable Development*. Springer, Singapore. doi:10.1007/978-981-13-0053-0_6

Kurniawan, A., Wirasembada, Y.C., Razaad, I.M.N., Novriansyah, A., Rafi, M., and Effendi, A.J. 2018. Polycyclic aromatic hydrocarbons in petroleum waste contaminated land: Review of microorganism growth, metabolism and biodegradation processes. *Journal of Environmental Science* 16(1): 9–24.

Kusumadirezza, I. 2023. Citarum yang tak Harum. https://www.iklimku.org/citarum-yang-tak-harum/

Li, B., Guan, Y., Zhang, N., Shi, R., and Qin, C. 2025. Assessing the industrial pollution risk potential to farmland soil environment based on spatialization: A single-year study in China. *Ecotoxicology and Environmental Safety* 303: 1–15. doi:10.1016/j.ecoenv.2025.118995

Magdoff, F., and van Es, H. 2021. Soil degradation: Erosion, compaction, and contamination. In: Magdoff, F., and van Es, H., Eds., *Buiding Soils for Better Crops, Ecological Management for Healthy Soils* (4th Edition). SARE Publisher, USA, 410p.

Marchiol, L., Fellet, G., Perosa, D., and Zerbi, G. 2007. Removal of trace metals by Sorghum bicolor and Helianthus annuus in a site polluted by industrial wastes: A field experience. *Plant Physiology and Biochemistry* 45(5): 379–387. doi:10.1016/j.plaphy.2007.03.018

McGrath, S.P. 1995. Chromium and nickel. In: Alloway, B.J., Ed., *Heavy Metals in Soils* (2nd Edition). Blackie, London, 368p.

Moeller, G., and Wall, G.R. 2017. *Zinc and Manganese in the Biogeochemical Cycles. Biogeochemical Cycles: A Global Approach*. Cambridge University Press, Cambridge.

Moldavan, L., Pimenowa, O., Prus, P., and Pimenow, S. 2024. Pollution problems in the economic agricultural sector: Evaluating the impact on natural resources and solutions for improvement. *Sustainability*, 16(24): 11294. doi:10.3390/su162411294

Mortvedt, J.J. 1996. Heavy metal contaminants in inorganic and organic fertilizers. *Fertilizer Research* 43: 55–61.

Mulyani, O., Joy, B., and Kurnia, D. 2022. The various forms of cow manure waste as adsorbents of heavy metals. *Applied Sciences* 12(11): 5763. doi:10.3390/app12115763

Murniasih, S., Santosa, S.J., and Roto, R. 2022. Assessment of multi-nutrients and heavy metals in inorganic fertilizers widely used by Indonesian farmers using NAA. *Indian Journal of Chemistry* 22(3): 666–682. doi:10.22146/ijc.69908

Notodarmojo, S. 2005. *Soil and Groundwater Pollution*. ITB Press, Indonesia, 490p.

Noviardi, R., Karuniawan, A., Sofyan, E.T., and Suryatmana, P. 2023. Heavy metals content in sweet potato (Ipomoea batatas L.) grown on soil contaminated by gold mine tailings with composted cow manure amendment. *Journal of Degraded and Mining Lands Management* 10(4): 4601–4610. doi:10.15243/jdmlm.2023.104.4601

O'Neill, P. 1995. Arsenic. In: Alloway, B.J., Ed., *Heavy Metals in Soils* (2nd Edition). Blackie, London, 105–121.

Pandolfo, E., Barra-Caracciolo, A., and Rolando, L. 2023. Recent Advances in Bacterial Degradation of Hydrocarbons. *Water* 15(375): 1–18. doi:10.3390/w15020375

Papoyan, A., and Kochian, L.V. 2004. Identification of Thlaspi caerulescencs genes that may be involved in heavy metal hyperaccumulation and tolerance. Characterization of a novel heavy metal transport ATPase. *Plant Physiology* 136(3): 3814–3823. doi:10.1104/pp.104.044503

Praveen, R., and Nagalakshmi, R. 2022. Review on bioremediation and phytoremediation techniques of heavy metals in contaminated soil from dump sites. *Materials Today: Proceedings* 68(5): 1562–1567. doi:10.1016/j.matpr.2022.07.190

Priya, A.K., Muruganandam, M., Ali, S.S., and Kornaros, M. 2023. Clean-up of heavy metals from contaminated soil by phytoremediation: A multidisciplinary and eco-friendly approach. *Toxics* 11(5): 422. doi:10.3390/toxics11050422

Rashid, A., Schutte, B.J., Ulery, A., Deyholos, M.K., Sanogo, S., Lehnhoff, E.A., and Beck, L. 2023. Heavy metal contamination in agricultural soil: Environmental pollutants affecting crop health. *Agronomy* 13(6): 1521. doi:10.3390/agronomy13061521

Rugh, C.L., Wilde, H.D., Stack, N.M., Thompson, D.M., Summers, A.O., and Meagher, R.B. 1996. Mercuric ion reduction and resistance in transgenic Arabidopsis thaliana plants expressing a modified bacterial merA gene. *Proceedings of the National Academy of Sciences of the United States of America* 93(8): 3182–3187. doi:10.1073/pnas.93.8.3182

Saidy, A.R., and Badruzsaufari, B. 2009. Relationship between concentration of Cr(VI) and soil chemical characteristics: Preliminary information for remediation of ex-mined lands in the South Kalimantan. *Jurnal Tanah Tropika* 14(2): 97–103.

Sitorus, S.R.P., and Pravitasari, A.E. 2017. Land degradation and landslide in Indonesia. *Sumatra Journal of Disaster, Geography and Geography Education* 1(2): 61–71.

Steinnes, E. 1995. Mercury. In: Alloway, B.J., Ed., *Heavy Metals in Soils* (2nd Edition). Blackie, London, 368p.

Varekamp, J.C., and Thomas, E. 1998. Volcanic sulfate emissions. *Journal of Volcanology and Geothermal Research* 177(4): 1077–1093.

Vaverkova, M.D., and Adamcova, D. 2014. Heavy metals uptake by select plant species in the landfill area of Stepanovice, Czech Republic. *Polish Journal of Environmental Studies* 23(6): 2265–2269. doi:10.15244/pjoes/26106

Wang, M.C., Chen, Y.T., Chen, S.H., Chang Chien, S.W., and Sunkara, S.V. 2012. Phytoremediation of pyrene contaminated soils amended with compost and planted with ryegrass and alfalfa. *Chemosphere* 87(3): 217–225. doi:10.1016/j.chemosphere.2011.12.063

Wolf, M.K., Wiesmeier, M., and Macholdt, J. 2023. Importance of soil fertility for climate-resilient cropping systems: The farmer's perspective. *Soil Security* 13: 100119. doi:10.1016/j.soisec.2023.100119

Wuri, R.L. 2024. BRIN Olah Limbah Tahu Jadi Biogas untuk Pembangkit Listrik. https://katadata.co.id/ekonomi-hijau/energi-baru/65cecb81dfcd8/brin-olah-limbah-tahu-jadi-biogas-untuk-pembangkit-listrik

Xun, Y., Feng, L., Li, Y., and Dong, H. 2017. Mercury accumulation plant Cyrtomium macrophyllum and its potential for phytoremediation of mercury polluted sites. *Chemosphere* 189: 161–170. doi:10.1016/j.chemosphere.2017.09.055

Yao, Z., Li, J., Xie, H., and Yu, C. 2012. Review on remediation technologies of soil contaminated by heavy metals. *Procedia Environmental Sciences* 16: 722–729. doi:10.1016/j.proenv.2012.10.099

Zhao, X., and Wang, D. 2010. Mercury in some chemical fertilizers and the effect of calcium superphosphate on mercury uptake by corn seedlings (Zea mays L.). *Journal of Environmental Sciences* 22(8): 1184–1188. doi:10.1016/S1001-0742(09)60236-9

Zinn, Y.L., de Faria, J.A., de Araujo, M.A., and Skorupa, A.L.A. 2020. Soil parent material is the main control on heavy metal concentrations in tropical highlands of Brazil, *Catena* 185: 104319. doi:10.1016/j.catena.2019.104319

Zulfikah, Z., Basir, M., and Isrun, I. 2014. Mercury (Hg) concentration in soil and plant tissue of water spinach (*Ipomoea Reptans*) given Kirinyu Bokashi (*Chromolaena Odorata* L.) in Poboya gold mining tailings waste, Palu City *e-J. Agrotekbis* 2(6): 587–595.

11 Soil Carbon Sequestration and Its Role in Agriculture

Mahmoud Abdelaziz, Nano Alemu Daba, Md Ashraful Alam, Kiya Adare Tadesse, Imtiaz Ahmed, and Jobira Defera Gonfa

11.1 INTRODUCTION

Soil carbon sequestration is also called "carbon farming" or "regenerative agriculture," which involves diverse methods that manage land, especially farmland, to increase the amount of carbon stored in soils. A soil system, therefore, refers to an intricate and ever-changing system that sustains core biological processes and plays a central role in the cycling of fundamental nutrients that are imperative for terrestrial living. Soils also play a role in climate control, for they are among the largest reservoirs of carbon through the absorption of CO_2 from the atmosphere. Soil organic carbon stock in the biosphere is estimated to be 2,110 gigatons, and of this, 73% is stored in the soil with an estimated stock of 1,550 gigatons. Soil organic carbon (SOC) as an emblematic index demonstrates a unique and crucial role in the health care systems of the land ecosystem that mediates the uses of fertile soil structure, environment, and fertility underpins. Carbon sequestration – Atmospheric CO_2 can also be reduced by storing carbon in the terrestrial carbon pool (Figure 11.1) The above discussion shows that: Carbon sequestration in soil is defined as the ability of biological and physical processes through which CO_2 is transferred from the atmospheric condition to the soil and stored for a long time (Anjum et al., 2022; Meena et al., 2020). Photosynthesis is the first point where the plant stores CO_2, and the second point transfers the CO_2 into the soil through the decomposition of plant residues and root secretion (Bhattacharyya et al., 2022), after that, microorganisms in the soil start to decompose and stabilize this organic matter, resulting in the formation of stable soil organic carbon compounds (Gunina & Kuzyakov, 2022). Carbon sequestration can affect (1) Mitigation of climate change: Soils serve as a main carbon sink, holding more carbon than both the atmosphere and terrestrial vegetation together (Wang & Huang, 2020). Through a synthesis of atmospheric CO_2 and storage of the compound in stable carbon pools that are associated with the soil, soil carbon sequestration is useful in minimizing greenhouse effects by limiting emissions (Kumar et al., 2020). (2) Soil health and productivity: Exports of soil organic matter (SOM), hence soil organic carbon (SOC), are crucial for maintaining

DOI: 10.1201/9781003714484-11

soil fertility and health. It enhances the physical properties of the soil, water holding capacity, nutrient cycling, and soil biotic components, which results in high yields of crops (Page et al., 2020; Rahman et al., 2019. (3) Food Security: Fertility of soil and maintenance is vital for crop yield to feed the increasing population in the world, as crops are the main source of food in the world (Batjes, 1996; Hinge et al., 2021). (4) Ecosystem Services: This revealed that soils rich in SOC content enhance several ecosystem values, from supporting the life of soil organisms and light filtration, water filtration, and check C thrashing, all advancing environmental privilege (Sanderman et al., 2017; Teague & Kreuter, 2020).

11.1.1 Formation of Soil Organic Matter

Microbial activity plays a pivotal role in the creation of soil organic matter (SOM). Microbes break down plant residues and other organic materials, transforming them into microbial biomass and necromass, both essential constituents of SOM (Kögel-Knabner, 2002). This microbial process is vital for securing organic carbon in the soil, as microbial by-products are more enduring and can establish stable connections with soil minerals (Kögel-Knabner, 2002). The chemical makeup of SOM influences its decomposition rates and capacity for carbon sequestration. Resistant compounds, such as lignin and tannins, decompose at a slower pace compared to easily degradable compounds like sugars and proteins. The inclusion of resistant carbon inputs in SOM can result in prolonged carbon storage in soils. Plant secondary compounds, like tannins, can interact with microbial necromass to generate durable organic matter, thereby enhancing carbon sequestration.

11.1.2 Soil Organic Carbon Dynamics

The dynamics of SOC are determined by an input/output system with inputs from plant residues, root exudates, microorganisms, and outputs through decomposition and mineralization (Figure 11.1). Soils receiving higher quality organic matter containing low C/N ratios have a faster rate of nutrient and carbon dioxide (Johnston et al., 2009. This interplay defines net gain or loss of SOC, which in consequence affects the agriculture capacity of the land/soil and the potential for carbon sequestration.

11.1.3 Role of Agriculture in the Carbon Cycle

Agriculture activities have an impact on the carbon cycle; it affects both carbon emission and carbon sequestration It is also considered one of the most important sources of greenhouse gases: carbon dioxide (CO_2), methane (CH_4), and nitrous oxide (N_2O) (Qiu et al., 2009; Benbi, 2013). There are different sources for those gases: CO_2 emissions because of organic matter decomposition in soils and soil-use changes, like transforming forests into farmland (Ngatia et al., 2021; Paustian et al., 1997). CH_4 is primarily emitted through enteric fermentation in ruminant livestock and manure management (Thakuri et al., 2020; Tongwane & Moeletsi, 2021). N_2O: Emissions created from the application of chemical and organic fertilizers, as well

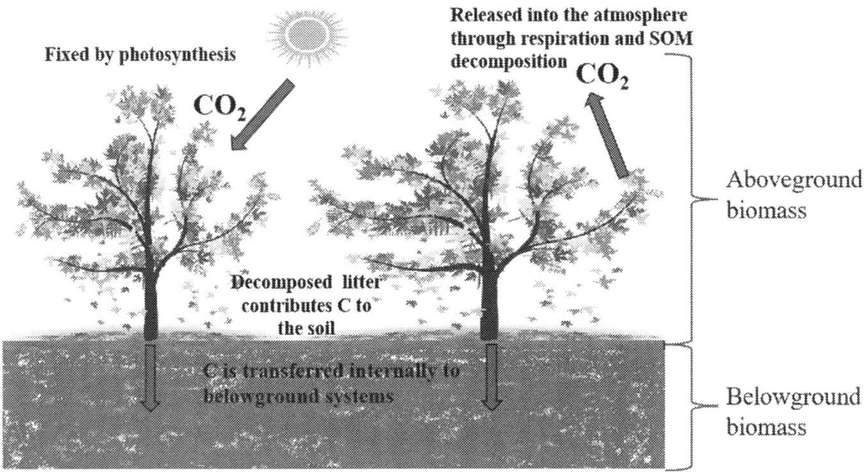

FIGURE 11.1 Carbon sequestration in plants.

as manure management (Hassan et al., 2022). Usage of Energy: Agricultural processes need substantial energy inputs, resulting in indirect CO_2 emissions from the use of machinery and other equipment powered by fossil fuels (Gao et al., 2022; Ozlu et al., 2022), soil organic carbon (SOC): Thus, roles of soils are; as a major carbon reservoir or carbon pool, steer carbon is assimilated from the atmosphere through plant photosynthesis. SOC sequestration involves loading plant residues, roots, and organic amendments to the ground where they are subjected to degradation by soil microbial activities (Macas & Camps Arbestain, 2010; Stephens et al., 2007). Land management practices and sustainable agricultural practices can enhance SOC sequestration. Cover cropping: sowing of crop residue during the off-planting season to enhance inputs to soil biomass and to protect the soil carbon (Boselli et al., 2020). Reduced Tillage: Minimizing soil disturbance to decrease carbon oxidation (CO_2) and loss (Derpsch et al., 2014; He et al., 2011). Crop rotation and diversification: Incorporating nitrogen-fixing crops and diversifying crop rotations to improve soil health and carbon storage (Zhang et al., 2021a,b). Agroforestry: planting trees and shrubs within agricultural systems for purposes of increasing biomass and carbon stock (Kay et al., 2019; Kumara et al., 2023). Through the following ways, soil organic amendments can contribute to changes in soil carbon storage: These increase the microbial biomass and the microbial activity in the soil (Liang et al., 2017), improving the water-stable soil aggregation (Mpeketula & Snapp, 2019), and incorporating resistant carbon (Smith, 2016). The chapter first define soil carbon sequestration and explains its importance. The chapter also analyzes the role of agriculture in the carbon cycle and evaluates sustainable land management practices. The chapter will also attempt to assess the potential and limitations of SOC sequestration and explore policy and investment strategies. Furthermore, it also highlights case studies and empirical evidence for future directions and research needs.

11.2 MECHANISMS OF SOIL CARBON SEQUESTRATION

11.2.1 Photosynthesis and Carbon Input into Soil

Photosynthesis is a process used by plants, algae, and certain bacteria to convert light energy from the sun into chemical energy in the form of glucose (sugar) and oxygen. It is an essential process that sustains life on Earth by providing food and oxygen. The overall equation for photosynthesis is as follows:

$$6CO_2 + 6H_2O + \text{light energy} \rightarrow C6H_{12}O_6 + 6O_2. \quad (11.1)$$

Photosynthesis is the main process through which atmospheric carbon dioxide (CO_2) is captured and transformed into organic carbon compounds in plants (Kell, 2012), The organic carbon produced during photosynthesis is transported underground into plant roots and the adjacent rhizosphere (root zone) (Kell, 2012; Panchal et al., 2022). Roots add carbon to the soil via different channels: (A) Root biomass: As roots expand and penetrate deeper into the soil, they enrich the soil's organic carbon (SOC) reserve through root turnover and decomposition. (B) Root exudates: Roots secrete diverse carbon-containing substances, like sugars, organic acids, and mucilage, into the rhizosphere. These exudates act as a supply of readily available carbon for soil microorganisms (Kell, 2012; Panchal et al., 2022). Root exudates and the microbial activity they foster in the rhizosphere can result in the creation of enduring soil organic matter via diverse mechanisms, including interactions with soil minerals, the formation of aggregates, and the generation of recalcitrant compounds (Kell, 2012; Panchal et al., 2022). The quantity of carbon that can be sequestered in the soil depends on how far the root system penetrates. Carrying carbon down to deeper layers with deeper root systems can make for improved long-term carbon storage. The composition of the microbial community present in the rhizosphere and the volume and characteristics of root exudates determine the destiny of carbon inputs to the soil and their capacity for enduring stabilization. The process of carbon input into the soil is begun by photosynthesis through the translocation of organic carbon compounds to the roots and their subsequent release into the rhizosphere as root exudates and biomass turnover. The depth and spread of the root system, and the interactions between roots, exudates, and soil microorganisms greatly influence the potential for long-term soil carbon sequestration.

11.2.2 Physical Mechanisms of Carbon Sequestration

It also enhances the stability as well as the quality of the contained SOC because the aggregation physically shields it from disintegration. SOC associates soil particles through organic matter and microbial products, thus forming compacts that minimize its availability to microbial decomposition. The studies indicate that the stability of soil aggregates, depending on the content of clay material and microbial metabolites, correlates with SOC stocks (Sekaran & Kumar, 2021). The distribution, accumulation, and deposition of organic matter in the deeper soil horizons are of great importance in sequestering carbon. However, it should be noted that as organic materials are buried, they are less easily accessible to microbial decay, leading to more long-term sequestration of carbon. Research shows that the depth of organic matter burial and its interaction with clay minerals helps to increase the stability of carbon stocks (Zhang et al., 2021a,b).

11.2.3 Biological Mechanisms of Carbon Sequestration

Around 25% of the existing biodiversity on our planet resides in soil, and the majority of such things are microbes, which happen to be varied and common on Earth (Heijden, 2008). The well-being of plants and soil relies on microorganisms (Paul, 2015; Wardle et al., 2004a,b). Many microorganisms serve a crucial part in biogeochemical cycling and general ecosystem function, among them fluxes of greenhouse gases like carbon dioxide, methane, and nitrous oxide (Singh et al., 2010; Whalen et al., 2022). The management and manipulating of soil and plant microbes can help draw down atmospheric CO_2 by bolstering soil C sequestration and stabilization (Ahmed et al., 2020; Buragohain, 2019; King, 2011; Vishwakarma et al., 2016), preserving soil and plant health, and performing an active part in ecological processes (notably fixing of N_2 and sequestration of C). through boosted crop yield and endurance.

Microbial action is at the heart of the processes of transformation and stabilization of SOC. Organic matter is decomposed with the help of microorganisms. The process also emits carbon dioxide and increases SOC by the biomass. Cells that are created during decomposition are more resilient to breakdown; therefore, they make up a significant portion of long-term carbon sequestration (De Deyn et al., 2008). The specific microbial composition depends on the quality of organic matter added and the characteristics of the soil they are in, and this determines the efficiency of sequestration of SOC. Soil microbial effects on SOC inputs are further amplified by plant and soil interactions. Root exudates are sugars, organic acids, and amino acids, and for soil microorganisms, they are a source of carbon, promoting microorganisms and the growth of microbial biomass. Mycorrhizal fungi interact with plant roots mutually, belonging to one of the most common networks of symbiotic associations that allow for the uptake of nutrients and for increasing SOC due to greater root mass and the deposition of organic matter. Such interactions have underlined the necessity of plant heterogeneity in the augmentation of carbon storage through optimizing microbial activities (Figure 11.2). Resistant organic carbon pools in soil have fine fractions with stable and persistent C pools and include mineral-associated organic matter (MAOM) and aggregated carbon (AggC). Low molecular weight compounds, which are part of MAOM, include lipids, polysaccharides, and tiny particles of highly decomposed plant and microbial origin. Conversely, several microbial and indigenous plant-driven processes produce AggC. There are two main ways that soil carbon sequestration happens: below-ground inputs, such as root litter and the discharge of root exudates, and above-ground plant inputs, such as plant litter and dissolved organic carbon (DOC). Retention of carbon and cycling processes include active, direct, or indirect engagement from soil microorganisms. According to Liang et al. (2019) and Kindler et al. (2006), direct mechanisms entail the incorporation of carbon into microbial biomass and necromass, potentially constituting a significant fraction of stable (SOC), estimated at around 50% and 80%. Carbon plants can appear as materials in microbes through anabolism that decomposes plant carbon and puts out extracellular carbon and CO_2 in catabolism. The necromass from the remaining

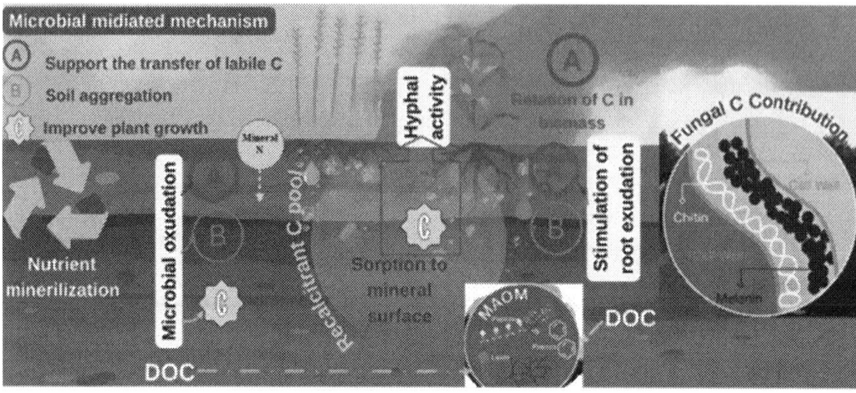

FIGURE 11.2 Microbial contributions to soil C sequestration and retention.

microbes can either be reinstated into the active microbial pool when it is utilized as a substrate for microbial metabolism, or it can be converted into more resilient C pools when it is incorporated into MAOM through adsorption onto soil minerals or AggC through microaggregation (Mason et al., 2023). Indirect effects include enhancing plant biomass in the ground and plant growth, stress tolerance, root-out ability, and other root exudation capabilities through microorganisms. With time, some of this biomass builds up in resistant C pools (Mason et al., 2023).

11.2.3.1 Role and Mechanisms of Bacteria for Carbon Sequestration

The development of the agricultural inoculant industry is projected to reach US$12.5 billion in income by 2027 (Mason et al., 2023). Understanding the contribution of soil microbes in soil carbon cycling is necessary to help the development of products that are capable of sequestering and keeping soil carbon through carbon allowances and other financially related tools (Mason et al., 2023).

The soil carbon cycle in the ecological process can be regulated by soil bacteria, which are the most abundant microbes in the soil. Soil carbon can also be used as an energy source for most bacteria in the soil, indicating a strong relationship between soil bacteria and carbon in the soil (Horner-Devine et al., 2004; Wardle et al., 2004a,b). For instance, the decomposition of SOM is positively driven by soil bacteria, which in turn contributes to the SOC pool (Treseder & Allen, 2000; Nave et al., 2010; Lange et al., 2015). As described by Yang et al. (2018), soil carbon cycling is mostly influenced by soil bacteria than other factors in the soils. Furthermore, the abundance of soil bacteria has a strong correlation with the soil C pool, showing >70% of the differences (Yang et al., 2018). Moreover, the extent of the SOC pool is reliant on the type of microbial involved in the organic matter decomposition (Burke, 2015; Doetterl et al., 2015; Lange et al., 2015). This is mainly due to the storage of SOC being directly influenced by the composition of soil microbes, such as the diversity and community of bacteria (Lange et al., 2015; Ling et al., 2017).

Figure 11.3 shows that CO_2 sequestration is mostly influenced by photosynthetic bacteria. They reduce ATP energy that accelerates the transformation of CO_2 to

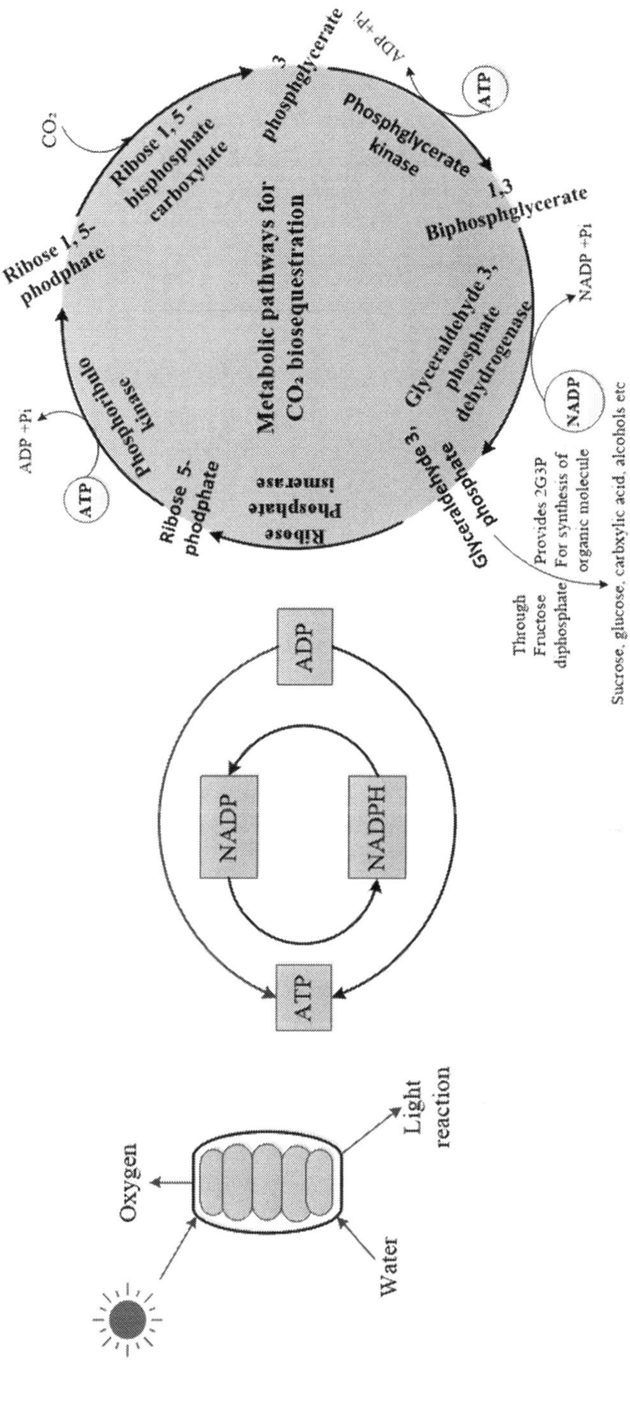

FIGURE 11.3 Process of soil bacteria for carbon sequestration.

bio-products (Angermayr et al., 2015). According to Bharti et al. (2014), bacteria can sequester CO_2 through ribulose-1,5-bisphosphate carboxylase and carbonic anhydrase. For example, bacteria that contain type I photosynthetic reaction, including heliobacteria and green-sulfur, often use Fe-S clusters as electron acceptors to support the reversing of the oxidative tricarboxylic acid cycle to carbon dioxide sequester. However, for autotrophic CO_2 fixation, the Calvin cycle is used (Tang et al., 2011). This sequestration of CO_2 by bacteria is also based on the environmental condition, where CO_2 for metabolism sequester typically with anaerobic bacteria (Agler et al., 2011). The catalytic action of gamma and zeta carbonic anhydrase enzymes is used to support CO_2 sequestration in anaerobic CO_2 sequestrating through bacteria (Shahid et al., 2022). Therefore, anaerobic sulfate-reducing bacteria are important to advance the alteration degree of CO_2 into solid minerals to improve a long-lasting C pool (Paul et al., 2017). A bacterium called *Bacillus altitudinis* bacteria is also effective in CO_2 sequestration (Nathan & Ammini, 2019). In general, soil bacteria significantly support the conversion of organic matter into CO_2 or soil carbon pools of terrestrial vegetation (Wang et al., 2021b). Therefore, it is vital to take into account the responses of bacterial diversity and ecological situations that determine the transformation process of organic carbon in the soil to realize the maximum level of CO_2 sequestration (Shahid et al., 2022).

11.2.3.2 Role and Mechanisms of Fungi for Carbon Sequestration

Root exudates represent the highest carbon sink of carbon input into the soil (Nguyen, 2009), with up to 20% of this carbon transferring into the rhizosphere through mycorrhizal fungi (Canarini et al., 2019). Lignin, traditionally considered a highly intricate molecule derived from plant materials, can be broken down by specialized lignolytic fungi during the early stages of decomposition (Klotzbücher et al., 2011). A rise in mesofauna and macrofauna enhances the occurrence of beneficial microbes for plants across the soil and improves soil structure through disturbance and reworking in the soil environment (Lavelle et al., 2006) (Figure 11.4). An increase in the relative dominance and abundance of beneficial fungi in agricultural ecosystems can positively impact ecosystem-wide carbon cycling, organic matter decomposition, carbon sink, plant production, and productivity, and ultimately crop yields (Figure 11.4). These benefits represent the key ecosystem services provided by agricultural soils (de Vries et al., 2013; Wagg et al., 2014). Under both active and abandoned agroecosystems, fungi take up approximately 50%–70% of the carbon from root exudates (Hannula et al., 2012; Hannula et al., 2017; Morri¨en et al., 2017), while they account for a tiny fraction of the overall microbial biomass. Arbuscular mycorrhizal fungi (AMF) have turnover rates similar to those of bacteria, while ectomycorrhizal fungi (EMF) and saprotrophic fungi exhibit slower turnover rates in comparison (Anthony et al., 2020; Hannula et al., 2019). In agroecosystems, fungal necromass accounts for about 40% of the SOC stock, while bacterial necromass contributes nearly about 20% (Liang et al., 2019). Emilia and Morriën (2022) emphasized that leveraging the natural ability of diverse fungal communities to sink carbon offers a dual advantage: The storage of the photosynthesized carbon helps mitigate climate change and boosts productivity by rebuilding organic matter in impoverished soils. In addition, Fontaine et al. (2011) proposed that fungi serve a key function in cellulose decomposition and

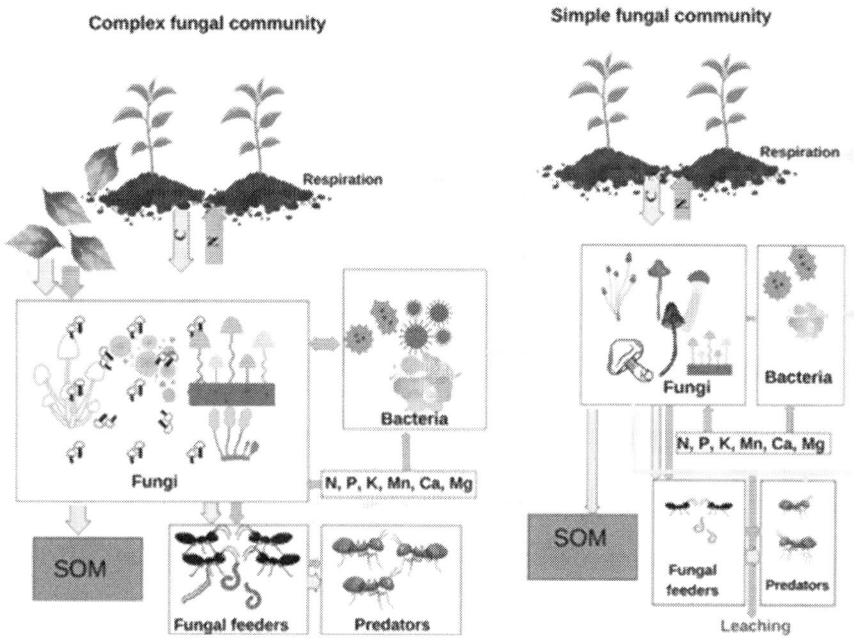

FIGURE 11.4 Role of fungi in soil organic carbon sequestration.

the induction of the priming effect (PE), adjusting their degradation based on nutrient availability.

The availability of nutrients in soil shows a dynamic relation with its fungal activity. Fungi will sequester nutrients and carbon when nutrient levels are abundant and low in plant demand (Frey et al., 2003). Fungi also release nutrients and carbon from soil organic matter under the condition of low nutrient availability. This adaptive mechanism coordinates the supply of soil minerals with plant nutrient uptake, suitable for mitigating nitrogen limitations and decreasing nitrogen losses through processes such as leaching and denitrification, as well as increasing organic nutrient and carbon accumulation.

Based on over 50 years of research, Mason et al. (2023) underscored the role of arbuscular mycorrhizal fungi (AMF) in sequestering carbon to less labile forms, including mineral-associated and aggregated carbon pools. Endophytic fungi have been identified as a promising supply of stable carbon soil and as being melanized. Mukasa Mugerwa and McGee (2016a) in a 14-week pot study, found melanized endophytic fungi had the potential to increase soil carbon content in a carbon-rich Alfisol by up to 17%.

Fungal groups, such as many of the Ascomycetes and Basidiomycetes, produce extensive vegetative mycelial networks, composed of thread-like hyphae in which carbon compounds are stored. These compounds contribute to the formation of recalcitrant carbon pools upon their death by processes like aggregation or sorption to mineral surfaces (Mason et al., 2023). However, AMFs, in particular, are crucial to soil carbon cycling and regenerative agriculture. They are obligate biotrophs that form symbiotic relationships with approximately 80% of vascular plants, and many of these are crops (Lewis, 2016). Understanding how they contribute to soil carbon sequestration provides an exciting opportunity to improve sustainable agriculture.

The symbiotic associations of fungi with plant roots increase their nutrient uptake, which improves plant growth, and consequently, fungi convert atmospheric CO_2 into plant biomass (Malyan et al., 2019). Apart from their role in sequestering carbon by assimilating carbon into their hyphae, soil fungi are also involved in carbon sequestration by producing recalcitrant decomposition products with the residence time of the order of years to centuries (Malyan et al., 2019; Mehar & Sundaramoorthy, 2018). During the decomposition of organic matter, these recalcitrant materials help provide the terrestrial ecosystem with a role as a carbon sink. Fungi contribute significantly toward lowering atmospheric CO_2 by sequestering carbon in multiple stable forms, thereby adding to mitigating climate change (Mehar & Sundaramoorthy, 2018).

11.2.3.3 Role and Mechanisms of Actinomycetes for Carbon Sequestration

Actinomycetes, referred to as filamentous microorganisms or Actinobacteria, are gram-positive), aerobic bacteria that provide spore and members of the Actinomycetales order (Strap, 2011). Actinobacteria contribute to soil biomass through the development of filamentous structures that might have a certain function in the retention and stabilization of soil C besides its aptitude to foster plant growth (Mason et al., 2023). These bacteria have a knack for helping break down SOM, including lignin, chitin, and pectin, and create an assortment of substances, including hydrolyzing enzymes (Gasmi et al., 2019). Therefore, actinomycetes are important for plant growth, climate change, and soil organic carbon sequestration. Pellegrini et al. (2022) also highlighted the need for research on these microorganisms.

11.2.4 Chemical Mechanisms of Carbon Sequestration

Stabilization of organic matter by chemical means is particularly important for the retention of carbon in the soil matrix. This process involves organic compounds being adsorbed and fixed with soil minerals, where they remain stable organomineral complexes that are resistant to microbial degradation. The stability of these complexes depends on the chemical reactivity of organic matter and the type of interacting soil minerals. It has been postulated from research that the presence of specific functional groups known to react with soil can increase the stability of the organic compounds in the matrix (Schmidt et al., 2011) (Figure 11.5).

11.3 FACTORS AFFECTING SOIL CARBON SEQUESTRATION

Soil organic carbon (SOC) originates from atmospheric CO_2, which is absorbed by plants through photosynthesis. The level of SOC is determined by the balance between carbon inputs and carbon losses from organic materials (Burke et al., 1989). Several factors influence the soil's capacity to store carbon, including soil properties, land management practices, climate, and types of vegetation (Table 11.1). In natural ecosystems, rainfall and temperature are key factors that influence the amount of plant biomass input and the decomposition of SOC for a specific soil type. These

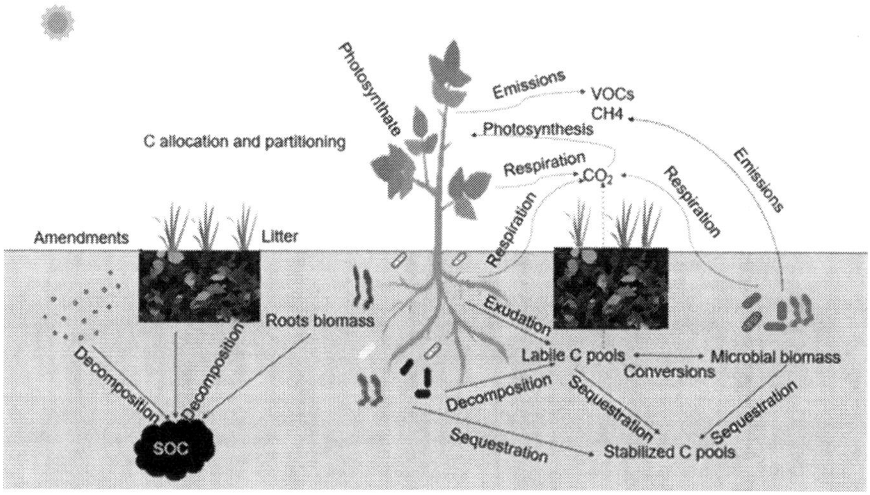

FIGURE 11.5 Mechanisms of soil carbon sequestration.

TABLE 11.1
Factors Influencing the Soil's Capacity to Store Carbon

Factor	Description	References
1. Soil properties		
Soil texture	Soils high in clay content can store carbon by preventing organic materials from breaking down.	Churchman et al. (2014)
Soil structure	Soils with good aggregation create microenvironments that help stabilize organic matter.	Dalal and Bridge (2020)
Soil pH	An ideal pH level of 6–7 enhances microbial activity and nutrient availability, which enhances carbon sequestration.	Bhattacharyya et al. (2022)
Soil moisture	Moist soil is essential for plant growth and microbial activity, both of which contribute carbon and help stabilize the soil.	Dignac et al. (2017)
Soil temperature	Mild temperatures boost microbial activity and carbon stabilization, while extreme temperatures inhibit both processes.	Knicker (2007)
2. Land management practice		
Tillage practices	Conservation tillage increases soil carbon levels by 57% compared to conventional tillage methods.	Sharma et al. (2016)
Crop rotation anddiversification	Rotating cover crops increases soil organic carbon by 20% compared to monoculture systems.	Duval et al. (2016)

(*Continued*)

TABLE 11.1 *(Continued)*

Factor	Description	References
Organic amendments	Adding organic materials such as compost and manure raises soil carbon levels. These organic additions help build up carbon reserves soil over time.	Diacono and Montemurro (2011)
Agroforestry	Incorporating trees into agricultural land enhances biomass and soil carbon storage. Agroforestry systems can sequester 1.5 mg of carbon per hectare each year.	Lorenz and Lal (2014)
Perennialcropping systems	Perennial crops such as switchgrass and miscanthus, with their deep root systems, are effective at storing soil carbon. These crops are particularly efficient at capturing carbon on less fertile landscapes.	Das et al. (2018)
3. Climate conditions		
Temperature andprecipitation	Climate has a significant effect on soil carbon sequestration, with warm, moist conditions enhancing plant growth and soil carbon levels.	Davidson and Janssens (2006)
Extreme weather events	Droughts, floods, and storms harm soil structure, and vegetation, and increase erosion, leading to a decrease in soil carbon sequestration.	Saco et al. (2021)
4. Vegetation types		
Grasslands	Grasslands are important carbon sinks because of their deep root systems and persistent ground cover.	Lorenz et al. (2018)
Forests	Forest biomass and soil retain a large amount of carbon. Depending on the species composition, age, and management practices, forest soils can store up to 2.0 mg of carbon per hectare per year.	Jandl et al. (2007)
Wetlands	The slow rate of decomposition makes wetlands among the most effective habitats for carbon sequestration. Wetlands, including peatlands, can capture up to 5.5 mg of carbon annually.	Iqbal and Shang (2019)

conditions establish the equilibrium SOC level. Equilibrium SOC levels are achieved when the input of organic carbon matches the losses. In agricultural systems, however, SOC turnover rates and equilibrium levels are additionally influenced by management practices. For instance, a significant amount of carbon captured through photosynthesis is lost during the harvest, with about 30%–50% of the fixed carbon in the form of aboveground dry matter commonly being taken away in cereal crops (Johnson et al., 2006). The remaining 50%–70% of the yearly fixed carbon consists of aboveground residues and belowground residues, such as root biomass. Accurately determining the contribution of each factor to soil carbon sequestration is challenging due to the complex interactions among these elements and the variability across different ecosystems and management strategies.

11.3.1 Conservation Tillage

Agricultural practices and fertilization schemes, as well as soil and climate conditions, influenced SOC inputs and turnover, contributing to variations in soil carbon pools (Wang et al., 2014) conservation tillage management was frequently opined with little modification in SOC pools to gain more insight, distinct fractions of SOC were determined impact with plowing management on soil organic carbon concentration (Blanco-Moure et al., 2016). Tillage methods might impact the unique state of SOM by altering SOM disintegrating paths (Ding et al., 2002). Prior studies revealed that tillage appeared to increase labile fractions of SOC. Hence, permitting the formation of resistant fractions of organic carbon and fixing organic carbon pools (Ramnarine et al., 2018). No tillage or reduced tillage are caused to foster root development and emit root exudates in soil (Dijkstra et al., 2021), leading to elevated rhizosphere carbon and building up labile fractions of carbon (Shrestha et al., 2015). Tillage habits have had an immediate effect on the flow and cycling of various SOC fractions in soil. So, conventional tillage has a considerable effect on the content of hydrophobic components in carbon. These impacts are strongly linked to straw management and cropping patterns (Shrestha et al., 2015).

11.3.2 Cropping Pattern

SOC concentration is being assessed primarily for its advantageous impact on soil physiochemical properties and crop yield security. Considering the shifting in soil organic carbon stocks due to climatic conditions, there is a new spotlight on examining agricultural fields soil's promise as an organic carbon store (Körschens et al., 2014; Poeplau et al., 2015). Long-term trials, which have been utilized for decades, present ongoing data on the impact of different strategies for the management of crop and soil variables (Smith et al., 2020). Factors on SOM entail a disparity between immobile carbon pools and biodegradable carbon pools that may be modified through management technologies, according to Körschens (2021), who stated for 68 European long-term experiments. The overall organic coil carbon content varies from 0.15% to 2.29%, with an average inert SOC content in sandy (0.32%) and black soil (1.46%). Ryegrass-sweet sorghum cropping pattern can generate a greater variety of biomass for bioreactors than straight solitary plants (Hill, 2017). In this vein, cyclic and biennial cropping should be promoted as methods of increasing the growth of crops. An effective cropping plan for biorefineries regularly involves biomass withdrawal, resulting in a trade-off between yields of biomass and carbon sequestration. Soil carbon stock changes are tightly related to carbon consumed from a combination of crop residues, root, and rhizodeposits-derived carbon (Gougoulias et al., 2014), and discarding biomass for biorefining in rotational and evergreen crops diminishes carbon input (Burks, 2013). Rotational and biennial cropping, on the other hand, can increase soil aggregate formation, leaving behind little SOC mineralization (Bansal et al., 2021). According to Chen et al. (2022), Single and/or perennial cropping may boost C sequestration and yield on a global scale. Versatile cropping is urged to generate biomass that meets the rising need for biorefining, however, inadequate biomass removal could erode soil carbon retains and ecosystem diversity (Sharma et al.,

2024). The demand worldwide for biomass production is going to skyrocket, driven by the growing need for green electricity and natural products (Antar et al., 2021). However, there is a desperate need for an effective cropping pattern serving a flexible that can strengthen yields sustainably (Von Cossel et al., 2019).

11.3.3 Organic Amendments (Manure, Compost, and Biochar)

It is well known that balanced fertilization with a combination of manure and chemical fertilizers may raise soil nutrients, crop growth, and crop yields to soil organic carbon concentrations. However, synthetic fertilization couldn't produce the same findings on its own. Inorganic fertilizers are known to induce lower soil pH and lead to environmental hazards for maize cultivation, including eutrophication and acidity (Lee et al., 2020). Adding manure is a more effective way to raise the SOC stocks than using inorganic fertilizers (Gai et al., 2018). The amount and types of organic materials used, the more times it can take for decomposition to occur, the type of irrigation water used, and the location can all affect the manure's quality and nutritional levels. Wang et al. (2021a) found that while manure inputs result in enhanced capture of carbon in fine soil components, their influence on stable carbon content was much bigger than that of inorganic fertilizers. Therefore, whereas inorganic fertilizers do not improve soil health, the incorporation of manure does. It also increases soil carbon sequestration. Because manure affects soil properties like biological indicators, microbial population works of art, the biomass of microbial species, earthworm populations in general, other microbes like enzymes, and soil physical parameters, including porosity, aggregation, compaction, and bulk density, it was shown to improve soil water relations. For example, a reduction in the use of organic materials also influences soil erosion and causes more compaction of soil (Bashir et al., 2021). It reduces soil aggregation, making it unable to stabilize soil structure, porosity, and water-holding capacity (WHC). Consequently, manure improves SOM, which is important for crop production and, as a result, helps to reduce global warming and create a win-win situation. Manure addition improves soil structure, which either directly or indirectly raises SOC.

Using compost is a reliable carbon sequestration method in crop fields. Prior researchers opined that it promotes SOC stocks and concentration (Ferreras et al., 2006; Lord & Sakrabani, 2019). Since 2015, the partial substitution of chemical fertilizers with organic ones has been used in crop fields, thus, SOC sequestration in Chinese croplands should rise (Guo et al., 2010; Li et al., 2020). Yet it is still unclear how much carbon may be trapped by farmland soil across China with a massive application of compost. In such scenarios, the importance of carbon sequestration urged by compost applied to the carbon neutrality aim until 2060 remains uncertain. China's net-zero carbon emissions are critical to reducing GWP, as it has the most carbon emissions. As an outcome, evaluating carbon sequestration created by compost usage throughout China, as well as its aid in carbon neutrality, is highly important.

Biochar can be utilized in soils to retain carbon as well as stabilize the carbon to which the crop has evolved (Joseph et al., 2021). Furthermore, owing to its carboxyl and phenolic component, biochar is charged negatively, increasing its surface charge negativity and cation exchange capacity (CEC). Therefore, composts or biochar have a chance to sequester carbon in soils while also eliminating CO_2 from the atmosphere

(Paramisparam et al., 2021). Biochar input may provide carbon to soil lasting decades, but longevity varies depending on other factors, including the type of Biochar, amounts, and soil's basic makeup (Jatav et al., 2021).

11.3.4 Agroforestry Systems

Agroforestry is based on the expectation that on-farm and off-farm plant cultivation can have an integral part in encouraging equitable land use and conservation of natural resources. This idea relies on the notion that land-use systems that are structurally and economically more intricate than the crop or tree solitary plants capture and utilize assets better, and also have greater structural variation, which leads to more rigid nutrient cycles. While above and belowground diversity promotes stability and resilience at site-specific levels, the systems enable them to connect with forest soil parameters at the landscape as well as watershed levels (Nair et al., 2009).

Agroforestry systems are thought to have the ability to trap carbon than grasslands or croplands. This assumption is based on the idea that planting trees into cropland and grassland will increase net aboveground and belowground carbon sequestration (Palm et al., 2004; Sharrow & Ismail, 2004). The consequence of the agroforestry system on soil organic carbon sequestration can be assessed by the amount and type of input items supplied by its planting materials, and also by soil capabilities such as form and agglomeration. For instance, if trees are permitted to develop on grassland, like free grazing land, some effects may be natural, especially variations in above- and belowground entire production, shifts in the root system in depth and dispersion, and modifications in numbers and its quality (Gillespie et al., 2021). This, in turn, could alter local and regional weather patterns (Herzfeld et al., 2021). Carbon sequestration takes place via processes involving humification, aggregation, biomass transfer into subsoil by long roots, and releasing of inorganic C into the soil as carbonates (Lal, 2001). Each of the pathways is prevalent with tree-based use of land systems.

11.3.5 Grazing Management

Grazing management that encompasses either native grassland or raised grazing land, acts as a key trap for CO_2 which impacts greatly the entire carbon cycle (Bracho et al., 2021; Paudel et al., 2023). Despite tropical forests, wherein nearly all of the carbon resides in above-ground plant life, herding land ecosystems contain upwards of 90 of their carbon reservoirs belowground (Schuman et al., 2002), providing carbon to be easily moved into more permanent storage in the soil due to soil carbon is longer-lasting than carbon tied with plant material. Soil carbon sequestration in grazing areas offers an ongoing approach to minimizing atmospheric GHG emissions (Sollenberger et al., 2019). Soil carbon status in grazing pastures is dictated by the amount and quality of plant-based carbon input from photosynthesis as well as carbon withdrawals via respiration. A prior study pointed out that these management practices boosting the creation of biomass from plants (e.g., N fertilization, adequate grazing management, or legume use) often provide carbon to the soil and, as a result, optimizing carbon sequestration (Conant et al., 2017; Henderson et al., 2015). In an overall survey of grasslands, Conant et al. (2017) noticed that soil carbon sensitivities to management were favorable, detrimental, or unchanged. Researchers ascribed the divergent findings to the delicate connection among management, rainfall, temperature, and soil structure.

The size and even trend of SOC storage through grazing can be strongly affected by grazing intensity as well as climatic scenarios (Zhan et al., 2020) because of plant community, soil microclimate, including microbes, The effects of grazing on carbon sequestration may differ significantly with mean precipitation of that area (Liang et al., 2018). Therefore, grazing intensity is regarded as an important factor for carbon sequestration.

11.4 QUANTIFYING CARBON SEQUESTRATION

In order to sustain agricultural systems and guide climate policy, it is critical to understand and evaluate carbon sequestration. This section investigates the key approaches for assessing soil organic carbon stock, modeling carbon dynamics, and integrating soil carbon into the life cycle assessments (LCAs) of agricultural systems.

11.4.1 Measuring Soil Organic Stock

The SOC is the carbon part of organic matter in the soil. Moreover, it plays an active role in CO_2 exchange in the atmosphere and secures global carbon balance (Xue et al., 2024). With unique advantages and constraints, soil organic carbon measures have been formulated and implemented. The most common method is direct soil sampling and subsequent laboratory analysis using dry combustion technique. This involves the collection of samples of the soil with varying strata and with the aid of the CHN; it analyses the comparative proportion of carbon, hydrogen, and nitrogen (Wang et al., 2022). Although this method is considered accurate but laborious and requires careful sampling to ensure representativeness across heterogeneous landscapes. Other indirect measurement techniques include remote sensing and infrared (IR) spectroscopy. These methods make it possible to estimate SOC over large-scale areas, making it less expensive and labor-intensive. Meanwhile, they are less precise than direct measurements and they frequently need to be calibrated using real data from soil samples (Viscarra Rossel et al., 2016). Challenges in measurement are as follows:

1. Spatial variability: Due to differences in soil type, topography, and land use, SOC levels can vary significantly over short distances even within the same field (Xin et al., 2016).
2. Temporal variation: Factors such as climate conditions, technology used in farming, and changing seasons, influenced SOC, and as such, frequent long-term evaluation is recommended (Nayak et al., 2019).
3. Depth of sampling: SOC is high at the surface soil, while deeper layers also have comparatively good amounts of carbon concentrations depending on ecosystems such as forests and garland. Deep soil sampling, however, is more expensive and challenging (Lorenz & Lal, 2014).

Despite these difficulties, determining SOC is still a vital approach to determining how well agricultural and natural ecosystems can sequester carbon.

11.4.2 Modeling Carbon Dynamics

In addition to direct measurements, to comprehend how soils function as carbon sinks over time, modeling soil organic dynamics is a key component. Models allow researchers to predict how land management practices, different environmental conditions, and climate change scenarios influence carbon sequestration (Paustian et al., 2016).

Many models are widely used to approximate the dynamics of soil carbon:

- RothC (Rothamsted Carbon Model): One of the most used models, predicts the organic carbon turnover in non-waterlogged soils under different land uses, including arable crops and grasslands (Coleman & Jenkinson, 1996). It clarifies how carbon is stabilized or released by separating soil carbon into different pools based on their decomposability.
- CENTURY Model: According to Liang et al. (2024) the CENTURY model utilizes the dynamics of carbon, nitrogen, phosphorus, and sulfur in a variety of ecosystems, including agricultural systems. It focuses on long-term dynamics and helps to predict how management practices like tillage, fertilization, and crop rotation affect soil carbon storage
- DayCent: It is a time-step daily form of the CENTURY model that allows for daily fluctuations in soil conditions, management practices, and weather resulting in more comprehensive simulations (Mathers et al., 2023). It is especially helpful for analyzing how examining the impact of agricultural practices on carbon sequestration greenhouse gas emissions in coming years.
- DNDC (DeNitrification-DeComposition): This model helps analyze trade-offs between environmental impacts and carbon storage because it incorporates carbon sequestration with GHG emissions (Wang et al., 2022).

These models require inputs such as crop type, soil characteristics, management practices, and prevailing climatic conditions, and by mimicking various processes, including plant growth, soil respiration, and organic matter decomposition, they provide approximate estimates of the potential of different systems to sequester carbon over time. Now, models help policymakers and farmers make long-term predictions to ensure that land management improves carbon sequestration.

Validation: One of the main challenges associated with applying the models is that to make sure that the models are accurate they have to crosscheck the predictions with the field data. However, further research, including long-term field testing and follow-up, is needed to increase the stability and applicability of these models to different ecosystems, and climates (Smith et al., 1997).

11.4.3 Life Cycle Assessment of Agriculture System

A method used to evaluate a system's effects on the environment is life cycle assessment, or LCA from the extraction of resources through the disposal of ultimate waste. To assess the carbon footprint of different farming systems and their role in mitigating greenhouse gas emissions, LCA can be used in the agriculture system as well (Feng et al., 2023).

Challenges exist in including soil carbon sequestration in LCA, especially in estimating how much carbon, and for how long, different soils store (Forfora et al., 2024). However, since SOC is tightly constrained by SOC, their combination is key for evaluating the agricultural system's equivalent carbon emissions and for determining whether the relevant activity is a carbon credit or debit.

Some of the most well-known practices that can assist in raising the quantity of carbon in soil include agroforestry, cover cropping, and reduced tillage. Having greater carbon sequestering potential, common LCA of systems employing these methods often show lower greenhouse gas emissions (Lal, 2020). There are, however, trade-offs with such techniques. However, high usage of nitrogen fertilizer to raise crop yields may result in high emission of N_2O, an important greenhouse gas that counterbalances the benefits of carbon sequestration. In agroforestry systems, trees provide ecosystem services such as increased biodiversity and retained water as they sequester carbon in soil and biomass through conventional farming systems, practices like monocropping and overtilling often reduce SOC levels resulting in increased total carbon footprint (Sileshi et al., 2023).

In the end, LCA provides a structure by which to understand the relationships between agricultural methods and their environmental effects. By adding SOC to LCA, agricultural systems can be made to optimize carbon sequestration and production contribution to mitigation of climate change.

11.5 CO-BENEFITS AND CHALLENGES

Around 2,344 Gt of organic carbon is found in soil globally which renders its biggest terrestrial SOC reservoir (Stockmann et al., 2013). The importance of SOC sequestration has gained and will continue to gain public attention due to the overwhelming positive co-benefits such as its significant contribution to maintaining some crucial soil functions and ecosystem services (Lal, 2004). There is increasing interest in the sequestration of SOC due to its important co-benefits that include the direct contribution to sustainably improved essential soil functions and ecosystem services (Lal, 2004). The co-benefits of sequestering carbon in soil encompass: It also improves food security, increases renewability of water, and encourages biodiversity (Lal et al., 2015). Carbon reservoirs in the soil are important for the worldwide carbon cycle (Nathanael et al., 2021). As organic substances are trapped in soil organic matter, SOC is retained in the earth, where it has beneficial effects on biological, chemical, and physical soil functions, impacting plant development and soil (micro-) biological processes (Rumpel et al., 2023). The sequestration of SOC, mostly implying a rise in its concentration, has been seen as important to soil fertility and agricultural productivity (Lal, 2016). Given its possible co-benefits, SOC sequestration has been presented as a win–win–win approach (Bossio et al., 2020). Currently, the majority of research concludes that an increase in SOC positively impacts soil water (quality and quantity), and soil physical characteristics, for example, bulk density and aggregation, nutrient retention, biological activity, and biodiversity (Rumpel et al., 2023). All these advantages increase soil performance and the services provided by the soil ecosystem (Rumpel et al., 2023). In addition, the array of co-benefits is extensive, encompassing wildlife habitats, water quality, and the beauty of landscapes (Feng et al., 2004), But with notable exceptions, the matter of co-benefits from sequester action garnered little to no attention.

The difficulties associated with carbon sequestration regarding other greenhouse gas (GHG) emissions, as well as water and nutrient needs (Rumpel et al., 2023), along with potential issues like N and P immobilization, water consumption, nutrient usage, financial investments, limited flexibility, biomass demands, reduced albedo, and greenhouse gas emissions, must be taken into account and tackled through sustainable management practices to restore and safeguard ecosystems (Rumpel et al., 2023). Consequently, considering variation in GHG emissions is crucial for assessing net reductions resulting from SOC sequestration at the field or project level (Rumpel et al., 2023), Feng et al. (2004) also noted that this nonpoint source nature of carbon sequestration is expected to result in several "co-benefits" of its carbon storage abilities, and the economical storage of carbon in agricultural soils is very encouraging. The expenses associated with measuring and monitoring the quantity of carbon sequestration in a field are expensive, which is why protocols to measuring and monitoring the quantity of carbon sequestration are still under development (Feng et al., 2004). Measurement of carbon stocks in soil, which is challenging, is less convenient along with its nonpoint source nature of carbon sequestration. Measuring and monitoring the quantity of carbon sequestered in a field or area is expensive, and the protocols for these processes are still under development (Feng et al., 2004). According to Abdullahi et al. (2018), the measurement of carbon stocks in soils can be difficult, time-consuming, and costly. 10% variations are hard to detect because of sampling errors, small-scale fluctuations, and also uncertainty in the measurement and analysis (Sparling et al., 2006). Soil carbon tends to increase annually by a relatively small amount, for example, between 0.25 and 1.0 t/ha (Ravindranath & Ostwald, 2008). Minor soil carbon gains or losses across various scales are difficult to account for due to methodological obstacles like monitor verification, samplings, and depths (Trumbore & Torn, 2003). Therefore, it is challenging to absolutely attribute these small modifications to a specific land use migrate, or approaches to management in each setting if they are noticeable (Abdullahi et al., 2018). However, the enhancement of carbon sequestration and retention potential of the soil is limited since the soil eventually gets to a certain steady state over time (Abdullahi et al., 2018). Carbon pools: Carbon sequestered within the soil is distributed among different pools of various residence times within the ecosystem. These are as follows:

1. Recalcitrant pool:
   The carbon in these pools often cycles over an extended period, from several hundreds to thousands of years (Abdullahi et al., 2018).
2. Active pool:
   Because of its quick disintegration, the carbon of this pool only stays in the soil for a brief period – usually between one day and one year.
3. Stable or humus pool:
   The sluggish breakdown of carbon in this pool conduces to greater turnover time; the resident duration usually lasts between one year and ten years (Abdullahi et al., 2018). For this reason, soil carbon sequestration is viewed

as a short-term strategy for carbon removal. The rate of carbon loss is influenced by various climatic, land use, and management factors (Abdullahi et al., 2018). (D) Separation: It is difficult to distinguish between carbon sequestered due to management practices or land use changes and carbon stored naturally in the soil (Abdullahi et al., 2018).

11.5.1 Improved Soil Health and Crop Production

SOC concentrations are a key variable in soil quality and positively linked to various soil functions that support, regulate, and provide ecosystem service (Rumpel & Chabbi, 2021). SOC is also recognized as essential for soil fertility as well as improving crop yields (Lal, 2016). These benefits include enhanced WHC and infiltrations, providing substrates to soil organisms, acting as a source of essential plant nutrients, and improving the stability of soil structure (Lal, 2004). Soil carbon sequestration also supports efforts to enhance the productivity of both forested and agricultural land. Carbon-sequestration technologies boost the amount of soil organic matter, improving its quality and production. This soil organic matter promotes water retention, soil structure, and access to nutrients while likewise offering an optimal setting for soil microbes (Lal, 2004). CS endeavors possess the potential to reclaim lost efficiency, particularly in farming systems that have forfeited 30%–55% of their original SOC pool following land conversion (Batjes, 1996). By using CS technologies, cultivated and poor soils might regain 50%–60% of their total amount of carbon (Lal, 2004). The study observed that considering the amount, type, and origin of SOM is a potent predictor of restoration since these traits are related to the function of ecosystems and soil health (Bieluczyk et al., 2023). Several soil functions and ecosystem benefits depend on SOC and its shifting patterns (Lal, 2016). Better soil health, in addition to raised water and access to nutrients, improves soil's resiliency to severe weather such as shortages and extreme heat, as well as offering disease-suppressive features (Lal, 2016). Protecting and enhancing soil health is integral to accomplishing the sustainable development goal (Lal, 2008; Lal, 2016).

Lal (2016) noted that the threshold level of SOC in the root zone should be between 1.5% and 2.0%. Maintaining SOC levels above this critical threshold is vital for (A) supporting soil structure and aggregation, which affect soil aeration and tilth; (B) improving water retention and efficiency, which enhances resilience to drought, heat waves, and sudden climate changes; (C) enhancing nutrient retention and efficiency, which reduces nonpoint source pollution, improves water quality, and prevents toxic algal blooms; (D) regulating rhizospheric processes, which drive elemental transformations and create disease-suppressive soils; and (E) reducing gaseous emissions (CO_2, CH_4, N_2O), thereby influencing atmospheric chemistry and mitigating climate change (Lal, 2016).

SOC dynamics play a significant role in global food and nutritional security by supplying essential macro- and micronutrients, which improve soil health (Lal, 2016). Organic matter also enhances soil's physical, chemical, and biological properties, leading to improved soil quality and higher crop yields. Increasing SOM in agricultural systems is vital to boosting soil health, raising sustainable crop yields, and storing CO_2 from the atmosphere. Activities that raise SOC levels also boost soil quality (Smith et al., 2020).

11.5.2 Potentials for Mitigating Climate Change

The goal of the Paris Agreement, which was made at the United Nations Framework Convention on Climate Change's 21st Conference of Parties (COP21), is to hold temperature rise to less than 2°C. Plus, it urges attempts that aim to minimize temperature rise to 1.5°C or less (Demenois et al., 2020).GHG emissions must be lowered to attain this aim, and technologies that can extract carbon dioxide from the air must be developed (Meinshausen et al., 2009; Lomax et al., 2015). Trying to achieve an overall rise in SOC stocks at a yearly rate of 4% or 0.4, the "Soils for Food Security and Climate" effort was started to address this problem. The goal of this effort is to mitigate the effects of climate change (Demenois et al., 2020). An estimated 13% of the global GHG emission each year, which was expected to be 53.5 Gt CO_2-eq. yr^{-1} in 2017, might be neutralized by the mitigation potential of soil carbon storing practices, which is estimated to reach 7 $GtCO_2$–eq.yr^{-1} (Fuss et al., 2019).

Since the Industrial Revolution, the concentration of CO_2 and other greenhouse gases in the atmosphere has risen (Table 11.2). Specifically, the concentration of CO_2 has surged from 280 ppmv in 1750 to 367 ppmv in 1999. Currently, it is increasing at a rate of 1.5 ppmv per year, which equates to 3.3 petagrams of carbon per year (Haywood & Schulz, 2007). During the same period, the concentration of atmospheric methane (CH_4) has risen from approximately 700–1,745 parts per billion by volume ppbv, with an annual increase of 7 ppbv. Likewise, the concentration of nitrous oxide N2O has grown from around 270 ppbv in 1750 to 314 ppbv, currently rising at a rate of 0.8 ppbv per year (Haywood & Schulz, 2007). The current radiative forcing values for these gases are 1.46 W/m^2 for CO_2, 0.5 W/m^2 for CH_4, and 0.15 W/m^2 for N_2O. The human-induced increase of GHGs in the atmosphere, along with the cumulative radiative forcing from all GHGs, has resulted in a rise in the average global surface temperature of 0.6°C since the late 19th century. The current warming rate is 0.17°C per decade (Haywood & Schulz, 2007). This observed increase in global mean temperature exceeds the critical threshold of 0.1°C per decade, beyond which ecosystems may struggle to adapt.

The ongoing rise in carbon dioxide CO_2 and other greenhouse gases GHGs in the atmosphere, primarily caused by human activities, is thought to be driving climate

TABLE 11.2
Change in the Atmospheric Concentration of Greenhouse Gases Since the Industrial Revolution at About 1750

Gas	Present Concentration	Percent Increase Since 1750	Present Rate of Increase (%/year)
Carbon dioxide (CO_2)	379 ppm	31	0.4
Methane (CH_4)	1,745 ppb	151	0.4
Nitrous oxide(N_2O)	314 ppb	17	0.25
Chlorofluorocarbons (CFCs)	268 ppt	—	Decreasing

Source: Modified from Haywood and Schulz (2007).
Note: ppm – parts per million, ppb – parts per billion, ppt – parts per trillion.

change and its associated impacts worldwide (Haywood & Schulz, 2007; Lal et al., 2007). This situation has sparked interest in creating strategies to reduce the accumulation of greenhouse gases and GHGs in the atmosphere. Of the 8.7 GtC emitted annually from human activities, only 3.8 GtC remains in the atmosphere (Denman et al., 2007; Lal, 2008). The unexplained difference of 4.9 GtC per year is thought to be stored in terrestrial environments, such as oceans, forests, and soils, which is often referred to as the "missing sink" (Battle et al., 2000; Denman et al., 2007). This understanding has led to increased interest in the potential of the terrestrial sector, including soil, to capture carbon in long-lasting reservoirs, thereby helping to decrease the amount of carbon in the atmosphere (Guo & Gifford, 2002; Lal, 2004; Post & Kwon, 2000; Stockmann et al., 2013).

Even small changes in SOC levels can have a significant impact on atmospheric carbon concentrations (Stockmann et al., 2013). For mitigating climate change, sequencing SOC can be vital (Balesdent & Arrouays, 1999). Compared to other approaches, storing carbon in soil is particularly beneficial since it is available and easily accessed. Based on recent forecasts, SOC may be causing 72%, 47%, and 9% of the mitigation potential in the biomes of forests, grasslands, and wetlands, respectively. Bossio et al. (2020) also indicated that this translates to soils in these biomes buffering or storing 1.2 Gt of CO_2e each year, 2.0 Gt CO_2e yearly, and 2.3 Gt CO2e annually, respectively. SOC sequestration has several benefits for mitigating climate change, including improved quality of water and air, greater biodiversity, and reduced soil erosion (Rumpel et al., 2023).

11.5.3 Barriers to the Adoption of Soil Organic Carbon Sequestration for Economic, Social, and Technical

Barriers to adopting SOC sequestration for economic, social, and technical The successful execution of SOC sequestration strategies can be delayed by a variety of economic, social, and technical social obstacles. According to Nganga et al. (2019), Mills et al. (2020), and Demenois et al. (2020), there are various difficulties to soil carbon management approaches, including one main hurdle:

Economic Barriers

1. Preliminary Costs: using Cover crop and decreased tillage are a pair of SOC sequestration tactics which might need starting with investments in seeds, equipment, or labor.
2. Unexpected Economic Refunds: Producers can be unwilling about using SOC techniques due to unknowns about long-term financial rewards, such as a rise crop yields for carbon credits Access to funding: Limited access to financial resources or credit can prevent farmers from investing in SOC practices.
3. Opportunity costs: Producers might need to decide between pursuing SOC methods and more lucrative farming strategies.
4. Market entry: Farmers may be less motivated to sequester SOC due to a lack of established carbon credit markets.

Social Barriers

1. Insufficient consciousness: Producers and landlords might not know the beneficial effects of storing carbon in soil or how to execute it well.
2. Cultural intolerance: conventional farming methods could be passionately built up, rising as opposed to new forms.
3. Group effects: Producers regularly depend on the guidance and expertise of their colleagues. If significant community members do not use SOC practices, others may be less likely to do so.
4. Perceived risk: Concerns about the risks associated with changing established practices can deter farmers from trying new methods.
5. Education and training: Restricted access to SOC methods, especially to remote areas, may restrict uptake efforts.

Technical Barriers

1. Knowledge gaps: There may be insufficient knowledge about the best practices for SOC sequestration, including site-specific recommendations.
2. Complexity of practices: Some SOC practices can be complex and require specific management skills that farmers may not possess.
3. Measurement and verification: Difficulty in measuring and verifying SOC levels can create challenges in demonstrating the effectiveness of practices and securing carbon credits.
4. Soil variability: Variability in soil types and conditions can make it challenging to implement a one-size-fits-all approach to SOC sequestration.
5. Lack of research and development: Insufficient research on the long-term impacts and benefits of SOC practices can limit the development of effective strategies.

11.6 CONCLUSIONS AND FUTURE DIRECTION

Carbon sequestration is a vital topic given that it is related to ecological health and farming soil fertility. Our findings reveal that agricultural practices have an immense effect on carbon sequestration. Soil carbon research has experienced hurdles due to regional and temporal disparities. Rotation of crops, diversified portfolios, organic amendments, agroforestry, and perpetual cropping systems all contribute to around 40% of carbon sequestration. A closer look suggests the common thread connecting these approaches to carbon sequestration is the role of organisms that break down organic materials. Microbial activity benefits the ecosystem by offering needed nutrients.

Through our study of soil carbon sequestration and its role in agriculture, we advise giving research aimed at grasping soil microbial activity, its function in storing carbon, and conditions that encourage this process top priority. Predictive modeling and isotope analysis are two methods that can provide important insights into future patterns and results.

ARTIFICIAL INTELLIGENCE DECLARATION

The authors have not used regenerative artificial intelligence tools during the writing the chapter.

BIBLIOGRAPHY

Abdullahi, A.C., Siwar, C., Shaharudin, M.I.I., & Anizan, I. (2018). Carbon sequestration in soils: The opportunities and challenges. *Carbon Capture, Utilization and Sequestration*, 1, 3–16.

Agler, M.T., Wrenn, B.A., Zinder, S.H., et al. (2011). Waste to bioproduct conversion with undefined mixed cultures: The carboxylate platform. *Trends in Biotechnology*, 29,70–78..

Ahmed, A. A. Q., Odelade, K. A., & Babalola, O. O. (2020). Microbial inoculants for improving carbon sequestration in agroecosystems to mitigate climate change. *Handbook of Climate Change Resilience*, 381–401.

Alkharabsheh, H.M., Seleiman, M.F., Battaglia, M.L., Shami, A., Jalal, R.S., Alhammad, B.A., Almutairi, K.F., & Al-Saif, A.M. (2021). Biochar and its broad impacts in soil quality and fertility, nutrient leaching and crop productivity: A review. *Agronomy*, 11(5), 993.

Angermayr, S.A., Gorchs Rovira, A., & Hellingwerf, K.J. (2015). Metabolic engineering of cyanobacteria for the synthesis of commodity products. *Trends in Biotechnology*, 33, 352–361.

Anjum, J., Sheikh, M.A., Tiwari, A., Sharma, S., & Kumari, B. (2022). Carbon Sequestration: An Approach to Sustainable Environment. In *Microbial and Biotechnological Interventions in Bioremediation and Phytoremediation* (pp. 425–444). Springer.

Antar, M., Lyu, D., Nazari, M., Shah, A., Zhou, X., & Smith, D.L. (2021). Biomass for a sustainable bioeconomy: An overview of world biomass production and utilization. *Renewable and Sustainable Energy Reviews*, 139, 110691.

Anthony, M.A., Crowther, T.W., Maynard, D.S., van den Hoogen, J., & Averill, C. (2020). Distinct assembly processes and microbial communities constrain Soil organic carbon formation. *One Earth*, 2, 349–360.

Balesdent, J., & Arrouays, D. (1999). Usage des terres et stockage de carbone dans les sols du territoire français. Une Estimation Des Flux Nets Annuels Pour La Période 1900–1999.

Bansal, S., Yin, X., Sykes, V., Lee, J., & Jagadamma, S. (2021). Soil aggregate-associated organic carbon and nitrogen response to long-term no-till crop rotation, cover crop, and manure application. *Soil Science Society of America Journal*, 85(6), 2169–2184.

Bashir, O., Ali, T., Baba, Z. A., Rather, G. H., Bangroo, S. A., Mukhtar, S. D., Naik, N., Mohiuddin, R., Bharati, V., & Bhat, R. A. (2021). Soil organic matter and its impact on soil properties and nutrient status. Microbiota and Biofertilizers, Vol 2: Ecofriendly Tools for Reclamation of Degraded Soil Environs (pp. 129–159). Cham: Springer International Publishing.

Basiru, S., & Hijri, M. (2024). Trade-off between soil organic carbon sequestration and plant nutrient uptake in arbuscular mycorrhizal symbiosis. *Fungal Biology Reviews*, 49, 100381.

Batjes, N.H. (1996). Total carbon and nitrogen in the soils of the world. *European Journal of Soil Science*, 47(2), 151–163.

Battle, M., Bender, M.L., Tans, P.P., White, J.W.C., Ellis, J.T., Conway, T., & Francey, R.J. (2000). Global carbon sinks and their variability inferred from atmospheric O2 and δ13C. *Science*, 287(5462), 2467–2470.

Benbi, D.K. (2013). Greenhouse gas emissions from agricultural soils: Sources and mitigation potential. *Journal of Crop Improvement*, 27(6), 752–772.

Berhane, M., Xu, M., Liang, Z., Shi, J., Wei, G., & Tian, X. (2020). Effects of long-term straw return on soil organic carbon storage and sequestration rate in North China upland crops: A meta-analysis. *Global Change Biology*, 26(4), 2686–2701.

Bharti, R.K., Srivastava, S., & Thakur, I.S. (2014). Proteomic analysis of carbon concentrating chemolithotrophic bacteria Serratia sp. for sequestration of carbon dioxide. *PLoS One*, 9, e91300.

Bhattacharyya, S.S., Ros, G.H., Furtak, K., Iqbal, H.M.N., & Parra-Saldívar, R. (2022). Soil carbon sequestration—An interplay between soil microbial community and soil organic matter dynamics. *Science of the Total Environment*, 815, 152928.

Bieluczyk, W., Asselta, F.O., Navroski, D., Gontijo, J.B., Venturini, A.M., Mendes, L.W., Simon, C.P., de Camargo, P.B., Tadini, A.M., Martin-Neto, L., et al. (2023). Linking above and belowground carbon sequestration, soil organic matter properties, and soil health in Brazilian Atlantic Forest restoration. *Journal of Environmental Management*, 344, 118573.

Blanco-Canqui, H., & Lal, R. (2004). Mechanisms of carbon sequestration in soil aggregates. *Critical Reviews in Plant Sciences*, 23(6), 481–504.

Blanco-Moure, N., Gracia, R., Bielsa, A.C., & López, M.V. (2016). Soil organic matter fractions as affected by tillage and soil texture under semiarid Mediterranean conditions. *Soil and Tillage Research*, 155, 381–389.

Boselli, R., Fiorini, A., Santelli, S., Ardenti, F., Capra, F., Maris, S.C., & Tabaglio, V. (2020). Cover crops during transition to no-till maintain yield and enhance soil fertility in intensive agro-ecosystems. *Field Crops Research*, 255, 107871.

Bossio, D.A., Cook-Patton, S.C., Ellis, P.W., Fargione, J., Sanderman, J., Smith, P., Wood, S., Zomer, R.J., Von Unger, M., Emmer, I.M., et al. (2020). The role of soil carbon in natural climate solutions. *Nature Sustainability*, 3(5), 391–398.

Bracho, R., Silveira, M.L., Boughton, R., Sanchez, J.M.D., Kohmann, M.M., Brandani, C.B., & Celis, G. (2021). Carbon dynamics and soil greenhouse fluxes in a Florida's native rangeland before and after fire. *Agricultural and Forest Meteorology*, 311, 108682.

Buragohain, P. (2019). Role of microbes on carbon sequestration. International Journal of Microbiology Research ISSN, 0975-5276.

Burke, D.J. (2015). Effects of annual and interannual environmental variability on soil fungi associated with an old-growth, temperate hardwood forest. *FEMS Microbiology Ecology*, 91(6), 185–188.

Burke, I.C., Yonker, C.M., Parton, W.J., Cole, C.V., Flach, K., & Schimel, D.S. (1989). Texture, climate, and cultivation effects on soil organic matter content in U.S. grassland soils. *Soil Science Society of America Journal*, 53(3), 800–805

Burks, J.L. (2013). *Eco physiology of Three Perennial Bioenergy Systems*. Purdue University.

Canarini, A., Kaiser, C., Merchant, A., Richter, A., Wanek, W., (2019). Root exudation of primary metabolites: Mechanisms and their roles in plant responses to environmental stimuli. *Frontiers in Plant Science*, 10, 157.

Change, I.P.O.C. (2007). Climate change 2007: The physical science basis. *Agenda*, 6(7), 333.

Chen, D., Wei, W., Daryanto, S., & Tarolli, P. (2020). Does terracing enhance soil organic carbon sequestration? A national-scale data analysis in China. *Science of the Total Environment*, 721, 137751.

Chen, J., Lærke, P.E., & Jørgensen, U. (2022). Land conversion from annual to perennial crops: A win-win strategy for biomass yield and soil organic carbon and total nitrogen sequestration. *Agriculture, Ecosystems & Environment*, 330, 107907.

Chen, W., Wolf, B., Zheng, X., Yao, Z., Butterbach-Bahl, K., Brueggemann, N., Liu, C., Han, S., & Han, X. (2011). Annual methane uptake by temperate semiarid steppes as regulated by stocking rates, aboveground plant biomass and topsoil air permeability. *Global Change Biology*, 17(9), 2803–2816.

Churchman, G. J., Noble, A., Bailey, G., Chittleborough, D., & Harper, R. (2014). Clay addition and redistribution to enhance carbon sequestration in soils. In *Soil Carbon* (pp. 327–335). Cham: Springer International Publishing.

Coleman, K., & Jenkinson, D. S. (1996). RothC-26.3-A model for the turnover of carbon in soil. In *Evaluation Of Soil Organic Matter Models: Using Existing Long-Term Datasets* (pp. 237–246). Berlin, Heidelberg: Springer Berlin Heidelberg.

Conant, R.T., Cerri, C.E.P., Osborne, B.B., & Paustian, K. (2017). Grassland management impacts on soil carbon stocks: A new synthesis. *Ecological Applications*, 27(2), 662–668.

Dalal, R. C., & Bridge, B. J. (2020). Aggregation and organic matter storage in sub-humid and semi-arid soils. In *Structure and Organic Matter Storage in Agricultural Soils* (pp. 263–307). CRC Press.

Das, S., Teuffer, K., Stoof, C.R., Walter, M.F., Walter, M.T., Steenhuis, T.S., & Richards, B.K. (2018). Perennial grass bioenergy cropping on wet marginal land: Impacts on soil properties, soil organic carbon, and biomass during initial establishment. *BioEnergy Research*, 11, 262–276.

Davidson, E.A., & Janssens, I.A. (2006). Temperature sensitivity of soil carbon decomposition and feedbacks to climate change. *Nature*, 440(7081), 165–173.

De Deyn, G.B., Cornelissen, J.H.C., & Bardgett, R.D. (2008). Plant functional traits and soil carbon sequestration in contrasting biomes. *Ecology Letters*, 11(5), 516–531.

De Vries, F.T., Bååth, E., Kuyper, T.W., & Bloem, J. (2009). High turnover of fungal hyphae in incubation experiments. *FEMS Microbiology Ecology*, 67, 389–396.

De Vries, F.T., Thébault, E., Liiri, M., Birkhofer, K., Tsiafouli, M.A., Bjørnlund, L., Jørgensen, H.B., Brady, M.V., Christensen, S., De Ruiter, P.C., D'Hertefeldt, T., Frouz, J., Hedlund, K., Hemerik, L., Hol, W.H.G., Hotes, S., Mortimer, S.R., Setälä, H., Sgardelis, S.P., Uteseny, K., Show, V., der Putten, W.H., Wolters, V., & Bardget, R.D. (2013). Soil food web properties explain ecosystem services across European land use systems. *Proceedings of the National Academy of Sciences of the United States of America*, 110, 14296–14301.

Deb, S., Bhadoria, P.B.S., Mandal, B., Rakshit, A., & Singh, H.B. (2015). Soil organic carbon: Towards better soil health, productivity and climate change mitigation. *Climate Change and Environmental Sustainability*, 3(1), 26–34.

Demenois, J., Torquebiau, E., Arnoult, M.H., Eglin, T., Masse, D., Assouma, M.H., Blanfort, V., Chenu, C., Chapuis-Lardy, L., Medoc, J.-M., et al. (2020). Barriers and strategies to boost soil carbon sequestration in agriculture. *Frontiers in Sustainable Food Systems*, 4, 37.

Denman, K.L., Brasseur, G., Chidthaisong, A., Ciais, P., Cox, P.M., Dickinson, R.E., Hauglustaine, D., Heinze, C., Holland, E., Jacob, D., Lohmann, U. & Ramachandran, S. (2007). *da Silva Dias {PL}*, ed. D. Qin, M. Manning, Z. Chen, M. Marquis, K.B. Averyt, M. Tignor, & H.L. Miller. Cambridge University Press.

Derner, J.D., & Schuman, G.E. (2007). Carbon sequestration and rangelands: A synthesis of land management and precipitation effects. *Journal of Soil and Water Conservation*, 62(2), 77–85.

Derpsch, R., Franzluebbers, A.J., Duiker, S.W., Reicosky, D.C., Koeller, K., Friedrich, T., Sturny, W.G., Sá, J.C.M., & Weiss, K. (2014). Why do we need to standardize no-tillage research? *Soil and Tillage Research*, 137, 16–22.

Diacono, M., & Montemurro, F. (2011). Long-term effects of organic amendments on soil fertility. *Sustainable Agriculture*, 2, 761–786.

Dignac, M.-F., Derrien, D., Barré, P., Barot, S., Cécillon, L., Chenu, C., Chevallier, T., Freschet, G.T., Garnier, P., Guenet, B., et al. (2017). Increasing soil carbon storage: Mechanisms, effects of agricultural practices and proxies. A review. *Agronomy for Sustainable Development*, 37, 1–27.

Dijkstra, F.A., Zhu, B., & Cheng, W. (2021). Root effects on soil organic carbon: A double-edged sword. *New Phytologist*, 230(1), 60–65.

Ding, G., Novak, J.M., Amarasiriwardena, D., Hunt, P.G., & Xing, B. (2002). Soil organic matter characteristics as affected by tillage management. *Soil Science Society of America Journal*, 66(2), 421–429.

Doetterl, S., Stevens, A., Six, J., Merckx, R., Oost, K.V., & Pinto, M.C. (2015). Soil carbon storage controlled by interactions between geochemistry and climate. *Nature Geoscience*, 8(10), 8–12.

Duval, M.E., Galantini, J.A., Capurro, J.E., & Martinez, J.M. (2016). Winter cover crops in soybean monoculture: Effects on soil organic carbon and its fractions. *Soil and Tillage Research*, 161, 95–105.

Emilia, Hannula S., & Morriën, Elly (2022). Will fungi solve the carbon dilemma?. *Geoderma*, 413, 115767.

Fan, J., Ding, W., Xiang, J., Qin, S., Zhang, J., & Ziadi, N. (2014). Carbon sequestration in an intensively cultivated sandy loam soil in the North China Plain as affected by compost and inorganic fertilizer application. *Geoderma*, 230, 22–28.

Feng, H., Kling, C.L., & Gassman, P.W. (2004). Carbon sequestration, co-benefits, and conservation programs. *Choices*, 19(3), 19–23.

Feng, S., Lakshmanan, P., Zhang, Y., Zhang, T., Liang, T., Zhang, W., Chen, X., & Wang, X. (2023). A comprehensive continental-scale analysis of carbon footprint of food production: Comparing continents around the world. *Journal of Cleaner Production*, 426, 138939.

Ferreras, L., Gomez, E., Toresani, S., Firpo, I., & Rotondo, R. (2006). Effect of organic amendments on some physical, chemical and biological properties in a horticultural soil. *Bioresource Technology*, 97(4), 635–640.

Field, C.B., Barros, V.R., Mastrandrea, M.D., Mach, K.J., Abdrabo, M.A.K., Adger, W.N., Anokhin, Y.A., Anisimov, O.A., Arent, D.J., Barnett, J., et al. (2015). Summary for policymakers. In *Climate Change 2014: Impacts, Adaptation, and Vulnerability. Part A: Global and Sectoral Aspects. Contribution of Working Group II to the Fifth Assessment Report of the Intergovernmental Panel on Climate Change* (pp. 1–32). Cambridge University Press.

Flannery, T. (2009). *Now or Never: Why We Must Act Now to End Climate Change and Create a Sustainable Future*. Open Road+ Grove/Atlantic.

Follett, R.F., & Schuman, G.E. (2005). Grazing land contributions to carbon sequestration. In *Grassland: A Global Resource* (pp. 265–277). Wageningen Academic.

Fontaine, S., Hénault, C., Aamor, A., Bdioui, N., Bloor, J.M.G., Maire, V., … & Maron, P.A. (2011). Fungi mediate long term sequestration of carbon and nitrogen in soil through their priming effect. *Soil Biology and Biochemistry*, 43(1), 86–96.

Forfora, N., Azuaje, I., Vivas, K.A., Vera, R.E., Brito, A., Venditti, R., Kelley, S., Tu, Q., Woodley, A., & Gonzalez, R. (2024). Evaluating biomass sustainability: Why below-ground carbon sequestration matters. *Journal of Cleaner Production*, 439, 140677.

Frank, S., Havlik, P., Soussana, J.-F., Levesque, A., Valin, H., Wollenberg, E., Kleinwechter, U., Fricko, O., Gusti, M., Herrero, M., et al. (2017). Reducing greenhouse gas emissions in agriculture without compromising food security? *Environmental Research Letters*, 12(10), 105004.

Frey, S.D., Six, J., & Elliott, E.T. (2003). Reciprocal transfer of carbon and nitrogen by decomposer fungi at the soilelitter interface. *Soil Biology and Biochemistry* 35, 1001–1004.

Fuss, C. B., Lovett, G. M., Goodale, C. L., Ollinger, S. V., Lang, A. K., & Ouimette, A. P. (2019). Retention of nitrate-N in mineral soil organic matter in different forest age classes. *Ecosystems*, 22(6), 1280–1294

Gai, X., Liu, H., Liu, J., Zhai, L., Yang, B., Wu, S., Ren, T., Lei, Q., & Wang, H. (2018). Long-term benefits of combining chemical fertilizer and manure applications on crop yields and soil carbon and nitrogen stocks in North China Plain. *Agricultural Water Management*, 208, 384–392.

Gao, H., Tian, H., Zhang, Z., & Xia, X. (2022). Warming-induced greenhouse gas fluxes from global croplands modified by agricultural practices: A meta-analysis. *Science of the Total Environment*, 820, 153288.

Gasmi, M., Kitouni, M., Carro, L., Pujic, P., Normand, P., & Boubakri, H. (2019). Chitinolytic actinobacteria isolated from an Algerian semi-arid soil: Development of an antifungal chitinase-dependent assay and GH18 chitinase gene identification. *Annals of Microbiology*, 69, 395–405.

Gillespie, L.M., Hättenschwiler, S., Milcu, A., Wambsganss, J., Shihan, A., & Fromin, N. (2021). Tree species mixing affects soil microbial functioning indirectly via root and litter traits and soil parameters in European forests. *Functional Ecology*, 35(10), 2190–2204.

Gougoulias, C., Clark, J.M., & Shaw, L.J. (2014). The role of soil microbes in the global carbon cycle: Tracking the below-ground microbial processing of plant-derived carbon for manipulating carbon dynamics in agricultural systems. *Journal of the Science of Food and Agriculture*, 94(12), 2362–2371.

Griggs, D.J., & Noguer, M. (2002). Climate change 2001: The scientific basis. Contribution of working group I to the third assessment report of the intergovernmental panel on climate change. *Weather*, 57(8), 267–269.

Gunina, A., & Kuzyakov, Y. (2022). From energy to (soil organic) matter. *Global Change Biology*, 28(7), 2169–2182.

Guo, J.H., Liu, X.J., Zhang, Y., Shen, J.L., Han, W.X., Zhang, W.F., Christie, P., Goulding, K.W.T., Vitousek, P.M., & Zhang, F.S. (2010). Significant acidification in major Chinese croplands. *Science*, 327(5968), 1008–1010.

Guo, L.B., & Gifford, R.M. (2002). Soil carbon stocks and land use change: A meta analysis. *Global Change Biology*, 8(4), 345–360.

Gurmu, G. (2019). Soil organic matter and its role in soil health and crop productivity improvement. *Forest Ecology and Management*, 7(7), 475–483.

Haile, S.G. (2007). Soil carbon sequestration and stabilization in tree-based pasture systems in Florida. 70(11).

Hannula, S. Emilia, Morriën, Elly, de Hollander, Mattias, van der Putten, Wim H., van Veen, Johannes A., & de Boer, Wietse. (2017). Shifts in rhizosphere fungal community during secondary succession following abandonment from agriculture. *ISME Journal*, 11(10), 2294–2304.

Hannula, S.E., Boschker, H.T.S., de Boer, W., & van Veen, J.A. (2012). 13C pulse-labeling assessment of the community structure of active fungi in the rhizosphere of a genetically starch-modified potato (Solanum tuberosum) cultivar and its parental isoline. *New Phytologist*, 194(3), 784–799.

Hannula, S.E., Kielak, A.M., Steinauer, K., Huberty, M., Jongen, R., De Long, J.R., Heinen, R., Bezemer, T.M. (2019). Time after time: Temporal variation in the effects of grass and forb species on soil bacterial and fungal communities. *mBio*, 10, 10-1128.

Hassan, M.U., Aamer, M., Mahmood, A., Awan, M.I., Barbanti, L., Seleiman, M.F., Bakhsh, G., Alkharabsheh, H.M., Babur, E., Shao, J., et al. (2022). Management strategies to mitigate N2O emissions in agriculture. *Life*, 12(3), 439.

Hawkins, H.J., Cargill, R.I.M., Van Nuland, M.E., Hagen, S.C., Field, K.J., Sheldrake, M., Soudzilovskaia, N.A., & Kiers, E.T. (2023). Mycorrhizal mycelium as a global carbon pool. *Current Biology*, 33, R560–R573.

Haywood, J., & Schulz, M. (2007). Causes of the reduction in uncertainty in the anthropogenic radiative forcing of climate between IPCC (2001) and IPCC (2007). *Geophysical Research Letters*, 34(20).

He, J., Li, H., Rasaily, R.G., Wang, Q., Cai, G., Su, Y., Qiao, X., & Liu, L. (2011). Soil properties and crop yields after 11 years of no tillage farming in wheat–maize cropping system in North China Plain. *Soil and Tillage Research*, 113(1), 48–54. https://doi.org/10.1016/j.still.2011.01.005

Heijden, V.D. (2008). The unseen majority: Soil microbes as drivers of plant diversity and productivity in terrestrial ecosystems. *Ecology Letters*, 11, 296–310.

Henderson, B.B., Gerber, P.J., Hilinski, T.E., Falcucci, A., Ojima, D.S., Salvatore, M., & Conant, R.T. (2015). Greenhouse gas mitigation potential of the world's grazing lands: Modeling soil carbon and nitrogen fluxes of mitigation practices. *Agriculture, Ecosystems \& Environment*, 207, 91–100.

Herzfeld, T., Heinke, J., Rolinski, S., & Müller, C. (2021). Soil organic carbon dynamics from agricultural management practices under climate change. *Earth System Dynamics*, 12(4), 1037–1055.

Hill, J.E. (2017). Quality and yield responses of seven warm season forage grasses to partial shading in a simulated silvopasture in East Texas. *Electronic Theses and Dissertations*. 125. https://scholarworks.sfasu.edu/etds/125

Hinge, G., Surampalli, R.Y., Goyal, M.K., Gupta, B.B., & Chang, X. (2021). Soil carbon and its associate resilience using big data analytics: For food Security and environmental management. *Technological Forecasting and Social Change*, 169, 120823.

Horner-Devine, M.C., Lage, M., Hughes, J.B., Bohannan, B.J. (2004). A taxa-area relationship for bacteria. *Nature*, 432(7018), 750.

Huang, X., Jia, Z., Guo, J., Li, T., Sun, D., Meng, H., Yu, G., He, X., Ran, W., Zhang, S., et al. (2019). Ten-year long-term organic fertilization enhances carbon sequestration and calcium-mediated stabilization of aggregate-associated organic carbon in a reclaimed Cambisol. *Geoderma*, 355, 113880.

Iqbal, A., & Shang, Z. (2019). Wetlands as a carbon sink: Insight into the Himalayan Region. In *Carbon Management for Promoting Local Livelihood in the Hindu Kush Himalayan (HKH) Region* (pp. 125–144). Cham: Springer International Publishing.

Jackson, N.A., Wallace, J.S., & Ong, C. K. (2000). Tree pruning as a means of controlling water use in an agroforestry system in Kenya. *Forest Ecology and Management*, 126(2), 133–148.

Jandl, R., Lindner, M., Vesterdal, L., Bauwens, B., Baritz, R., Hagedorn, F., Johnson, D.W., Minkkinen, K., & Byrne, K.A. (2007). How strongly can forest management influence soil carbon sequestration? *Geoderma*, 137(3–4), 253–268.

Jansson, C., Vogel, J., Hazen, S., Brutnell, T., & Mockler, T. (2018). Climate-smart crops with enhanced photosynthesis. *Journal of Experimental Botany*, 69(16), 3801–3809.

Janzen, H.H. (2011). What place for livestock on a re-greening earth? *Animal Feed Science and Technology*, 166, 783–796.

Jatav, H.S., Rajput, V.D., Minkina, T., Singh, S.K., Chejara, S., Gorovtsov, A., Barakhov, A., Bauer, T., Sushkova, S., Mandzhieva, S., et al. (2021). Sustainable approach and safe use of biochar and its possible consequences. *Sustainability*, 13(18), 10362.

Johnson, J.M.F., Allmaras, R.R., & Reicosky, D.C. 2006. Estimating source C from crop residues, roots and rhizodeposits using the national grain-yield database. *Agronomy Journal*. 98, 622–636

Johnston, A.E., Poulton, P.R., & Coleman, K. (2009). Soil organic matter: Its importance in sustainable agriculture and carbon dioxide fluxes. *Advances in Agronomy*, 101, 1–57.

Joseph, S., Cowie, A.L., Van Zwieten, L., Bolan, N., Budai, A., Buss, W., Cayuela, M.L., Graber, E.R., Ippolito, J.A., Kuzyakov, Y., et al. (2021). How biochar works, and when it doesn't: A review of mechanisms controlling soil and plant responses to biochar. *Gcb Bioenergy*, 13(11), 1731–1764.

Kallenbach, C.M., Frey, S.D., & Grandy, A.S. (2016). Direct evidence for microbial-derived soil organic matter formation and its ecophysiological controls. *Nature Communications*, 7(1), 13630.

Kay, S., Rega, C., Moreno, G., den Herder, M., Palma, J.H.N., Borek, R., Crous-Duran, J., Freese, D., Giannitsopoulos, M., Graves, A., et al. (2019). Agroforestry creates carbon sinks whilst enhancing the environment in agricultural landscapes in Europe. *Land Use Policy*, 83, 581–593.

Kell, D.B. (2012). Large-scale sequestration of atmospheric carbon via plant roots in natural and agricultural ecosystems: Why and how. *Philosophical Transactions of the Royal Society B: Biological Sciences*, 367(1595), 1589–1597.

Kindler, R., Miltner, A., Richnow, H., & Kastner, M. (2006). Fate of gram-negative bacterial biomass in soil—mineralization and contribution to SOM. *Soil Biology and Biochemistry*, 38, 2860–2870.

King, G.M. (2011). Enhancing soil carbon storage for carbon remediation: Potential contributions and constraints by microbes. *Trends in Microbiology*, 19, 75–84.

Kirby, K.R., & Potvin, C. (2007). Variation in carbon storage among tree species: Implications for the management of a small-scale carbon sink project. *Forest Ecology and Management*, 246(2–3), 208–221.

Klotzbücher, Thimo, Kaiser, Klaus, Guggenberger, Georg, Gatzek, Christiane, & Kalbitz, Karsten (2011). A new conceptual model for the fate of lignin in decomposing plant litter. *Ecology*, 92(5), 1052–1062.

Knicker, H. (2007). How does fire affect the nature and stability of soil organic nitrogen and carbon? A review. *Biogeochemistry*, 85(1), 91–118.

Koerner, S.E., & Collins, S.L. (2014). Interactive effects of grazing, drought, and fire on grassland plant communities in North America and South Africa. *Ecology*, 95(1), 98–109.

Kögel-Knabner, I. (2002). The macromolecular organic composition of plant and microbial residues as inputs to soil organic matter. *Soil Biology and Biochemistry*, 34(2), 139–162.

Körschens, M. (2021). Long-Term Field Experiments (LTEs)—Importance, overview, soil organic matter. In *Exploring and Optimizing Agricultural Landscapes* (pp. 215–231). Cham: Springer International Publishing

Körschens, M., Albert, E., Baumecker, M., Ellmer, F., Grunert, M., Hoffmann, S., Kismanyoky, T., Kubat, J., Kunzova, E., Marx, M., et al. (2014). Humus und Klimaänderung-Ergebnisse aus 15 langjährigen Dauerfeldversuchen. *Archives of Agronomy and Soil Science*, 60(11), 1485–1517.

Kumar, S.S., Mahale, A.G., & Patil, A.C. (2020). Mitigation of climate change through approached agriculture-soil carbon sequestration (a review). *Current Journal of Applied Science and Technology*, 39(33), 47–64.

Kumara, K., Pal, S., Chand, P., & Kandpal, A. (2023). Carbon sequestration potential of agroforestry systems in Indian agricultural landscape: A meta-analysis. *Ecosystem Services*, 62, 101537.

Lal, R. (2001). Potential of desertification control to sequester carbon and mitigate the greenhouse effect. *Climatic Change*, 51(1), 35–72.

Lal, R. (2003). Soil erosion and the global carbon budget. *Environment International*, 29(4), 437–450.

Lal, R. (2004). Soil carbon sequestration impacts on global climate change and food security. *Science*, 304(5677), 1623–1627.

Lal, R. (2008). Carbon sequestration. *Philosophical Transactions of the Royal Society B: Biological Sciences*, 363(1492), 815–830.

Lal, R. (2014). Societal value of soil carbon. *Journal of Soil and Water Conservation*, 69(6), 186A–192A.

Lal, R. (2016). Soil health and carbon management. *Food and Energy Security*, 5(4), 212–222.

Lal, R. (2020). Managing soils for negative feedback to climate change and positive impact on food and nutritional security. *Soil Science and Plant Nutrition*, 66(1), 1–9.

Lal, R., Follett, R.F., Stewart, B.A., & Kimble, J.M. (2007). Soil carbon sequestration to mitigate climate change and advance food security. *Soil Science*, 172(12), 943–956.

Lal, R., Negassa, W., & Lorenz, K. (2015). Carbon sequestration in soil. *Current Opinion in Environmental Sustainability*, 15, 79–86.

Lange, M., Eisenhauer, N., Sierra, C.A., Bessler, H., Engels, C., & Griffiths, R.I. (2015). Plant diversity increases soil bacterial activity and soil carbon storage. *Nature Communications*, 6, 6707.

Lavelle, P., Decaëns, T., Aubert, M., Barot, S., Blouin, M., Bureau, F., Margerie, P., Mora, P., & Rossi, J.-P. (2006). Soil invertebrates and ecosystem services. *European Journal of Soil Biology*. 42, S3–S15.

Lee, E.K., Zhang, X., Adler, P.R., Kleppel, G.S., & Romeiko, X.X. (2020). Spatially and temporally explicit life cycle global warming, eutrophication, and acidification impacts from corn production in the US Midwest. *Journal of Cleaner Production*, 242, 118465.

Lewis, J. (2016). *Mycorrhizal Fungi, Evolution and Diversification of Encyclopedia of Evolutionary Biology*. Academic Press.

Li, C., Xiong, Y., Huang, Q., Xu, X., & Huang, G. (2020). Impact of irrigation and fertilization regimes on greenhouse gas emissions from soil of mulching cultivated maize (*Zea mays* L.) field in the upper reaches of Yellow River, China. *Journal of Cleaner Production*, 259. https://doi.org/10.1016/j.jclepro.2020.120873

Liang, C., Amelung, W., Lehmann, J., & Kastner, M. (2019). Quantitative assessment of microbial necromass contribution to soil organic matter. *Global Change Biology*, 25, 3578–3590.

Liang, C., Schimel, J.P., & Jastrow, J.D. (2017). The importance of anabolism in microbial control over soil carbon storage. *Nature Microbiology*, 2(8), 1–6.

Liang, M., Chen, J., Gornish, E.S., Bai, X., Li, Z., & Liang, C. (2018). Grazing effect on grasslands escalated by abnormal precipitations in inner Mongolia. *Ecology and Evolution*, 8(16), 8187–8196.

Liang, S., Sun, N., Meersmans, J., Longdoz, B., Colinet, G., Xu, M., & Wu, L. (2024). Impacts of climate change on crop production and soil carbon stock in a continuous wheat cropping system in southeast England. *Agriculture, Ecosystems & Environment*, 365, 108909.

Liang, Y., Al-Kaisi, M., Yuan, J., Liu, J., Zhang, H., Wang, L., Cai, H., & Ren, J. (2021). Effect of chemical fertilizer and straw-derived organic amendments on continuous maize yield, soil carbon sequestration and soil quality in a Chinese Mollisol. *Agriculture, Ecosystems \& Environment*, 314, 107403.

Ling, N., Chen, D., Guo, H., Wei, J., Bai, Y., & Shen, Q. (2017). Differential responses of soil bacterial community to long-term n and p inputs in a semi-arid steppe. *Geoderma*, 292,25–33.

Lomax, G., Lenton, T.M., Adeosun, A., & Workman, M. (2015). Investing in negative emissions. *Nature Climate Change*, 5(6), 498–500.

Lord, R., & Sakrabani, R. (2019). Ten-year legacy of organic carbon in non-agricultural (brownfield) soils restored using green waste compost exceeds 4 per mille per annum: Benefits and trade-offs of a circular economy approach. *Science of the Total Environment*, 686, 1057–1068.

Lorenz, K., & Lal, R. (2014). Soil organic carbon sequestration in agroforestry systems. A review. *Agronomy for Sustainable Development*, 34, 443–454.

Lorenz, K., Lal, R., Lorenz, K., & Lal, R. (2018). Carbon sequestration in grassland soils. In *Carbon Sequestration in Agricultural Ecosystems* (pp. 175–209). Cham: Springer International Publishing.

Luo, Z., Wang, E., & Sun, O.J. (2010). Can no-tillage stimulate carbon sequestration in agricultural soils? A meta-analysis of paired experiments. *Agriculture, Ecosystems & Environment*, 139(1–2), 224–231.

Ma, Z., Shrestha, B.M., Bork, E.W., Chang, S.X., Carlyle, C.N., Döbert, T.F., Sobrinho, L.S., & Boyce, M.S. (2021). Soil greenhouse gas emissions and grazing management in northern temperate grasslands. *Science of the Total Environment*, 796, 148975.

Macas, F., & Camps Arbestain, M. (2010). Soil carbon sequestration in a changing global environment. *Mitigation and Adaptation Strategies for Global Change*, 15, 511–529.

Malyan, S.K. et al. (2019). Role of fungi in climate change abatement through carbon sequestration. In A. Yadav, S. Singh, S. Mishra, & A. Gupta (eds.). *Recent Advancement in White Biotechnology Through Fungi*. Fungal Biology. Springer.

Mason, A.R.G., Salomon M.J., Lowe, A.J., & Cavagnaro, T.R. (2023). Microbial solutions to soil carbon sequestration. *Journal of Cleaner Production*, 417, 137993.

Mathers, C., Black, C.K., Segal, B.D., Gurung, R.B., Zhang, Y., Easter, M.J., Williams, S., Motew, M., Campbell, E.E., Brummitt, C.D., et al. (2023). Validating DayCent-CR for cropland soil carbon offset reporting at a national scale. *Geoderma*, 438, 116647.

Meena, R.S., Kumar, S., & Yadav, G.S. (2020). Soil carbon sequestration in crop production. In *Nutrient Dynamics for Sustainable Crop Production* (pp. 1–39). Singapore: Springer.

Mehar, S.K., & Sundaramoorthy, S. (2018). Carbon sequestration and the significance of soil fungi in the process. In P. Gehlot, & J. Singh (eds.). *Fungi and their Role in Sustainable Development: Current Perspectives*. Springer.

Meinshausen, M., Meinshausen, N., Hare, W., Raper, S.C.B., Frieler, K., Knutti, R., Frame, D.J., & Allen, M.R. (2009). Greenhouse-gas emission targets for limiting global warming to 2 C. *Nature*, 458(7242), 1158–1162.

Miajee, M.J.H. (2020). Estimation of soil organic carbon stock by fixed depth and equivalent soil mass methods in fine and coarse textured soils with cereal and grass based cropping history. University of Helsinki.

Mills, J., Ingram, J., Dibari, C., Merante, P., Karaczun, Z., Molnar, A., Sánchez, B., Iglesias, A., & Ghaley, B.B. (2020). Barriers to and opportunities for the uptake of soil carbon management practices in European sustainable agricultural production. *Agroecology and Sustainable Food Systems*, 44(9), 1185–1211.

Morriën, Elly, Hannula, S. Emilia, Snoek, L. Basten, Helmsing, Nico R., Zweers, Hans, de Hollander, Mattias, Soto, Raquel Luján, Bouffaud, Marie-Lara, Buée, Marc, Dimmers, Wim, Duyts, Henk, Geisen, Stefan, Girlanda, Mariangela, Griffiths, Rob I., Jørgensen, Helene-Bracht, Jensen, John, Plassart, Pierre, Redecker, Dirk, Schmelz, Rűdiger M., Schmidt, Olaf, Thomson, Bruce C., Tisserant, Emilie, Uroz, Stephane, Winding, Anne, Bailey, Mark J., Bonkowski, Michael, Faber, Jack H., Martin, Francis, Lemanceau, Philippe, de Boer, Wietse, van Veen, Johannes A., & van der Putten, Wim H. (2017). Soil networks become more connected and take up more carbon as nature restoration progresses. *Nature Communications*, 8(1), 14349.

Mpeketula, P.M.G., & Snapp, S.S. (2019). Structural stability conditions soil carbon gains from compost management and rotational diversity. *Soil Science Society of America Journal*, 83(1), 203–211.

Mugerwa, T. M., & McGee, P. A. (2016a). Potential effect of melanised endophytic fungi on levels of organic carbon within an Alfisol. *Soil Research*, *55*(3), 245–252.

Mugerwa, T.M., Mcgee, P.A. (2016b). Potential effect of melanised endophytic fungi on levels of organic carbon within an Alfisol. *Soil Research*, 55(3), 245–252.

Nair, P.K.R., Nair, V.D., Kumar, B.M., & Haile, S.G. (2009). Soil carbon sequestration in tropical agroforestry systems: A feasibility appraisal. *Environmental Science & Policy*, 12(8), 1099–1111.

Nathan, V.K., & Ammini, P. (2019). Carbon dioxide sequestering ability of bacterial carbonic anhydrase in a mangrove soil microcosm and its bio-mineralization properties. *Water Air & Soil Pollution*, 230, 192.

Nathanael, A. J., Kannaiyan, K., Kunhiraman, A. K., Ramakrishna, S., & Kumaravel, V. (2021). Global opportunities and challenges on net-zero CO 2 emissions towards a sustainable future. *Reaction Chemistry & Engineering*, 6(12), 2226–2247.

Nave, L.E., Vance, E.D., Swanston, C.W., & Curtis, P.S. (2010). Harvest impacts on soil carbon storage in temperate forests. *Forest Ecology and Management*, 259(5), 857–866.

Nayak, A.K., Rahman, M.M., Naidu, R., Dhal, B., Swain, C.K., Nayak, A.D., Tripathi, R., Shahid, M., Islam, M.R., & Pathak, H. (2019). Current and emerging methodologies for estimating carbon sequestration in agricultural soils: A review. *Science of the Total Environment*, 665, 890–912.

Nganga, S.K., Jalang'o, D.A., & Girvetz, E.H. (2019). Soil carbon enhancing practices: A systematic review of barriers and enablers of adoption. *SN Applied Sciences*, 1(12), 1726.

Ngatia, L.W., Moriasi, D., Grace III, J.M., Fu, R., Gardner, C.S., & Taylor, R.W. (2021). Land use change affects soil organic carbon: An indicator of soil health. Environmental Health. *Intech Open*, 2021 (January 2021), 1–15.

Nguyen, C. (2009). Rhizodeposition of organic c by plant: Mechanisms and controls. In E. Lichtfouse, M. Navarrete, P. Debaeke, S. Véronique, & Alberola, C. (eds.). *Sustainable Agriculture* (pp. 97–123). Springer.

Ozlu, E., Arriaga, F.J., Bilen, S., Gozukara, G., & Babur, E. (2022). Carbon footprint management by agricultural practices. *Biology*, 11(10), 1453.

Page, K.L., Dang, Y.P., & Dalal, R.C. (2020). The ability of conservation agriculture to conserve soil organic carbon and the subsequent impact on soil physical, chemical, and biological properties and yield. *Frontiers in Sustainable Food Systems*, 4, 31.

Palm, C., Tomich, T., Van Noordwijk, M., Vosti, S., Gockowski, J., Alegre, J., & Verchot, L. (2004). Mitigating GHG emissions in the humid tropics: Case studies from the Alternatives to Slash-and-Burn Program (ASB). *Environment, Development and Sustainability*, 6, 145–162.

Panchal, P., Preece, C., Peñuelas, J., & Giri, J. (2022). Soil carbon sequestration by root exudates. *Trends in Plant Science*, 27(8), 749–757.

Paramisparam, P., Ahmed, O.H., Omar, L., Ch'ng, H.Y., Johan, P.D., & Hamidi, N.H. (2021). Co-application of charcoal and wood ash to improve potassium availability in tropical mineral acid soils. *Agronomy*, 11(10), 2081.

Paudel, S., Gomez-Casanovas, N., Boughton, E.H., Chamberlain, S.D., Wagle, P., Peterson, B.L., Bajgain, R., Starks, P.J., Basara, J., Bernacchi, C.J., et al. (2023). Intensification differentially affects the delivery of multiple ecosystem services in subtropical and temperate grasslands. *Agriculture, Ecosystems & Environment*, 348, 108398.

Paul, E. (2015). *Soil Microbiology, Ecology and Biochemistry*. Elsevier.

Paul, V.G., Wronkiewicz, D.J., & Mormile, M.R. (2017). Impact of elevated CO_2 concentrations on carbonate mineral precipitation ability of sulfate-reducing bacteria and implications for CO_2 sequestration. *Applied Geochemistry*, 78, 250–271.

Paustian, K., Andren, O., Janzen, H.H., Lal, R., Smith, P., Tian, G., Tiessen, H., Van Noordwijk, M., & Woomer, P.L. (1997). Agricultural soils as a sink to mitigate CO2 emissions. *Soil Use and Management*, 13, 230–244.

Paustian, K., Lehmann, J., Ogle, S., Reay, D., Robertson, G.P., & Smith, P. (2016). Climate-smart soils. *Nature*, 532(7597), 49–57.

Pellegrini, M., Djebaili, R., Pagnani, G., & del Gallo, M. (2022). Role of actinomycetes in mitigating the impact of climate change. Mechanisms of action and perspectives. In R.Z. Sayyed, & V.G. Uarrota (eds.). *Secondary Metabolites and Volatiles of PGPR in Plant-Growth Promotion*. Springer. https://doi.org/10.1007/978-3-031-07559-9_9

Poeplau, C., Kätterer, T., Bolinder, M.A., Börjesson, G., Berti, A., & Lugato, E. (2015). Low stabilization of aboveground crop residue carbon in sandy soils of Swedish long-term experiments. *Geoderma*, 237, 246–255.

Post, W.M., & Kwon, K.C. (2000). Soil carbon sequestration and land-use change: Processes and potential. *Global Change Biology*, 6(3), 317–327.

Qiu, J., Li, C., Wang, L., Tang, H., Li, H., & Van Ranst, E. (2009). Modeling impacts of carbon sequestration on net greenhouse gas emissions from agricultural soils in China. *Global Biogeochemical Cycles*, 23(1), GB1007.

Rahman, G. K. M. M., Rahman, M. M., Alam, M. S., Kamal, M. Z., Mashuk, H. A., Datta, R., & Meena, R. S. (2019). Biochar and organic amendments for sustainable soil carbon and soil health. *Carbon and Nitrogen Cycling in Soil* (pp. 45–85). Singapore: Springer Singapore.

Ramachandran Nair, P.K., Mohan Kumar, B., & Nair, V.D. (2009). Agroforestry as a strategy for carbon sequestration. *Journal of Plant Nutrition and Soil Science*, 172(1), 10–23.

Ramnarine, R., Voroney, R. P., Dunfield, K. E., & Wagner-Riddle, C. (2018). Characterization of the heavy, hydrolysable and non-hydrolysable fractions of soil organic carbon in conventional and no-tillage soils. *Soil and Tillage Research*, 181, 144–151.

Ravindranath, N.H., & Ostwald, M. (2008). Generic methods for inventory of carbon pools. In *Carbon Inventory Methods Handbook for Greenhouse Gas Inventory, Carbon Mitigation and Roundwood Production Projects* (pp. 99–111). Dordrecht: Springer Netherlands.

Ren, F., Misselbrook, T.H., Sun, N., Zhang, X., Zhang, S., Jiao, J., Xu, M., & Wu, L. (2021). Spatial changes and driving variables of topsoil organic carbon stocks in Chinese croplands under different fertilization strategies. *Science of the Total Environment*, 767, 144350.

Rossel, R.A.V., Behrens, T., Ben-Dor, E., Brown, D.J., Demattê, J.A.M., Shepherd, K.D., Shi, Z., Stenberg, B., Stevens, A., Adamchuk, V., et al. (2016). A global spectral library to characterize the world's soil. *Earth-Science Reviews*, 155, 198–230.

Rumpel, C., & Chabbi, A. (2021). Managing soil organic carbon for mitigating climate change and increasing food security. *Agronomy*, 11(8), 1553.

Rumpel, C., Amiraslani, F., Bossio, D., Chenu, C., Cardenas, M. G., Henry, B., Espinoza, A. F., Koutika, L.-S., Ladha, J., Madari, B. E., & et al. (2023). Studies from global regions indicate promising avenues for maintaining and increasing soil organic carbon stocks: The Scientific and Technical Committee of the 4 per 1000 initiative. *Regional Environmental Change*, 23(1), 8.

Saco, P.M., McDonough, K.R., Rodriguez, J.F., Rivera-Zayas, J., & Sandi, S.G. (2021). The role of soils in the regulation of hazards and extreme events. *Philosophical Transactions of the Royal Society B*, 376(1834), 20200178.

Sanderman, J., Hengl, T., & Fiske, G.J. (2017). Soil carbon debt of 12,000 years of human land use. *Proceedings of the National Academy of Sciences*, 114(36), 9575–9580.

Schmidt, M.W.I., Torn, M.S., Abiven, S., Dittmar, T., Guggenberger, G., Janssens, I.A., Kleber, M., Kögel-Knabner, I., Lehmann, J., Manning, D.A.C., et al. (2011). Persistence of soil organic matter as an ecosystem property. *Nature*, 478(7367), 49–56.

Schuman, G.E., Janzen, H.H., & Herrick, J.E. (2002). Soil carbon dynamics and potential carbon sequestration by rangelands. *Environmental Pollution*, 116(3), 391–396.

Scotti, R., Bonanomi, G., Scelza, R., Zoina, A., & Rao, M.A. (2015). Organic amendments as sustainable tool to recovery fertility in intensive agricultural systems. *Journal of Soil Science and Plant Nutrition*, 15(2), 333–352.

Sekaran, U., & Kumar, S. (2021). Responses of soil carbon storage, compaction, and biological properties under no-till and conventional-till systems. *Conservation Agriculture: A Sustainable Approach for Soil Health and Food Security: Conservation Agriculture for Sustainable Agriculture*, (pp. 359–378). Singapore: Springer Singapore.

Shahid, M.K., Kashif, A., Rout, P.R., & Choi, Y. (2022). An overview of soil bacteria for CO_2 sequestration. In S.K. Nayak, B. Baliyarsingh, I. Mannazzu, A. Singh, & B.B. Mishra (eds.). *Advances in Agricultural and Industrial Microbiology*. Springer.

Sharma, C., Pathak, P., & Gautam, S. (2024). Transforming agri-crop residue biomass for value addition: An innovative strategy toward resilient circular economy. In *Handbook of Biomass* (pp. 1517–1542). Singapore: Springer Nature Singapore.

Sharma, P., Abrol, V., Sharma, K.R., Sharma, N., Phogat, V.K., & Vikas, V. (2016). Impact of conservation tillage on soil organic carbon and physical properties – A review. *International Journal of Bio-Resource and Stress Management*, 7(1), 151–161.

Sharrow, S.H., & Ismail, S. (2004). Carbon and nitrogen storage in agroforests, tree plantations, and pastures in western Oregon, USA. *Agroforestry Systems*, 60, 123–130.

Shrestha, B.M., Singh, B.R., Forte, C., & Certini, G. (2015). Long-term effects of tillage, nutrient application and crop rotation on soil organic matter quality assessed by NMR spectroscopy. *Soil Use and Management*, 31(3), 358–366.

Shrestha, G., & Stahl, P.D. (2008). Carbon accumulation and storage in semi-arid sagebrush steppe: Effects of long-term grazing exclusion. *Agriculture, Ecosystems & Environment*, 125(1–4), 173–181.

Sileshi, G.W., Dagar, J.C., Nath, A.J., & Kuntashula, E. (2023). Agroforestry as a climate-smart agriculture: Strategic interventions, current practices and policies. In *Agroforestry for Sustainable Intensification of Agriculture in Asia and Africa* (pp. 589–640). Singapore: Springer Nature Singapore.

Singh, B.K., Bardgett, R.D., Smith, P., & Reay, D.S. (2010). Microorganisms and climate change: Terrestrial feedbacks and mitigation options. *Nature Reviews Microbiology*, 8, 779–790.

Smith, P. (2016). Soil carbon sequestration and biochar as negative emission technologies. *Global Change Biology*, 22(3), 1315–1324.

Smith, P., Smith, J.U., Powlson, D.S., McGill, W.B., Arah, J.R.M., Chertov, O.G., Coleman, K., Franko, U., Frolking, S., Jenkinson, D.S., et al. (1997). A comparison of the performance of nine soil organic matter models using datasets from seven long-term experiments. *Geoderma*, 81(1–2), 153–225.

Smith, P., Soussana, J.-F., Angers, D., Schipper, L., Chenu, C., Rasse, D.P., Batjes, N.H., Van Egmond, F., McNeill, S., Kuhnert, M., et al. (2020). How to measure, report and verify soil carbon change to realize the potential of soil carbon sequestration for atmospheric greenhouse gas removal. *Global Change Biology*, 26(1), 219–241.

Sollenberger, L.E., Kohmann, M.M., Dubeux Jr, J.C.B., & Silveira, M.L. (2019). Grassland management affects delivery of regulating and supporting ecosystem services. *Crop Science*, 59(2), 441–459.

Sparling, G.P., Wheeler, D., Vesely, E.-T., & Schipper, L.A. (2006). What is soil organic matter worth? *Journal of Environmental Quality*, 35(2), 548–557.

Stephens, B.B., Gurney, K.R., Tans, P.P., Sweeney, C., Peters, W., Bruhwiler, L., Ciais, P., Ramonet, M., Bousquet, P., Nakazawa, T., et al. (2007). Weak northern and strong tropical land carbon uptake from vertical profiles of atmospheric CO2. *Science*, 316(5832), 1732–1735.

Stocker, T. F., Qin, D., Plattner, G.-K., Tignor, M. M. B., Allen, S. K., Boschung, J., Nauels, A., Xia, Y., Bex, V., & Midgley, P. M. (2013). Summary for policymakers. *Climate change 2013: The physical science basis. Contribution of working group I to the fifth assessment report of the Intergovernmental Panel on Climate Change*, 3, 159–204.

Stockmann, U., Adams, M., Crawford, J.W., Field, D.J., Henakaarchchi, N., Jenkins, M., Minasny, B., McBratney, A.B., de Courcelles, V.D., Singh, K., Wheeler, I., Abbott, L., Angers, D.A., Baldock, J., Bird, M., Brookes, P.C., Chenu, C., Jastrowh, J.D., Lal, R., … Zimmermann, M. (2013). The knowns, known unknowns and unknowns of sequestration of soil organic carbon. *Agriculture, Ecosystems and Environment*, 164, 80–99.

Strap, J.L. (2011). Actinobacteria–plant interactions: A boon to agriculture. In *Bacteria in Agrobiology: Plant Growth Responses* (pp. 285–307). Berlin, Heidelberg: Springer Berlin Heidelberg.

Tang, K.-H., Tang, Y., & Blankenship, R. (2011). Carbon metabolic pathways in phototrophic bacteria and their broader evolutionary implications. *Frontiers in Microbiology*, 2, 165.

Teague, R., & Kreuter, U. (2020). Managing grazing to restore soil health, ecosystem function, and ecosystem services. *Frontiers in Sustainable Food Systems*, 4, 534187.

Thakuri, S., Baskota, P., Khatri, S.B., Dhakal, A., Chaudhary, P., Rijal, K., & Byanju, R.M. (2020). Methane emission factors and carbon fluxes from enteric fermentation in cattle of Nepal Himalaya. *Science of the Total Environment*, 746, 141184.

Thompson, N., Hughes, M., Nuworsu, E., Reeling, C., Armstrong, J., Langemeier, M., DeLay, N., & Foster, K. (2021). *Opportunities and Challenges Associated with "carbon farming" for US Row-crop Producers*. Purdue University Center for Commercial Agriculture.

Tongwane, M.I., & Moeletsi, M.E. (2021). Provincial cattle carbon emissions from enteric fermentation and manure management in South Africa. *Environmental Research*, 195, 110833.

Torsvik, V., & Øvreås, L. (2002). Bacterial diversity and function in soil: From genes to ecosystems. *Current Opinion in Microbiology*, 5(3), 240.

Treseder, K.K., & Allen, M.F. (2000). Mycorrhizal fungi have a potential role in soil carbon storage under elevated CO2 and nitrogen deposition. *New Phytologist*, 147(1), 189–200.

Trumbore, S.E., & Torn, M.S. (2003). Soils and the global carbon cycle. Notes from the Underground. Publication of a NATO/ASI Workshop. E.A. Holland (Ed.), Soils and Global Change, NATO Advanced Study Institute.

Vishwakarma, K., Sharma, S., Kumar, N., Upadhyay, N., Devi, S., & Tiwari, A. (2016). Contribution of microbial inoculants to soil carbon sequestration and sustainable agriculture. In *Microbial Inoculants in Sustainable Agricultural Productivity. Vol. 2: Functional Applications* (pp. 101–113). New Delhi: Springer India.

Von Cossel, M., Wagner, M., Lask, J., Magenau, E., Bauerle, A., Von Cossel, V., Warrach-Sagi, K., Elbersen, B., Staritsky, I., Van Eupen, M., et al. (2019). Prospects of bioenergy cropping systems for a more social-ecologically sound bioeconomy. *Agronomy*, 9(10), 605.

Wagg, C., Bender, S.F., Widmer, F., & van der Heijden, M.G.A. (2014). Soil biodiversity and soil community composition determine ecosystem multifunctionality. *Proceedings of the National Academy of Sciences of the United States of America*, 111(14), 5266–5270.

Wang, C., Amon, B., Schulz, K., & Mehdi, B. (2021a). Factors that influence nitrous oxide emissions from agricultural soils as well as their representation in simulation models: A review. *Agronomy*, 11(4), 770.

Wang, C., Sun, Y., & Chen, H.Y.H. et al. (2021b). Effects of elevated CO2 on the C:N stoichiometry of plants, soils, and microorganisms in terrestrial ecosystems. *Catena*, 201, 105219.

Wang, J., Wang, K., Wang, X., Ai, Y., Zhang, Y., & Yu, J. (2018). Carbon sequestration and yields with long-term use of inorganic fertilizers and organic manure in a six-crop rotation system. *Nutrient Cycling in Agroecosystems*, 111, 87–98.

Wang, J., Zhao, X., Ouyang, X., Zhao, L., Wang, W., Zuo, C., Zhang, Z., Zhou, H., Watson, A., & Li, Y. (2022a). The role of herbivores in the grassland carbon budget for Three-Rivers Headwaters region, Qinghai--Tibetan Plateau, China. *Grassland Research*, 1(3), 207–219.

Wang, Q., Wang, Y., Wang, Q., & Liu, J. (2014). Impacts of 9 years of a new conservational agricultural management on soil organic carbon fractions. *Soil and Tillage Research*, 143, 1–6.

Wang, S., & Huang, Y. (2020). Determinants of soil organic carbon sequestration and its contribution to ecosystem carbon sinks of planted forests. *Global Change Biology*, 26(5), 3163–3173.

Wang, Y., Tao, F., Yin, L., & Chen, Y. (2022). Spatiotemporal changes in greenhouse gas emissions and soil organic carbon sequestration for major cropping systems across China and their drivers over the past two decades. *Science of the Total Environment*, 833, 155087.

Wardle, D.A., Bardgett, R.D., Klironomos, J.N., Setala, H., Van Der Putten, W.H., & Wall, D.H., (2004a). Ecological linkages between aboveground and belowground biota. *Science*, 304, 1629–1633.

Wardle, D.A., Yeates, G.W., Williamson, W.M., Bonner, K.I., & Barker, G.M. (2004b). Linking aboveground and belowground community: The indirect influence of aphid species identity and diversity on a three trophic level soil food web. *Oikos*, 107(2), 283–294.

Whalen, E.D., Grandy, A.S., Sokol, N.W., Keiluweit, M., Ernakovich, J., Smith, R.G., & Frey, S.D. (2022). Clarifying the evidence for microbial- and plant-derived soil organic matter, and the path toward a more quantitative understanding. *Global Change Biology*, 28, 7167–7185.

Wiesmeier, M., Hübner, R., Spörlein, P., Geuß, U., Hangen, E., Reischl, A., Schilling, B., von Lützow, M., & Kögel-Knabner, I. (2014). Carbon sequestration potential of soils in southeast Germany derived from stable soil organic carbon saturation. *Global Change Biology*, 20(2), 653–665.

Xin, Z., Qin, Y., & Yu, X. (2016). Spatial variability in soil organic carbon and its influencing factors in a hilly watershed of the Loess Plateau, China. *Catena*, 137, 660–669.

Xue, T., Zhang, L., Yang, F., Cheng, C., Wang, J., Sang, Q., Yang, S., Zhao, X., & Zhang, L. (2024). Carbon sink and soil organic carbon sequestration mechanisms in vineyards. *Journal of Cleaner Production*, 469, 143217.

Yang, Y., Dou, Y., & An, S. (2018). Testing association between soil bacterial diversity and soil carbon storage on the Loess Plateau. *Science of the Total Environment*, 626, 48–58.

Zhan, T., Zhang, Z., Sun, J., Liu, M., Zhang, X., Peng, F., Tsunekawa, A., Zhou, H., Gou, X., & Fu, S. (2020). Meta-analysis demonstrating that moderate grazing can improve the soil quality across China's grassland ecosystems. *Applied Soil Ecology*, 147, 103438.

Zhang, K., Maltais-Landry, G., & Liao, H.-L. (2021a). How soil biota regulate C cycling and soil C pools in diversified crop rotations. *Soil Biology and Biochemistry*, 156, 108219.

Zhang, Y., Kalkhajeh, Y.K., Hu, W., Tian, K., Yu, D., Huang, B., et al. (2024). Effective indicators and drivers of soil organic matter in intensive orchard production systems. *Soil and Tillage Research*, 238, 105999.

Zhang, Y., Wang, J., & Feng, Y. (2021b). The effects of biochar addition on soil physicochemical properties: A review. *Catena*, 202, 105284.

12 Role of Nanoparticles and Nano-Based Approaches for Sustainable Agriculture

Soma A. El Mogy and Neama A. Shalaby

12.1 INTRODUCTION

Nanoparticles (NPs) have become a suitable medium for continuous culture because of their unique characteristics. NPs possess a large ratio of surface area to volume, enabling improved interactions with plant and soil elements. They can improve soil nutrient uptake, minimize the requirement for traditional agrochemicals, and mitigate the impact of environmental stress on crops (Usman et al., 2020). This chapter presented a thorough examination of present circumstances and breakthroughs in the use of NPs in sustainable cultivation and biosecurity. We investigated the roles of NPs in agriculture, exploring their potential advantages and risks. We delved into the difficulties associated with their safe and responsible utilization, examining recent research on the effects of NPs on soil health, plant growth, and food quality. Additionally, we assessed their potential impact on human and environmental health.

NPs can serve an ample scope of activities in agriculture, including nutrient delivery, insect and disease management, and soil quality enhancement. NPs can be used as effective delivery vehicles for nutrients, fertilizers, and insecticides, lowering their environmental impact and increasing their performance. For example, silica and oxide NPs were proven to promote nutrient uptake as well as crop growth and yield (Gogos et al., 2012; Raliya & Tarafdar, 2013; Soares et al., 2012).

In addition, NPs can also control pests and diseases in crops. Silver and copper nanoparticles are efficient against a diversity of plant diseases, including bacteria, fungi, and viruses. NPs can also promote soil quality by enhancing water retention, reducing erosion, and promoting microbial activity. For instance, the use of NPs such as nano-clay and nanohydroxyapatite has been shown to promote soil structure and fertility. NPs have been proven to promote soil quality through a diversity of processes, including erosion reduction and increased microbial activity. The ability of NPs to reduce erosion can be attributed to their impact on soil structure and stability. NPs have been found to enhance soil aggregation, leading to the creation of enduring soil clumps that resist erosion by wind or water. They also help to improve water retention in soil, lowering the danger of runoff and erosion.

DOI: 10.1201/9781003714484-12

Additionally, NPs have been shown to enhance microbial activity in soil, which can further contribute to soil health and fertility. Several conducted research studies indicated that NPs are beneficial to soil quality (Gogos et al., 2012; Schlich & Hund-Rinke, 2015). However, it is important to consider the solubility problem associated with NPs in agricultural applications. To resolve this issue, different solutions have been investigated, including the encapsulating of NPs within protective coatings or matrices, surface modification techniques, and formulation into nanocomposites, to enhance their solubility and dispersibility in agricultural natural systems (Rashad et al., 2022; Usman et al., 2020).

The use of NPs in cultivation offers several potential benefits, such as increased crop yield, reduced agrochemical use, and improved soil health (P. Wang et al., 2016). NPs can boost plant development and increase effectiveness through improvements in nutrient uptake and minimizing the impact of environmental stresses. Additionally, the use of NPs can reduce the need for traditional agrochemicals, leading to a decrease in their environmental impact (Khodakovskaya et al., 2012a).

However, Concerns have been raised about the potential dangers of utilizing NPs in agriculture toward the ecosystem and human well-being. NPs can interact with soil components and potentially alter soil microbial communities, with potential consequences for soil health and ecosystem functioning.

The use of NPs in cultivation raises several challenges regarding their safe and responsible use. The lack of standardized methods for NPs synthesis and characterization makes it difficult to compare results across studies, hindering our understanding of the potential dangers to the environment and human health (Nowack et al., 2013). Figure 12.1. Demonstrates an outline of the benefits and risks of NPs in agriculture.

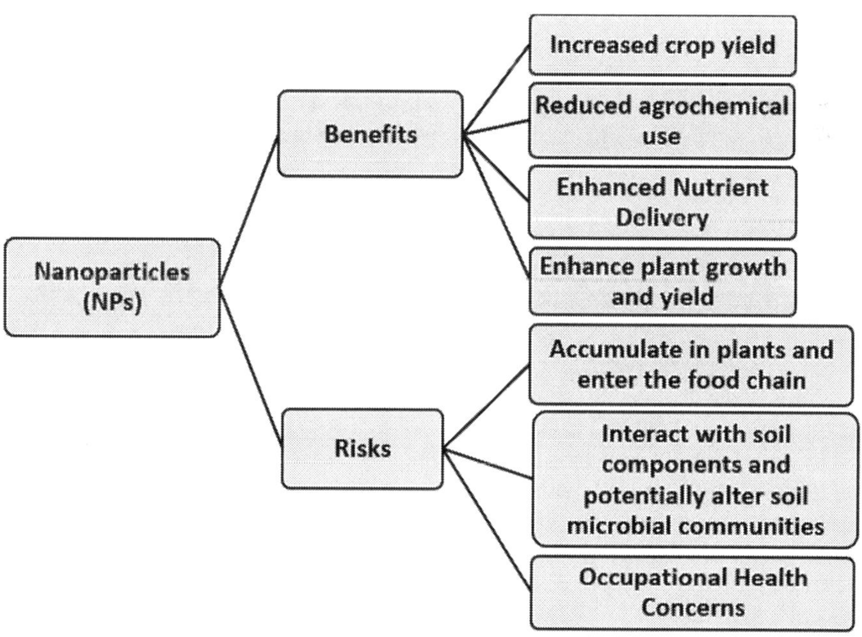

FIGURE 12.1 Scheme of benefits and risks associated with using NPS in agriculture.

12.2 HISTORICAL BACKGROUND OF NANOPARTICLES IN AGRICULTURE

The use of nanoparticles in cultivation is a fairly new subject, with most research dating back to the previous 20 years. The capacity of NPs to advance crop growth and production, minimize the use of traditional agrochemicals, and ameliorate environmental stressors has sparked increased interest in their application in sustainable agriculture. The oldest known usage of nanotechnology in agriculture dates back to the early 2000s. Researchers began researching the potential of NPs to stimulate cultivation and yield. In 2003, researchers from the University of California presented a study demonstrating that gold NPs may penetrate plant roots and leaves and aggregate in plant tissues without inflicting any visible harm. This study set the path for future research into the possible applications of NPs in agriculture. Since then, various research studies have been undertaken to look into the possible benefits of NPs in cultivation. These NPs have been shown to progress nutrient uptake, mitigate the effects of environmental stresses, and manage pests and diseases in crops. In previous years, there was increasing concern about the use of NPs in agriculture due to their potential to boost crop development and output while decreasing the environmental effect of standard cultivation operations.

NPs are emerging as a transformative tool in agriculture, offering advantageous like enhanced crop growth, improved nutrient-use efficiency, and effective pest and disease management. Their application as nanofertilizers and nanopesticides enables targeted, slow-release delivery, reducing the environmental impact of traditional agrochemicals (Bongaarts, 2021). Studies have shown promising results, such as increased crop yields and stress tolerance, with examples like nano-silica boosting tomato yields by 50%. However, concerns persist about the potential environmental and healthiness risks of NPs (R. Liu & Lal, 2015). They can accumulate in plants, enter the food chain, and disrupt soil microbial populations critical for nutrient cycling and plant health. The long-term effects of NPs on soil ecosystems, plant-associated microorganisms, and human health remain underexplored. While nanotechnology holds promise for sustainable agriculture, further research is desired to address its risks and optimize its benefits for precision farming practices

12.3 SYNTHESIS AND CHARACTERIZATION OF NANOPARTICLES

We will explore the fundamental properties of NPs at the nanoscale through their synthesis and characterization. NPs, ranging in size from 1 to 100 nanometers, show distinct qualities that vary considerably from larger materials, such as heightened reactivity, greater strength, and changed optical properties. We will examine different synthesis methods, such as mechanical milling for top-down approaches and chemical vapor deposition and sol-gel processes for bottom-up techniques. Every approach offers unique benefits for regulating particle size, shape, and composition.

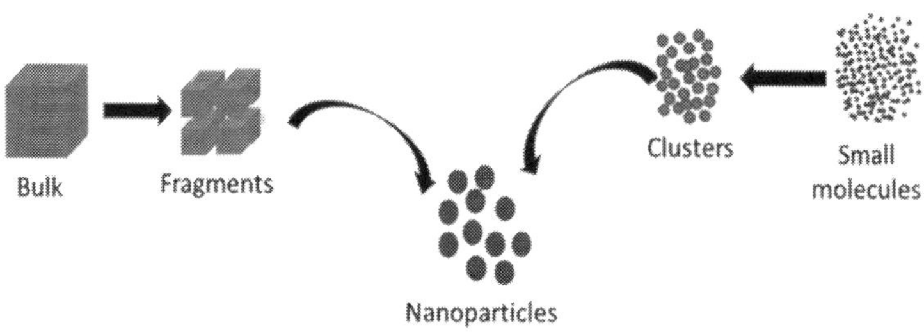

FIGURE 12.2 Top-down and bottom-up approaches of nanoparticles.

12.3.1 Synthesis of Nanoparticles

NPs can be produced through a variation of techniques, commonly divided into top-down and bottom-up methods. Figure 12.2 shows a simplified overview of the process.

12.3.1.1 Bottom-Up Method

The bottom-up, or constructive, method of nanoparticle synthesis involves building materials from the atomic level, gradually forming clusters and ultimately nanoparticles. Several prominent techniques are employed for this purpose, including sol-gel processes, spinning methods, chemical vapor deposition, pyrolysis, and biosynthesis.

12.3.1.1.1 Sol-Gel

The sol is a liquid phase containing suspended solids in a colloidal solution. The solid macromolecule is immersed in a solvent to form the gel. Due to its simplicity and broad applicability, the sol-gel method is a leading bottom-up technique for nanoparticle creation. This wet chemical process utilizes a chemical solution as a precursor to form a cohesive network of discrete particles. Metal oxides and metal chlorides are commonly used as source materials in the sol-gel process (Ramesh, 2013). Subsequently, the precursor is dispersed within a liquid medium using techniques such as agitation, stirring, or sonication, creating a system containing both liquid and solid phases. NPs are then extracted via separation processes like sedimentation, filtration, or centrifugation, and finally dried to eliminate residual moisture.

12.3.1.1.2 Spinning

NPs are produced using a spinning disc reactor (SDR), where synthesis occurs in a controlled environment. This reactor features a rotating disc that enables precise regulation of temperature and other physical variables during the process. To ensure stability, the system operates under an inert atmosphere – such as nitrogen – to exclude oxygen and minimize unwanted chemical interactions (Mohammadi et al., 2014). As the disc rotates at controlled speeds, a liquid solution – typically containing the precursor and water – is introduced into the reactor. The disc's rotation facilitates the aggregation of atoms or molecules, leading to the formation of NPs. These NPs are subsequently collected, dried, and precipitated to isolate the final product. The properties of the resulting NPs produced via this method are highly dependent on operational factors. These include the liquid flow rate, the rotational speed of the disc, the ratio of liquid to precursor, the location of the feed, and the surface characteristics of the disc itself.

12.3.1.1.3 Chemical Vapor Deposition

Chemical vapor deposition (CVD) is a process in which a gaseous precursor is used to create a thin film on a substrate. In this method, gaseous reactants are combined within a reaction chamber, often at ambient temperature, to facilitate the deposition process. When this gas mixture comes into contact with a heated substrate, a chemical reaction occurs, resulting in the formation of a solid material on the substrate's surface (Labus et al., 2023). The reaction leads to the creation of a thin layer of the desired product on the substrate, which is then collected and utilized. The substrate temperature is a key parameter in the CVD process. Advantages of CVD include the production of highly pure, uniform, durable, and robust NPs. However, CVD also has limitations, such as the requirement for specialized and costly equipment, as well as the generation of potentially hazardous gaseous by-products (Labus et al., 2023).

12.3.1.1.4 Pyrolysis

Pyrolysis stands out as a widely adopted technique in industrial settings for the large-scale manufacturing of NPs. This method involves the thermal decomposition of a precursor by applying heat, often through combustion. The precursor, which can be in liquid or vapor form, is introduced into a high-temperature furnace under elevated pressure through a small opening, where it undergoes combustion. The resulting NPs are then extracted from the exhaust stream through air classification of the combustion gases and by-products. In some pyrolysis setups, lasers or plasma are employed instead of flames to achieve the high temperatures necessary for efficient vaporization of the precursor material (D'Amato et al., 2013). The pyrolysis method offers several advantages, including its simplicity, effectiveness, cost-efficiency, continuous operation, and potential for high-volume production.

12.3.1.1.5 Biosynthesis

Biosynthesis is an eco-friendly method to produce NPs that are both non-toxic and able to be broken down naturally (Kuppusamy et al., 2016). This method employs various organisms like bacteria, plant extracts, and fungi, as well as precursor materials, to create NPs for bioreduction and capping, as opposed to traditional chemicals. The NPs that are produced through biosynthesis have special and improved qualities that are useful in biomedical fields.

12.3.1.2 Top-Down Method

The top-down, or subtractive, approach to nanoparticle synthesis involves reducing bulk materials into nanoscale particles. Common techniques employed include mechanical milling, nanolithography, laser ablation, sputtering, and thermal decomposition.

12.3.1.2.1 Mechanical Milling

Mechanical milling is among the most prevalent top-down techniques for the production of diverse NPs. It is used both to reduce the size of materials and to mill and post-anneal nanoparticles during synthesis, often performed under an inert atmosphere to process various elements (Prasad Yadav et al., 2012). The factors that influenced mechanical milling include plastic deformation causing changes in particle shape, fracture causing particles to become smaller, and cold-welding causing particles to become larger.

12.3.1.2.2 Nanolithography

Nanolithography focuses on the fabrication of structures at the nanoscale, typically with at least one dimension ranging from 1 to 100 nanometers. Various techniques within nanolithography include optical lithography, electron-beam lithography, multiphoton lithography, nanoimprint lithography, and scanning probe lithography (Pimpin & Srituravanich, 2012). Lithography, in general, refers to the process of shaping or structuring a light-sensitive material by selectively removing portions of it to achieve the intended design or pattern. Nanolithography has the key benefit of being able to create structures ranging from individual NPs to clusters with specific shapes and sizes. Complex equipment and cost are some of the drawbacks.

12.3.1.2.3 Laser Ablation Synthesis in Solution

Laser Ablation Synthesis in Solution (LASiS) is a common technique for producing nanoparticle colloidal solutions in various solvents. A laser beam creates a plasma plume in a liquid solution surrounding a metal, leading to the production of NPs (Amendola & Meneghetti, 2009). LASiS provides a reliable top-down alternative to chemical reduction for creating metal NPs. This method is considered environmentally friendly, as it allows for the sustainable production of NPs in both organic solvents and water without needing stabilizing agents or chemicals. LASiS involves using a laser beam to strike a solid target submerged in a liquid, leading to the creation of NPs through the condensation of the resulting plasma plume.

12.3.1.2.4 Sputtering

Sputtering is a process where NPs are deposited onto a substrate by ejecting atoms from a target material through bombardment with energetic ions. The process usually involves depositing a thin film of the material, followed by annealing to induce nanoparticle formation. Factors such as the initial film thickness, annealing temperature and duration, and the nature of the substrate significantly affect the resulting size and shape of the NPs (Zeng et al., 2018).

12.3.1.2.5 Thermal Decomposition

Thermal decomposition is a process in which chemical bonds within a compound are broken down by heat, representing an endothermic chemical reaction (Nessim, 2010). The decomposition temperature is the specific temperature at which a substance undergoes chemical decomposition. NPs can be formed through this process by carefully controlling the decomposition of a metallic compound at specific temperatures, inducing a chemical reaction that leads to the formation of the desired NPs and other by-products.

12.3.2 Characterization of Nanoparticles

The potential and application of a nanoparticle are determined by its distinct characteristics, requiring thorough characterization to understand and optimize its performance. Several techniques are employed to analyze key properties, including size, surface area, composition, surface morphology, surface charge, crystallography, and concentration. These properties are crucial for determining how the nanoparticle will behave and interact in its intended application. Figure 12.3 illustrates the diverse measurement techniques used for nanoparticle characterization, providing a visual guide to the methods available for a comprehensive analysis.

Nanoparticle characterization is essential for determining their properties and potential applications. Key characteristics include particle size and distribution, which are commonly assessed using electron microscopy (SEM and TEM) for individual particles and laser diffraction for bulk samples (Marsalek, 2014). For particles in liquid, photon correlation spectroscopy and centrifugation are employed, while a Scanning Mobility Particle Sizer is preferred for gaseous particles. Surface area, crucial for nanoparticle performance, is typically determined by Brunauer–Emmett–Teller (BET) analysis; nuclear magnetic resonance spectroscopy is used for liquid samples (V. Sharma & Rao, 2013).

Chemical composition, which affects purity and performance, is measured using X-ray photoelectron spectroscopy, with alternative methods like mass spectrometry for chemically broken-down particles (Bzdek et al., 2011). Nanoparticle shape and surface structure, whether spherical, cylindrical, crystalline, or amorphous, are analyzed via electron microscopy (Hodoroaba et al., 2014). Surface charge, dictating particle interactions, is measured using zeta potentiometers for solutions and Differential

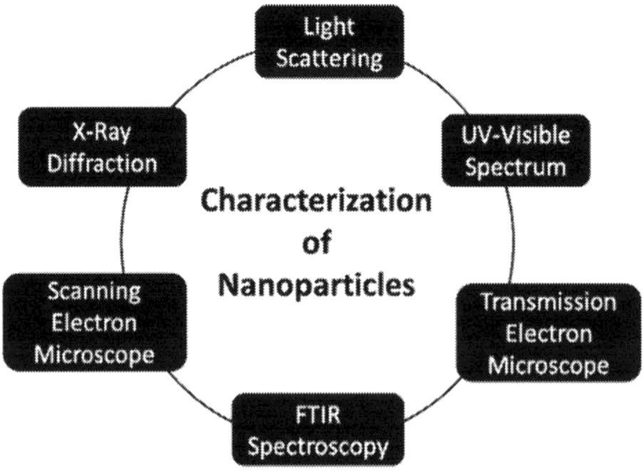

FIGURE 12.3 Techniques for nanoparticle characterization.

Mobility Analyzers for gaseous samples (Marsalek, 2014). Crystallinity, the arrangement of atoms, is determined through powder X-ray, electron, or neutron diffraction. Finally, concentration levels are measured using a Condensation Particle Counter.

12.4 NANOTECHNOLOGY POTENTIAL IN SUSTAINABLE AGRICULTURAL DEVELOPMENT

Nanotechnology is a rapidly advancing field with the potential to revolutionize agriculture and promote sustainable development. By enabling precise nutrient management, it can significantly enhance productivity while minimizing environmental impact (Gruère, 2012; Mukhopadhyay, 2014), as well as surveil long-term water quality and pesticide usage in cultivation development (Ram et al., 2014). Nanotechnology can contribute to agriculture's long-term viability by improving agricultural practice efficiency, minimizing the environmental impact of agrochemicals, and increasing crop nutritional value (Elbasyoni et al., 2023).

Nanotechnology research plays a vital role in advancing agriculture and ensuring sustainable development. Various nanomaterials, such as nanotubes, fullerenes, and biosensors, are being applied in the food industry for diverse purposes. These include controlled delivery systems for nutrients and pesticides, nanofiltration for improved water management, and nanosensors for monitoring soil quality and food safety. These innovations enhance productivity while addressing environmental challenges (Sabir et al., 2014). This technology has been successful in managing agricultural resources, delivering herbal drugs, and conserving soil fertility. Additionally, biomass and plant by-products, as well as food processing and packaging systems, are consistently being evaluated, along with risk assessment (Mukhopadhyay, 2014). The interaction of NPs with plants is a complex process that involves several mechanisms affecting plant growth and development. The mechanism of NPs interaction with plant was indicated in Figure 12.4.

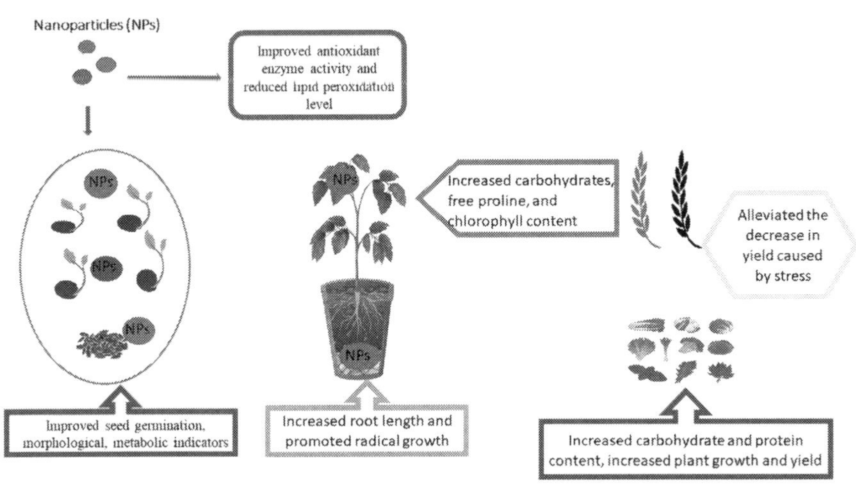

FIGURE 12.4 Mechanism of nanoparticles interaction with plant.

While nanotechnology has the potential to greatly improve cultivation sustainability, there are worries about the harm to the well-being of people and the natural surroundings. Nanopesticides, nanofertilizers, nano-biosensors, and nanoherbicides are prominent examples of nanotechnology applications in agriculture, offering innovative and sustainable solutions to enhance agricultural productivity by improving resource efficiency and reducing negative environmental impacts, as discussed below.

12.4.1 Nano-Agrochemicals

Ensuring food security and nutrition is crucial for long-term agricultural development. Furthermore, the drastic decrease in agricultural output due to global climate change has also impacted sustainable natural resource utilization, urban growth, and ecological issues (Aouada & de Moura, 2015; Panpatte et al., 2016; Pramanik & Pramanik, 2016). The most important issues in modern agriculture are excessive fertilizer use and chemical accumulation. This constitutes a significant risk to the world's food supply. Because of this, modern agriculture requires the usage of nano agrochemicals (Campos et al., 2014; Pulimi & Subramanian, 2016). The integration of nanotechnology in agriculture through nanofertilizers and nanopesticides plays a vital role in achieving sustainable crop production. These innovative tools enhance nutrient delivery and pest management, contributing significantly to environmentally friendly farming practices and improved agricultural productivity (Mastronardi et al., 2015). Notably, nano-agrochemicals encompass a range of products, including pesticides, fertilizers, and soil conditioners at the nanoscale, among others.

12.4.1.1 Nanopesticides

Nanopesticides are nanotechnology-based agricultural inputs that can increase pest control efficiency and effectiveness. Traditional insecticides can be inefficient and can have negative effects on both environmental ecosystems and human well-being (Kookana et al., 2014). Nanopesticides can help address these issues by enhancing pesticide distribution and lowering the amount required. Besides synthetic pesticides being commonly used, biobased pesticides have a distinctive role in controlling insects and pests. The field of biopesticides has been revolutionized by the introduction of new technology through nanotechnology. The creation of eco-friendly nanopesticides that are non-toxic is essential in this aspect. These nanopesticides have metal NPs that exhibit significantly better antibacterial, antifungal, and antipathogenic properties (Saeed, 2021). Soil microorganisms have a crucial function in plants by carrying out various ecological and biological functions. Fusarium wilt negatively impacts lettuce and tomatoes in various countries. Using NPs has been suggested as a different approach effectively manage plant diseases. Ghidan et al. (Ghidan et al., 2016) produced CuO NPs and evaluated the impact of different dosages on the green peach aphid in a controlled greenhouse setting.

Several studies confirm the effectiveness of nanopesticides in agriculture, highlighting their advantages over conventional pesticides. For instance, a comprehensive analysis revealed that nanopesticides demonstrated an overall efficacy increase of 31.5% against target organisms compared to their non-nanoscale counterparts, with a notable 18.9% improvement observed in field trials (D. Wang et al., 2022). Another study indicated that responsive nanopesticides are, on average, 24% more efficient than traditional pesticides, showcasing their potential in sustainable agricultural practices (D. Wang et al., 2022). Furthermore, studies indicate that nanopesticides can significantly decrease environmental toxicity, posing 43.1% less risk to non-target organisms. They also help minimize the premature degradation of active ingredients, thereby enhancing their efficacy and reducing potential environmental harm (Yousef et al., 2023). These findings collectively underscore the promising role of nanopesticides in enhancing pest control efficacy while promoting environmental sustainability. Overall, nanopesticides can be more effective than regular pesticides. Their small size and distinct features may enhance their capacity to enter plant tissues and target pests. This approach can boost the effectiveness of pest management while reducing the quantity of insecticides required.

12.4.1.2 Nanofertilizers

Nanofertilizers are nanotechnology-based agriculture inputs that can improve the efficiency of fertilizer usage and reduce its environmental impact. Traditional fertilizers can be ineffective, with only a tiny portion of the administered fertilizer being absorbed by plants (Pesavento et al., 2021). Nanofertilizers enhance nutrient absorption by delivering nutrients in a targeted and gradual manner, reducing the need for frequent applications and minimizing environmental waste. Some examples of nutrients were demonstrated in Table 12.1.

Nanofertilizers can improve soil health and reduce fertilizer needs, but concerns exist about their safety and potential environmental impacts due to nanoparticle accumulation in soil and water (Giraldo et al., 2019). As a result, nano fertilizers must be rigorously evaluated and regulated to assure their safety and reduce their environmental impact.

12.4.2 Nano-Biosensors

Nano-biosensors are a new technology that combines nanotechnology with biosensors. Biosensors are analytical instruments that detect and quantify biological and chemical molecules. In contrast, biosensors use nanomaterials like NPs, nanowires, and nanotubes to enhance their level of sensitivity, selectivity, and accuracy. Nano-biosensors have diverse uses, involving tests for medical conditions, environmental observation, and food security. Nano-biosensors can potentially improve food safety by detecting pollutants, pathogens, and other dangerous compounds in real-time (Marvier, 2009). An example of how nano-biosensors can be used in food security is the detection of pathogens found in food like Salmonella, *E. coli*, and Listeria. Traditional techniques of detecting these diseases need time-consuming and labor-intensive processes that can take days to generate results. Nano-biosensors, on the other hand, can detect these pathogens in minutes or even seconds, allowing for the rapid identification of tainted food and the prevention of foodborne illness outbreaks. Nano-biosensors can also detect other pollutants in food, including pesticides and heavy metals. These pollutants can harm human health and lead to chronic health problems. Nano-biosensors can detect these toxins in very low quantities, allowing us to assure that food is safe to consume.

TABLE 12.1
Types of Nutrients and Their Chemical Forms and Uses

Types	Examples	Chemical Forms of Elements Absorbed by Plants	Uses	Crop, Concentration	References
Primary Nutrients					
Nitrogenous	Ammonium chloride, ammonium sulfate, and amide nitrogen	NO^{-3}, NH^{+4}	Used to address the nitrogen shortage in the land.	Most plants	Tantawy et al. (2012)
Phosphorus	Available phosphate	H_2PO^{-4}, HPO_4^{-3}, PO_4^{-3}	Used for meeting the phosphor deficiency in the land	Most plants	Tantawy et al. (2012)
Potassium	Muriate potassium chloride and potash sulfate. Sulfate of potassium	K^-	Necessary for healthy plant growth	Arabidopsis 50 mg/L	Tantawy et al. (2012)
Secondary Nutrients					
Magnesium (Mg)	Magnesium (Mg)	Mg^{+2}	Enzymes control the movement of nutrients within the plant.	Wheat 0.5 g/L	Cai et al. (2018)
Calcium (Ca)	Calcium (Ca)	Ca^{+2}	It serves as a potent initiator in numerous enzyme reactions.	Chili pepper 0.5, 1, 2, 3 g/L	Brackhage et al. (2013)
Sulfur (S)	Sulfur (S)	$SO4^{-2}$	Essential for Chloroplasts, Nitrogen Fixation, and Protein Synthesis	Wheat (50–800 ppm)	Hong et al. (2015), Karimi and Mohsenzadeh (2016), P.-H. Liu et al. (2017), and Tantawy et al. (2012)

Micronutrients					
Boron (B)	B	BO_3^{-3}	It serves a variety of functions within plants.	Pomegranate 34 mg	Davarpanah et al. (2016), Hong et al. (2015), and Liu et al. (2017)
Copper (Cu)	(Cu)	Cu^{+}, Cu^{+2}	It is necessary for the production of lignin.	Lettuce 130, 660 mg/Kg	Hong et al. (2015) and Liu et al. (2017)
Iron (Fe)	(Fe)	Fe^{+3}	Iron is an enzyme cofactor.	Lettuce 10, 20 mg/L	Liu et al. (2017)
Manganese (Mn)	(Mn)	Mn^{+2}	Mn plays an important role in chloroplast development.	(5–20 ppm) (0.05–1 ppm)	Liu et al. (2017) and Prashantha Kumar et al. (2016)
Molybdenum (Mo)	(Mo)	$Moo4^{-2}$	Mo is a key cofactor in amino acid synthesis and contributes to nitrogen metabolism.		
Zinc (Zn)	(Zn)	Zn^{+2}	Zinc is essential for various enzymes to function properly. It has a significant role in DNA transcription.	Maize 10 ppm	Adebisi et al. (2017)

Nano biosensors can be utilized for both food safety and food quality management. They can detect changes in the composition and qualities of food, like variations in pH, temperature, and moisture content. This can assist food manufacturers in ensuring that their products are of consistent quality and satisfy regulatory standards. Nano-biosensors can also be utilized for precision agriculture. Precision cultivation is the utilization of technology to maximize crop yield while minimizing waste. Nano-biosensors are capable of monitoring soil nutrients, moisture levels, and other factors that impact the growth of crops. This data can be used to enhance irrigation and fertilizer efficiency, leading to higher crop production and reduced environmental harm (González-Díaz et al., 2019).

12.4.3 Nanoherbicides

The most effective way to control weeds involves eliminating their seeds in the soil and preventing germination under favorable conditions. Nanoherbicides can effectively integrate into the soil, providing a sustainable method to eradicate weeds without leaving harmful residues and combating weed species resistant to traditional herbicides. Weeds often thrive and spread through underground structures like tubers and robust root systems. Plowing fields with disease and manually removing weeds can lead to the spread of unwanted plants to healthy areas. Using nano application remains consistent, whether from a nanosized active ingredient or creating a nanosized formulation with an adjuvant. If the herbicide is connected to a smart delivery system, it will only be used as needed based on the agricultural field's conditions. Soils contaminated with weeds result in reduced agricultural productivity. Enhancements in herbicide performance with nanotechnology could lead to higher crop yields without causing harm to farm workers responsible for manually removing weeds in the absence of herbicide application (Ram et al., 2014).

12.5 PLANT-GROWTH-ENHANCING NPS

Many studies have discovered NPs that may promote plant development through an unknown method because they are neither a source of macronutrients or micronutrients nor a transporter of nutrients. The sections that follow provide a detailed description of these NPs.

12.5.1 Titanium Dioxide

Titanium Dioxide (TiO_2) NP's sizes range from 5 to 30 nm and are commonly made and employed throughout the globe (Jomini et al., 2015). There is a high demand for these NPs made from titanium photocatalytic capabilities. Because of this characteristic, they are utilized to make pigments (Sang et al., 2014). These NPs increase plant growth and photosynthesis (see Table 12.2). Several researchers have observed its good influence on photosynthetic efficacy and enzyme activities, such as nitrogenase, consequently improving nitrogen supply (N) to crop plants via biological nitrogen fixation, and eventually plant development.

TABLE 12.2
Impacts of TiO_2 NPs on Plant Growth and Development

Effect	Description	References
Salinity Reduction	TiO_2 NPs reduced soil salinity in broad beans, improving growth and nitrogen fixation.	Abdel Latef et al. (2017)
Growth Enhancement	TiO_2 NPs enhanced enzyme activity, chlorophyll production, nitrogen uptake, and stress tolerance.	Lyu et al. (2017)
Pest Control	TiO_2 residues in plants acted as an efficient insecticide by destroying pests.	Lyu et al. (2017)
Spinach Yield Increase	Weekly Ti-NPs spray (2.5 g/L) doubled fresh and dry weight compared to controls.	Yang et al. (2007)
Optimal Concentration for Growth	Lower Ti-NP doses (0.5 g/L) enhanced growth; higher doses inhibited it.	Song et al. (2012)
Nitrogen Metabolism Stimulation	Increased nitrate reductase activity and enzymatic functions (e.g., glutamate dehydrogenase).	Mishra et al. (2014)
Seed Germination & Growth	Boosted germination, root, and shoot growth in various crops like canola, tomato, and wheat.	Raliya and Tarafdar (2013)
Photosynthesis Improvement	Enhanced Rubisco activity, light absorption, and chloroplast longevity for better photosynthesis.	Yang et al. (2006)
Antioxidant Enzyme Activity	Increased catalase, peroxidase, and superoxide dismutase activity to protect against stress.	Hong et al. (2015)
Water-Limited Conditions	Ti-NPs (0.01–0.03%) improved wheat growth and yield under drought stress.	Gupta and Tripathi (2011)

12.5.2 Carbon Nanotubes

Carbon nanotubes, encompassing both single-walled and multi-walled variants, have emerged as promising tools in agriculture due to their distinctive characteristics. These nanomaterials can boost seed germination, enhance plant development, and improve photosynthetic efficiency. They facilitate better water and nutrient absorption by influencing cellular processes and activating specific proteins involved in water transport. However, excessive concentrations can lead to adverse effects on plant health. The impact of CNTs varies significantly depending on factors such as plant type, growth phase, CNT type, and dosage. To fully leverage their potential, further studies are necessary to elucidate the underlying mechanisms and validate practical applications in agricultural settings (see Table 12.3). Numerous studies have highlighted the beneficial effects of multi-walled carbon nanotubes on seed

TABLE 12.3
Effects of Carbon Nanotubes on Plant Growth

Effect	Description	References
Seed germination	MWCNTs enhanced seed germination in barley, soybean, wheat, and other crops.	Lahiani et al. (2013)
Water uptake & nutrient absorption	CNT surface charges activated aquaporins, improving water and nutrient (Ca and Fe) absorption.	Khodakovskaya et al. (2012a), Lahiani et al. (2013), and Tiwari et al. (2013)
Photosynthesis enhancement	SWCNTs boosted photosynthetic efficiency by three times and improved electron transport speed.	Giraldo et al. (2014)
Gene activation	MWCNTs triggered genes for cell division (CycB), cell wall formation (NtLRX1), and aquaporins.	Khodakovskaya et al. (2009)
Root growth & division	CNTs enhanced root elongation in rye, onion, cucumber, and corn plants.	Tripathi and Sarkar (2014)
Plant yield improvement	MWCNTs improved tomato yield, biomass, water retention, and medicinal properties.	Khodakovskaya et al. (2012b)
Stress tolerance	CNTs mitigated salinity effects and improved drought resistance in plants.	Lahiani et al. (2013)
Enzyme activity enhancement	Increased peroxidase and dehydrogenase activity in roots and leaves.	Wang et al. (2012) and Tripathi and Sarkar (2014)
Phytotoxic effects at high doses	Elevated CNT levels inhibited growth in radish, lettuce, cucumber, and other crops.	Begum and Fugetsu (2012), and Tiwari et al. (2013)

germination across various crops, ultimately resulting in enhanced plant development (Khodakovskaya, et al., 2012b; Tiwari et al., 2013; Villagarcia et al., 2012). The proposed explanation for the improved plant growth is that the surface charges on carbon nanotubes activate water channel proteins, thereby increasing the efficiency of water and essential nutrient uptake, such as calcium and iron, which in turn boosts seed germination and overall plant development (Khodakovskaya et al., 2012a; Lahiani et al., 2013; Tiwari et al., 2013).

12.5.3 Silver (Ag NPs)

Based on the data that is currently available, there have been limited research studies regarding the impact of Ag NPs. The results, whether good, neutral, or bad, are determined by the specific plant, the particle size, and the quantity used (Krishnaraj et al., 2012). Lately, a study on Arabidopsis demonstrated that decahedral Ag NPs had the most significant effect on root growth promotion (RGP) among the three types tested. Nevertheless, the SNPs had no impact on RGP but led to increased levels of anthocyanin accumulation in Arabidopsis plants (Syu et al., 2014). Spherical Ag NPs had the highest yield of copper/zinc superoxide dismutase, while decahedral

Ag NPs had the lowest yield. Furthermore, there are three distinct kinds and forms of Ag NPs that regulate protein synthesis, gene expression related to cell formation, inhibition of aminocyclopropane-1-carboxylic acid (ACC) – a factor that promotes root elongation in Arabidopsis plants, and triggers gene activation. Ag NPs have been shown to reduce the expression of ACC synthase 7 and ACC oxidase 2 genes, suggesting that Ag NPs can act as inhibitors of ethylene perception and may interfere with ethylene biosynthesis. Similarly, Feizi et al. (Feizi et al., 2011) reported that Ag NPs stimulated root growth of *Crocus sativus* by inhibiting ethylene signaling. Recently, Gruyer et al. (Gruyer et al., 2014) observed a growth in root length in barley, while it was suppressed in lettuce. Furthermore, Yin et al. (2012) examined the impact of Ag NPs on the seedling growth of 11 wetland plant species (*Lolium multiflorum, Panicum virgatum, Carex lurida, C. scoparia, C. vulpioidea, C. crinita, Eupatorium fistulosum, Phytolaca americana, Scirpus cyperinus, Lobelia cardinalis, and Juncus effusus*) and observed enhanced germination in just *E. fistulosum*. In a different hydroponic experiment, the germination rate of *Bacopa monnieri* showed a notable increase when biologically produced Ag NPs were used (Krishnaraj et al., 2012). Furthermore, the identical therapies enhanced the synthesis of proteins and carbohydrates, while decreasing the total phenolic content and the activity of catalase and peroxidase. Comparable findings were documented in *Brassica juncea*, common bean, and maize (Sharma et al., 2012).

12.5.4 Silicon Dioxide (SiO_2 NPs)

The growth and development of plants start with the germination of seeds, which is then followed by the elongation of roots and the emergence of shoots as initial indicators of development. Although silica is not necessary for plant growth, it plays a critical role in the germination and growth process by impacting the metabolism of other elements like (c), (N), and (P) (Brackhage et al., 2013; Schaller et al., 2011, 2012, 2013). Numerous studies concluded that the impact of SiO_2 NPs on seed sprouting varies according to their concentration. Smaller amounts enhanced the sprouting of tomato seeds (Siddiqui & Al-Whaibi, 2014). Likewise, SiO_2 NPs enhanced maize seed germination through increased availability of nutrients (Suriyaprabha et al., 2012). Bao-shan et al. (Bao-shan et al., 2004) demonstrated that the external use of SiO_2 NPs boosted the growth and quality of Changbai larch seedlings (Larix olgen sis). Haghighi et al. (Haghighi et al., 2012) and Siddiqui et al. (Siddiqui & Al-Whaibi, 2014) stated in tomatoes that SiO_2 NPs enhanced seed germination and increased the antioxidant system in squash plants when exposed to NaCl stress. SiO_2 NPs improve plant growth and proline levels in leaves when plants are exposed to salinity stress. SiO_2 particles increased plants' resistance to environmental stress by improving levels of proline, amino acids, nutrients, and antioxidant enzymes (Haghighi et al., 2012; Siddiqui & Al-Whaibi, 2014). Wang et al. (Wang et al., 2014) reported that rice root growth was greatly enhanced by Silica coated with quantum dots. SiO_2 NPs improved plant growth and development by increasing gas exchange and chlorophyll fluorescence attributes (Siddiqui & Al-Whaibi, 2014). Similarly, Noji et al. (Noji et al., 2010) stated that combining a Nano-mesoporous silica compound (SBA) with photosystem II (PSII) led to consistent activity in the photosynthetic oxygen-evolving reaction, showing light-induced electron transfer from water to quinone molecules. This indicated that the PSII-SBA combination could have characteristics related to light detection and mimicry of photosynthesis.

12.5.5 Zinc Oxide

Zinc is an essential micronutrient. It affects farming in soils abundant in calcium carbonate, which acts as an antacid. Soil with a high content of calcium carbonate is a significant and abundant resource for farming in the Mediterranean area of the globe. The main micronutrient problem often addressed in soils is zinc deficiency, typically treated with (ZnO) and $ZnSO_4$. Zinc oxide nanoparticles applied as composts could result in zinc residues and lower bioavailability in soils abundant in calcium carbonate. ZnO NPs, which are a type of metal oxide, have become popular over the past 23 years because of their various uses in optics, electronics, and biological systems (Sirelkhatim et al., 2015). Various shapes of ZnO NPs such as nanoplates, nanoboxes, nanorods, nanospheres, tripods, hexagonal, and nanotubes are formed through different manufacturing techniques, similar to other NPs (Siddiqi et al., 2018). ZnO NPs can be produced through different methods such as physical, chemical, and biological ways, but the safest, cheapest, and most environmentally friendly option is the green approach involving plants. The lack of toxicity of ZnO NPs makes them suitable for various applications in medicine, as preservatives, and as antibacterial agents. ZnO NPs are tiny and possess a large surface-to-volume ratio, enabling increased contact with microorganisms. This sets apart ZnO NPs from the majority of zinc particles in their antibacterial properties. It can stimulate reactive oxygen species (ROS) production, leading to cell death by overwhelming its antioxidant capabilities. ZnO NPs generate reactive oxygen species in wheat, reducing the levels of chlorophyll (Javed et al., 2019).

12.6 NANOPARTICLES AND THEIR POTENTIAL ECOTOXICOLOGICAL EFFECTS

NPs have garnered significant interest due to their unique properties and diverse potential applications across sectors like agriculture. Nonetheless, as with any emerging technology, concerns arise about their potential environmental impact. This section will go over the probable ecotoxicological impacts of nanoparticles and their implications for the environment.

12.6.1 Sources of Nanoparticle Pollution

NPs can infiltrate the setting from a diversity of sources, inclusive of industrial and agricultural activities, and consumer products. Nanoparticles can be released during the manufacturing, usage, and disposal of NP-containing items. Once released, NPs can be carried via air and water and accumulated in soils, sediments, and biota (Handy et al., 2011).

12.6.2 Ecotoxicological Effects of Nanoparticles

NPs' small size and unique features can cause different ecotoxicological consequences than larger particles. The potential ecological effects of NPs on the setting and living beings are being researched; however, some of the known effects include the following:

12.6.2.1 Changes in Nutrient Availability

NPs can alter the opportunity of nutriment in the soil, affecting plant growth and ecosystem function. Studies revealed that NPs can adsorb nutrients such as phosphate and potassium, limiting their availability for plants (Gogos et al., 2012). This can result in reduced plant development and productivity, with cascade impacts on the ecosystem.

12.6.2.2 Toxicity to Organisms

NPs can be harmful to living things like plants, animals, and bacteria. NPs' toxicity is influenced by several parameters, such as their volume, formation, surface charge, and chemical composition (Muraisi et al., 2022). According to studies, NPs can cause oxidative strain, DNA damage, and inflammation in living organisms, resulting in poor health and survival outcomes (Alarifi et al., 2017; Ansari et al., 2019).

12.6.2.3 Bioaccumulation and Biomagnification

NPs can accumulate in living organisms, including plants and animals, and can potentially biomagnify through the food chain. This can result in high concentrations of NPs in top predators, which can have implications for their health and survival (Zhu et al., 2008). According to studies, some NPs, such as silver NPs, can accumulate in fish tissues and have harmful effects (Kakakhel et al., 2021).

12.6.2.4 Effects on Soil and Water Quality

NPs can affect both soil and water quality, leading to repercussions for ecosystem functioning. Studies have demonstrated that NPs can alter soil physicochemical parameters, for example, pH and organic matter concentration, influencing nutrient availability and plant growth. NPs in water can influence the pH, conductivity, and nutrient content, which can have an impact on aquatic life (Hafez et al., 2022).

12.7 FUTURE PERSPECTIVES

Nanotechnology has grown significantly in recent years, and it will certainly continue to have a big impact on cultivation and food systems in the future. Here are some possible future applications of nanotechnology in cultivation and food.

12.7.1 Maximizing the Usage of Nano Fertilizers

As the global population grows, farmers face increased pressure to produce more food with limited resources. Nanofertilizers can improve crop yield by enhancing nutrient uptake and targeted distribution while reducing the overall amount of fertilizer needed. This greater efficiency can result in better resource usage and cost-effectiveness in agricultural systems.

12.7.2 Developing Novel Nano Pesticides

Current pesticides may impact the environment and human health, necessitating safer and more effective alternatives. Nanopesticides have the potential to be more targeted and effective, therefore lowering the need for vast amounts of pesticides.

12.7.3 Nano-Biosensors in Food Safety and Quality

Nanobiosensors can improve food safety by detecting pollutants, pathogens, and dangerous compounds in real time. As technology advances, it may be possible to create handheld devices that can be used in the field to identify foodborne viruses and other contaminants.

Nano-biosensors improve food quality control by detecting changes in composition and characteristics, ensuring consistency and regulatory compliance. As technology advances, it may be possible to create sensors that can detect a broader range of qualities, allowing for more accurate and precise quality control. Nanotechnology can improve food packaging and preservation by creating materials that prevent rotting and contamination. Nanomaterials, for example, can be utilized to make antimicrobial coatings for food packaging, preventing hazardous bacteria from growing.

12.7.4 Precision Agriculture Through Nano Biosensors

Nano-biosensors can track soil nutrients, moisture levels, and other elements that impact crop growth. As technology advances, it may be possible to create sensors that can be integrated with precision agricultural systems, allowing for more targeted and efficient resource utilization.

Nevertheless, nanotechnology, like any other developing technology, raises worries about its possibilities for influence regarding the surroundings and mortal health. Additional studies are necessary to fully comprehend these concerns and build suitable regulatory structures to guarantee safety and ethical use of nanotechnology in agriculture and food systems.

12.8 CONCLUSION

Nanotechnology can alter cultivation in food manufacturing by increasing crop yields, improving food safety, and lowering environmental impact. It can be utilized in nanofertilizers, nanopesticides, and nano-biosensors to improve crop management efficiency and decrease toxic chemicals. NPs can be used in the wrapping and conservation of food to improve shelf life, and avoid contamination. Concerns have been expressed concerning NPs' ecotoxicological consequences, their ability to penetrate the food chain, and the need for responsible regulation and control. The equitable distribution of the benefits of nanotechnology, particularly in developing nations, is vital. Nanofertilizers created utilizing nanotechnology are advantageous to plant growth and development, but their unrestricted and unregulated use can harm soil and plant health. Proper nanoparticle management can preserve soil health, ensure human and environmental health, and boost crop yield. As technology progresses, weighing possible benefits and risks is critical while ensuring safe and responsible development and use.

ARTIFICIAL INTELLIGENCE DECLARATION

The authors declare that they have not used generative artificial intelligence in the writing of this chapter or in the creation of images, graphics, tables, or their corresponding captions.

REFERENCES

Abdel Latef, A. A. H., Srivastava, A. K., El-Sadek, M. S. A., Kordrostami, M., & Tran, L. P. (2017). Titanium dioxide nanoparticles improve growth and enhance tolerance of broad bean plants under saline soil conditions. *Land Degradation & Development*, *29*(4), 1065–1073. https://doi.org/10.1002/ldr.2780

Adebisi, J. A., Agunsoye, J. O., Bello, S. A., Ahmed, I. I., Ojo, O. A., & Hassan, S. B. (2017). Potential of producing solar grade silicon nanoparticles from selected agro-wastes: A review. *Solar Energy*, *142*, 68–86. https://doi.org/10.1016/j.solener.2016.12.001

Alarifi, S., Ali, D., & Alkahtani, S. (2017). Oxidative stress-induced DNA damage by manganese dioxide nanoparticles in human neuronal cells. *BioMed Research International*, *2017*, 1–10. https://doi.org/10.1155/2017/5478790

Amendola, V., & Meneghetti, M. (2009). Laser ablation synthesis in solution and size manipulation of noble metal nanoparticles. *Physical Chemistry Chemical Physics*, *11*(20), 3805. https://doi.org/10.1039/b900654k

Ansari, M. O., Parveen, N., Ahmad, M. F., Wani, A. L., Afrin, S., Rahman, Y., Jameel, S., Khan, Y. A., Siddique, H. R., Tabish, M., & Shadab, G. G. H. A. (2019). Evaluation of DNA interaction, genotoxicity and oxidative stress induced by iron oxide nanoparticles both in vitro and in vivo: Attenuation by thymoquinone. *Scientific Reports*, *9*(1). https://doi.org/10.1038/s41598-019-43188-5

Aouada, F. A., & de Moura, M. R. (2015). Nanotechnology applied in agriculture: Controlled release of agrochemicals. In *Nanotechnologies in Food and Agriculture* (pp. 103–118). Springer International Publishing. https://doi.org/10.1007/978-3-319-14024-7_5

Bao-Shan, L., Shao-Qi, D., Chun-Hui, L., Li-Jun, F., Shu-Chun, Q., & Min, Y. (2004). Effect of TMS (nanostructured silicon dioxide) on growth of Changbai larch seedlings. *Journal of Forestry Research*, *15*(2), 138–140. https://doi.org/10.1007/bf02856749

Begum, P., & Fugetsu, B. (2012). Phytotoxicity of multi-walled carbon nanotubes on red spinach (*Amaranthus tricolor* L) and the role of ascorbic acid as an antioxidant. *Journal of Hazardous Materials*, *243*, 212–222. https://doi.org/10.1016/j.jhazmat.2012.10.025

Bongaarts, J. (2021). FAO, IFAD, UNICEF, WFP and WHO the state of food security and nutrition in the world 2020. Transforming food systems for affordable healthy dietsFAO, 2020, 320 p. *Population and Development Review*, *47*(2), 558. https://doi.org/10.1111/padr.12418

Brackhage, C., Schaller, J., Bäucker, E., & Dudel, E. G. (2013). Silicon availability affects the stoichiometry and content of calcium and micro nutrients in the leaves of common reed. *Silicon*, *5*(3), 199–204. https://doi.org/10.1007/s12633-013-9145-3

Bzdek, B. R., Zordan, C. A., Luther, G. W., & Johnston, M. V. (2011). Nanoparticle chemical composition during new particle formation. *Aerosol Science and Technology*, *45*(8), 1041–1048. https://doi.org/10.1080/02786826.2011.580392

Cai, L., Liu, M., Liu, Z., Yang, H., Sun, X., Chen, J., Xiang, S., & Ding, W. (2018). MgONPs can boost plant growth: Evidence from increased seedling growth, morpho-physiological activities, and Mg uptake in tobacco (Nicotiana tabacum L.). *Molecules*, *23*(12), 3375. https://doi.org/10.3390/molecules23123375

Campos, E. V. R., de Oliveira, J. L., Fraceto, L. F., & Singh, B. (2014). Polysaccharides as safer release systems for agrochemicals. *Agronomy for Sustainable Development*, *35*(1), 47–66. https://doi.org/10.1007/s13593-014-0263-0

D'Amato, R., Falconieri, M., Gagliardi, S., Popovici, E., Serra, E., Terranova, G., & Borsella, E. (2013). Synthesis of ceramic nanoparticles by laser pyrolysis: From research to applications. *Journal of Analytical and Applied Pyrolysis*, *104*, 461–469. https://doi.org/10.1016/j.jaap.2013.05.026

Davarpanah, S., Tehranifar, A., Davarynejad, G., Abadía, J., & Khorasani, R. (2016). Effects of foliar applications of zinc and boron nano-fertilizers on pomegranate (*Punica granatum* cv. Ardestani) fruit yield and quality. *Scientia Horticulturae*, *210*, 57–64. https://doi.org/10.1016/j.scienta.2016.07.003

Elbasyoni, I. S., Morsy, S., Abdelghany, A. M., Naser, M., Mashaheet, A. M., Abdallah, A. M., Hafez, M., Frels, K., & Baenziger, P. S. (2023). Nebraska winter wheat unexpected flowering in Egypt: New improvement opportunities. *Agronomy Journal*, *115*(2), 698–712. https://doi.org/10.1002/agj2.21243

Feizi, H., Rezvani Moghaddam, P., Shahtahmassebi, N., & Fotovat, A. (2011). Impact of bulk and nanosized titanium dioxide (TiO2) on wheat seed germination and seedling growth. *Biological Trace Element Research*, *146*(1), 101–106. https://doi.org/10.1007/s12011-011-9222-7

Ghidan, A. Y., Al-Antary, T. M., & Awwad, A. M. (2016). Green synthesis of copper oxide nanoparticles using *Punica granatum* peels extract: Effect on green peach Aphid. *Environmental Nanotechnology, Monitoring & Management*, *6*, 95–98. https://doi.org/10.1016/j.enmm.2016.08.002

Giraldo, J. P., Landry, M. P., Faltermeier, S. M., McNicholas, T. P., Iverson, N. M., Boghossian, A. A., Reuel, N. F., Hilmer, A. J., Sen, F., Brew, J. A., & Strano, M. S. (2014). Plant nanobionics approach to augment photosynthesis and biochemical sensing. *Nature Materials*, *13*(4), 400–408. https://doi.org/10.1038/nmat3890

Giraldo, J. P., Wu, H., Newkirk, G. M., & Kruss, S. (2019). Nanobiotechnology approaches for engineering smart plant sensors. *Nature Nanotechnology*, *14*(6), 541–553. https://doi.org/10.1038/s41565-019-0470-6

Gogos, A., Knauer, K., & Bucheli, T. D. (2012). Nanomaterials in plant protection and fertilization: Current state, foreseen applications, and research priorities. *Journal of Agricultural and Food Chemistry*, *60*(39), 9781–9792. https://doi.org/10.1021/jf302154y

González-Díaz, A., Ojeda-Morales, M. E., Hernández-Rivera, M. A., Córdova-Bautista, Y., Díaz-Flores, L. L., López-Lázaro, J. D. L. S., & Álvarez-Ramírez, J. G. (2019). Effect of biofertilizers application on the growth of Eucalyptus grandis seedlings under greenhouse conditions. *Journal of Plant Nutrition*, *42*(19), 2560–2576. https://doi.org/10.1080/01904167.2019.1655040

Gruère, G. P. (2012). Implications of nanotechnology growth in food and agriculture in OECD countries. *Food Policy*, *37*(2), 191–198. https://doi.org/10.1016/j.foodpol.2012.01.001

Gruyer, N., Dorais, M., Bastien, C., Dassylva, N., & Triffault-Bouchet, G. (2014). Interaction between silver nanoparticles and plant growth. *Acta Horticulturae*, *1037*, 795–800. https://doi.org/10.17660/actahortic.2014.1037.105

Gupta, S. M., & Tripathi, M. (2011). A review of TiO2 nanoparticles. *Chinese Science Bulletin*, *56*(16), 1639–1657. https://doi.org/10.1007/s11434-011-4476-1

Hafez, M., Abdallah, A. M., Mohamed, A. E., & Rashad, M. (2022). Influence of environmental-friendly bio-organic ameliorants on abiotic stress to sustainable agriculture in arid regions: A long term greenhouse study in northwestern Egypt. *Journal of King Saud University – Science*, *34*(6), 102212. https://doi.org/10.1016/j.jksus.2022.102212

Haghighi, M., Afifipour, Z., & Mozafarian, M. (2012). The alleviation effect of silicon on seed germination and seedling growth of tomato under salinity stress. *Journal of Fruit and Ornamental Plant Research*, *76*(1), 119–126. https://doi.org/10.2478/v10032-012-0008-z

Handy, R. D., Cornelis, G., Fernandes, T., Tsyusko, O., Decho, A., Sabo-Attwood, T., Metcalfe, C., Steevens, J. A., Klaine, S. J., Koelmans, A. A., & Horne, N. (2011). Ecotoxicity test methods for engineered nanomaterials: Practical experiences and recommendations from the bench. *Environmental Toxicology and Chemistry*, *31*(1), 15–31. https://doi.org/10.1002/etc.706

Hodoroaba, V., Rades, S., & Unger, W. E. S. (2014). Inspection of morphology and elemental imaging of single nanoparticles by high-resolution SEM/EDX in transmission mode. *Surface and Interface Analysis*, *46*(10–11), 945–948. https://doi.org/10.1002/sia.5426

Hong, J., Rico, C. M., Zhao, L., Adeleye, A. S., Keller, A. A., Peralta-Videa, J. R., & Gardea-Torresdey, J. L. (2015). Toxic effects of copper-based nanoparticles or compounds to lettuce (*Lactuca sativa*) and alfalfa (*Medicago sativa*). *Environmental Science: Processes & Impacts*, *17*(1), 177–185. https://doi.org/10.1039/c4em00551a

Javed, Z., Dashora, K., Mishra, M. D., Fasake, V., & Srivastva, A. (2019). Effect of accumulation of nanoparticles in soil health – a concern on future. *Frontiers in Nanoscience and Nanotechnology*, *5*(2). https://doi.org/10.15761/fnn.1000181

Jomini, S., Clivot, H., Bauda, P., & Pagnout, C. (2015). Impact of manufactured TiO2 nanoparticles on planktonic and sessile bacterial communities. *Environmental Pollution*, *202*, 196–204. https://doi.org/10.1016/j.envpol.2015.03.022

Kakakhel, M. A., Wu, F., Sajjad, W., Zhang, Q., Khan, I., Ullah, K., & Wang, W. (2021). Long-term exposure to high-concentration silver nanoparticles induced toxicity, fatality, bioaccumulation, and histological alteration in fish (*Cyprinus carpio*). *Environmental Sciences Europe*, *33*(1). https://doi.org/10.1186/s12302-021-00453-7

Karimi, J., & Mohsenzadeh, S. (2016). Effects of silicon oxide nanoparticles on growth and physiology of wheat seedlings. *Russian Journal of Plant Physiology*, *63*(1), 119–123. https://doi.org/10.1134/s1021443716010106

Khodakovskaya, M., Dervishi, E., Mahmood, M., Xu, Y., Li, Z., Watanabe, F., & Biris, A. S. (2009). Carbon nanotubes are able to penetrate plant seed coat and dramatically affect seed germination and plant growth. *ACS Nano*, *3*(10), 3221–3227. https://doi.org/10.1021/nn900887m

Khodakovskaya, M. V., de Silva, K., Biris, A. S., Dervishi, E., & Villagarcia, H. (2012a). Carbon nanotubes induce growth enhancement of tobacco cells. *ACS Nano*, *6*(3), 2128–2135. https://doi.org/10.1021/nn204643g

Khodakovskaya, M. V., Kim, B., Kim, J. N., Alimohammadi, M., Dervishi, E., Mustafa, T., & Cernigla, C. E. (2012b). Carbon nanotubes as plant growth regulators: Effects on tomato growth, reproductive system, and soil microbial community. *Small*, *9*(1), 115–123. https://doi.org/10.1002/smll.201201225

Kookana, R. S., Boxall, A. B. A., Reeves, P. T., Ashauer, R., Beulke, S., Chaudhry, Q., Cornelis, G., Fernandes, T. F., Gan, J., Kah, M., Lynch, I., Ranville, J., Sinclair, C., Spurgeon, D., Tiede, K., & den Brink, P. J. (2014). Nanopesticides: Guiding principles for regulatory evaluation of environmental risks. *Journal of Agricultural and Food Chemistry*, *62*(19), 4227–4240. https://doi.org/10.1021/jf500232f

Krishnaraj, C., Jagan, E. G., Ramachandran, R., Abirami, S. M., Mohan, N., & Kalaichelvan, P. T. (2012). Effect of biologically synthesized silver nanoparticles on *Bacopa monnieri* (Linn.) Wettst. Plant growth metabolism. *Process Biochemistry*, *47*(4), 651–658. https://doi.org/10.1016/j.procbio.2012.01.006

Kuppusamy, P., Yusoff, M. M., Maniam, G. P., & Govindan, N. (2016). Biosynthesis of metallic nanoparticles using plant derivatives and their new avenues in pharmacological applications – An updated report. *Saudi Pharmaceutical Journal*, *24*(4), 473–484. https://doi.org/10.1016/j.jsps.2014.11.013

Labus, M., Matyasik, I., & Ziemianin, K. (2023). Thermal decomposition processes in relation to the type of organic matter, mineral and maceral composition of menilite shales. *Energies*, *16*(11), 4500. https://doi.org/10.3390/en16114500

Lahiani, M. H., Dervishi, E., Chen, J., Nima, Z., Gaume, A., Biris, A. S., & Khodakovskaya, M. V. (2013). Impact of carbon nanotube exposure to seeds of valuable crops. *ACS Applied Materials & Interfaces*, *5*(16), 7965–7973. https://doi.org/10.1021/am402052x

Liu, P.-H., Wen, M., Tan, C.-S., Navlani-García, M., Kuwahara, Y., Mori, K., Yamashita, H., & Chen, L.-J. (2017). Surface plasmon resonance enhancement of production of H2 from ammonia borane solution with tunable Cu2–xS nanowires decorated by Pd nanoparticles. *Nano Energy*, *31*, 57–63. https://doi.org/10.1016/j.nanoen.2016.10.064

Liu, R., & Lal, R. (2015). Potentials of engineered nanoparticles as fertilizers for increasing agronomic productions. *Science of the Total Environment*, *514*, 131–139. https://doi.org/10.1016/j.scitotenv.2015.01.104

Lyu, S., Wei, X., Chen, J., Wang, C., Wang, X., & Pan, D. (2017). Titanium as a beneficial element for crop production. *Frontiers in Plant Science*, *8*. https://doi.org/10.3389/fpls.2017.00597

Marsalek, R. (2014). Particle size and zeta potential of ZnO. *APCBEE Procedia*, *9*, 13–17. https://doi.org/10.1016/j.apcbee.2014.01.003

Marvier, M. (2009). Pharmaceutical crops in California, benefits and risks: A review. In *Sustainable Agriculture* (pp. 191–201). Springer. https://doi.org/10.1007/978-90-481-2666-8_13

Mastronardi, E., Tsae, P., Zhang, X., Monreal, C., & DeRosa, M. C. (2015). Strategic role of nanotechnology in fertilizers: Potential and limitations. In *Nanotechnologies in Food and Agriculture* (pp. 25–67). Springer International Publishing. https://doi.org/10.1007/978-3-319-14024-7_2

Mishra, V., Mishra, R. K., Dikshit, A., & Pandey, A. C. (2014). Interactions of nanoparticles with plants. In *Emerging Technologies and Management of Crop Stress Tolerance* (pp. 159–180). Elsevier. https://doi.org/10.1016/b978-0-12-800876-8.00008-4

Mohammadi, S., Harvey, A., & Boodhoo, K. V. K. (2014). Synthesis of TiO2 nanoparticles in a spinning disc reactor. *Chemical Engineering Journal*, *258*, 171–184. https://doi.org/10.1016/j.cej.2014.07.042

Mukhopadhyay, S. S. (2014). Nanotechnology in agriculture: Prospects and constraints. *Nanotechnology, Science and Applications*, 63. https://doi.org/10.2147/nsa.s39409

Muraisi, L., Hariyadi, D. M., Athiyah, U., & Pathak, Y. (2022). Eco-friendly nanotechnology in agriculture: Opportunities, toxicological implications, and occupational risks. In *Sustainable Nanotechnology* (pp. 287–296). Wiley. https://doi.org/10.1002/9781119650294.ch18

Nessim, G. D. (2010). Properties, synthesis, and growth mechanisms of carbon nanotubes with special focus on thermal chemical vapor deposition. *Nanoscale*, *2*(8), 1306. https://doi.org/10.1039/b9nr00427k

Noji, T., Kamidaki, C., Kawakami, K., Shen, J.-R., Kajino, T., Fukushima, Y., Sekitoh, T., & Itoh, S. (2010). Photosynthetic oxygen evolution in mesoporous silica material: Adsorption of Photosystem II Reaction Center Complex into 23 nm nanopores in SBA. *Langmuir*, *27*(2), 705–713. https://doi.org/10.1021/la1032916

Nowack, B., David, R. M., Fissan, H., Morris, H., Shatkin, J. A., Stintz, M., Zepp, R., & Brouwer, D. (2013). Potential release scenarios for carbon nanotubes used in composites. *Environment International*, *59*, 1–11. https://doi.org/10.1016/j.envint.2013.04.003

Panpatte, D. G., Jhala, Y. K., Shelat, H. N., & Vyas, R. V. (2016). Nanoparticles: The next generation technology for sustainable agriculture. In *Microbial Inoculants in Sustainable Agricultural Productivity* (pp. 289–300).India: Springer. https://doi.org/10.1007/978-81-322-2644-4_18

Pesavento, M., Zeni, L., De Maria, L., Alberti, G., & Cennamo, N. (2021). SPR-optical fiber-molecularly imprinted polymer sensor for the detection of furfural in wine. *Biosensors*, *11*(3), 72. https://doi.org/10.3390/bios11030072

Pimpin, A., & Srituravanich, W. (2012). Review on micro- and nanolithography techniques and their applications. *Engineering Journal*, *16*(1), 37–56. https://doi.org/10.4186/ej.2012.16.1.37

Pramanik, S., & Pramanik, G. (2016). Nanotechnology for sustainable agriculture in India. In *Nanoscience in Food and Agriculture 3* (pp. 243–280). Springer International Publishing. https://doi.org/10.1007/978-3-319-48009-1_10

Prasad Yadav, T., Manohar Yadav, R., & Pratap Singh, D. (2012). Mechanical milling: A top down approach for the synthesis of nanomaterials and nanocomposites. *Nanoscience and Nanotechnology*, *2*(3), 22–48. https://doi.org/10.5923/j.nn.20120203.01

Prashantha Kumar, T. K. M., Mandlimath, T. R., Sangeetha, P., Revathi, S. K., & Ashok Kumar, S. K. (2016). Selective removal of nitrate and phosphate from wastewater using nanoscale materials. In *Nanoscience in Food and Agriculture 3* (pp. 199–223). Springer International Publishing. https://doi.org/10.1007/978-3-319-48009-1_8

Pulimi, M., & Subramanian, S. (2016). Nanomaterials for soil fertilisation and contaminant removal. In *Nanoscience in Food and Agriculture 1* (pp. 229–246). Springer International Publishing. https://doi.org/10.1007/978-3-319-39303-2_8

Raliya, R., Nair, R., Chavalmane, S., Wang, W.-N., & Biswas, P. (2015). Mechanistic evaluation of translocation and physiological impact of titanium dioxide and zinc oxide nanoparticles on the tomato (*Solanum lycopersicum* L.) plant. *Metallomics*, *7*(12), 1584–1594. https://doi.org/10.1039/c5mt00168d

Raliya, R., & Tarafdar, J. C. (2013). ZnO nanoparticle biosynthesis and its effect on phosphorous-mobilizing enzyme secretion and gum contents in Clusterbean (*Cyamopsis tetragonoloba* L.). *Agricultural Research*, *2*(1), 48–57. https://doi.org/10.1007/s40003-012-0049-z

Ram, P., Vivek, K., & Kumar, S. P. (2014). Nanotechnology in sustainable agriculture: Present concerns and future aspects. *African Journal of Biotechnology*, *13*(6), 705–713. https://doi.org/10.5897/ajbx2013.13554

Ramesh, S. (2013). Sol-Gel synthesis and characterization of nanoparticles. *Journal of Nanoscience*, *2013*, 1–8. https://doi.org/10.1155/2013/929321

Rashad, M., Hafez, M., Popov, A. I., & Gaber, H. (2022). Toward sustainable agriculture using extracts of natural materials for transferring organic wastes to environmental-friendly ameliorants in Egypt. *International Journal of Environmental Science and Technology*, *20*(7), 7417–7432. https://doi.org/10.1007/s13762-022-04438-8

Sabir, S., Arshad, M., & Chaudhari, S. K. (2014). Zinc Oxide nanoparticles for revolutionizing agriculture: Synthesis and applications. *The Scientific World Journal*, *2014*, 1–8. https://doi.org/10.1155/2014/925494

Saeed, A. (2021). Applications of nanobiotechnology in plant sciences. *Biomedical Journal of Scientific & Technical Research*, *35*(1). https://doi.org/10.26717/bjstr.2021.35.005632

Sang, L., Zhao, Y., & Burda, C. (2014). TiO2Nanoparticles as functional building blocks. *Chemical Reviews*, *114*(19), 9283–9318. https://doi.org/10.1021/cr400629p

Schaller, J., Brackhage, C., & Dudel, E. G. (2012). Silicon availability changes structural carbon ratio and phenol content of grasses. *Environmental and Experimental Botany*, *77*, 283–287. https://doi.org/10.1016/j.envexpbot.2011.12.009

Schaller, J., Brackhage, C., Gessner, M. O., Bäuker, E., & Gert Dudel, E. (2011). Silicon supply modifies C:N:P stoichiometry and growth of *Phragmites australis*. *Plant Biology*, *14*(2), 392–396. https://doi.org/10.1111/j.1438-8677.2011.00537.x

Schaller, J., Brackhage, C., Paasch, S., Brunner, E., Bäucker, E., & Dudel, E. G. (2013). Silica uptake from nanoparticles and silica condensation state in different tissues of *Phragmites australis*. *Science of the Total Environment*, *442*, 6–9. https://doi.org/10.1016/j.scitotenv.2012.10.016

Schlich, K., & Hund-Rinke, K. (2015). Influence of soil properties on the effect of silver nanomaterials on microbial activity in five soils. *Environmental Pollution*, *196*, 321–330. https://doi.org/10.1016/j.envpol.2014.10.021

Sharma, P., Bhatt, D., Zaidi, M. G. H., Saradhi, P. P., Khanna, P. K., & Arora, S. (2012). Silver nanoparticle-mediated enhancement in growth and antioxidant status of *Brassica juncea*. *Applied Biochemistry and Biotechnology*, *167*(8), 2225–2233. https://doi.org/10.1007/s12010-012-9759-8

Sharma, V., & Rao, L. J. M. (2013). An overview on chemical composition, bioactivity and processing of leaves of *Cinnamomum tamala*. *Critical Reviews in Food Science and Nutrition*, *54*(4), 433–448. https://doi.org/10.1080/10408398.2011.587615

Siddiqi, K. S., Ur Rahman, A., & Tajuddin, Husen A. (2018). Properties of zinc oxide nanoparticles and their activity against microbes. *Nanoscale Research Letters*, *13*(1). https://doi.org/10.1186/s11671-018-2532-3

Siddiqui, M. H., & Al-Whaibi, M. H. (2014). Role of nano-SiO2 in germination of tomato (*Lycopersicum esculentum* seeds Mill.). *Saudi Journal of Biological Sciences*, *21*(1), 13–17. https://doi.org/10.1016/j.sjbs.2013.04.005

Sirelkhatim, A., Mahmud, S., Seeni, A., Kaus, N. H. M., Ann, L. C., Bakhori, S. K. M., Hasan, H., & Mohamad, D. (2015). Review on zinc oxide nanoparticles: Antibacterial activity and toxicity mechanism. *Nano-Micro Letters*, *7*(3), 219–242. https://doi.org/10.1007/s40820-015-0040-x

Soares, J. R., Cantarella, H., & Menegale, M. L. de C. (2012). Ammonia volatilization losses from surface-applied urea with urease and nitrification inhibitors. *Soil Biology and Biochemistry*, *52*, 82–89. https://doi.org/10.1016/j.soilbio.2012.04.019

Song, G., Gao, Y., Wu, H., Hou, W., Zhang, C., & Ma, H. (2012). Physiological effect of anatase TiO2 nanoparticles on Lemna minor. *Environmental Toxicology and Chemistry*, *31*(9), 2147–2152. https://doi.org/10.1002/etc.1933

Suriyaprabha, R., Karunakaran, G., Yuvakkumar, R., Rajendran, V., & Kannan, N. (2012). Silica nanoparticles for increased silica availability in Maize (*Zea mays* L) Seeds under hydroponic conditions. *Current Nanoscience*, *8*(6), 902–908. https://doi.org/10.2174/157341312803989033

Syu, Y., Hung, J.-H., Chen, J.-C., & Chuang, H. (2014). Impacts of size and shape of silver nanoparticles on Arabidopsis plant growth and gene expression. *Plant Physiology and Biochemistry*, *83*, 57–64. https://doi.org/10.1016/j.plaphy.2014.07.010

Tantawy, A. S., Abdel-Mawgoud, A. M. R., El-Nemr, M. A., & Sassine, Y. N. (2012). Combined effects of different forms of phosphorus associated with potassium on performance of squash crop. *Acta Horticulturae*, *936*, 321–326. https://doi.org/10.17660/actahortic.2012.936.41

Tiwari, D. K., Dasgupta-Schubert, N., Villaseñor Cendejas, L. M., Villegas, J., Carreto Montoya, L., & Borjas García, S. E. (2013). Interfacing carbon nanotubes (CNT) with plants: Enhancement of growth, water and ionic nutrient uptake in maize (*Zea mays*) and implications for nanoagriculture. *Applied Nanoscience*, *4*(5), 577–591. https://doi.org/10.1007/s13204-013-0236-7

Tripathi, S., & Sarkar, S. (2014). Influence of water soluble carbon dots on the growth of wheat plant. *Applied Nanoscience*, *5*(5), 609–616. https://doi.org/10.1007/s13204-014-0355-9

Usman, M., Farooq, M., Wakeel, A., Nawaz, A., Cheema, S. A., Rehman, H. Ur, Ashraf, I., & Sanaullah, M. (2020). Nanotechnology in agriculture: Current status, challenges and future opportunities. *Science of the Total Environment*, *721*, 137778. https://doi.org/10.1016/j.scitotenv.2020.137778

Villagarcia, H., Dervishi, E., de Silva, K., Biris, A. S., & Khodakovskaya, M. V. (2012). Surface chemistry of carbon nanotubes impacts the growth and expression of water channel protein in tomato plants. *Small*, *8*(15), 2328–2334. https://doi.org/10.1002/smll.201102661

Wang, A., Zheng, Y., & Peng, F. (2014). Thickness-controllable silica coating of CdTe QDs by reverse microemulsion method for the application in the growth of rice. *Journal of Spectroscopy*, *2014*, 1–5. https://doi.org/10.1155/2014/169245

Wang, D., Saleh, N. B., Byro, A., Zepp, R., Sahle-Demessie, E., Luxton, T. P., Ho, K. T., Burgess, R. M., Flury, M., White, J. C., & Su, C. (2022). Nano-enabled pesticides for sustainable agriculture and global food security. *Nature Nanotechnology*, *17*(4), 347–360. https://doi.org/10.1038/s41565-022-01082-8

Wang, P., Lombi, E., Zhao, F.-J., & Kopittke, P. M. (2016). Nanotechnology: A new opportunity in plant sciences. *Trends in Plant Science*, *21*(8), 699–712. https://doi.org/10.1016/j.tplants.2016.04.005

Wang, X., Han, H., Liu, X., Gu, X., Chen, K., & Lu, D. (2012). Multi-walled carbon nanotubes can enhance root elongation of wheat (*Triticum aestivum*) plants. *Journal of Nanoparticle Research*, *14*(6). https://doi.org/10.1007/s11051-012-0841-5

Yang, F., Hong, F., You, W., Liu, C., Gao, F., Wu, C., & Yang, P. (2006). Influences of nano-anatase TiO_2 on the nitrogen metabolism of growing spinach. *Biological Trace Element Research*, *110*(2), 179–190. https://doi.org/10.1385/bter:110:2:179

Yang, F., Liu, C., Gao, F., Su, M., Wu, X., Zheng, L., Hong, F., & Yang, P. (2007). The improvement of spinach growth by nano-anatase TiO2 treatment is related to nitrogen photoreduction. *Biological Trace Element Research*, *119*(1), 77–88. https://doi.org/10.1007/s12011-007-0046-4

Yin, L., Colman, B. P., McGill, B. M., Wright, J. P., & Bernhardt, E. S. (2012). Effects of silver nanoparticle exposure on germination and early growth of eleven wetland plants. *PLoS ONE*, *7*(10), e47674. https://doi.org/10.1371/journal.pone.0047674

Yousef, H. A., Fahmy, H. M., Arafa, F. N., Abd Allah, M. Y., Tawfik, Y. M., El Halwany, K. K., El-Ashmanty, B. A., Al-Anany, F. Sh., Mohamed, M. A., & Bassily, M. E. (2023). Nanotechnology in pest management: Advantages, applications, and challenges. *International Journal of Tropical Insect Science*, *43*(5), 1387–1399. https://doi.org/10.1007/s42690-023-01053-z

Zeng, J., Ji, X., Ma, Y., Zhang, Z., Wang, S., Ren, Z., Zhi, C., & Yu, J. (2018). 3D graphene fibers grown by thermal chemical vapor deposition. *Advanced Materials*, *30*(12). https://doi.org/10.1002/adma.201705380

Zhu, H., Han, J., Xiao, J. Q., & Jin, Y. (2008). Uptake, translocation, and accumulation of manufactured iron oxide nanoparticles by pumpkin plants. *Journal of Environmental Monitoring*, *10*(6), 713. https://doi.org/10.1039/b805998e

13 Wastewater Re-Using in Irrigation for Agricultural Sustainability and Electricity Production

Lamiaa Ibrahim Ahmed, Hanan S Khalefa, and Ghada Abd-Elmonsef Mahmoud

13.1 INTRODUCTION

The wastewater comprises a variety of organic and inorganic substances, poisonous metals, hazardous chemicals, aromatic compounds, dyes, and so on. According to Jamrah et al. (2023), these hazardous compounds pose a significant threat to natural resources and can disrupt their regular functioning. Natural ecosystems and biogeochemical cycles are essential to all life on Earth, and their restoration is a top priority in light of the current crisis. Excessive contaminants can disrupt the biogeochemical cycles, which are driven by the metabolic activities of microbial populations (Liu et al., 2016). The majority of the harmful contaminants in our water supply and on our farmland come from man-made sources, such as mining, pesticides, and other industrial effluents (Ungureanu et al., 2020). There are many methods currently in use for treating wastewater, including trickling filters, activated sludge process, membrane separation, fluidized aerobic bioreactors, and many more. However, these technologies are not viable for widespread field use because of their high capital, operation, and maintenance expenses (Vaz et al., 2024).

Regardless of weather predictions, by 2100, world temperatures will have risen 3.5% over preindustrial levels. Because of the devastating effects of drought and wildfires on the various ecological compartments, water extraction is an absolute need (Ungureanu et al., 2020). The need for environmentally friendly methods of treating wastewater and discovering new ways to recover valuable resources is growing in the manufacturing industry. Globally, almost 70% of the withdrawals of fresh water are attributed to agricultural uses (Lakshmi and Reddy, 2017). Worldwide, wastewater treatment (WWT) is a problem that no one technique can solve because of variations in size, contaminants, and local conditions. Traditional WWT plants primarily employ mechanical means to remove suspended particles and activated sludge to lower biological oxygen demand (Wang et al., 2017). An essential step in avoiding the eutrophication of water sources downstream, including rivers and lakes,

DOI: 10.1201/9781003714484-13

is the biodegradation process, which entails the disintegration of both inorganic and organic components, including phosphorus and nitrogen compounds (Manganiello et al., 2024).

Conventional wastewater treatment methods often employ physical, chemical, and biological processes (Shabir et al., 2023). Price, practicality, efficiency, dependability, environmental impact, operation, sludge production, pre-treatment requirements, and the creation of potentially dangerous by-products are some of the limitations of each treatment. Furthermore, neither the wastewater nor its treatment is enriched by the current methods. One viable solution to the problems of pollution, energy shortages, and resource depletion is sustainable wastewater treatment. According to Li et al. (2014), a sustainable treatment process should have the following goals: reduced resource consumption, energy-neutral operation, consistent treatment results, high effluent quality to satisfy demands for water reclamation and reuse, balanced economic production and investment, robust social fairness, and little environmental effect. Using microbial electrolysis cells (MECs) or microbial fuel cells (MFCs), microbial electrochemical technologies (METs) offer the potential to treat wastewater biologically while producing energy. These devices oxidize organic materials and move electrons to anodic surfaces by using microorganisms as biocatalysts., resulting in bioelectricity (Khater et al., 2024) Due to their environmental friendliness, ease of production, and great efficiency, green fuels are seen as a potential long-term energy solution (Saravanan et al., 2020).

Fossil fuels account for 86% of global energy production, including natural gas and petroleum, investigating energy and hydrogen cogeneration in MECs could be a viable alternative for decentralized hydrogen production and a cost-effective strategy for treating wastewater. Greenhouse gas (GHG) emissions from burning fossil fuels have increased, contributing to climate change and pollution that harms ecosystems and the health of humans and other animals. In addition to not being renewable, it won't be enough to fulfill the world's anticipated 20,000 Mtoe growths in energy demand in 2050 (Mittal et al., 2022). Hydrogen gas has a high calorific value (120–142 MJ kg^{-1}) in comparison to other gaseous fuels like CH4 (50 MJ kg^{-1}), ethanol (26.8 MJ kg^{-1}), and petrol (44 MJ kg^{-1}), is an efficient vehicle fuel, and is environmentally friendly with zero or nearly negligible CO_2 emissions (Safari et al., 2022). Unfortunately, fossil fuels such as coal, natural gas, and oil are used to produce almost all the hydrogen gas that is produced. Hydrogen generation from renewable and biological sources is emerging as an attractive option for clean, sustainable energy production, with the added benefit of lowering process energy demands, in response to rising global temperatures and falling oil supplies caused by the combustion of fossil fuels (Banu et al., 2020). Electrochemical, biochemical, photochemical, and thermochemical processes are among the many methods that can manufacture hydrogen. A potential bio-electrochemical device for producing hydrogen is the MEC, which uses the oxidation of organic substances to spark the progress of hydrogen in a cathode chamber chemically. Producing biohydrogen from biowastes (e.g., sewage sludge, agricultural by-products, and other biodegradable wastes) has dual purposes: treating biowastes and recovering bioenergy. According to Nawaz et al. (2022), Microorganisms act as catalysts in MFCs, which are bio-electrochemical devices that transform chemical energy from both organic and inorganic substrates into electrical current. Multiple fuel cells employ a variety of cathode, anode, and proton exchange membranes for a broad array of purposes, including bio-hydrogen production, bio-sensors, treatment of wastewater, and power generation (Kumar et al., 2017).

13.2 TREATMENT OF WASTEWATER

Wastewater treatment occurs at multiple levels: primary (chemical and physic-chemical), secondary (biological and chemical), and tertiary (chemical and physical). Physical and mechanical pretreatments are also included in this hierarchy (Vaz et al., 2024) (Figure 13.1).

13.2.1 Primary Methods

This method primarily treats raw wastewater by eliminating floating, suspended, and bulk particles. As a rule, these procedures involve advanced oxidation, adsorption, coagulation, membrane filtering, and flocculation. Both industrial and residential wastewater can have their color and other pollutants eliminated using these methods (Kishor et al., 2021). During this stage of treatment, the entering wastewater had a BOD reduction of 20%–30% and a total suspended particle reduction of 50%–60%. Using these methods can be an ordeal due to their high costs, lengthy processing times, and, in some cases, the secondary pollution they cause – sludge (Kishor et al., 2021).

13.2.2 Secondary Treatment

In the secondary treatment, organic dissolved elements that were not eliminated by the first treatment are removed by the secondary treatment. Using biological wastewater treatment technologies resulted in no harm to the surroundings at a low cost. This approach is the most practical and effective alternative to traditional methods for reducing the harmfulness of wastewater toxins by the use of aquatic macrophytes

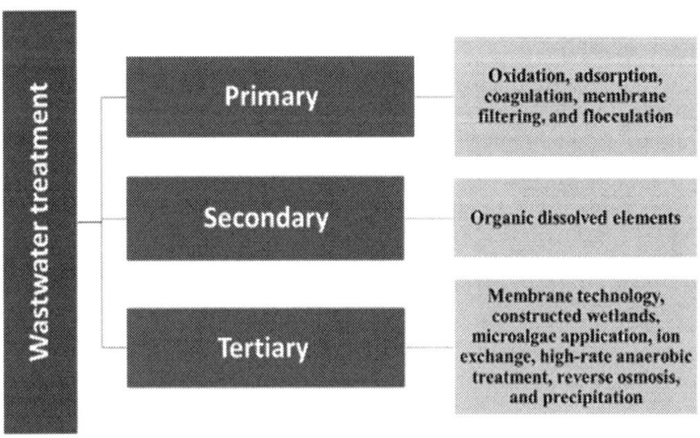

FIGURE 13.1 Different types of wastewater treatment.

and microorganisms (Kishor et al., 2021). Bacteria, fungi, and algae have been utilized by researchers in multiple areas to clean nearly every kind of wastewater. Kumar et al. (2020) and Kishor et al. (2021) found that these microbial communities can decolorize, detoxify, and break down pollutants in wastewater. They can also absorb heavy metals. The biological oxygen demand (BOD) and suspended solids can be reduced by about 85% with the help of secondary treatment. The method also eliminates carbonaceous pollutants and separates the biological sludge from the clean water. The residual water undergoes nitrification or denitrification after carbon and nitrogen have been removed. Biological wastewater treatment methods that have been linked to the removal of micropollutants include nitrification, solid retention time (SRT), and hydraulic retention time (HRT) (Helbling et al., 2012). After the primary and secondary treatments, the solid waste is subjected to a gravity-thickening tank that receives a constant flow of air. Two further tanks, one for centrifugal dewatering and the other for lime stabilization, complete the processing of the solid waste. Once handled, solid waste can be utilized for a variety of purposes, including landfilling, fertilizer production, and construction. Many additional technologies for treating wastewater have been developed and are now being implemented in reactors on a larger scale than activated sludge. According to Kishor et al. (2021), a 99% elimination capability for harmful chemicals and pollutants can be achieved through the integration of numerous biological and physicochemical processes.

13.2.3 Tertiary Treatment

Tertiary treatment systems are often used to describe advanced wastewater treatment systems. Compared to secondary treatment approaches, these advanced methods aim to generate quality high-quality effluents (Fito and Van Hulle, 2021). Nevertheless, the use of advanced technologies is subjective and is dictated by the efficacy of secondary treatment and the quality and characteristics of the treated water. There are a variety of modern ways of treating wastewater, including membrane technology, oxidation, adsorption, constructed wetlands, filtration, ion exchange, microalgae application, reverse osmosis, high-rate anaerobic treatment, and flocculation/precipitation (Mahmoud and Bagy, 2018).

13.3 REUSE OF WASTEWATER IRRIGATION

13.3.1 Recent Guidelines for the Safe Reuse

To control the risks to public health that result from utilizing wastewater for irrigation, there exist regulations for the safe reuse of wastewater irrigation to allow for the practical use of this water. Regarding protecting the environment and public health, the new standard is more stringent than the old one (Jaramillo and Restrepo, 2017). In many nations throughout the world, the WHO, EPA, and FAO standards serve as the foundation for the creation of rules. Therefore, these rules are recommended as a solution if no national guidelines exist in any country. Several particular water qualities that are important for preserving soil fertility, crop production, crop quality, and environmental preservation are included in irrigation water quality standards (Altabbaa et al., 2023). Guidelines for the capitalization of wastewater in agriculture stated that they typically contain the following physicochemical parameters: suspended solids, chemical and biological oxygen demand, heavy metals, turbidity, nutrients, salinity, pH, electrical conductivity (EC), and sodium absorption rate (SAR) in addition to the microbiological parameters (Salmonella, *E. coli*, fecal coliforms, fecal enterococci, Shigella, and nematode eggs). Distinct countries have distinct regulations that define limitations for fecal coliforms (cfu 100 mL^{-1}), total coliforms (cfu 100 mL^{-1}), *Escherichia coli* (cfu 100 mL^{-1}), and nematode eggs (number L^{-1}). These regulations also specify which crops and/or soils, depending on their quality, can be irrigated with wastewater. Public health is concerned about intestinal protozoans (*Cryptosporidium* and *Giardia*), helminths (*Taenia* and *Ascaris*), and bacterial pathogens (*Shigella* sp., *Salmonella* sp., *Escherichia coli*, *Legionella* sp., and *Vibrio cholerae*) worldwide. The biggest danger of transmission by wastewater reuse is posed by waterborne viruses such as rotavirus, adenovirus, HEV, and HAV (WHO, 2006). Therefore, health and environmental authorities believe that the most dependable way to assess the quality of water and the effectiveness of systems of water treatment is to utilize microbiological indicators of fecal contamination (Chibuike et al., 2014). The new water reuse minimum requirements regulation for agricultural irrigation was published by the European Commission on May 25, 2020. It went into effect on June 26, 2023, and is anticipated to encourage and facilitate water reuse throughout the European Union (European Commission (EC) Regulation EU, 2020/741). The specifications needed for the quality of WW utilized in crop irrigation are shown in Table 13.1. Apart from the parameters mentioned in Table 13.1, EU, 2020/741 stipulated the following for the four effluent quality classes: For irrigation of fields, use intestinal nematodes and aerosolization (helminth eggs) at <=1 egg L^{-1}, *Legionella* sp. at <1,000 CFU L^{-1}. The United Nations 2030 Agenda

TABLE 13.1
Reclaimed Water Quality Standards for Use in Agricultural Irrigation (EU, 2020/741)

Minimum Quality Class	Aim	E.coli [No./100 mL]	Quality Requirements: Biological Oxygen Demand [mg L^{-1}]	Total Suspended Solids [mg L^{-1}]	Turbidity [NTU]
A	Secondary treatment, disinfection, and filtration	<10	<10	<10	<5
B	Disinfection and secondary treatment	Less than or equal to 100	Following Directive 91/271/EEC		–
C	Disinfection and secondary treatment	1,000			–
D	Disinfection and secondary treatment	10,000			–

for Sustainable Development's sixth goal, which is to ensure that water and sanitation are available and managed for the world's population, as well as to significantly increase water recycling and safe water reuse worldwide, will be achieved with the help of the standard's implementation. By 2025, European farms may be capable of reusing about 6,000 million m^3 of water annually with the support of stronger regulations and financial incentives (EU, 2020/741).

13.3.2 Reuse of Treated Wastewater for Agriculture

Nearly 80% of all freshwater consumption goes toward agricultural irrigation, and that number will rise to 15% by 2030, causing water shortages in already-scarce areas. Thus, it is crucial to utilize alternate water sources. Reusing wastewater is one of the most important decisions to address the water shortage (Vaz et al., 2024). When wastewater leaves a household treatment facility, it undergoes a series of treatments, including basic disinfection and secondary treatment, before being recycled. This process is known as treated wastewater (TW). For arid regions in particular, water resource expansion through TW reuse has the potential to be a major alternative option (Hashem and Qi, 2021; Kesari et al., 2021). Because it is both economically and environmentally beneficial, treating wastewater and then using it in agricultural production are a widespread practice around the world (Manganiello et al., 2024). It has the potential to lessen or do away with the requirement to apply costly chemical fertilizers to the ground. Because a wide variety of nutrients, including phosphorus, nitrogen, potassium, and sulfur, are present in wastewater. Wastewater is a valuable resource for increasing agricultural productivity at a low cost because plants can readily absorb the large amounts of nitrogen and phosphorus found in it. On the other hand, there are risks associated with reusing wastewater in agriculture. Soil salinity increased, pathogenic microbes were found, and contaminants such as heavy metals were released. Reusing treated wastewater (TW) will solve two problems at once: the scarcity of water for agricultural development and the health risks and pollution caused by the careless removal of raw sewage (Ungureanu et al., 2020; Kesari et al., 2021). Therefore, to improve the treated water's quality, avoid environmental contamination, and protect human health from related risks, a cutting-edge technique for treating wastewater should be implemented before discharging wastewater into agricultural land, rivers, and soils (Liu et al., 2023).

A significant decrease in the effect of runoff on receiving regions could be achieved by recycling wastewater following sufficient treatment. There are several pros and cons of irrigating with wastewater. Pedrero et al. (2018) listed many key advantages, including protecting the livelihoods of millions of farmers, reducing the demand for freshwater, achieving ecological balance, bolstering the economy, decreasing fertilizer consumption, and preserving natural water sources. Groundwater, soil, and crops are gradually poisoned with toxic metals and microorganisms as a result of wastewater irrigation, which poses significant health hazards (Ungureanu et al., 2020). The use of various wastewaters for crop irrigation, including industrial effluents, sewage liquid sludge, and agricultural and municipal wastewaters, leads to soil deposition of heavy metals, which in turn causes crops to absorb these metals, the major entry point for heavy metals into aquatic resources. Results from agricultural wastewater treatment vary widely, with a range of 1.56%–6.66% worldwide (Ungureanu et al., 2020). Worldwide, almost 10% of the population buys food grown using irrigation water from wastewater systems (WHO, 2006). It is possible that the use of treated or untreated wastewater in farming can damage the environment, especially the soil. Published research has shown that soil physicochemical characteristics can change. Also, according to Becerra-Castro et al. (2015), agricultural wastewater reuse can change soil microbiome structure and quantity. When untreated wastewater is used for irrigation, it raises soil salinity, seals the land, and then causes sodium to accumulate. This causes erosion of the soil and a decrease in its permeability, which in turn prevents crops from receiving nutrients from the soil. Reusing wastewater in agriculture is not without its drawbacks, the most significant of which are the potential for increased health hazards and environmental damage. To find a positive use for it, treated wastewater (TW) is defined as wastewater from municipalities that have undergone treatment to achieve certain water quality standards. Eliminating potentially hazardous chemicals and biological components from wastewater, as well as lowering SS and OM, allows the treatment to achieve its primary goal of providing TW with an acceptable and vulnerable level of threat to public health and the environment (Wang et al., 2017). This opens up a lot of possibilities for water reuse in the future, whether in developed or developing nations. TW can remain polluted even after treatment and can enter the food chain of all living things, therefore recycling it could be bad for the environment and people's health (Lyu et al., 2016). Yet, with good irrigation management, the environmental and public health effects of TW can be mitigated (Helmecke et al., 2020; Hashem and Qi, 2021). Alcalde-Sanz and Gawlik (2017), and ISO (2015) all declare that treated wastewater's safe reutilization in agriculture is guaranteed by applying the standards and quality criteria in conjunction with advancements in treatment technologies. If we were to summarize the primary water parameters used to assess and control irrigation water quality according to FAO (2003), we would have the following.

13.3.2.1 Salinity

To determine if TW is suitable for irrigation, it is necessary to know the amount and types of salts that are present. The salt level in treated wastewater (TW) is often greater than that in WW, albeit this difference is source and treatment-type dependent (Qadir et al., 2015). Tropical water typically has a salinity level that is 1.5–2 times greater than that of freshwater in the study applied by Chen et al. (2013). Soil salinization, osmication, and structural changes caused by TW's high salt content can worsen, resulting in lower yields. El-gallal et al. (2016) came to a similar conclusion, stating that salinity poses the greatest ecological risks associated with TW irrigation in arid regions. Considerations such as choosing salt-tolerant crops, choosing an irrigation system, scheduling irrigation, leaching, installing drainage facilities, ensuring sufficient land leveling, and groundwater depth, and using wastewater in combination with freshwater can help reduce the salinity problem (Hashem and Qi, 2021).

13.3.2.2 Alkalinity

There is soluble salt in wastewater, which may cause an increase in sodium ions or soil salinization. According to Mutengu et al. (2007), soils irrigated with TW experience severe permeability problems due to the increased sodium, calcium, and magnesium levels, which in turn raise the soil alkalinity and SAR. However, Ofori et al. (2024) reported that sodium levels increased in soil irrigated with TWW significantly but did not induce soil sodicity. Soil electrical conductivity (EC) values amplified by 12.4% and 84.2%, while soil aquifer retention ratio (SAR) improved by 64.5% and 145.8%, respectively, when irrigated with TW in contrast to the control treatments (groundwater and drinking water irrigation) according to the study performed by Pan et al. (2012). According to Morugán-Coronado et al. (2011), SAR is the most reliable metric for estimating soil physicochemical alterations, hence monitoring its value is essential. Soil crust formation, which happens when water with a high SAR content is irrigated, can be reduced by using chemical amendments like gypsum, organic manure, or straw as an organic matter (OM), as well as by spraying the soil with drippers or mini sprinklers set to a low discharge rate and giving them enough time to run.

13.3.2.3 Nutrients from Crops in TWW

Nutrients, including P, N, and K, along with micronutrients like Mn, Fe, Cu, and Zn, and a considerable quantity of OM are found in wastewater and TW, which are crucial for the growth of plants (Mahmoud et al., 2023a; Mahmoud et al., 2024). According to De Carlo et al. (2020), the nitrogen, potassium, and phosphorus content of secondary treated municipal wastewater varied from 20 to 60 mg L^{-1}, from 10 to 30 mg L^{-1}, and from 6 to 15 mg L^{-1}, respectively. As a result, TW is a fertilizer-rich water source that can improve the fertility of the soil and boost the yield of crops. Elgallal et al. (2016) warned that improper TW irrigation management could cause plants to absorb more nutrients than they need, which could have back drawbacks on both human health and the environment. Reducing economic output and food nutrient quality (Chen et al., 2013) and possibly affecting the microbial communities of soil (Becerra-Castro et al., 2015) could result from providing plants with nitrogen over their specific requirements through wastewater. Consistent with this, Chiou (2008) noted that post-secondary treatment TW nitrogen levels in Taiwan were high (15–20 mg L^{-1}), ruling out its suitability for rice farming. To determine how much fertilizer a plant needs, it is necessary to subtract the amounts of nutrients found in soil and TW. This makes nutrient estimations all the more important.

13.3.2.4 Heavy Metal Pollution

The appropriate treatment processes effectively remove heavy metals present in municipal wastewater, including copper (Cu), molybdenum (Mo), Arsenic (As), Cadmium (Cd), nickel (Ni), Chromium (Cr), zinc (Zn), and Lead (Pb), until their concentrations are within the permissible limits or are comparable to those in freshwater (Ibrahim et al., 2020; Mahmoud et al., 2023b). The presence of HM in TW and its subsequent accumulation in soils poses a threat, according to Arunakumara et al. (2013). This is particularly true when industrial wastewater or a combination of industrial and domestic wastewater is used as a source (Liang et al., 2011). When utilizing this water for an extended period, it is essential to consider the possibility of the previously mentioned heavy metals entering the food chain (Martínez-Cortijo and Ruiz-Canales, 2018; Mahmoud, 2021). Soils that were watered with TW for 20 years showed a high concentration of HM in their upper layers (Hashem and Qi, 2021). Chen et al. (2013) showed that the concentration of certain heavy metals in treated water was similar to that in groundwater (GW), with the exception of Zn Because of the high concentration of HM in the wastewater; these metals can accumulate in crops and soils, which can be dangerous for both people and animals. Consequently, it is important to keep an eye on the levels of HM in the water, soil, and crops while employing TW irrigation. The maximum amounts of trace elements in irrigation water were recommended by FAO (1992) to be around 0.2, 2.0, 5.0, 0.01, 0.2, 5.0, and 0.2 mg L^{-1}, for Mn, Zn, Fe, Cd, Cu, Pb, and Ni, respectively. One way to lessen TW's negative effects and make it more extensively used is through efficient and suitable irrigation management (Hashem and Qi, 2021; Abeed et al., 2022).

13.3.3 Impact of TW Irrigation on the Properties of Soil

13.3.3.1 Physical Properties

Soil physicochemical and microbiological characteristics are changed by wastewater irrigation (Elgallal et al., 2016). Urban wastewater irrigation, as pointed out by Ababsa et al. (2020), increases hydraulic conductivity and soil porosity. Soil OM buildup was the explanation for the rise in soil bulk density observed by Coppola et al. (2004) in soils that were watered with wastewater. On the other hand, Abd-Elwahed (2018) discovered that soil that was watered with wastewater for an extended period had a little lower bulk density. According to Guo et al. (2017), the electrical conductivity and water content of the soil were found to be enhanced by TW irrigation. In a study conducted in the Middle East, where water is scarce, Farhadkhani et al. (2018) compared the use of treated WWI with freshwater irrigation in terms of soil properties and quality of several crops. In control plots, there were no discernible effects on the WWI-related variables. SAR and EC values were somewhat greater in WWI regions. In areas where water is scarce, the study found that treated WWI could be a viable alternative for watering root and leaf crops.

13.3.3.2 Chemical Properties

Wastewater irrigation's effects on soil chemical properties, including salinity and pH, have been the subject of several investigations. Soil salt saturation and infiltration difficulties are the two main indicators of salinity, which are measured by EC and SAR, respectively (Gharaibeh et al., 2016). Soil physicochemical properties were not significantly changed by using TW in irrigation, according to a study by Farhadkhani et al. (2018), except slightly higher levels of EC and SAR in TW soil samples compared to freshwater. According to Petousi et al. (2019), soil sodium levels increased after the irrigation period when TW was used as an irrigation method, but they decreased during the wet season. The primary consequence of TW irrigation, according to Salgado-Méndez et al. (2019), was an elevation of electrical conductivity in the bottom layers of soil. Furthermore, after two years, they found that Al, Fe, and Zn had accumulated in the topsoil layer two to eight times faster than before. According to Gu et al. (2019), the topsoil pH level was found to be higher after TW irrigation as compared to the control treatment. Soils that were watered with TW for around ten years had a high SAR value, as pointed out by Pan et al. (2012). Contrarily, TW irrigation did not significantly affect soil salinization according to Yang et al. (2006). Additional research in China found the same thing (Wang et al., 2017). According to Erel et al. (2019), soil salinity was shown to be greater when treated with TW as opposed to freshwater. Neither the tree's output nor the concentrations of chloride nor sodium in its leaves were affected. However, they did point out that soil physicochemical properties could be negatively affected by long-term TW irrigation.

13.3.3.3 Biological Properties

13.3.3.3.1 Soil Enzymes

Enzyme activity and other microbiological characteristics react quickly to changes in soil management, fertilization methods, and HM content, suggesting that they may be useful as a soil quality index. Because soil microbes create soil organic carbon, which increases water-retaining capacity and soil fertility, they are crucial for disintegrating organic materials, enriching the soil, cycling nutrients, and building the soil's structure (Hashem and Qi, 2021). The metabolic reactions required for microbial life to function are catalyzed by soil enzymes. Soil microbes are essential for a variety of reasons, including their ability to change nutrients, increase nutrient accessibility, and serve as sensors for soil physicochemical conditions and microbial states. According to Qu et al. (2020), "Receiving soils' long-term sustainability can be in danger" if some important enzymes are inhibited, which could affect the rate of elemental cycling.

Organic contaminants, HM, salts, and nutrients are all present in TW. Soil enzyme activity could be altered by all of these. According to Chen et al. (2014), adding TW to soil could increase its enzyme activity. Furthermore, it was noted by Li et al. (2019) that TW irrigation has the potential to enhance plant development and yields by increasing soil enzyme activity. Soil biochemical and physical-chemical properties were improved, according to Adrover et al. (2012), who also found an effect on alkaline phosphatase and glucosidase activity. According to research by Chen et al. (2013), soil-biological activities were enhanced after long-term irrigation with TW at two locations in California and Beijing. Activities of soil enzymes (alkaline, invertase, phosphatase dehydrogenase, catalase, and urease) were enhanced and soil microbial biomass was raised when treated with TW irrigation, according to a Beijing experiment by Pan et al. (2012). In comparison to freshwater, they discovered that TW increased the enzyme activity by an average of 36.7% in the top 0–20 cm of soil.

13.3.3.3.2 Microbial Biomass

Among the many crucial soil biological characteristics is microbial biomass. The biophysical integration of organic matter (OM) with soil-solid, water-based, and gaseous phases is one of the important ecosystem processes regulated by this parameter. Consequently, shifts in microbial biomass and activity may signal deterioration in soil health and have a profound impact on some soil processes (Hashem and Qi, 2021). According to Guo et al. (2017), the soil Proteobacteria population was shown to rise in soils treated with TW, but Acido-bacteria numbers were found to increase in soils treated with clean water. Moreover, they determined that electrical conductivity, total phosphorus, and ammonia nitrogen (NH_4+-N) were the key variables that significantly affected the structure of the microbial community in soil when treated with TW irrigation. According to Mkhinini et al. (2020), soils that received 20 years of TW irrigation showed a large increase in microbial biomass, while soils that were irrigated for shorter periods showed a significant decrease. According to Li et al. (2019), a 40-year study about TW irrigation's effects on soil microbes found no statistically significant difference in the amount or composition of soil microbes when irrigated with either TW or groundwater. Furthermore, it was found by Ibekwe et al. (2018) that soils irrigated with fresh water and TW did not differ significantly in terms of microbial diversity. Nevertheless, it was noted that a majority of the sequences linked to the nitrifying bacteria, bacteria that fix nitrogen, fecal indicator bacteria, carbon degraders, denitrifying bacteria, and potential pathogens were more frequent in aquatic environments other than freshwater. Consequently, TW could harbor bacteria that are very active in several soil processes and could also contain some harmful microbes.

13.3.3.4 Fertility Status

The chemical, physical, and soil fertility are all enhanced when TW is used for irrigation and plant nourishment. Soil organic matter (OM) and nutrients could be applied using TW irrigation, leading to increased crop yields. Soils irrigated with TW had much higher effective P, total N, and total K contents than those irrigated with freshwater, according to numerous studies (Hashem and Qi, 2021; Christou et al., 2024). How TW irrigation affects soil and crop levels of potentially harmful elements (PTEs) has been the subject of multiple investigations. According to Zhang et al. (2018) and Hashem and Qi (2021), topsoil layers irrigated with TW had a higher concentration of PTE than those irrigated with freshwater. After 35 years of TW irrigation, Gu et al. (2019) found that the PTE contents in soil were slightly affected. They also noted that the soil quality might improve and the OM content would increase significantly with long-term TW irrigation. The HM (Pb, Zn, Cu, Co, Ni, and Cd) in the soil was shown to increase due to TW irrigation, according to Fang et al. (2017). A major roadblock was the increase in accessible HM, which hurt soil properties that may have affected soil quality. Soil HM concentrations in TW and freshwater-irrigated soils did not vary significantly over the years, according to Yang et al. (2011). According to Dotaniya et al. (2018), HM levels in sewage water may be minimal, but prolonged use of this water could lead to soil HM accumulation. Using TW as a supplementary treatment to irrigate tomatoes in a field trial, Lu et al. (2016) showed that the soil and crop HM content was significantly lower than the Chinese national limits for food hygiene permission values and soil environmental quality. Soil and agricultural contamination would not result from using wastewater treated with secondary treatment. Soil HM concentrations were shown to be dependent on irrigation duration (long or short) and HM concentrations in irrigation water, according to earlier research.

13.3.4 Impact of Plant Irrigation with Wastewater

TW is a reliable water source that may provide a lot of nutrients, which can greatly boost crop yields (Christou et al., 2024). Because sludge may contain undesirable components such as HM and pollutants, it is important to exercise caution when using wastewater for agricultural irrigation (Khawla et al., 2019; Hashem and Qi, 2021). A wide range of crops have been successfully irrigated using wastewater in multiple experimental studies. Furthermore, Libutti et al. (2018) assessed how wastewater irrigation (WWI) affected soil parameters and the yield of broccoli and tomatoes. In order to compare the outcomes, the study took into account different irrigation regimes, including those that used treated effluent and fresh water. Nitrate and Sodium ion levels were marginally greater than the acceptable standard values for vegetables, and the results demonstrated elevated soil chemical values under WWI in comparison to freshwater irrigation. Based on the study's findings, water-scarce Mediterranean regions can safely employ treated wastewater for crop production. The potential risks to human health from eating vegetables grown in wastewater were assessed by Farhadkhani et al. (2020) in Middle Eastern nations. Under a specific set of circumstances in water-scarce regions, the study found that using treated wastewater for vegetable crops did not pose a significant risk of campylobacter infections to human health. In addition, a three-year study found that the nectarines treated with TW had greater levels of antioxidant chemicals, total phenolic, and quality metrics than freshwater-treated fruits. This is likely owing to the abundance of nutrients in TW. Although there was a decrease in fruit quantity due to wastewater treatment, the increase in fruit weight more than made up for it (Pedrero et al., 2018). Huang et al. (2007) assessed the impact of fresh water and TW on the development of soybeans and maize in Beijing through the use of pot experiments. Results showed that after receiving TW therapy, maize and soybean yields were much higher. In their study, Mahmoudi et al. (2020) looked at how okra growth characteristics and production were affected by irrigation with TW and fresh water under subsurface drip irrigation (SSDI) and surface drip irrigation (SDI). The results showed that okra's yield attributes and growth factors were positively affected by TW. The studies also showed that when compared to freshwater, TW produced the best agronomic performance for okra. Depending on the results of Li et al. (2019), who compared TW irrigation, to groundwater irrigation, it is likely that TW irrigation can completely replace groundwater irrigation in terms of yield and growth for cucumber, tomato, and eggplant. Compared to sorghum irrigated with freshwater, sorghum irrigated with TW resulted in higher energy yield, ethanol, and production of dry biomass (Campi et al., 2016). Trees that were watered with TW without fertilization did not suffer any negative effects. According to Erel et al. (2019), indicating that TW provides plants with enough nutrients. In comparison to the freshwater treatment, the TW treatments also produced a greater yield.

Consistent with Petousi et al. (2019), tertiary TW improved the growth and production of young grapevines compared to freshwater, but secondary TW reduced fruit safety. Alves et al. (2017) found no significant variations in the microbial quality of bananas watered with freshwater and TW over two years, even when chemical fertilizers were also used. When tomatoes and broccoli were watered with TW, the results were comparable (Chojnacka et al., 2020). The results showed that microbial contamination on tomato surfaces was unrelated to the source of the irrigation waters. Additionally, there was no statistically significant difference in bacterial contamination between tomatoes irrigated with TW and those irrigated with tap water. The tomato surface that was irrigated with tap water also tested positive for Cryptosporidium infections, though. Bacterial markers were not different in washed and unwashed samples from crops irrigated with tap water compared to crops irrigated with TW in a greenhouse trial (Sheikh, 2015). The results of a field testing, however, showed that total coliform levels were greater in the TW-grown vegetables than in the tap-water group. Several factors could explain the observed differences in crop performance when irrigated with TW. These include the properties of the wastewater, the kind and species of crop, the plant's capacity to thrive in situations with nutrient deficits, and the plant's sensitivity to environmental and climate factors (Egbuikwem et al., 2020). Finally, Ofori et al. (2024) mentioned that potassium levels and Nitrate-nitrogen increased in soil irrigated with TWW, while decreasing in soil irrigated with tap water. Minimizing the risk of wastewater reuse in agriculture can be achieved through proper wastewater management, which includes choosing crops and irrigation strategies that are suitable, in addition to keeping an eye on the amounts and dissemination of trace metallic elements in plants and soil (Hussain and Qureshi, 2020).

13.3.5 Microbial Fuel Cell and Microbial Electrolysis Cell

Emerging as environmentally friendly technologies, MECs can remove pollutants from wastewater while generating bioelectricity, a renewable energy source that relies on specific microorganisms. This process produces low levels of greenhouse gas and has less impact on the environment (Yang et al., 2020). Zhao et al. (2020) stated that MEC is a considerable and useful technique for the production of biohydrogen due to its high efficiency of production and low consumption of energy. Many technical terms have been coined to describe the process by which microbial electrolysis cells produce biohydrogen, including microbial electrochemical cells, electrohydrogenesis, bio-electrochemical systems, electrochemically assisted hydrogen production, MES, and bio-catalyzed electrolysis (Cardéna et al., 2019). "Microbial electrolysis cell" is the most common and precise word used thus far.

The traditional MEC uses two chambers, the cathode and the anode chambers, with the anaerobic conditions applied to each. A cation exchange membrane (CEM) or an anion exchange membrane (AEM) splits the two chambers of the proton pump. Wastewater organic components are first broken down into carbon dioxide (CO2) and biomass at the anode by electrochemically active bacteria (EAB). In the next step, hydrogen gas is produced when electrons at the cathode decrease protons. But if the anode doesn't produce enough protons and electrons, the cathode might not be able to reduce as much. According to Khater et al. (2024), commercializing MECs and achieving efficient and sustainable H_2 production depend on enhancing the bioelectrocatalytic activity at the anode.

In terms of anode catalysts, the most common and often producing highly effective microbial anodes have been carbon-based electrodes, such as activated graphite, carbon, and carbon felt. According to Arun et al. (2023), these structures are cross-linked, which helps the particles and anodes form a strong bond. Stainless steel and other metallic anodes, especially when formed into foam, are appealing alternatives to other materials for making microbial anodes (Rousseau et al., 2020). Microorganisms that produce energy from organic and inorganic chemicals require biocompatibility, porosity to adhere to and colonize, and the large specific surface area (Selihin and Tay, 2022). If you want to get a lot of hydrogen out of MECs, you have to make sure that non-electrogenic bacteria don't thrive. 2-Bromomethanesulfonate (BES) and lumazine are two examples of chemical-based inhibitors. Chemicals like this drive up the overall cost of the process, therefore, finding cheaper alternatives is essential (Varanasi et al., 2019).

The catalyst and cathode base material, however, would significantly affect MEC performance. Titanium, graphite, or any other electrically conductive material could serve as the cathode's foundation. On simple carbon electrodes, the hydrogen evolution reaction (HER) is a laborious and sometimes hazardous procedure to produce hydrogen (Kundu et al., 2013). Despite their widespread use in the past, platinum catalysts are no longer recommended due to their high cost, unsustainable nature, and potential for sulfide poisoning (Logan et al., 2008). For MEC cathodes, the best materials to employ are stainless steel and nickel, owing to their availability, efficacy, and low cost. Also, some stainless-steel alloys have lower corrosion weight losses than other metals and alloys, therefore, it's probable that they'll be preferred as MEC cathodes (Rivera et al., 2019). Electrodes and PEMs, two components of MFCs, must be efficient, inexpensive, and ecologically friendly (Table 13.2). In addition to being precise and efficient, fabrication procedures should also be eco-friendly.

13.3.5.1 Microbial Metabolism

A variety of microorganisms, such as fungi and bacteria, have been revealed to be quite beneficial in MFCs. It is possible to employ either pure or mixed cultures of bacteria. *Pseudomonas aeruginosa, Escherichia coli, Geobacter sulfurreducens, Shewanella putrefaciens*, and other bacteria are frequently employed in the anode (Sharma and Kundu, 2010). In the sulfate-laden wastewater, MFC was used as a biocatalyst using a mixed consortium that included *Clostridium*, *Tetrathiobacter*, and *Desulfovibrio* (Kumar et al., 2020). Wastewater from wood hydrothermal treatment was treated in the MFC by a consortium, including *Paenibacillus* and *Thermoanaero* bacteria. After using the consortia from municipal wastewater, the generation power was significantly boosted from 70 to 360 mW m^{-2} (Toczyłowska-Mamińska et al., 2018).

13.3.5.2 Electron Transfer Mechanism

Extracellular electron transfer (EET) is the mechanism by which electrons are moved from microorganisms to an anode. Direct electron transfer and mediated electron transfer are two different possible routes. Direct electron transfer involves the establishment of physical contact between the cell membrane of microbe and electrode without using dissolved redox mediator. Species with complex networks of outer membrane cytochrome that can carry out direct electron transfer have been the subject of much research. Two such species are Geobacter and Shewanella. When microbes can't transfer electrons naturally because they don't come into contact with the electrode surface, a redox mediator helps by transferring electrons from the bacterial cells to the anode, but in the process, it gets oxidized. Later on, the oxidized mediator can transport electrons. According to Aiyer (2020), an ideal mediator would have a positive redox potential to aid electron transfer, be easily soluble in the anolyte, be able to traverse the cell membrane of microbes with ease, and not be poisonous to bacteria.

TABLE 13.2
Sources and Advantages of Some Waste-Derived MFC Components

Waste-Derived MFC Components	Source	Advantages	References
Biochar (BC)	BC is produced from various wastes under thermal treatments and oxygen-free conditions.		Chakraborty et al. (2020), Tomczyk et al. (2020), Naveenkumar and Senthilkumar (2021), and Kamali et al. (2022)
	1. BC from wood and crop residues	• Specific surface area higher than 300 m^2 g^{-1}. • Used in MFCs as anode, cathode, and proton exchange membranes. • Modified MFC has 95% elimination of chemical oxygen demand, with a power density of 6 W m^{-3}. • BC electrodes are of low cost compared to carbon electrodes, also, the power density reaches to 156.8 mW m^{-2}.	
	2. BC from coconut shell	• Elevated specific surface area of 400 m^2 g^{-1}. • Suitable anode. • Used as an electrode in MFCs with the addition of Cu to coconut shell BC to increase functioning, surface areas, and porosity.	
	3. BC from algae	• Low specific surface areas (2.7–14 m^2 g^{-1}).	
	4. BC from corn stover	• Has zinc content of 89 mg kg^{-1}, iron (3.3 g kg^{-1}). • Has high surface areas (56 m^2 g^{-1})	
	5. BC from swine manure	• Has Cu (1.3), and Zn (1.9 g kg^{-1}).	
	6. Graphitic BC	• Improved air cathode MFC (AC-MFC) performance • Have a high power density of 146.7 mW m^{-2}.	

Carbon nanotubes (CNTs)	• Possess exceptional stability, electrical conductivity, and specific surface area. • Suitable for uses where a high electron transfer capacity is needed. • Promote the oxygen reduction reactions (ORRs) • Traditional techniques for producing carbon nanotubes (CNTs) remain costly and challenging to apply in practical settings.		Yang et al. (2007), Delord et al. (2017)
Graphite-based materials	Graphite-based materials from b algae, hardwood, and agricultural residues.	• Because of its exceptional electron transfer rate, chemical inertness, high specific surface area, low cost, high conductivity, and mechanical and chemical stability, it is proposed as an ideal electrode. • They showed effective treatment of dairy effluents with an electrical energy output of 0.9–1.6 W m^{-3}. • It was able to remove more than 90% of total organic carbon (TOC), biological oxygen demand (BOD), nitrogen, sulfate, and phosphorus. • Certain MFCs employed graphitic electrodes	Kamal et al. (2020) and Marassi et al. (2022)
Carbon nitride-based materials	Mushroom-derived g-C3N4, waste toner powder.	• It has superior qualities, including high stability, outstanding electrical properties, and excellent visible light activity. • Seen as an inexpensive material with great promise for manufacturing MFC components. • A superactive cathode was produced by modifying iso-reticular metal-organic framework-3 (IRMOF-3) with g-C3N4. This new, hierarchically nitrogen-doped porous carbon structure was nitrogen-doped.	Yang et al. (2020) and Zhong et al. (2020)
Diamond-based materials	• Diamonds show extremely high over-potentials for the evolution of oxygen and hydrogen according to the inner-sphere mechanism, resulting in an inert surface for chemical species adsorption. This creates a broad electrochemical potential window that is perfect for electrochemical reactions. • Electrodes made of boron-doped diamond (BDD) are frequently utilized in electrochemical processes due to their strong nature, broad potential range, low background current, and excellent biocompatibility. • Complex wastewater has been successfully treated electrochemically with their help		Li et al. (2021) and Pierpaoli et al. (2021)

13.3.5.3 Factors Affecting the Process

Anaerobic systems' performance may be affected by several issues, such as reactor configuration, organic loading rate, pH, salinity, ionic strength, inhibitory chemicals, and temperature (Karadag et al., 2015). By influencing hydrogenase activity, metabolic pathways, bacterial community structure, and pH value of the liquid phase within the reactor plays a crucial role in improving hydrogen generation (Tomczak et al., 2018). Several studies found that pH values between 5.0 and 6 were ideal for dark fermentative hydrogen production (Angeriz-Campoy et al., 2015). Accordingly, Tremouli et al. (2017) inspected the impact of temperature (ranging from 24°C to 35°C), pH (ranging from 6 to 9), and salinity (ranging from 0 to 6.7 g KCl L^{-1}) on a dual chamber MFC for wastewater treatment. They discovered that up to a salinity level of 4.1 gL^{-1}, the COD elimination efficiency of 70% and the highest power generation of 66 mW m^{-2} were constant. When the salinity level was raised from 4.1 to 6.7 gL^{-1}, the power production efficiency dropped by 92% and the COD removal efficiency by 25.3%. The highest power density rose to 50.6 mW when the anolyte's pH was elevated from 6 to 9 m^{-2}, a 37% increase. The maximum power density also rose from 24°C to 35°C, increasing 64% to 59 mW m^{-2}. Over a wide range of pH levels and temperatures, the COD removal effectiveness maintained a consistent 75%–80%. In light of this, Vaez and Zilouei (2020) used paper mill wastewater in a two-phase method of anaerobic digestion and dark fermentation to assess the impact of starting pH, temperature, and substrate concentration on the efficiency of biohydrogen and biomethane synthesis. At an initial pH of 5, the hydrogen yield was at its highest (35.1 mL H_2 g^{-1} COD initial), while at 55°C and an initial concentration of 5 g COD L^{-1}, the methane production was at its highest (553.8 mL CH_4 g^{-1} COD initial). When using mixed cultures as inoculum, HRT is an additional critical component that impacts the composition of the microbial population. When other pretreatment methods fail to inhibit H_2-consuming bacteria, including homoacetogens, propionate makers, and hydrogenotrophic methanogens, this parameter becomes very important (Akhlaghi et al., 2020). Since H_2–producing bacteria proliferate at a faster pace than H_2-consuming bacteria, a reduction in HRT can cause them to become the majority population in the reactor (Bundhoo and Mohee, 2016).

According to Karaosmanoglu-Gorgeç and Karapinar (2019), research on the production of biohydrogen in packed-bed reactors has typically used simple sugars like glucose and sucrose as substrates. The results have shown hydrogen contents ranging from 9.5% to 85% in the produced gas, a hydrogen production rate between 0.01 and 7.4 L H_2 $h^{-1}L^{-1}$, and a yield between 0.01 and 3.9 mol H_2 mol^{-1} substrate. Hydrogen has been produced by the dark fermentation process employing APBR on a variety of substrates; however, the majority of these substrates have been derived from simple sugars like glucose and sucrose. Hydrogen and volatile fatty acids are valuable by-products of dark fermentation that can be used in scale-up systems to remediate wastewater and carbohydrate-rich waste (Ghimire et al., 2015).

13.3.5.4 Types of MFC

There are three generations of MFCs, according to Ieropoulos et al. (2005). The first generation used synthetic redox mediators in combination with microorganisms, such as gram-negative *E. coli*. Sulfate and sulfur dioxide were used by Gen-2 in

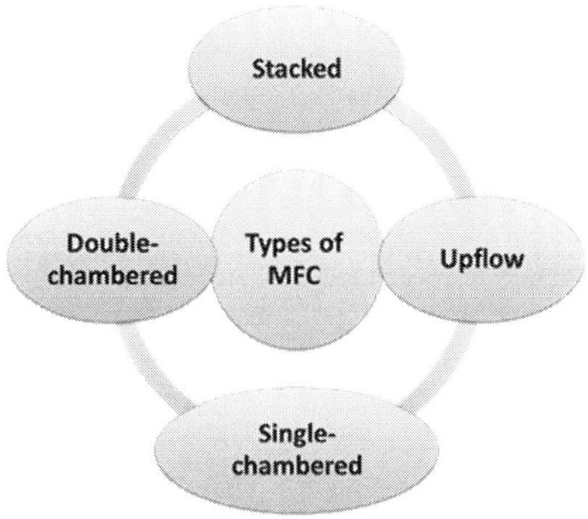

FIGURE 13.2 Four categories of microbial fuel cells.

conjunction with sulfate reducers such as Desulfovibrio desulfuricans. Among the Anodophilic species, Geobacter sulfurreducens served as the basis for Gen-3 MFCs, which did not employ soluble mediators. Kumar et al. (2017) further divided MFCs into four categories: double-chambered, single-chambered, upflow, and stacked, according to the design and construction methods (Figure 13.2).

Single-unit Single-Chambered MFCs Have the Anode and Cathode positioned at opposing ends of a single chamber (Park and Zeikus, 2003).

There is direct contact between the electrolyte and one side of the cathode and air on the other. The most significant disadvantage of single chamber MEC designs is that microbes such as methanogens, electrogens, and homoacetogens devour the produced hydrogen, particularly during long-term operation, despite the designs' simplicity and cost-effectiveness. Biofouling, microbial adulteration, and the inversion of oxygen flow from the cathode to the anode are some of the drawbacks, though (Wang et al., 2017).

13.3.5.4.1 Two-Chamber MFC

In a two-chamber MFC, with a proton exchange membrane separating the two reaction chambers, one serves as the anode and the other as the cathode. This design promotes better hydrogen recovery and purity while preventing methanogenic bacteria from consuming hydrogen. The arrangement that maintains the separation of the cathodic and anodic chamber liquid components is its primary benefit. To encourage microbial development, an anodic chamber is kept in an oxygen-free environment. To oxidize organic matter through metabolic processes, microorganisms cling to the anodic surface and conduct bioelectrochemical reactions, resulting in the formation of a biofilm at the anode. As a result, the solution gains protons, electrons, and carbon dioxide. A proton exchange membrane passes protons from the anodic to the cathodic chambers. Under aerobic conditions, the cathodic chamber produces water as a result of oxygen-accepting electrons (Borja-Maldonado et al., 2022; Arun et al., 2024; Khater et al., 2024).

13.3.5.4.2 Up-Flow MFC

This cylindrical design is a newer type of MFC. According to Ghorab et al. (2022), it has a proton exchange membrane that splits the cathode chamber from the anode chamber, which is located at the bottom of the device. It is common practice to pump the substrate from the chamber's base through the anode and up through the cathode. Since the output power generated is less than the energy expenditure of pumping the substrate, upflow MFCs may be more easily scaled for use in wastewater treatment than other MFC power applications (Zhou et al., 2013; Khater et al., 2024).

13.3.5.4.3 Stack MFC

Typically, stacked MFCs consist of numerous MFCs arranged in series or parallel to boost power output; nevertheless, the total voltage of the combined MFCs will be less than the sum of their voltages (Borja-Maldonado et al., 2022). Compared to MFCs connected in series, those stacked parallelly reportedly generate more power and have a faster bioelectrochemical reaction rate. As a result, improved wastewater treatment with parallel-stacked MFCs achieves a higher COD removal (Kumar et al., 2017).

A minor reduction in the voltage generated in the MFC led to a large fall in the biohydrogen production stability of the combined MFC and MEC system. For MECs to receive steady, high voltages, energy storage circuits based on capacitors show promise. To improve MEC biohydrogen output, the capacitors could be charged in parallel by several MFCs and discharged in series. So, to determine if it is possible to use paper mill effluent (PME) in an anaerobic packed bed biofilm reactor (APBR) continuously without pretreatment or modification, Namdarimonfared et al. (2023) conducted an assessment. The greatest hydrogen concentration of 41.5% was achieved after 4 hours of heat-reactive treatment (HRT). At 2 hours of HRT, the maximum rate of hydrogen synthesis was 6.21 L H_2 d^{-1} L^{-1}, and at 8 hours of HRT, the maximum yield was 0.96 mmol H_2 g^{-1} COD. The hydrogen production rate was positively affected for all HRTs except the shortest one as the OLR increased from 27.35 to 547 g COD L^{-1} d^{-1} with a decrease in HRT from 24 to 1.2 hours. In addition, the effectiveness of COD removal decreased from 25.2% to 6.2%. The conversion of organic input to useful metabolites, like VFAs, in the effluent, caused the low value of COD removal efficiency. According to Jiangbo et al. (2023), the primary technological challenge in biohydrogen generation is the low efficiency of hydrogen synthesis due to volatile fatty acid stress. Two kinds of microbes, Bacillus cereus A1 and *Brevundimonas naejangsanensis* B1, worked together to form hydrogen-producing biofilms that were spatially organized layer by layer. Compared to suspended fermentation (a hydrogen production increase of 103.27%) and mixed biofilms (MBs) fermentation (a hydrogen production increase of 66.50%), spatially ordered biofilms (SBs) considerably enhanced the cooperation efficiency of the two strains. At an inoculation volume ratio of A1:B1 = 1:1, the maximum hydrogen accumulation of 1,755 mL L^{-1} was achieved, and the mixed ratios of the two strains were further improved. After five rounds of fermentation using the biofilms, the performance of hydrogen production remained consistent, with 1,736 mL L^{-1} of hydrogen buildup on average. To increase the stability and cooperation efficiency of microbial consortiums consisting of two species, this paper proposed a method for producing hydrogen from biofilms layer by layer. That relies on the functioning bacteria's spatially organized distribution. Lastly, the percentages of COD removal from wastewater using various MFCs were as follows: >90% for Upflow MFC (He et al., 2006), 98.0% for Lu et al., 2008, 88.0% for Single chamber MFC (Sawasdee and Pisutpaisal, 2016), 64.0% for double chamber (Guo et al., 2016), 8.76% for dual chamber (Penteado et al., 2016), 75–80% for dual chamber (Mohanakrishna et al., 2019; Namdarimonfared et al., 2023), and 90.0% for stacked MFC (Das et al., 2020).

13.4 FUTURE RESEARCH PROSPECT

Reusing wastewater can result in new revenue streams, which can lower construction costs and increase the operational efficiency of water treatment facilities. In the future, a lot of environmentally friendly technologies with enhanced wastewater treatment methods will be released. Agricultural irrigation uses up almost 80% of freshwater use; by 2030, that percentage will increase to 15%, creating water shortages in already-scarce locations. Consequently, it is essential to use alternative water sources. Reusing wastewater is one of the most important decisions you can make to address the water shortage. Therefore, it is critical to determine the extent of pollution brought on by harmful substances and to identify the contaminants that are causing contamination in agriculture. Toxic pollutant treatment potential should be investigated in existing plant diversity to investigate advanced.

13.5 CONCLUSION

This chapter aims to provide an overview of the techniques applied globally for wastewater treatment in sustainable agriculture. The aforementioned highlights the present condition of affairs and possible forthcoming concerns with the repurposing of wastewater in agriculture and the utilization of natural agriculture-based wastewater treatment methods. Moreover, it clears the effects of using this treated water on the soil properties, fertility, and whole plant growth. At the same time, MEC is a newly developed ecologically friendly technology that can also produce bioelectricity, a renewable energy source that depends on particular microorganisms, and remove pollutants from wastewater. This technique has less of an environmental impact and produces minimal amounts of greenhouse gas. MEC's low energy consumption and excellent production efficiency make it a significant and useful technology for the manufacture of biohydrogen. Numerous technical words, such as "microbial electrochemical cells," "electrohydrogenesis," "electrochemically assisted hydrogen production," "bio-catalyzed electrolysis," "bio-electrochemical systems," and "MES," have been used to characterize the biohydrogen synthesis in microbial electrolysis cells.

ARTIFICIAL INTELLIGENCE DECLARATION

The authors have not used Regenerative Artificial Intelligence tools during the writing of this chapter.

REFERENCES

Ababsa, N., Kribaa, M., Tamrabet, L., Addad, D., Hallaire, V., & Ouldjaoui, A. 2020. A long-term effects of wastewater reuse on hydro physicals characteristics of grassland grown soil in semi-arid Algeria. *Journal of King Saud University Science*, 32, 1004–1013.

Abd-Elwahed, M.S. 2018. Influence of long-term wastewater irrigation on soil quality and its spatial distribution. *Annals of Agriculture Sciences*, 63, 191–199.

Abeed, A.H.A., Mahdy, R.E., Alshehri, D., Hammami, I., Eissa, M.A., Abdel Latef, A.A., & Mahmoud, G.A.-E. 2022. Induction of resilience strategies against biochemical deteriorations prompted by severe cadmium stress in sunflower plants when Trichoderma and bacterial inoculation were used as biofertilizers. *Frontiers in Plant Science*, 13, 1004173.

Adrover, M., Farrús, E., Moyà, G., & Vadell, J. 2012. Chemical properties and biological activity in soils of Mallorca following twenty years of treated wastewater irrigation. *Journal of Environmental Management*, 95, S188–S192.

Aiyer, K.S. 2020. How does electron transfer occur in microbial fuel cells? *World Journal of Microbiology Biotechnology*, 36, 1–9.

Akhlaghi, N., & Najafpour-Darzi, G. (2020). A comprehensive review on biological hydrogen production. *International Journal of Hydrogen Energy*, 45(43), 22492–22512.

Alcalde-Sanz, L., & Gawlik, B.M. 2017. Minimum quality requirements for water reuse in agricultural irrigation and aquifer recharge. Towards a water reuse regulatory instrument at EU level. Available online: https://committee.iso.org/files/live/users/aj/bc/fe/tc282contributor%40iso.org/files/EU%20JRC%20report%20Water%20Reuse%20Minimum%20Quality%20Requirements%202017.pdf (accessed on 27 February 2021).

Altabbaa, S., Mann, N.A., Chauhan, N., Utkarsh, K., Thakur, N., & Mahmoud, G.A.-E. 2023. Era connecting nanotechnology with agricultural sustainability: Issues and challenges. *Nanotechnology for Environmental Engineering*, 8, 481–498.

Alves, P.F.S., Dos Santos, S.R., Kondo, M.K., Mizobutsi, G.P., Caldeira, L.A., Antunes, A.B., & De Oliveira, G.F. 2017. Banana fertigation with treated sanitary wastewater: Postharvest and microbiological quality. *Semina: Ciências Agrárias*, 38, 1229–1240.

Angeriz-Campoy, R., Álvarez-Gallego, C.J., & Romero-García, L.I. 2015. Thermophilic anaerobic co-digestion of organic fraction of municipal solid waste (OFMSW) with food waste (FW): Enhancement of bio-hydrogen production. *Bioresource Technology*, 194, 291–296.

Arun, J., Panneer, S. S., Kirubanandam, G. P., Packiyadoss, P., Sivaprasad, S., Rangarajan, G., Quynh, H. L., & Arivalagan, P. 2023. New insights into microbial electrolysis cells (MEC) and microbial fuel cells (MFC) for simultaneous wastewater treatment and green fuel (hydrogen) generation. *Fuel*, 355(2024) 129530. https://doi.org/10.1016/j.fuel.2023.129530

Arun, J., SundarRajan, P., Pavithra, K.G., Priyadharsini, P., Shyam, S., Goutham, R., & Pugazhendhi, A. 2024. New insights into microbial electrolysis cells (MEC) and microbial fuel cells (MFC) for simultaneous wastewater treatment and green fuel (hydrogen) generation. *Fuel*, 355, 129530.

Arunakumara, K., Walpola, B.C., & Yoon, M.-H. 2013. Current status of heavy metal contamination in Asia's rice lands. *Reviews in Environmental Science and Bio-Technology* 12, 355–377.

Banu, J.R., Kavitha, S., Kannah, R.Y., Bhosale, R.R., & Kumar G. 2020. Industrial wastewater to biohydrogen: Possibilities towards successful biorefinery route. *Bioresource Technology*, 298, 122378.

Becerra-Castro, C., Lopes, A.R., Vaz-Moreira, I., Silva, E. F., Manaia, C. M., & Nunes, O. C. 2015. Wastewater reuse in irrigation: A microbiological perspective on implications in soil fertility and human and environmental health. *Environment International*, 75, 117–135.

Borja-Maldonado, F., Ángel, M., & Zavala, L. 2022. Contribution of configurations, electrode and membrane materials, electron transfer mechanisms, and cost of components on the current and future development of microbial fuel cells. *Heliyon*, 8, e09849.

Bundhoo, M.A.Z., & Mohee, R. 2016. Inhibition of dark fermentative bio-hydrogen production: A review. *International Journal of Hydrogen Energy*, 41, 6713–6733.

Campi, P., Navarro, A., Palumbo, A.D., Modugno, F., Vitti, C., & Mastrorilli, M. 2016. Energy of biomass sorghum irrigated with reclaimed wastewaters. *European Journal of Agronomy*, 76, 176–185.

Carde̒na, R., Cercado, B. & Buitrón, G. 2019. Microbial electrolysis cell for biohydrogen production. *Biohydrogen*, 2019, 159–185.

Chakraborty, I., Das, S., Dubey, B.K., & Ghangrekar, M.M. 2020. Novel low-cost proton exchange membrane made from sulphonated biochar for application in microbial fuel cells. *Materials Chemistry and Physics*, 239, 122025.

Chen, W., Lu, S., Jiao, W., Wang, M., & Chang, A.C. 2013. Reclaimed water: A safe irrigation water source? *Environmental Development*, 8, 74–83.

Chen, W.P., Lu, S.D., Zhang, W.L., Yi, L., & Jiao, W. 2014. Ecological risks and sustainable utilization of reclaimed water and wastewater irrigation. *Acta Ecologica Sinica*, 34, 163–172.

Chibuike, G.U., Obiora, S.C., & Chibuike, G.U. 2014. Heavy metal polluted soils: Effect on plants and bioremediation methods. *Applied and Environmental Soil Science*, 752708, 1–12.

Chiou, R.-J. 2008. Risk assessment and loading capacity of reclaimed wastewater to be reused for agricultural irrigation. *Environmental Monitoring and Assessment*, 142, 255–262.

Chojnacka, K., Witek-Krowiak, A., Moustakas, K., Skrzypczak, D., Mikula, K., & Loizidou, M. 2020. A transition from conventional irrigation to fertigation with reclaimed wastewater: Prospects and challenges. *Renewable and Sustainable Energy Reviews*, 130, 109959.

Christou, A., Beretsou, V.G., Iakovides, I.C., Karaolia, P., Michael, C., Benmarhnia, T., Chefetz, B., Donner, E., Gawlik, B.M., Lee, Y., Lim, T.T., Lundy, L., Maffettone, R., Rizzo, L., Topp, E., & Fatta-Kassinos, D. 2024. Sustainable wastewater reuse for agriculture. *Nature Reviews Earth & Environment*, 5, 504–521.

Coppola, A., Santini, A., Botti, P., Vacca, S., Comegna, V., & Severino, G. 2004. Methodological approach for evaluating the response of soil hydrological behavior to irrigation with treated municipal wastewater. *Journal of Hydrology*, 292, 114–134.

Das, I., Ghangrekar, M.M., Satyakam, R., Srivastava, P., Khan, S., & Pandey, H.N. 2020. On-site sanitary wastewater treatment system using 720-L stacked microbial fuel cell: Case study. *Journal of Hazardous, Toxic, Radioactive Waste*, 24, 04020025.

De Carlo, L., Battilani, A., Solimando, D., & Caputo, M.C. 2020. Application of time-lapse ERT to determine the impact of using brackish wastewater for maize irrigation. *Journal of Hydrology*, 582, 124465.

Delord, B., Neri, W., Bertaux, K., Derre, A., Ly, I., & Mano, N. 2017. Carbon nanotube fiber mats for microbial fuel cell electrodes. *Bioresource Technology*, 243, 1227–1231.

Dotaniya, M.L., Rajendiran, S., Meena, V.D., Coumar, M.V., Saha, J.K., Kundu, S., & Patra, A.K. 2018. Impact of long-term application of sewage on soil and crop quality in Vertisols of Central India. *Bulletin Environmental Contamination Toxicology*, 101, 779–786.

Egbuikwem, P.N., Mierzwa, J.C., & Saroj, D.P. 2020. Assessment of suspended growth biological process for treatment and reuse of mixed wastewater for irrigation of edible crops under hydroponic conditions. *Agricultural Water Management*, 231, 106034.

Elgallal, M., Fletcher, L., & Evans, B. 2016. Assessment of potential risks associated with chemicals in wastewater used for irrigation in arid and semiarid zones: A review. *Agricultural Water Management*, 177, 419–431.

Erel, R., Eppel, A., Yermiyahu, U., Ben-Gal, A., Levy, G., Zipori, I., Schaumann, G.E., Mayer, O., & Dag, A. 2019. Long-term irrigation with reclaimed wastewater: Implications on nutrient management, soil chemistry, and olive (Olea europaea L.) performance. *Agricultural Water Management*, 213, 324–335.

European Commission (EC). 2020. *Regulation EU 2020/741 of the European Parliament and of the Council, on Minimum Requirements for Water Reuse*. European Commission: Brussels, Belgium. Available online: https://eur-lex.europa.eu/legal-content/EN/TXT/PDF/?uri=CELEX:32020R0741&from=EN (accessed on 1 September 2020).

Fang, W., DeLapp, R.C., Kosson, D.S., Sloot, V.D., & Liu, J. 2017. Release of heavy metals during long-term land application of sewage sludge compost: Percolation leaching tests with repeated additions of compost. *Chemosphere*, 169, 271–280.

FAO. 1992. *Waste-Water Treatment and Use in Agriculture; Soils Bull. No. 47*. Rome, Italy: FAO.

FAO. 2003. *User's Manual for Irrigation with Treated Wastewater*. FAO Regional Office for the Near East: Cairo, Egypt.

Farhadkhani, M., Nikaeen M., Hadi M., Gholipour S., & Yadegarfar G. 2020. Campylobacter risk for the consumers of wastewater-irrigated vegetables based on field experiments. *Chemosphere*, 251, 126408.

Farhadkhani, M., Nikaeen, M., Yadegarfar, G., Hatamzadeh, M., Pourmohammadbagher, H., Sahbaei, Z., & Rahmani, H.R. 2018. Effects of irrigation with secondary treated wastewater on physicochemical and microbial properties of soil and produce safety in a semi-arid area. *Water Research*, 144, 356–364.

Fito, J., & Van Hulle, S. W. (2021). Wastewater reclamation and reuse potentials in agriculture: Towards environmental sustainability. *Environment, Development and Sustainability*, 23(3), 2949–2972.

Gharaibeh, M.A., Ghezzehei, T.A., Albalasmeh, A.A., & Main, Z.A. 2016. Alteration of physical and chemical characteristics of clayey soils by irrigation with treated wastewater. *Geoderma*, 276, 33–40.

Ghimire, A., Frunzo, L., Pirozzi, F., Trably, E., Escudie, R., & Lens, P.N.L. 2015. A review on dark fermentative biohydrogen production from organic biomass: Process parameters and use of by-products. *Applied Energy*, 144, 73–95.

Ghorab, R.E.A., Pugazhendi, A., Jamal, M.T., Jeyakumar, R.B., Godon, J.J., & Mathew, D.K. 2022. Tannery wastewater treatment coupled with bioenergy production in upflow microbial fuel cell under saline condition. *Environmental Research*, 212, 113304.

Gu, X., Xiao, Y., Yin, S., Liu, H., Men, B., Hao, Z., Qian, P., Yan, H., Hao, Q., & Niu, Y. 2019. Impact of Long-term reclaimed water irrigation on the distribution of potentially toxic elements in soil: An in-situ experiment study in the North China Plain. *International Journal of Environmental Research Public Health*, 16, 649.

Guo, W., Andersen, M.N., Qi, X.-B., Li, P., Li, Z.-Y., Fan, X.-Y., & Zhou, Y. 2017. Effects of reclaimed water irrigation and nitrogen fertilization on the chemical properties and microbial community of soil. *Journal of Integrative Agriculture*, 16, 679–690.

Guo, X., Zhan, Y., Chen, C., Cai, B., Wang, Y., & Guo, S. 2016. Influence of packing material characteristics on the performance of microbial fuel cells using petroleum refinery wastewater as fuel. *Renew Energy*, 87, 437–444.

Hashem, M.S., & Qi, X. 2021. Treated wastewater irrigation—A review. *Water*, 13, 1527.

He, Z., Wagner, N., Minteer, S.D., & Angenent, L.T. 2006. An upflow microbial fuel cell with an interior cathode: Assessment of the internal resistance by impedance spectroscopy. *Environmental Science & Technology* 40, 5212–5217.

Helmecke, M., Fries, E., & Schulte, C. 2020. Regulating water reuse for agricultural irrigation: Risks related to organic micro-contaminants. *Environmental Sciences Europe*, 32, 1–10.

Helbling, D. E., Johnson, D. R., Honti, M., & Fenner, K. 2012. Micropollutant biotransformation kinetics associate with WWTP process parameters and microbial community characteristics. *Environmental Science & Technology*, 46(19), 10579–10588.

Huang, Z., Miao, Z., Hou, L., Jiao, Z., & Ma, M. 2007. Effect of irrigation time and mode with reclaimed water on growth and quality of crops. *Journal of Agro Environment Science*, 26, 2257–2261.

Hussain, M.I., & Qureshi, A.S. 2020. Health risks of heavy metal exposure and microbial contamination through consumption of vegetables irrigated with treated wastewater in Dubai. *UAE Environmental Science and Pollution Research*, 27, 11213–11226.

Ibekwe, A., Gonzalez-Rubio, A., & Suarez, D. 2018. Impact of treated wastewater for irrigation on soil microbial communities. *Science of the Total Environment*, 622, 1603–1610.

Ibrahim, A.B.M., Zidan, A.S.A., Aly, A.A.M., Mosbah, H.K., & Mahmoud, G.A.-E. 2020. Mesoporous cadmium sulfide nanoparticles derived from a new cadmium anthranilato complex: Characterization and induction of morphological abnormalities in pathogenic fungi. *Applied Organometallic Chemistry*, 34, e5391.

Ieropoulos, I.A., Greenman, J., Melhuish, C., & Hart, J. 2005. Comparative study of three types of microbial fuel cell. *Enzyme and Microbial Technology*, 37, 238–245.

ISO. 2015. *Guidelines for Treated Wastewater Use for Irrigation Projects—Part 1–3; ISO 16075-1-32015*. Beuth Verlag: Berlin, Germany.

Jamrah, A., Al-Zghoul, T.M., & Darwish, M.M. 2023. A comprehensive review of combined processes for olive mill wastewater treatments. *Case Studies in Chemical and Environmental Engineering*, 8, 100493.

Jaramillo, M.F., & Restrepo, I. 2017. Wastewater reuse in agriculture: A review about its limitations and benefits. *Sustainability*, 9, 1734.

Jiangbo, L., Wang, K., Wang, S., & Su, H. 2023. Spatially-ordered layer-by-layer biofilms of a two-species microbial consortium promote hydrogen production. *Renewable Energy*, 215, 118905.

Kamal, A.S., Othman, R., & Jabarullah, N.H. 2020. Preparation and synthesis of synthetic graphite from biomass waste: A review. *Systematic Reviews in Pharmacy*, 11, 881–894.

Kamali, M., Sweygers, N., Al-Salem, S., Appels, L., Aminabhavi, T.M., & Dewil, R. 2022. Biochar for soil applications-sustainability aspects, challenges and prospects. *Chemical Engineering Journal*, 428, 131189.

Karadag, D., Köroğlu, O.E., Ozkaya, B., & Cakmakci, M. 2015. A review on anaerobic biofilm reactors for the treatment of dairy industry wastewater. *Process Biochemistry*, 50(2), 262–271. https://doi.org/10.1016/j.procbio.2014.11.005

Karaosmanoglu-Gorgeç, F., & Karapinar, I. 2019. Production of biohydrogen from waste wheat in continuously operated UPBR: The effect of influent substrate concentration. *International Journal of Hydrogen Energy*, 44(32), 17323–17333. https://doi.org/10.1016/j.ijhydene.2018.12.213

Kesari, K.K., Soni, R., Jamal, O.M.S., Tripathi, P., Lal, G.A., Jha, N.K., Siddiqui, M.H., Tripathi, V., & Ruokolainen, J. 2021. Wastewater treatment and reuse: A review of its applications and health implications. *Water Air Soil & Pollution*, 232, 208.

Khater, D.Z., Amin, R.S., Fetohi, A.E., El-Khatib, K.M., & Mahmoud, G.A-E. 2024. Microbial fuel cells: Bio-battery for environmental bioremediation and bioelectricity generation. In: Prasad, R. (ed.), *Environmental Materials and Waste (Second Edition): Circular Economy and Pollution Abatement* (pp. 813–833). Elsevier. https://doi.org/10.1016/B978-0-443-22069-2.00019-X

Khawla, K., Besma, K., Enrique, M., & Mohamed, H. 2019. Accumulation of trace elements by corn (Zea mays) under irrigation with treated wastewater using different irrigation methods. *Ecotoxicology and Environmental Safety*, 170, 530–537.

Kishor, R., Purchase, D., Saratale, G.D., Saratale, R.G., Ferreira, L.F.R., Bilal, M., & Bharagava, R.N. 2021. Ecotoxicological and health concerns of persistent coloring pollutants of textile industry wastewater and treatment approaches for environmental safety. *Journal of Environmental Chemical Engineering*, 9, 105012.

Kumar, R., Singh, L., & Kalia, V.C. 2017. Waste biomass management – A holistic approach. *Waste Biomass Management – A Holistic Approach*, 2017, 1–392. https://doi.org/10.1007/978-3-319-49595-8

Kumar, S.S., Kumar, V., Gnaneswar Gude, V., Malyan, S.K., & Pugazhendhi, A. 2020. Alkalinity and salinity favor bioelectricity generation potential of Clostridium, Tetrathiobacter and Desulfovibrio consortium in Microbial Fuel Cells (MFC) treating sulfate-laden wastewater. *Bioresource Technology*, 306, 123110.

Kundu, A., Sahu, J.N., Redzwan, G., & Hashim, M.A. 2013. An overview of cathode material and catalysts suitable for generating hydrogen in microbial electrolysis cell. *International Journal of Hydrogen Energy*, 38, 1745–1757.

Lakshmi, K.S., & Reddy, M.A. 2017. Wastewater-an overview. *International Journal of Contemperary Research Review*, 8, 20254–20262.

Li, B., Cao, Y., Guan, X., Li, Y., Hao, Z., Hu, W., & Chen, L. 2019. Microbial assessments of soil with a 40-year history of reclaimed wastewater irrigation. *Science of the Total Environment*, 651, 696–705.

Li, H., Zhou, K., Cao, J., Wei, Q., Lin, C.T., & Pei, S.E. 2021. A novel modification to boron-doped diamond electrode for enhanced, selective detection of dopamine in human serum. *Carbon New York*, 171, 16–28.

Li, W-W., Yu, H-Q., & He, Z. 2014. Towards sustainable wastewater treatment by using microbial fuel cells-centered technologies. *Energy & Environmental Science*, 7, 911–924.

Liang, J., Chen, C., Song, X., Han, Y., & Liang, Z. 2011. Assessment of heavy metal pollution in soil and plants from Dunhua sewage irrigation area. *International Journal of Electrochemical Science*, 6, 5314–5324.

Libutti, A., Gatta, G., Gagliardi, A., Vergine, P., Pollice, A., Beneduce, L., Disciglio, G., & Tarantino, E. 2018. Agro-industrial wastewater reuse for irrigation of a vegetable crop succession under Mediterranean conditions. *Agricultural Water Management*, 196, 1–14.

Liu, L., Chou, T.Y., Lee, C.Y., Lee, D.J., Su, A., & Lai, J.Y. 2016. Performance of freshwater sediment microbial fuel cells: Consistency. *International Journal of Hydrogen Energy*, 41:4504–4508. https://doi.org/10.1016/j.ijhydene.2015.07.139

Liu, X., Zhang, L., Yang, F., & Zhou, W. 2023. Determining reclaimed water quality thresholds and farming practices to improve food crop yield: A meta-analysis combined with random forest model. *Science of the Total Environment*, 862, 160774.

Logan, B.E., Call, D., Cheng, S., Hamelers, H.V.M., Sleutels, T.H.J.A., Jeremiasse, A.W. 2008. Microbial electrolysis cells for high yield hydrogen gas production from organic matter. *Environmental Science & Technology*, 42, 8630–8640.

Lu, N., Zhou, S., Zhuang, L., Zhang, J., & Ni, J. 2008. Electricity generation from starch processing wastewater using microbial fuel cell technology. *Biochemical Engineering Journal*, 43, 246–251. https://doi.org/10.1016/j.bej.2008.10.005

Lu, S., Wang, J., & Pei, L. 2016. Study on the effects of irrigation with reclaimed water on the content and distribution of heavy metals in soil. *International Journal of Environmental Research and Public Health*, 13, 298.

Lyu, S., Chen, W., Zhang, W., Fan, Y., & Jiao, W. 2016. Wastewater reclamation and reuse in China: Opportunities and challenges. *Journal of Environmental Sciences*, 39, 86–96.

Mahmoud, G.A-E. 2021. Microbial scavenging of heavy metals using bioremediation strategies. In Kumar, V., Prasad, R., & Kumar, M. (eds.), *Rhizobiont in Bioremediation of Hazardous Waste* (pp. 265–289). Springer: Singapore. https://doi.org/10.1007/978-981-16-0602-1_12

Mahmoud, G.A-E., Abeed, A.H.A., Mostafa, H.H.A., & Abdel Monsef, O. 2023a. Responses of pea (Pisum sativum L.) to single and consortium bio-fertilizers in clay and newly reclaimed soils. *Plants*, 12, 3931. https://doi.org/10.3390/plants12233931

Mahmoud, G.A-E., & Bagy M.M.K. 2018. Microbial degradation of petroleum hydrocarbons. In: Kumar, V., Kumar, M., & Prasad, R. (eds.), *Microbial Action on Hydrocarbons* (pp. 299–320). Springer: Singapore. https://doi.org/10.1007/978-981-13-1840-5_12

Mahmoud, G.A.E., Hefzy, M., & Zahran, M.M.A.A. 2024. Synergistic effects of microbial gibberellic acid and vitamins on onion (Allium cepa L.) Yield, and quality in low-fertility soil. *Journal of Soil Science and Plant Nutrition*, 24(3), 5342–5359. https://doi.org/10.1007/s42729-024-01911-w

Mahmoud, G.A-E., Thakur, N., & Hefzy, M. 2023b. Responses of marine fungi to heavy metal contamination. In Soni, R., Suyal, D.C., Morales-Oyervides, L., & Fouillaud, M. (eds.), *Current Status of Marine Water Microbiology* (pp. 263–277). Springer: Singapore. https://doi.org/10.1007/978-981-99-5022-5_11

Mahmoudi, M., Khelil, M.N., Ghrib, R., Douh, B., & Boujelben, A. 2020. Assessement of growth and yield of okra (Abelmoschus esculentus) under surface and subsurface drip irrigation using treated waste water. *International Journal of Recycling of Organic Waste in Agriculture*, 9, 349–356.

Manganiello, V., Zucaro, R., & Dono G. 2024. The use of purified wastewater for irrigation: Possible strategies in the Capitanata area (Apulia, Italy). *Agricultural Water Management*, 291, 108614.

Marassi, R.J., López, M.B.G., Queiroz, L.G., Silva, D.C.V.R., da Silva, F.T., & de Paiva, T.C.B. 2022. Efficient dairy wastewater treatment and power production using graphite cylinders electrodes as a biofilter in microbial fuel cell. *Biochemical Engineering Journal*, 178, 108283.

Martínez-Cortijo, J., & Ruiz-Canales, A. 2018. Effect of heavy metals on rice irrigated fields with waste water in high pH Mediterranean soils: The particular case of the Valencia area in Spain. *Agricultural Water Management*, 210, 108–123.

Mittal, V., Talapatra, K.N. & Ghosh, U.K. 2022. A comprehensive review on biodiesel production from microalgae through nanocatalytic transesterification process: Lifecycle assessment and methodologies. *International Nano Letters*, 12, 351–378.

Mkhinini, M., Boughattas, I., Alphonse, V., Livet, A., Gıustı-Mıller, S., Bannı, M., & Bousserrhıne, N. 2020. Heavy metal accumulation and changes in soil enzymes activities and bacterial functional diversity under long-term treated wastewater irrigation in East Central region of Tunisia (Monastir governorate). *Agricultural Water Management*, 235, 106150.

Mohanakrishna, G., Al-Raoush, R.I., Abu-Reesh, I.M., & Pant, D. 2019. A microbial fuel cell configured for the remediation of recalcitrant pollutants in soil environment. *RSC Advances*, 9, 41409–41418.

Morugán-Coronado, A., García-Orenes, F., Mataix-Solera, J., Arcenegui, V., & Mataix-Beneyto, J. 2011. Short-term effects of treated wastewater irrigation on Mediterranean calcareous soil. *Soil and Tillage Research*, 112, 18–26.

Mutengu, S., Hoko, Z., & Makoni, F.S. 2007. An assessment of the public health hazard potential of wastewater reuse for crop production. A case of Bulawayo city, Zimbabwe. *Physics Chemistry of the Earth Parts A/B/C*, 2007(32), 1195–1203.

Namdarimonfared, M., Hamid, Z., & Hadiseh, T. (2023). Biological hydrogen production from paper mill effluent via dark fermentation in a packed bed biofilm reactor. *Fuel*, 338, 127231. https://doi.org/10.1016/j.fuel.2022.127231

Naveenkumar, M., & Senthilkumar, K. 2021. Microbial fuel cell for harvesting bio-energy from tannery effluent using metal mixed biochar electrodes. *Biomass Bioenergy*, 149, 106082.

Nawaz, A., Haq, I., Qaisar, K., Gunes, B., Raja, S.I., & Mohyuddin, K. 2022. Microbial fuel cells: Insight into simultaneous wastewater treatment and bioelectricity generation. *Process Safety and Environmental Protection*, 161, 357–373.

Ofori, S., Abebrese, D.K., & Ružicková, I., Wanner, J. 2024. Reuse of treated wastewater for crop irrigation: Water suitability, fertilization potential, and impact on selected soil physicochemical properties. *Water*, 16, 484.

Pan, N., Chen, W.-P., Jiao, W.-T., Zhao, Z.-M., & Hou, Z.-A. 2012. Soil salinity in greenland irrigated with reclaimed water and risk assessment. *Huan Jing Ke Xue Huanjing Kexue*, 33, 4088.

Park, D.H., & Zeikus, J.G. 2003. Improved fuel cell and electrode designs for producing electricity from microbial degradation. *Biotechnology and Bioengineering*, 81, 348–355.

Pedrero, F., Camposeo, S., Pace, B., Cefola, M., & Vivaldi, G.A. 2018. Use of reclaimed wastewater on fruit quality of nectarine in Southern Italy. *Agricultural Water Management*, 203, 186–192.

Penteado, E.D., Fernandez-Marchante, C.M., Zaiat, M., Canizares, P., Gonzalez E.R., & Rodrigo, M.A. 2016. Influence of sludge age on the performance of MFC treating winery wastewater. *Chemosphere*, 151, 163–170.

Petousi, I., Daskalakis, G., Fountoulakis, M., Lydakis, D., Fletcher, L., Stentiford, E., & Manios, T. 2019. Effects of treated wastewater irrigation on the establishment of young grapevines. *Science of the Total Environment*, 658, 485–492.

Pierpaoli, M., Szopinska, M., Wilk, B.K., Sobaszek, M., Łuczkiewicz, A., & Bogdanowicz, R. 2021. Electrochemical oxidation of PFOA and PFOS in landfill leachates at low and highly boron-doped diamond electrodes. *Journal of Hazardous Materials*, 403, 123606.

Qadir, M., Mateo-Sagasta, J., Jiménez, B., Siebe, C., Siemens, J., & Hanjra, M.A. 2015. Environmental risks and cost-effective risk management in wastewater use systems. *Wastewater*, 55–72. https://doi.org/10.1007/978-94-017-9545-6_4

Qu, Y., Tang, J., Li, Z., Zhou, Z., Wang, J., Wang, S., & Cao, Y. 2020. Soil enzyme activity and microbial metabolic function diversity in soda saline–Alkali rice paddy fields of Northeast China. *Sustainability*, 12, 10095.

Rivera, I., Schroder, U., & Patil, S.A. 2019. Microbial electrolysis for biohydrogen production: Technical aspects and scale-up experiences. *Microbial Electrochemical Technology* 2019, 871–898. https://doi.org/10.1016/B978-0-444-64052-9.00036-4

Rousseau, R., Etcheverry, L., Roubaud, E., Basseguy, R., Delia, M-L., & Bergel, A. 2020. Microbial electrolysis cell (MEC): Strengths, weaknesses and research needs from electrochemical engineering standpoint. *Applied Energy*, 257, 113938.

Safari, M., Tondro, H., & Zilouei, H. 2022. Biohydrogen production from diluted-acid hydrolysate of rice straw in a continuous anaerobic packed bed biofilm reactor. *International Journal of Hydrogen Energy*, 47, 5879–5890.

Salgado-Méndez, S., Gilabert-Alarcón, C., Daesslé, L.W., Mendoza-Espinosa, L., Avilés-Marín, S., & Stumpp, C. 2019. Short-term effects on agricultural soils irrigated with reclaimed water in Baja California. *México City Bulletin of Environmental Contamination Toxicology*, 102, 829–835.

Saravanan, A., Karishma, S., Kumar, P.S., Yaashikaa, P.R., Jeevanantham, S., & Gayathri, B. 2020. Microbial electrolysis cells and microbial fuel cells for biohydrogen production: Current advances and emerging challenges. *Biomass Conversion and Biorefinery*, 13(10), 8403–8423.

Sawasdee, V., & Pisutpaisal, N. 2016. Simultaneous pollution treatment and electricity generation of tannery wastewater in air-cathode single chamber MFC. *International Journal of Hydrogen Energy*, 41, 15632–15637.

Selihin, N.M., & Tay, M.G. 2022. A review on future wastewater treatment technologies: Micro-nanobubbles, hybrid electro-Fenton processes, photocatalytic fuel cells, and microbial fuel cell. *Water Science & Technology*, 85(1), 319–341.

Shabir, S., Ilyas, N., Saeed, M., Bibi, F., Sayyed, R.Z., & Almalki, W.H. 2023. Treatment technologies for olive mill wastewater with impacts on plants. *Environmental Research*, 216, 114399.

Sharma, V., Kundu, P.P. 2010. Biocatalysts in microbial fuel cells. *Enzyme and Microbial Technology*, 2010(47), 179–188. https://doi.org/10.1016/j.enzmictec.2010.07.001

Sheikh, B. 2015. Recycled Water for Irrigation of Edible Crops (pp. 1–48). Available online: http://www.denverwater.org/docs/assets/D37B8FE5-155D-01CB-0C6B47FCC5B59221/recycled-water-edible-crop-white-paper.pdf (accessed on 18 March 2021).

Toczyłowska-Mamińska, R., Szymona, K., & Kloch, M. 2018. Bioelectricity production from wood hydrothermal-treatment wastewater: Enhanced power generation in MFC-fed mixed wastewaters. *Science of the Total Environment*, 634, 586–594. https://doi.org/10.1016/j.scitotenv.2018.04.002

Tomczak, W., Ferrasse, J.H., Giudici-Orticoni, M.T., & Soric, A. 2018. Effect of hydraulic retention time on a continuous biohydrogen production in a packed bed biofilm reactor with recirculation flow of the liquid phase. *International Journal of Hydrogen Energy* 2018, 43. https://doi.org/10.1016/j.ijhydene.2018.08.094

Tomczyk, A., Sokołowska, Z. & Boguta, P. 2020. Biochar physicochemical properties: Pyrolysis temperature and feedstock kind effects. *Reviews in Environmental Science and Bio/Technology*, 19, 191–215. https://doi.org/10.1007/s11157-020-09523-3

Tremouli, A., Martinos, M., & Lyberatos, G. 2017. The effects of salinity, pH and temperature on the performance of a microbial fuel cell. *Waste and Biomass Valorization*, 8, 2037–2043.

Ungureanu, N., Vlăduţ, V., & Voicu, G. 2020. Water scarcity and wastewater reuse in crop irrigation. *Sustainability*, 12, 9055.

Vaez, E., & Zilouei, H. 2020. Towards the development of biofuel production from paper mill effluent. *Renewable Energy*, 146, 1408–1415. https://doi.org/10.1016/j.renene.2019.07.059

Varanasi, J.L., Veerubhotla, R., Pandit, S., & Das, D. 2019. Biohydrogen production using microbial electrolysis cell: Recent advances and future prospects. *Microbial Electrochemical Technologies*, 2019, 843–869.

Vaz, T., Margarida, M.J., Quina, R.C., & Martins, G.J. 2024. Olive mill wastewater treatment strategies to obtain quality water for irrigation: A review. *Science of the Total Environment*, 931, 172676.

Wang, J., Song, X., Wang, Y, Zhao, Z., Wang, B., & Yan, D. 2017. Effects of electrode material and substrate concentration on the bioenergy output and wastewater treatment in air-cathode microbial fuel cell integrating with constructed wetland. *Ecological Engineering*, 99, 191–198.

World Health Organization (WHO). 2006. Guidelines for the safe use of wastewater, excreta and greywater (Vol. 4). In *Wastewater Use in Agriculture*. WHO: Geneva, Switzerland.

Yang, H.C., Chao, M.W., Chou, C.J., Wang, K.H., & Hu, C. 2020. Mushroom waste-derived g-C3N4 for methyl blue adsorption and cytotoxic test for Chinese hamster ovary cells. *Materials Chemistry and Physics*, 244, 122715.

Yang, J., Chen, T., Lei, M., Liu, H., Wu, W., & Zhou, J. 2011. Assessing the effect of irrigation with reclaimed water: The soil and crop pollution risk of heavy metals. *Journal of Natural Resources*, 2011(26), 209–217.

Yang, J., Li, W., Li, Q., Wu, S., Yu, B., & Jing, X. 2007. Oxygen adsorption by carbon nanotubes and its application in radiotherapy. *IET Nanobiotechnology*, 1, 10–14.

Yang, L., Yang, P., Ren, S., & Wang, C. 2006. Experimental studies on effects of reclaimed water irrigation on soil physicochemical properties. *Journal of Soil and Water Conservation*, 2006(2), 82–85.

Zhang, S., Yao, H., Lu, Y., Shan, D., & Yu, X. 2018. Reclaimed water irrigation effect on agricultural soil and maize (Zea mays L.) in northern China. *Clean Soil Air Water*, 46, 1800037.

Zhao, N., Liang, D, Meng, S., & Li, X. 2020. Bibliometric and content analysis on emerging technologies of hydrogen production using microbial electrolysis cells. *International Journal of Hydrogen Energy*, 45(58), 33310–33324.

Zhong, K., Wang, Y., Wu, Q., You, H., Zhang, H., & Su, M. 2020. Highly conductive skeleton graphitic-C3N4 assisted Fe-based metal-organic frameworks derived porous bimetallic carbon nanofiber for enhanced oxygen-reduction performance in microbial fuel cells. *Journal of Power Sources*, 467, 228313.

Zhou, M., Wang, H, Hassett, D.J., & Gu, T. 2013. Recent advances in microbial fuel cells (MFCs) and microbial electrolysis cells (MECs) for wastewater treatment, bioenergy and bioproducts. *Journal of Chemical Technology & Biotechnology*, 88, 508–518.

Index

Pages in *italics* refer to figures and pages in **bold** refer to tables.

T

U

W

Y

Z

Made in the USA
Las Vegas, NV
19 September 2026

1e6b6937-8591-42d2-92c5-892799084483R01